Evaluation of Energy Efficiency and Flexibility in Smart Buildings

Evaluation of Energy Efficiency and Flexibility in Smart Buildings

Editor

Alessia Arteconi

MDPI • Basel • Beijing • Wuhan • Barcelona • Belgrade • Manchester • Tokyo • Cluj • Tianjin

Editor
Alessia Arteconi
Università Politecnica delle Marche
Italy
KU Leuven
Belgium

Editorial Office
MDPI
St. Alban-Anlage 66
4052 Basel, Switzerland

This is a reprint of articles from the Special Issue published online in the open access journal *Energies* (ISSN 1996-1073) (available at: https://www.mdpi.com/journal/energies/special_issues/evaluation_efficiency_flexibility_buildings).

For citation purposes, cite each article independently as indicated on the article page online and as indicated below:

LastName, A.A.; LastName, B.B.; LastName, C.C. Article Title. *Journal Name* **Year**, *Volume Number*, Page Range.

ISBN 978-3-03943-849-5 (Hbk)
ISBN 978-3-03943-850-1 (PDF)

Contents

About the Editor

Alessia Arteconi (Associate Professor) has been an Associate Professor at Università Politecnica delle Marche (Italy) since December 2018 and she teaches HVAC systems and Renewable Energy Systems in Buildings. Since October 2019, she has been working as a part-time Associate Professor at KU Leuven (Belgium). She spent several periods abroad in other institutions: she was a visiting researcher at the University of Ulster—Northern Ireland in 2011, TU/e—The Netherlands in 2012–2013, and KU Leuven—Belgium from 2013 to 2019, working on the topic of demand-side management in the built environment. She has taken part in national and international research projects. She is the author of several papers published in international journals and presented at international conferences. Her research topics are:

(i) heat pumps and thermal storage;

(ii) demand-side management and smart cities;

(iii) refrigeration and cryogenics.

Preface to "Evaluation of Energy Efficiency and Flexibility in Smart Buildings"

Energy use in the residential sector accounts for about 40% of the total energy use both in Europe and the US. In particular, heating and cooling in buildings represent a high share of the overall energy use. Nevertheless, buildings are considered a resource in power systems thanks to the high energy flexibility they provide. Indeed, in buildings, there are several deferrable loads (e.g., laundry machines and dishwashers) and thermostatically controlled loads (TCL), such as heat pumps, refrigerators, and air conditioners. The latter technologies, together with the properties of the building envelope, contain various forms of storage, which can be used to alter the electric load without affecting the quality of the energy service. The energy flexibility provided by buildings is paramount to mitigating the upcoming challenges of future power systems, and its exact definition and quantification have a central role. Furthermore, the EU's Energy Performance of Buildings Directive pushes towards new and better-performing buildings—nearly zero energy buildings (nZEB)—where energy efficiency and energy flexibility are essential to achieve the required performance targets. Given this premise, the sector of heating and cooling in buildings is promising for the application of demand-side management (DSM) strategies aimed at modifying the final user's electricity demand on the basis of electricity grid needs. The relevance of DSM is related to the growing share of renewable energy sources (RES) in the generation mix and consequently to the necessity of integrating them and of adapting the energy demand to their intermittent and unpredictable production. DSM technologies can be used to activate the energy flexibility of buildings. They can be divided into three main categories: (i) energy-efficient end-use devices; (ii) additional equipment, systems, and controls to enable load shaping (e.g., energy storage); and (iii) communication systems between end-users and external parties, for example, demand response (DR) programs. The identification of the different technologies within the aforementioned categories, their technical details, and the effects of their application at the system or even country level is paramount. Different possibilities and analysis are collected in this book to shed light on this very interesting topic.

Alessia Arteconi

Editor

Article

Triggering Optimal Control of Air Conditioning Systems by Event-Driven Mechanism: Comparing Direct and Indirect Approaches

Junqi Wang [1,2,*], Rundong Liu [1,2], Linfeng Zhang [3], Hussain Syed ASAD [4] and Erlin Meng [1,2]

1 School of Environmental Science and Engineering, Suzhou University of Science and Technology, Suzhou 215009, China; ruhaning@163.com (R.L.); m20_njnu@126.com (E.M.)

2 National and Local Joint Engineering Laboratory of Municipal Sewage Resource Utilization Technology, Suzhou University of Science and Technology, Suzhou 215009, China

3 Institute of Geotechnical Engineering, Southeast University, Nanjing 211189, China; becharles@hnu.edu.cn

4 Department of Architecture and Civil Engineering, City University of Hong Kong, Hong Kong 999077, China; sahussain2-c@my.cityu.edu.hk

* Correspondence: junqi.alan.wang@outlook.com

Received: 10 September 2019; Accepted: 9 October 2019; Published: 12 October 2019

Abstract: Real-time optimal control of air conditioning (AC) is important, and should respond to the condition changes for an energy efficient operation. The traditional optimal control triggering mechanism is based on the "time clock" (called time-driven), and has certain drawbacks (e.g., delayed or unnecessary actions). Thus, an event-driven optimal control (EDOC) was proposed. In previous studies, the part-load ratio (PLR) of chiller plants was used as events to trigger optimal control actions. However, PLR is an indirect indicator of operation efficiency, which could misrepresent the system coefficient of performance (SCOP). This study thus proposes to directly monitor the SCOP deviations from the desired SCOP values. Two events are defined based on transient and cumulative SCOP deviations, which are systematically investigated in terms of energy performance and robustness. The PLR-based and SCOP-based EDOC are compared, in which energy saving and optimal control triggering time are analyzed. Results suggest that SCOP-based EDOC has better energy performance compared with PLR-based EDOC, but the frequent event triggering might happen due to the parameter uncertainty. For actual applications, the SCOP-based EDOC can be recommended when the ideal SCOP model is available with the properly-handled uncertainty. Nevertheless, the PLR-based EDOC could still be a more practical option to replace the traditional TDOC considering its acceptable energy performance and better robustness.

Keywords: real-time optimal control; system coefficient of performance; event-driven optimal control

---

1. Introduction

If the air conditioning (AC) system experiences identical operation condition all of the time, the engineer can set up the optimal settings and maintain the system energy efficiency easily. However, the fact is that the operation condition of AC systems involves many random variables over time, e.g., weather and occupancy. This makes real-time optimal control important. Each time the AC system experiences significant changes, the control system should respond and reset the previous optimal settings. Thus, the optimal control of the AC system should be repeated over the operation period.

Central AC systems contribute towards a large portion of a building's energy consumption [1,2]. Optimal control has been considered as a powerful measure to improve the operating efficiency of AC systems [3–7], where an objective function is optimized through optimizing the control set-points or operation modes (e.g., chiller sequence). Although many advanced optimization algorithms (e.g., reinforcement learning and model-predictive control [8]) and sophisticated modelling techniques (e.g., grey-box models or data-driven models [9] were developed, the mechanism to trigger the optimal control actions is still simple. In fact, an efficient triggering of optimal control can enhance the energy efficiency with the same resource consumption [10].

In most current practices, the control actions are triggered periodically based on a time "clock". This mechanism is thus termed as a time-driven optimal control (TDOC) [11,12]. In the TDOC, the time interval between two neighboring optimizations is called "optimal control frequency". Typical optimal control frequency ranges from a few minutes to a few hours [13]. The TDOC scheme has been widely used due to its simplicity and effectiveness. For example, Kusiak, Li and Tang [14] performed an hourly optimization of the set-points of a supply air pressure and temperature, leading to a 7.66% energy saving. Huang, Zuo and Sohn [15] optimized the condenser water set-point for an existing chiller plant, and the hourly optimization offered a 9.67% energy saving.

However, since AC systems always experience stochastic changes in their operation conditions, such as weather and occupancy changes [10], using a periodic optimization mechanism, TDOC will have inherent drawbacks, e.g., stochastic (or aperiodic) changes cannot be captured promptly or correctly. Consequently, the TDOC may lead to delayed or unnecessary control actions, which would degrade the expected optimization performance. To deal with this issue, an event-driven optimal control (EDOC) was recently developed by Wang et al. for AC optimal control [11], where control actions are triggered only when pre-defined events occur. In the study of Wang et al. [11], two events were defined based on the part-load ratio (PLR), because PLR has great impact on the AC optimal control [16]. One event was defined as the significant change of the chiller plant PLR, and one was defined as the chiller sequence change, which is also triggered by the variation of PLR. Thus, this EDOC was titled as a PLR-based EDOC. Numerical studies showed that the PLR-based EDOC was able to achieve a better energy efficiency (0.4–2.6% higher), and simultaneously reduce the computational cost by 60–84% when compared with a traditional TDOC approach.

Many studies use PLR as an indicator to optimize the AC operations, e.g., optimal chiller loading [17]. However, for triggering the optimal control, the real and direct trigger of optimal control is the system coefficient of performance (SCOP) [18]. SCOP links the system cooling/heating output to the system power [19,20]. Only when the SCOP deviates from the optimal value, an optimal control action is needed. Since the relationship between the PLR and SCOP is nonlinear [12,21], the PLR cannot reflect the SCOP precisely. Using PLR variation to trigger would lead to non-optimal actions. For instance, when the PLR has a significant change (e.g., 10%) compared with the last optimization time instant, it is possible that the system operation efficiency remains at its ideal level, but the PLR-based EDOC strategy will trigger optimization, leading to unnecessary control actions. When the PLR has a little change (e.g., 2%), it is possible that the system operation efficiency deviates largely from its ideal level, but the optimal control is not triggered, leading to a degraded operation efficiency.

For a more efficient EDOC, events could be directly defined based on the control objective(s). For instance, Xu et al. [22] investigated a PMV-based event-triggered mechanism for building energy management. The control objective is to satisfy the thermal comfort of occupants, while minimizing the building electricity cost. Event 1 was defined as the equality between the predicted and actual occupied time, and Event 2 was defined by a thermal comfort range. The two events were directly linked to the control objective, and the simulation results show that the energy cost can be saved with the reduced power demand. Therefore, to optimize the operation efficiency, events could be directly defined based on the SCOP. SCOP has been widely used to evaluate the performance of AC system operations [23,24]. A higher SCOP indicates that less power is required when the same cooling/heating capacity is provided. Thus, the SCOP of AC systems should be maintained at its ideal value to minimize the energy use. Accordingly, events can be defined such that the optimal control is triggered when the SCOP is lower than its ideal value.

However, using the SCOP to trigger the AC optimal control has not been systematically investigated in terms of its energy performance and robustness. This study aims to develop the SCOP-based EDOC for AC systems, and systematically compares SCOP-based and PLR-based EDOC, regarding their merits and demerits. Since the ideal SCOP varies with the working condition [22,23], an artificial neural network (ANN) is developed to predict the ideal SCOP under various conditions. The methodology of SCOP-based EDOC is firstly illustrated, including events definitions and a SCOP prediction model. Then, the EDOC approaches are evaluated by testing in a case AC system. The control performance and applicability of different EDOC approaches are compared. Discussions and concluding remarks are given at the end.

2. Methodology

2.1. Overview

The real-time optimal control mainly consists of two steps (see Figure 1): (1) Triggering optimal control; (2) solving the optimal control problem using a certain search algorithm subject to operation constraints. Finally, optimal control settings will be sent to the AC system to supervise the operation. In this study, the triggering of optimal control will be studied, and three triggering approaches will be discussed, i.e., The time-driven approach, part-load ratio (PLR)-based event-driven, optimal control (EDOC) approach and the system coefficient of performance (SCOP)-based EDOC approach.

A case commercial AC system is simulated in the Transient System Simulation Tool (TRNSYS) with validated component models. The optimal control codes and basic local control codes are programmed in MATLAB. Actual weather and cooling load data are used as the simulation inputs. The actual system operation is simulated through co-simulation between TRNSYS and MATLAB (Figure 2). Typical weather and load profiles are tested with each case in 24 h. The typical load profiles are identified using a PPA-based K-means clustering approach (see Appendix D). At last, the simulation results are compared to evaluate the performance.

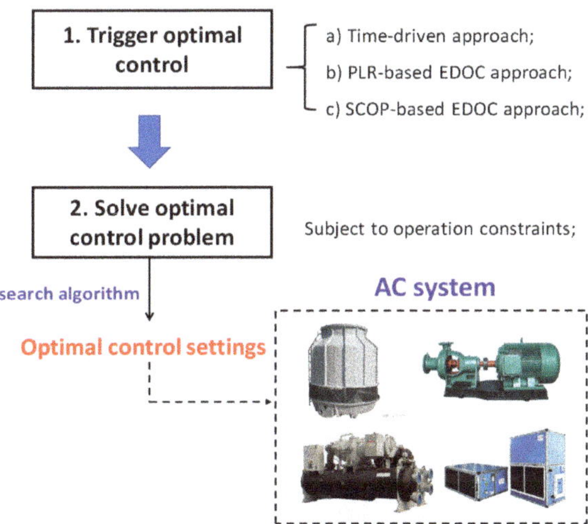

Figure 1. Overview of the research.

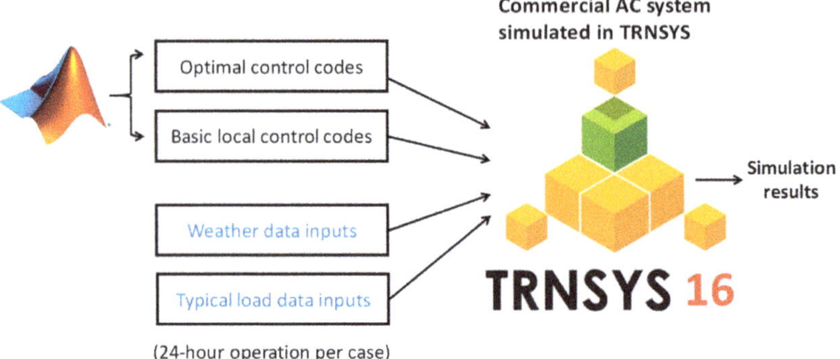

Figure 2. Simulation diagram and software.

2.2. Optimal Control Triggering Approaches

2.2.1. PLR-based EDOC Approach

For PLR-based EDOC, "PLR significant change" and "chiller operating number change" are used as the events. The PLR is a continuous variable, and a threshold is critical for the event definition [11]. Till now, there is no selection or calculation method on this PLR variation threshold. The American Society of Heating, Refrigerating and Air-Conditioning Engineers (ASHRAE) handbook's heating, ventilation, and air conditioning (HVAC) applications (in Section 3.2 of Chapter 42) suggests an example of 10% for the threshold of chilled-water load change. For a certain system, [10] suggested that the threshold of PLR variation can be customized for a better energy performance. This study will choose a suitable threshold from 5% to 10%.

2.2.2. SCOP-based EDOC Approach

The basic idea of the SCOP-based EDOC is to control the deviation between the current SCOP and its ideal SCOP within an acceptable range. The SCOP is defined by

$$SCOP = Q_{sys,tot} / P_{sys,tot} \tag{1}$$

where $Q_{sys,tot}$ is the system total cooling capacity (kW), and $P_{sys,tot}$ is the system total power of all the major equipment (kW).

Two events are defined to capture the transient and cumulative SCOP deviations.

Event 1: The transient SCOP deviation (which deviates from its desired value) is larger than σ_{tra}.

$$e_1 := \{[SCOP_{idl}(\tau) - SCOP(\tau)] > \sigma_{tra}\} \tag{2}$$

where τ is a time index; $SCOP_{idl}$ is the reference of SCOP; and σ_{tra} is a predefined threshold.

Event 1 actually defines an unacceptable SCOP curve, denoted by the curve $SCOP_2$ in Figure 3, the value of which is calculated by $SCOP_2(\tau) = SCOP_{idl}(\tau) - \sigma_{tra}$. Event 1 occurs at every moment when the SCOP curve crosses the $SCOP_2$ curve downward, which indicates the SCOP becoming unacceptable, and thus the optimization should be taken immediately to improve the SCOP.

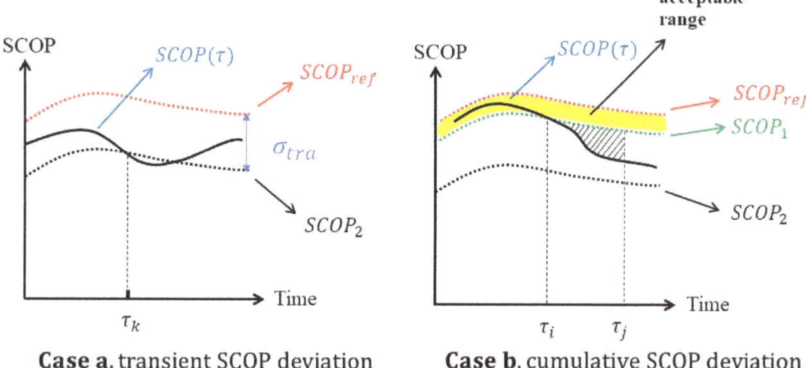

Figure 3. Two events of SCOP deviations.

It is possible that the SCOP is lower than a certain value in a long period, but does not cross down the unacceptable curve ($SCOP_2$). In this situation, if no optimal control is taken, the operation efficiency cannot be guaranteed in the long run. Therefore, an event dedicated for the cumulative SCOP deviation is defined as below:

Event 2: The cumulative SCOP deviation is larger than σ_{cum}.

$$e_2 := \{\int_{\tau_i}^{\tau_j}[SCOP_1(\tau) - SCOP(\tau)]d\tau > \sigma_{cum} \\ and\ SCOP(\tau) \in [SCOP_2(\tau), SCOP_1(\tau)]\ \forall\tau \in (\tau_i, \tau_j)\} \tag{3}$$

where τ_i and τ_j are start and end time of a continuous period; $SCOP_1$ is a bound to define the acceptable SCOP range; and σ_{cum} is a predefined threshold.

Note that in Event 2, to exclude the model uncertainty in the calculation of the cumulative SCOP deviation, $SCOP_1$ is used instead of $SCOP_{idl}$ in the integration. The curves $SCOP_1$ and $SCOP_{idl}$, as shown in Figure 3b, define a range in which the SCOP is acceptable. The calculation of the cumulative deviation should start only when the SCOP goes into the range $[SCOP_2, SCOP_1]$. The calculation will stop, and the cumulative deviation value will be reset to zero when Event 2 or Event 1 occurs.

In the event detection, the SCOP deviation is calculated based on the measured data that are collected at each sampling time [10]. Considering the discrete-time sampling, "SCOP·mins" is used to approximate the integration in Equation (3). One minute is used as the discretization time interval to approximate the integration, as one minute is small enough for sampling time intervals in buildings. The cumulative SCOP deviation is represented by SCOP·mins in Equation (4), where the integration is approximated by a summation.

$$SCOP{\cdot}mins(\tau_i, \tau_j) = \int_{\tau_i}^{\tau_j}[SCOP_1(\tau) - SCOP(\tau)]d\tau \approx \sum_{\tau_i}^{\tau_j}[SCOP_1(\tau) - SCOP(\tau)] \tag{4}$$

2.3. Ideal SCOP Model

To develop the Artificial Neural Network (ANN)-based SCOP ideal ($SCOP_{idl}$) model, important variables affecting the SCOP should firstly be identified. A crucial step is to select the input variables, since the inclusion of unimportant variables may bring in redundant information and decrease the ANN model accuracy [25].

Traditional techniques, e.g., correlation coefficient, standard regression coefficient and the products of these two coefficients, are inadequate to handle correlated data [26]. Thus, advanced techniques are required, such as variance-decomposition-based, variable-transformation-based and machine-learning-based techniques [26,27]. The random forest (RF) algorithm is selected because

RF can handle highly correlated variables, avoid overfitting and improve the prediction accuracy. Moreover, the variable importance can be represented by "%IncMSE" from RF [28]. The calculation details of "%IncMSE" are shown in Appendix A.

To develop the prediction model of $SCOP_{idl}$, the ANN algorithm is used since ANN has demonstrated its capability in handling complex relationships in building energy fields [29,30]. A typical three-layer feed-forward ANN (including the input layer, the hidden layer and the output layer) is used in this study (see Figure 4). The identified important variables are used as the ANN model inputs, while the $SCOP_{idl}$ is the ANN output. The ANN toolbox in MATLAB is used, and the Levenberg–Marquardt algorithm (see "trainlm" of MATLAB) is adopted in the training. A different number of hidden neurons are tested to select a suitable number, where the mean squared error (MSE) is used for the model evaluation.

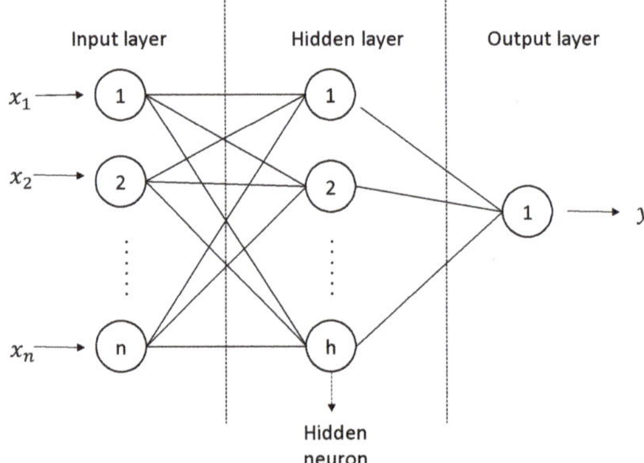

Figure 4. The structure of the Artificial Neural Network (ANN).

3. Case Study: System, Problem Formulation and Simulation

The selected case air-conditioning system serves a super-tall office building in Hong Kong. As Hong Kong is a subtropical region, only cooling is considered in the case study.

3.1. System Description

The schematic diagram is shown in Figure 5, which is a typical primary-constant and secondary-variable chilled water system. The specifications of its main components are presented in Table 1. Four critical temperatures are controlled by PI controllers, namely the supply cooling water temperature (T_{scw}), the supply chilled water temperature at the primary ($T_{schw,prm}$) and secondary loops ($T_{schw,sec}$), and the supply air temperature (T_{sa}). T_{scw} is controlled through varying the cooling tower fan frequency; $T_{schw,prm}$ is controlled by modulating the refrigerant flow rate; $T_{schw,sec}$ is controlled by modulating the water flow rate; T_{sa} is controlled by AHUs. The set-points of the above four critical temperatures are optimized and reset in real-time to achieve the best energy efficiency.

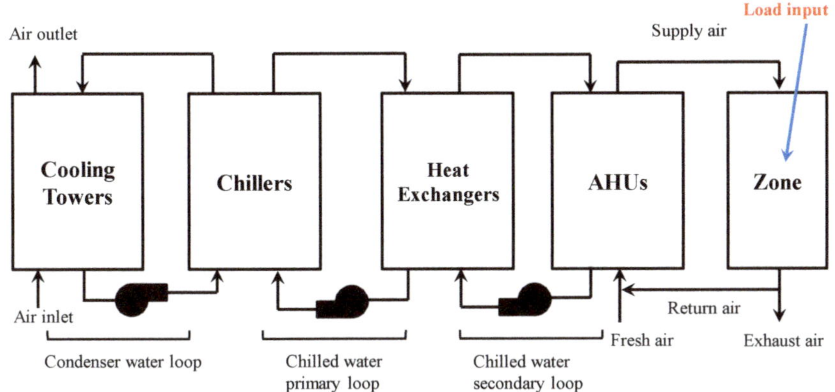

Figure 5. Schematic of the air-conditioning (AC) system.

Table 1. Specifications of the main components.

Equipment	Number	Capacity (kW)	Rated Power (kW)	$M_{w,ev}$ (L/s)	$M_{w,cd}$ (L/s)
Water-cooled Chiller	6	7230	1346	345	410
Equipment	Number	Heat rejection capacity (kW)	Rated power (kW)	M_w (L/s)	M_a (m^3/s)
Cooling tower - type A	6	5234	152	250	157.2
Cooling tower - type B	5	4061	120	194	127.0
Equipment	Type	Efficiency (%)	Rated power (kW)	M_w (L/s)	Head (m)
Condenser water pump	Constant speed	83.6	202	410	41.6
Primary chilled-water pump	Constant speed	84.5	126	345	31.6
Secondary chilled-water pump	Variable speed	84.2	163	345	41.4

Note: M_a is the air flow rate; M_w is the water flow rate; *ev* and *cd* stand for evaporator and condenser respectively.

3.2. Optimal Control Problem Formulation

For all-electric AC systems without thermal storage, the minimization of the energy use can be simplified as the minimization of the system total power at each time instance [16]. In this case study, the decision variables that significantly affect the energy efficiency of the system operation are T_{scw}, $T_{schw,prm}$, $T_{schw,sec}$ and T_{sa}. Thus, the real-time optimal control problem is formulated as

$$\left(T^*_{scw}, T^*_{schw,prm}, T^*_{schw,sec}, T^*_{sa}\right) = argmin\ P_{sys,tot}\left(T_{scw}, T_{schw,prm}, T_{schw,sec}, T_{sa}\right) \tag{5}$$

where $P_{sys,tot} = P_{ch,tot} + P_{ct,tot} + P_{pump,tot} + P_{AHU,tot}$.

The system total power can be written as a function of these four decision variables

$$P_{sys,tot} = f\left(T_{scw}, T_{schw,prm}, T_{schw,sec}, T_{sa}\right) \tag{6}$$

where $f(\cdot)$ is always a nonlinear function. The decision variables are limited in their feasible ranges, which are treated as operational constraints (shown in Equations (7)–(10)).

Table 2 shows the values of the operational constraints.

$$T_{scw,lower} \leq T_{scw} \leq T_{scw,upper} \tag{7}$$

$$T_{schw,prm} \leq T_{schw,prm} \leq T_{schw,prm,upper} \tag{8}$$

$$T_{schw,sec} \leq T_{schw,sec} \leq T_{schw,sec,upper} \tag{9}$$

$$T_{sa,lower} \leq T_{sa} \leq T_{sa,upper} \tag{10}$$

Table 2. Operational constraints of the case AC system [11].

Name	Value		
	Spring	Summer	Autumn
$[T_{scw,lower}, T_{scw,upper}]$	[20, 28] °C	[28, 35] °C	[24, 30] °C
$[T_{schw,prm,lower}, T_{schw,prm,upper}]$		[5, 8] °C	
$[T_{schw,sec,lower}, T_{schw,sec,upper}]$		[6.5, 11.5] °C	
$[T_{sa,lower}, T_{sa,upper}]$		[12, 18] °C	

3.3. Simulation Platform

A dynamic simulation platform integrating TRNSYS and MATLAB was constructed to test the performance of the proposed optimal control approaches. The platform layout was shown in Figure 6. The used TRNSYS models were verified by measured data with acceptable model accuracies. The model details are presented in Table 3. The control logics presented in Section 2 were coded in MATLAB. The simulation duration for each run was 24 h, with a simulation time step of 30 s. Several typical loads and weather data are simulated. The computer used to perform the simulation has a six-core processor (3.60 GHz) with 16 GB RAM. Since cooling load and weather data are crucial for the AC optimal control, actual load and weather data were input to the TRNSYS simulation to mimic the actual operations. The airflow rate was calculated based on the cooling load and the enthalpy difference between the supply air (the temperature set-point with 95% relative humidity) and the room air (set as 25 °C with 50% relative humidity).

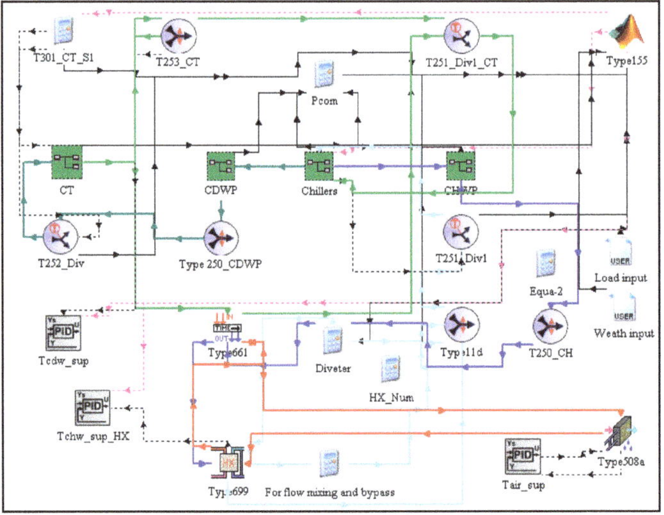

Figure 6. Transient System Simulation Tool (TRNSYS) model (screenshot).

Table 3. TRNSYS model and description.

Name	Model Type	Description
Chillers	Grey-box model (dynamic)	Verified by the measured data (see Chapter 3.2 and Chapter 5 of [31]); Model accuracy: ±10%.
Cooling towers	Grey-box model (dynamic)	Verified by the measured data (see Chapter 3.2 and Chapter 5 of [31]); Model accuracy: ±6.9% for predicting the heat reject load and ±1% for predicting outlet water temperature.
Variable-speed Pumps	Polynomial model (steady-state)	Verified by the measured data (see Chapter 3.2 and Chapter 5 of [31]); Model accuracy: the pump model can predict exactly as the catalogue data.
AHU	Physical model (dynamic)	TRNSYS Model: Type 508a
Heat exchanger	Physical model (dynamic)	TRNSYS Model: Type 699
PID Controller	Physical model (dynamic)	TRNSYS Model: Type 23; The PID settings: P = – 0.95 and I = 35 s [11]

3.4. Ideal SCOP Model

To develop an ANN model for predicting the ideal SCOP, important variables that have significant influence on the SCOP were firstly identified. Basically, important variables that affect the SCOP are similar in typical AC systems that contain chillers, cooling towers, pumps and AHUs/Coils [16,32]. The variables include the outdoor dry/wet-bulb temperature (T_{db}/T_{wb}), the chilled water supply temperature (T_{schw}), the condensing water return temperature (T_{scw}), the chilled water mass flow rate at the primary/secondary loop ($M_{w,prm,pump}/M_{w,sec,pump}$), the PLR, the cooling tower fan frequency ($Freq_{ct,fan}$) and the enthalpy driving force in the operation of cooling towers "Δh" [33].

To evaluate the variable importance, the RF algorithm was implemented in the software R using the package of "randomForest" [34] and the results were plotted in Figure 7. The variable PLR has the highest importance, while the $M_{w,sec,pump}$ has the lowest importance. As discussed in [35], less important variables (e.g., % IncMSE lower than 10%) can be omitted. Therefore, $M_{w,sec,pump}$ was not used in the ideal SCOP model, and totally eight variables were used as the ANN model inputs.

In the ANN model, 70% of the data set was used for training, 15% of the data set was used for validation and the additional 15% of the data set was used to test whether the model can perform well. Since the optimal number of the hidden layer neurons depends upon the specific problem, this study tested a different number of hidden neurons (from 2 to 20). The MSEs of different numbers of hidden neurons were plotted in Figure 8. Figure 8. shows that when the number of neurons increased, the MSE decreased. However, the reduction in MSE became insignificant as the number of neurons was greater than 15. Thus, "15" was selected as the optimal number of hidden neurons. The R-value of the ANN model with 15 hidden neurons was 0.96052 (see Figure 9), with R values of training, validation and test are 0.96007, 0,95783 and 0.9656 respectively, which showed a good prediction accuracy.

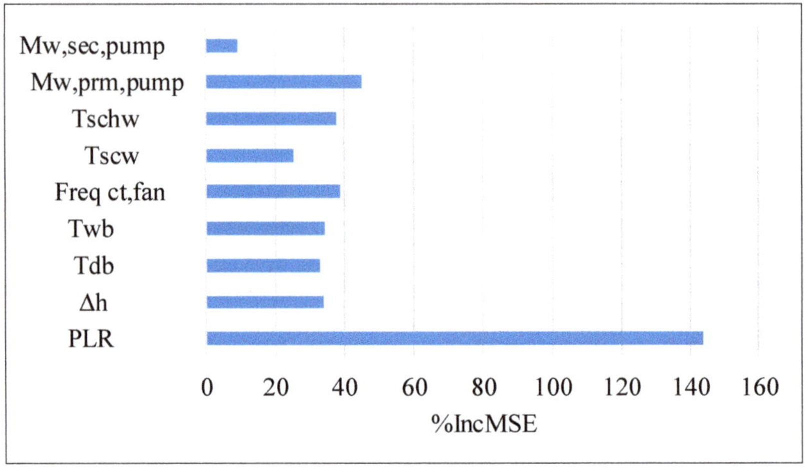

Figure 7. Results of random forest (higher %IncMSE means more important).

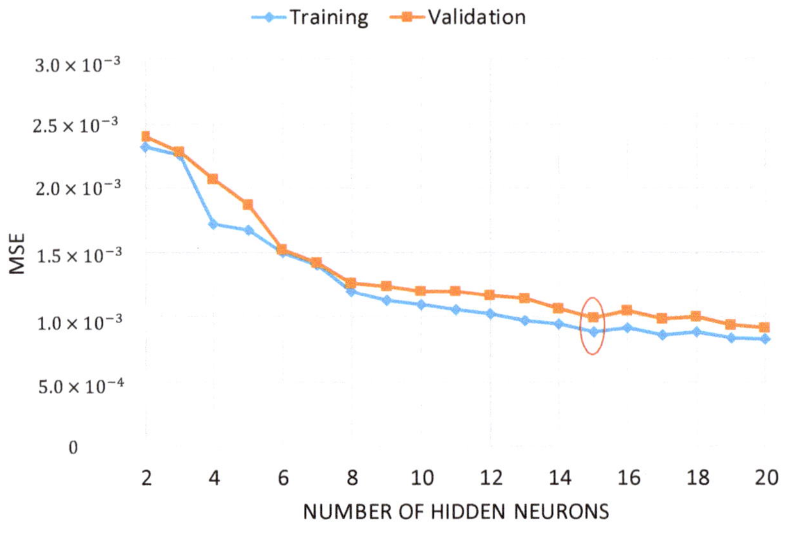

Figure 8. Number of ANN hidden neurons vs. mean squared error (MSE).

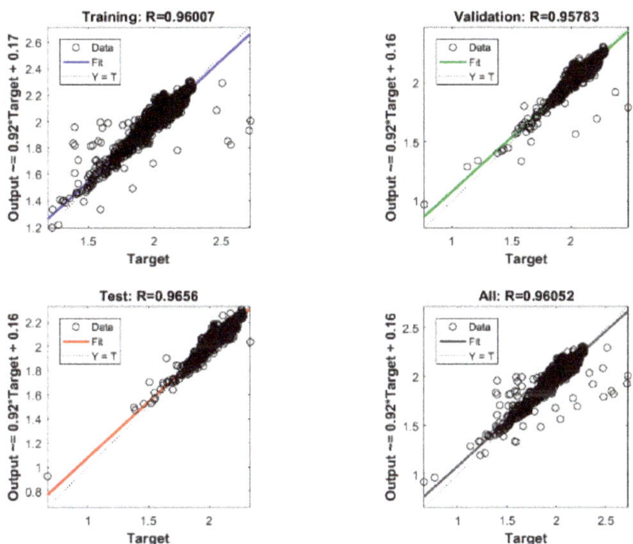

Figure 9. Regression plots of the ANN model (with "15" hidden neurons).

3.5. Event Threshold

For the event "PLR significant change", 7% was selected as the PLR change threshold based on a previous study on the same AC system [11]. The threshold of SCOP deviation can be calculated based on the value of $SCOP_{idl}$ and an expected energy saving percentage (as shown in Appendix C Equation (A8)). For instance, an expected energy saving percentage can be specified by users (e.g., 10%), and the $SCOP_{idl}$ can be obtained from the developed ANN model. After the threshold is calculated, it should be checked that whether the calculated threshold is feasible or not based on the SCOP deviations in historical data. The threshold selection should also consider the uncertainties in the $SCOP_{idl}$ model and SCOP measurement in terms of the application concerns.

The thresholds of the SCOP deviation are set as 0.2057 (transient) and 2.057 (cumulative). The selection details are presented as follows: The targeted energy saving is set as 10%, based on handbook [16] and a previous study [10]. To help the threshold selection, the histogram of $SCOP_{idl}$ and SCOP deviation are plotted in Figure 10. The base case keeps all the four temperature set-points as constant. The SCOP deviation is the difference between SCOP of the base case and $SCOP_{idl}$. From Figure 10, the mean of $SCOP_{idl}$ is 2.0544. Using a $SCOP_{idl}$ of 2.0544 and an expected energy saving of 10.0%, the threshold of SCOP deviation is 0.2057, based on Equation (A8) (Appendix C).

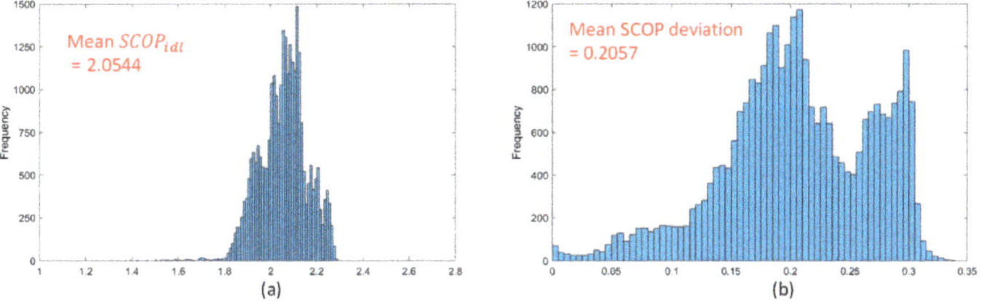

Figure 10. (**a**) Histogram of $SCOP_{idl}$; (**b**) Histogram of SCOP deviation.

Since "0.2057" is within the range of SCOP deviation and the most frequent case (Figure 10b), it is a feasible threshold value.

For uncertainties, the value 0.01 was used for as the model uncertainty of $SCOP_{idl}$, which was the averaged value over the training and validation Mean Squared Error (MSE) of the ANN model. The measurement uncertainty for the case study was calculated in Appendix B, where sensor uncertainties were added to the simulation-generated operation data. As a negative value is considered, "–0.0694" was used for measurement uncertainty. Thus, the combined uncertainty is "–0.0794" ("–0.01"+"–0.0694"), and $SCOP_1$ can be set as "$SCOP_{idl} - 0.0794$".

In comparison, the threshold 0.2057 is nearly three times greater the combined uncertainty, which can largely prevent the uncertainty-caused event triggering. Thus, 0.2057 is used as the threshold of transient SCOP deviation. The cumulative SCOP deviation depends on both the time duration and the SCOP deviation. The deviation for cumulative SCOP was considered as half of the transient SCOP deviation threshold (i.e., 0.2057/2), or a 5% energy saving equivalently. The time duration was selected as 20 minutes for a demonstration. Thus, the threshold of cumulative SCOP deviation is 2.057.

4. Results and Discussions

4.1. Energy Performance

The optimal control performance is evaluated by the energy saving percentage. The energy saving percentage is computed based on the base case (i.e., constant control settings without optimal control). The energy saving percentages of EDOC range from 5.83% to 12.06% for ten different load clusters. In general, as can be seen from Figure 11, SCOP-based EDOC outperforms PLR-based EDOC, and EDOC outperforms TDOC (1 h per optimal control).

The biggest energy performance difference can be 3%, as shown in load cluster no. 3. The possible reason of such superior performance is that the SCOP-based events can directly represent the system operation efficiency without mismatch compared with the PLR-based events. By continuous monitoring and tracking of the transient and cumulative SCOP deviations, a better timing of optimal control can be achieved. As a result, the system operation efficiency is always maintained at or near the ideal level. The next section will discuss the triggering time of optimal control in details.

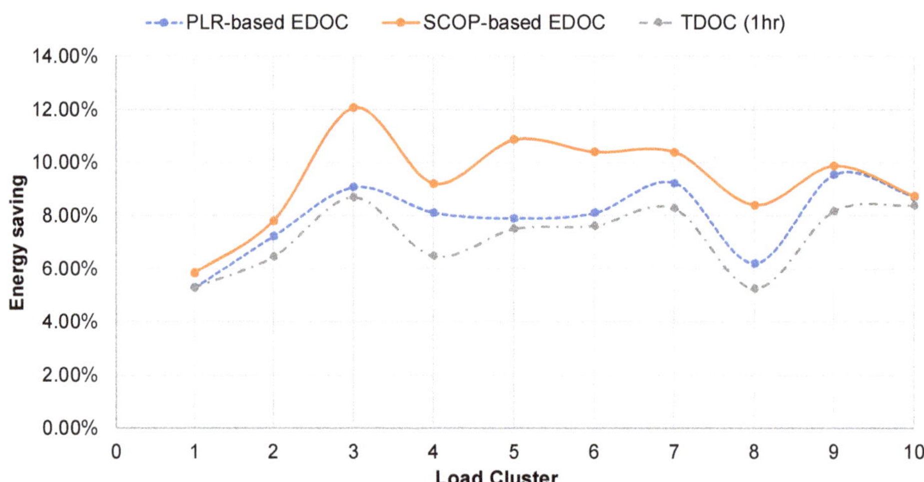

Figure 11. Energy saving percentages.

4.2. Triggering Time of Optimal Control

It was observed that (see Table 4), in load cluster 1, both TDOC and EDOC approaches triggered seven times control actions. Thus, load cluster 1 is the best for analyzing the effect of different optimal control triggering times.

In Figure 12, the optimal control time of EDOC is plotted. The Y axis refers to the time of the optimal control action, and the Y axis is updated each time when this optimal control is triggered. The X axis is the simulation time step in minutes. From Figure 12 (see dashed circles), SCOP-based EDOC tends to react in advance compared to the PLR-based EDOC. The possible reason is because SCOP is a more comprehensive energy efficiency index than the PLR index. SCOP-based EDOC can capture the operation variation earlier than the PLR-based EDOC, leading to a better energy performance. Moreover, inside the solid circles of Figure 12, the SCOP-based EDOC did not take response, while this PLR-based EDOC triggered optimal control. In such circumstances, the PLR varies, but the SCOP still remains at the ideal level, which is unnecessary to trigger optimal control. The SCOP-based EDOC could avoid such unnecessary actions, while the PLR-based EDOC cannot (which is a demerit).

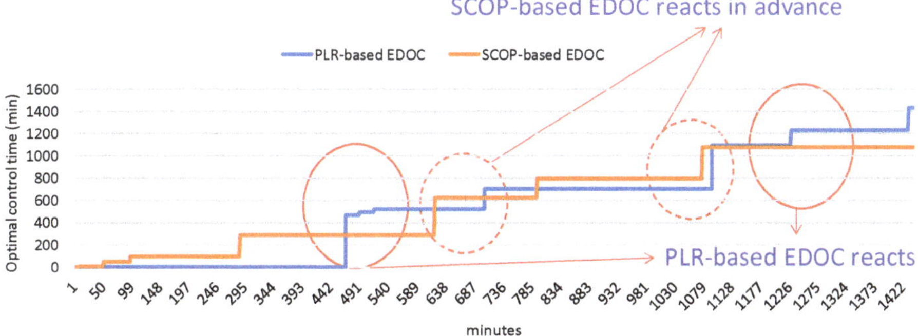

Figure 12. Optimal control time of EDOC (load cluster 1).

The total optimal control times of EDOC are shown in Table 4, which reflects the frequency of optimal control. Based on literatures [15,36,37], it can be found that 1 h per optimal control is the most common optimal control frequency used for TDOC. Thus, this study uses 1 h as the benchmark. Then, less than 24 times can be regarded as an acceptable triggering frequency for one day. On average, PLR-based EDOC (22.7 times per day) satisfies the condition, while SCOP-based EDOC (26.4 times per day) triggers more actions than the benchmark. It should be noted that, in load cluster no. 8, the SCOP-based EDOC performed 61 times of optimal control. The reason for such frequent triggering is possibly due to the uncertainty in "SCOP". As the parameter, SCOP is calculated based on the power meter and the load measurement, the combined uncertainty of power and load could be significant. Besides, the model of ideal SCOP also has model uncertainty. When the SCOP value fluctuated at the pre-defined threshold value, the optimal control action can be triggered frequently. From this, a demerit of SCOP-based EDOC is that frequent optimal control may be triggered compared with PLR-based EDOC. This problem could be mitigated by adding an additional time interval constraint between two adjacent optimal control actions (e.g., at least 10 mins between two control actions). Although SCOP-based EDOC obtains the highest energy saving, in terms of the energy saving performance, triggering frequency and robustness of optimal control, PLR-based EDOC can still be recommended to replace traditional TDOC for practical applications.

Table 4. Optimal control triggering frequency of EDOC.

Load Cluster	PLR-Based EDOC	SCOP-Based EDOC
1	7	7
2	16	21
3	31	19
4	13	37
5	32	25
6	29	27
7	28	11
8	15	61
9	29	36
10	27	20
Average optimal control triggering frequency	22.7	26.4

4.3. Analyses of SCOP-based Events

As both transient and cumulative SCOP deviations are used (Section 2.2.2), the event triggering times of the two events are analyzed in this section. From Figure 13, e_1 and e_2 were triggered differently under different load clusters, e_1 sometimes dominates, and e_2 sometimes reacts more. For example, in load cluster C1 and C2, only e_2 was triggered. In load cluster C3–C10, e_1 and e_2 both appeared. If only using e_1 in load cluster C1 and C2, the SCOP-based EDOC may not work. This validates the idea of combing e_1 and e_2 to accommodate broader operation conditions, since one event may fail sometimes.

Regarding the individual energy performance of the single event (e_1 and e_2), load cluster C7 was used as an illustration, where two events are triggered almost equally. Simulation results show: Using only e_1 gives 9.84% of energy saving, and using only e_2 gives 9.97% of energy saving. In Figure 11, combining e_1 and e_2 outputs 10.38% of energy saving. It can be seen that combining e_1 and e_2 will be beneficial for improving the energy performance of EDOC. The reason is because the single event may not capture the critical operation variations in a comprehensive manner.

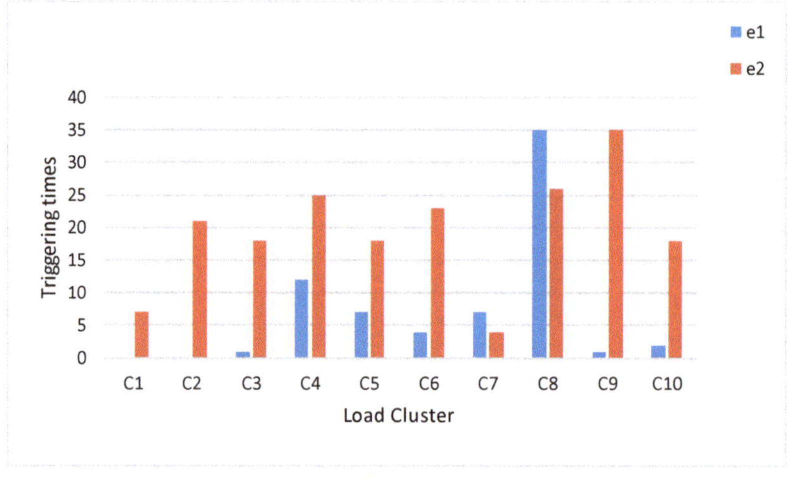

(e_1: transient SCOP deviation, e_2: cumulative SCOP deviation, C1: load cluster 1.)

Figure 13. Triggering times of SCOP-based events.

5. Conclusions

This study investigates the optimal control triggering of EDOC in AC systems. Optimal control triggering approaches are classified into direct and indirect approaches based on the used triggering indicator. In the direct SCOP-based EDOC, two events have been defined to capture both the transient and cumulative SCOP deviations, which maintains the SCOP at the ideal level in a closed-loop manner. Simulated case studies suggest that direct SCOP-based EDOC achieved a higher energy saving than indirect PLR-based EDOC in all the typical load scenarios. The main reason is due to a more direct performance indicator, i.e., SCOP, that can reflect the system performance precisely and timely. However, the development of an accurate ideal SCOP model requires sufficient operation data and extensive work, which could restrict its implementation [8]. The robustness issues caused by the uncertainty of SCOP should also be paid attention to. Thus, in terms of actual applications, the SCOP-based EDOC can be recommended when the ideal SCOP model is available with a properly-handled uncertainty issue. However, the PLR-based EDOC could still be a more practical option to replace traditional TDOC, which features an easy development process, better robustness and acceptable energy performance.

Author Contributions: Conceptualization, J.W.; Data curation, R.L. and L.Z.; Formal analysis, H.S.A. and E.M.; Funding acquisition, J.W.; Investigation, R.L.; Methodology, J.W.; Resources, L.Z.; Supervision, J.W.; Validation, H.S.A. and E.M.

Funding: This research was funded by Natural Science Foundation of Jiangsu Province (BK20190942), Suzhou University of Science and Technology Research projects funding for the introducing talents (331911204), Comprehensive curriculum reform program (2018KJZG-09), and Suzhou science and technology plan projects (SNG2017052).

Acknowledgments: The authors would acknowledge to the Staffs of City University of Hong Kong for their supports.

Conflicts of Interest: The authors declare no conflict of interest.

Nomenclature

	Variables/Abbreviations	Subscripts	
SCOP	system coefficient of performance	scw	supply cooling water
EDOC	event-driven optimal control	schw	supply chilled water
TDOC	time-driven optimal control	sa	supply air
TLP	typical load profile	prm	primary
PLR	part-load ratio	sec	secondary
SCOP	system coefficient of performance	wb	wet-bulb
s	event threshold	db	dry-bulb
T	Temperature (°C)	sys	system
P	Power (kW)	tot	total
Mw	water flow rate (L/s)	ct	cooling tower
Freq	Frequency (Hz)	ch	chiller
PAA	Piecewise Aggregate Approximation	pump	pump
Δh	enthalpy difference between the saturated (at cooling tower inlet water temperature) and incoming air of cooling tower (kJ/kg)	fan	fan
		ref	reference value
		lower	lower bound
		upper	upper bound

Appendix A. Variable Importance Measures ("%IncMSE")

In the random forest algorithm, the variable importance is represented by "%IncMSE".

$$\%IncMSE(v_j) = 100\% \times \left[MSE(-v_j) - MSE\right]/MSE \tag{A1}$$

where "$MSE(-v_j)$" stands for the MSE if v_j is not used in the prediction. A higher $\%IncMSE$ suggests that the variable v_j is more important.

Appendix B. SCOP Measurement Uncertainty

This part is to calculate the measurement uncertainty of SCOP. The equations for SCOP are shown in (A2)–(A4), where Q_{sys} is the system cooling capacity, c_w and m_w are the water specific heat capacity and water flow rate (others please see the nomenclature). Assuming the sensors/meters are well-calibrated, the reported accuracies of sensors/meters from a chiller design guide [38] can be used to quantify the measurement uncertainty of SCOP.

$$SCOP = Q_{sys}/P_{sys} \quad (A2)$$

$$Q_{sys} = c_w m_w (T_{CHWR} - T_{CHWS}) \quad (A3)$$

$$P_{sys} = P_{ch} + P_{ct} + P_{pump} + P_{AHU} \quad (A4)$$

Considering the measurement uncertainties in the cooling capacity and power, the Equation (A2) can be re-written as Equation (A5), where the subscript mea means the measured value, $\Delta P_{sys,mea}$, $\Delta Q_{sys,mea}$ and $\Delta SCOP_{mea}$ are the measurement uncertainties for the power, cooling capacity and SCOP.

$$SCOP_{mea} = Q_{sys,mea}/P_{sys,mea} \quad (A5)$$

where $P_{sys,mea} = P_{sys} + \Delta P_{sys,mea}$; $Q_{sys,mea} = Q_{sys} + \Delta Q_{sys,mea}$; $SCOP_{mea} = SCOP + \Delta SCOP_{mea}$;

Since multiple uncertainties are involved in the SCOP calculation, uncertainty shift can be used to integrate the variable uncertainties for easy analyses [39]. The measurement uncertainty of power consumption can be easily quantified by the power meter uncertainty. For the measurement uncertainty of the cooling capacity, the measured cooling capacity is represented by Equation (A6), and the measurement uncertainty can be calculated by Equation (A7), as demonstrated by [39].

$$Q_{sys,mea} = c_w m_{w,mea}\left(T_{CHWR,mea} - T_{CHWS,mea}\right) \quad (A6)$$

$$\begin{aligned} \Delta Q_{sys,mea} &= Q_{sys,mea} - Q_{sys} \\ &= \Delta m_{w,mea}\left(T_{CHWR,mea} - T_{CHWS,mea}\right) + m_{w,mea}\left(\Delta T_{CHWR,mea} - \Delta T_{CHWS,mea}\right) \end{aligned} \quad (A7)$$

With $m_{w,mea} = m_w + \Delta m_{w,mea}$; $T_{CHWR,mea} = T_{CHWR} + \Delta T_{CHWR,mea}$; $T_{CHWS,mea} = T_{CHWS} + \Delta T_{CHWS,mea}$; where $\Delta m_{w,mea}$, $\Delta T_{CHWR,mea}$, $\Delta T_{CHWS,mea}$ are the measurement uncertainties for the corresponding variables.

Based on measurement accuracies in Table A1, the measurement uncertainties of cooling capacity and power can be calculated. The simulated operation data is considered as the measured value. As presented in Table A2, the maximum positive and negative relative uncertainties for $\Delta SCOP_{mea}$ are +3.73% and -3.95%, with the absolute uncertainties of "0.0708" and "-0.0694" respectively.

Table A1. Typical measurement accuracy [38].

Measured Variable	Reported Accuracy
Water flow	±1% *of full scale*
Water temperature	±0.1°C
Electrical power	1% *of reading*

Table A2. Measurement uncertainties of SCOP (worst cases).

$\Delta m_{w,mea}$	$(\Delta T_{CHWR,mea} - \Delta T_{CHWS,mea})$	$\Delta Q_{sys,mea}$	$\Delta P_{sys,mea}$	Relative SCOP Measurement Uncertainty	Absolute SCOP Measurement Uncertainty
+1% of full scale	+0.2°C	197 kW	−38.01 kW (−1% of reading)	+3.73%	+0.0708
−1% of full scale	-0.2°C	−197 kW	+38.01 kW (+1% of reading)	-3.95%	-0.0694

Appendix C. Energy Saving Estimation

The energy saving percentage at a time instant can be estimated using Equation (A8) by substituting the value of $SCOP(\tau)$ and $SCOP_{idl}(\tau)$.

$$\frac{\Delta P(\tau)}{P(\tau)} = \frac{[P(\tau) - P_{idl}(\tau)]}{P(\tau)} = [SCOP_{idl}(\tau) - SCOP(\tau)]/SCOP_{idl}(\tau) \quad \text{(A8)}$$

where $P(\tau) = \frac{load(\tau)}{SCOP(\tau)}$, $P_{idl}(\tau) = \frac{load(\tau)}{SCOP_{idl}(\tau)}$, P is power, idl is the ideal value, τ is the time instant.

Appendix D. Load Clustering

A K-means clustering method is proposed to generate typical load profiles (TLPs) in this study. Figure A1a shows a general procedure of the K-means clustering with a Piecewise Aggregate Approximation (PAA) transformation. To cluster the load profile (a time-series data), the distance between two profiles should be computed for measuring the similarity. However, computing the distance on raw time-series data is difficult and slow. Therefore, the approximation is normally carried out to reduce the computational difficulties [40]. Many representation techniques can be used to transform the raw time-series data, such as Discrete Fourier Transformation [41], Discrete Wavelet Transformation [42], Single Value Decomposition [43], PAA [44], etc. The PAA was used in this paper due to its simplicity and fast calculation speed [44].

In the PAA method, the original load profile is firstly segmented into equal-distance pieces (Figure A1b). Then, the mean of each piece is used to approximate the original segment (Figure A1c). This approximation greatly reduces the data dimension, while the fundamental characteristics in the original time-series data are still captured. After the PAA transformation, the Euclidean distance between two load profiles is calculated, and the standard K-means clustering algorithm is applied. The initialization of the K-means is important as it affects the final result [45]. A different K value will be tested to find a suitable one. A realistic testing range for the K value is between 2 and $\sqrt{n}$ [46], where "n" is the number of data samples. Then, the "furthest first initialization" is used, which starts at a random point as the first cluster center, and adding more cluster centers which are furthest from the existing ones [47]. Based on the clustering result, one representative profile from each cluster will be selected to form the TLPs.

The measured cooling load data of the case building in year 2013 was used. In total 214 daily load profiles from spring season to autumn season were used, which covered typical cooling seasons for sub-tropical regions like Hong Kong. The clustered loads were given in Figure A2a, where the similar load profiles were grouped together as a load cluster. In each cluster, the load profile closest to the cluster centroid was selected to constitute the TLPs (see Figure A2b). These TLPs and their associated weather data (from Hong Kong Observatory) were used as the simulation inputs.

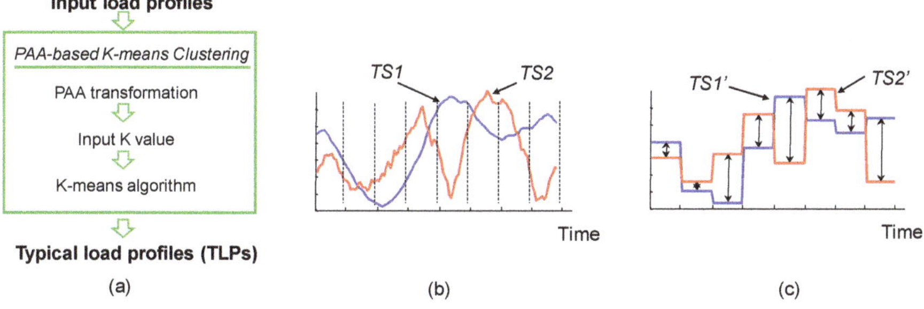

Figure A1. (**a**) Process of generating typical load profiles (TLPs); (**b**) two original load profiles; (**c**) two load profiles after PAA transformation.

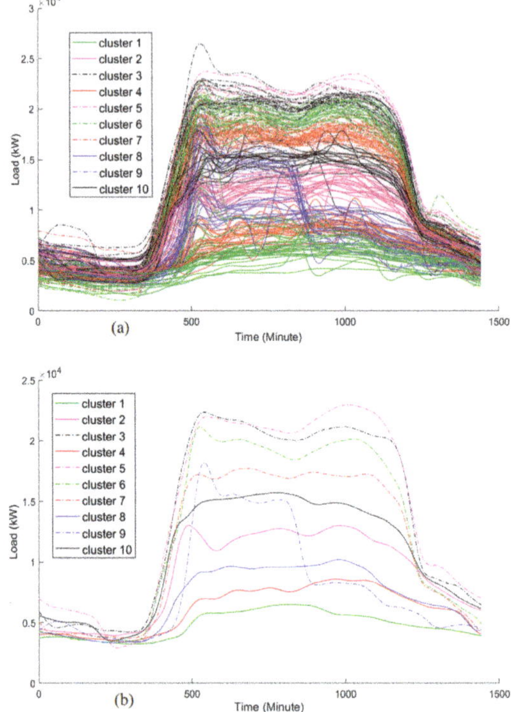

Figure A2. (**a**) Clustered load profiles; (**b**) Selected typical load profiles.

Table A3. Load and weather data of ten TLPs.

Load Cluster.	Load (kW)			Tdb (°C)			Twb (°C)		
	Mean	Max	Min	Mean	Max	Min	Mean	Max	Min
C1#	4833	6495	3268	21.5	23.5	20.6	18.8	19.7	17.4
C2	9072	13038	3954	23.7	25.2	22.1	22.9	23.6	21.5
C3	13998	22361	4212	29.7	32.9	27.6	26.6	28.1	25.9
C4	6164	8593	3437	29.2	32.5	27.3	26.2	27.4	25.3
C5	14492	22996	2923	29.7	33.1	27.4	26.1	27.9	24.9
C6	12646	21177	3367	28.2	31.0	26.4	25.3	26.0	24.4
C7	11465	17715	3740	27.8	29.7	26.8	24.0	25.1	22.8
C8	7037	10241	3355	29.2	31.9	27.2	26.1	27.1	25.1
C9	8067	18175	3475	31.2	34.2	28.4	25.1	26.8	23.5
C10	10628	15714	3487	24.8	27.3	22.9	21.5	22.5	19.5

(#: 'C1' means the load cluster 1.).

References

1. Ahmad, M.W.; Mourshed, M.; Yuce, B.; Rezgui, Y. Computational intelligence techniques for HVAC systems: A review. *Build. Simul.* **2016**, *9*, 359–398. [CrossRef]
2. Liu, Z.; Song, F.; Jiang, Z.; Chen, X.; Guan, X. Optimization based integrated control of building HVAC system. *Build. Simul.* **2014**, *7*, 375–387. [CrossRef]

3. Sun, Y.; Huang, G. Recent Developments in HVAC System Control and Building Demand Management. *Curr. Sustain. Renew. Energy Rep.* **2017**, *4*, 15–21. [CrossRef]
4. Kim, J.-H.; Seong, N.-C.; Choi, W. Modeling and Optimizing a Chiller System Using a Machine Learning Algorithm. *Energies* **2019**, *12*, 2860. [CrossRef]
5. Zhang, S.; Cheng, Y.; Fang, Z.; Lin, Z. Dynamic Control of Room Air Temperature for Stratum Ventilation Based on Heat Removal Efficiency: Method and Experimental Validations. *Build. Environ.* **2018**, *140*, 107–118. [CrossRef]
6. Zhang, S.; Cheng, Y.; Fang, Z.; Huan, C.; Lin, Z. Optimization of room air temperature in stratum-ventilated rooms for both thermal comfort and energy saving. *Appl. Energy* **2017**, *204*, 420–431. [CrossRef]
7. Wang, S.; Ma, Z. Supervisory and optimal control of building HVAC systems: A review. *HVAC&R Res.* **2008**, *14*, 3–32.
8. Atam, E. New Paths Toward Energy-Efficient Buildings: A Multiaspect Discussion of Advanced Model-Based Control. *IEEE Ind. Electron. Mag.* **2016**, *10*, 50–66. [CrossRef]
9. Asad, H.S.; Yuen, R.K.K.; Liu, J.; Wang, J. Adaptive modeling for reliability in optimal control of complex HVAC systems. *Build. Simul.* **2019**, *12*, 1095–1106. [CrossRef]
10. Wang, J.; Jia, Q.; Huang, G.; Sun, Y. Event-driven optimal control of central air-conditioning systems: Event-space establishment. *Sci. Technol. Built Environ.* **2018**, *24*, 839–849. [CrossRef]
11. Wang, J.; Huang, G.; Sun, Y.; Liu, X. Event-driven optimization of complex HVAC systems. *Energy Build.* **2016**, *133*, 79–87. [CrossRef]
12. Wang, J.; Huang, G.; Zhou, P. Event-driven optimal control of complex HVAC systems based on COP mins. *Energy Procedia* **2017**, *105*, 2372–2377. [CrossRef]
13. Darby, M.L.; Nikolaou, M.; Jones, J.; Nicholson, D. RTO: An overview and assessment of current practice. *J. Process Control* **2011**, *21*, 874–884. [CrossRef]
14. Kusiak, A.; Li, M.; Tang, F. Modeling and optimization of HVAC energy consumption. *Appl. Energy* **2010**, *87*, 3092–3102. [CrossRef]
15. Huang, S.; Zuo, W.; Sohn, M.D. Improved cooling tower control of legacy chiller plants by optimizing the condenser water set point. *Build. Environ.* **2017**, *111*, 33–46. [CrossRef]
16. ASHRAE 2015. Chapter 42–Supervisory control strategies and optimization. In *ASHRAE Handbook–HVAC Applications*; ASHRAE Inc.: New York, NY, USA, 2011.
17. Saeedi, M.; Moradi, M.; Hosseini, M.; Emamifar, A.; Ghadimi, N. Robust optimization based optimal chiller loading under cooling demand uncertainty. *Appl. Therm. Eng.* **2019**, *148*, 1081–1091. [CrossRef]
18. Braun, M.R.; Walton, P.; Beck, S.B.M. Illustrating the relationship between the coefficient of performance and the coefficient of system performance by means of an R404 supermarket refrigeration system. *Int. J. Refrig.* **2016**, *70*, 225–234. [CrossRef]
19. Huang, S.; Zuo, W.; Sohn, M.D. Amelioration of the cooling load based chiller sequencing control. *Appl. Energy* **2016**, *168*, 204–215. [CrossRef]
20. Yan, C.; Gang, W.; Niu, X.; Peng, X.; Wang, S. Quantitative evaluation of the impact of building load characteristics on energy performance of district cooling systems. *Appl. Energy* **2017**, *205*, 635–643. [CrossRef]
21. Yu, F.W.; Chan, K.T. Economic benefits of optimal control for water-cooled chiller systems serving hotels in a subtropical climate. *Energy Build.* **2010**, *42*, 203–209. [CrossRef]
22. Xu, Z.; Hu, G.; Spanos, C.J.; Schiavon, S. PMV-based event-triggered mechanism for building energy management under uncertainties. *Energy Build.* **2017**, *152*, 73–85. [CrossRef]
23. Du, Z.; Jin, X.; Fang, X.; Fan, B. A dual-benchmark based energy analysis method to evaluate control strategies for building HVAC systems. *Appl. Energy* **2016**, *183*, 700–714. [CrossRef]
24. Fang, X.; Jin, X.; Du, Z.; Wang, Y. The evaluation of operation performance of HVAC system based on the ideal operation level of system. *Energy Build.* **2016**, *110*, 330–344. [CrossRef]
25. Ding, Y.; Zhang, Q.; Yuan, T.; Yang, F. Effect of input variables on cooling load prediction accuracy of an office building. *Appl. Therm. Eng.* **2018**, *128*, 225–234. [CrossRef]
26. Bi, J. A review of statistical methods for determination of relative importance of correlated predictors and identification of drivers of consumer liking. *J. Sens. Stud.* **2012**, *27*, 87–101. [CrossRef]
27. Tian, W. A review of sensitivity analysis methods in building energy analysis. *Renew. Sustain. Energy Rev.* **2013**, *20*, 411–419. [CrossRef]

28. Archer, K.J.; Kimes, R.V. Empirical characterization of random forest variable importance measures. *Comput. Stat. Data Anal.* **2008**, *52*, 2249–2260. [CrossRef]
29. Wang, Z.; Srinivasan, R.S. A review of artificial intelligence based building energy use prediction: Contrasting the capabilities of single and ensemble prediction models. *Renew. Sustain. Energy Rev.* **2017**, *75*, 796–808. [CrossRef]
30. Zhao, L.; Liu, Z.; Mbachu, J. Energy Management through Cost Forecasting for Residential Buildings in New Zealand. *Energies* **2019**, *12*, 2888. [CrossRef]
31. Ma, Z. Online Supervisory and Optimal Control of Complex Building Central Chilling Systems. Ph.D. Thesis, The Hong Kong Polytechnic University, Hong Kong, China, 2008.
32. Yoon, Y.R.; Moon, H.J. Energy Consumption Model with Energy Use Factors of Tenants in Commercial Buildings Using Gaussian Process Regression. *Energy Build.* **2018**, *168*, 215–224. [CrossRef]
33. Chang, C.; Shieh, S.; Jang, S.; Wu, C.; Tsou, Y. Energy conservation improvement and ON–OFF switch times reduction for an existing VFD-fan-based cooling tower. *Appl. Energy* **2015**, *154*, 491–499. [CrossRef]
34. Breiman, L.; Cutler, A.; Liaw, A.; Wiener, M. *Breiman and Cutler's Random Forests for Classification and Regression*; 2015; The R Project for Statistical Computing; Available online: https://www.researchgate.net/publication/304378707_Package_\T1\textquoteleftrandomForest\T1\textquoteright_Breiman_and_Cutler\T1\textquoterights_random_forests_for_classification_and_regression (accessed on 9 September 2019).
35. Yu, F.; Ho, W.; Chan, K.; Sit, R. Critique of operating variables importance on chiller energy performance using random forest. *Energy Build.* **2017**, *139*, 653–664. [CrossRef]
36. Wei, X.; Kusiak, A.; Li, M.; Tang, F.; Zeng, Y. Multi-objective optimization of the HVAC (heating, ventilation, and air conditioning) system performance. *Energy* **2015**, *83*, 294–306. [CrossRef]
37. Mossolly, M.; Ghali, K.N.; Ghaddar, N. Optimal control strategy for a multi-zone air conditioning system using a genetic algorithm. *Energy* **2009**, *34*, 58–66. [CrossRef]
38. EDR. *Chilled Water Plant Design Guide*; EDR, 2009; Available online: https://www.google.com.hk/url?sa=t&rct=j&q=&esrc=s&source=web&cd=1&ved=2ahUKEwj19YKenJblAhVBMN4KHVqDATQQFjAAegQIAhAC&url=http%3A%2F%2Fwww.taylor-engineering.com%2FWebsites%2Ftaylorengineering%2Fimages%2Fguides%2FEDR_DesignGuidelines_CoolToolsChilledWater.pdf&usg=AOvVaw3bjt6a4g9we8XKcaXHHeN (accessed on 9 September 2019).
39. Liao, Y.; Huang, G.; Sun, Y.; Zhang, L. Uncertainty analysis for chiller sequencing control. *Energy Build.* **2014**, *85*, 187–198. [CrossRef]
40. Yildiz, B.; Bilbao, J.; Dore, J.; Sproul, A. Recent advances in the analysis of residential electricity consumption and applications of smart meter data. *Appl. Energy* **2017**, *208*, 402–427. [CrossRef]
41. Li, X.; Li, F.; Chen, S.; Li, Y.; Zou, Q.; Wu, Z.; Lin, S. An Improved Commutation Prediction Algorithm to Mitigate Commutation Failure in High Voltage Direct Current. *Energies* **2017**, *10*, 1481. [CrossRef]
42. Li, S.; Wen, J. A model-based fault detection and diagnostic methodology based on PCA method and wavelet transform. *Energy Build.* **2014**, *68*, 63–71. [CrossRef]
43. Lü, X.; Lu, T.; Kibert, C.J.; Viljanen, M. A novel dynamic modeling approach for predicting building energy performance. *Appl. Energy* **2014**, *114*, 91–103. [CrossRef]
44. Keogh, E.; Chakrabarti, K.; Pazzani, M.; Mehrotra, S. Dimensionality reduction for fast similarity search in large time series databases. *Knowl. Inf. Syst.* **2001**, *3*, 263–286. [CrossRef]
45. Steinley, D.; Brusco, M.J. Initializing k-means batch clustering: A critical evaluation of several techniques. *J. Classif.* **2007**, *24*, 99–121. [CrossRef]
46. Ren, X.; Yan, D.; Hong, T. Data mining of space heating system performance in affordable housing. *Build. Environ.* **2015**, *89*, 1–13. [CrossRef]
47. Jin, X.; Han, J. K-means clustering. In *Encyclopedia of Machine Learning*; Springer: Berlin, Germany, 2011; pp. 563–564.

Article

A Fuzzy Control Strategy Using the Load Forecast for Air Conditioning System

Jing Zhao * and Yu Shan

Tianjin Key Lab of Indoor Air Environmental Quality Control, School of Environmental Science and Engineering, Tianjin University, Tianjin 300072, China; shanyu@tju.edu.cn

* Correspondence: zhaojing@tju.edu.cn; Tel.: +86-022-8740-2072

Received: 9 December 2019; Accepted: 19 January 2020; Published: 21 January 2020

Abstract: The energy consumption of air-conditioning systems is a major part of energy consumption in buildings. Optimal control strategies have been increasingly developed in building heating, ventilation, and air-conditioning (HVAC) systems. In this paper, a load forecast fuzzy (LFF) control strategy was proposed. The predictive load based on the SVM method was used as the input parameter of the fuzzy controller to perform feedforward fuzzy control on the HVAC system. This control method was considered as an effective way to reduce energy consumption while ensuring indoor comfort, which can solve the problem of hysteresis and inaccuracy in building HVAC systems by controlling the HVAC system in advance. The case study was conducted on a ground source heat pump system in Tianjin University to validate the proposed control strategy. In addition, the advantages of the LFF control strategy were verified by comparing with two feedback control strategies, which are the supply water temperature (SWT) control strategy and the room temperature fuzzy (RTF) control strategy. Results show that the proposed LFF control strategy is capable not only to ensure the minimum indoor temperature fluctuations but also decrease the total energy consumption.

Keywords: load forecast fuzzy (LFF) control; SVM method; building HVAC system; time delay effect; optimal control strategy

---

1. Introduction

Buildings energy systems account for about one-third of the global energy consumption [1]. In China, the total energy consumption of HVAC systems is expected to account for 65% of residential buildings by 2020 [2]. Optimal control strategies have been increasingly developed in building HVAC systems [3]. The energy consumption of building energy systems can be greatly reduced by developing effective control strategies for building HVAC systems [4].

Some scholars have focused on the air-conditioning control strategies. Yordanova et al. [5] designed a fuzzy controller for temperature and humidity control. This method uses fuzzy control to ensure indoor comfort and reduce energy consumption. Wang et al. [6] proposed a direct load control strategy to optimize the distribution of set values for local and global refrigerators by setting adaptive effect functions, which saves energy while ensuring indoor comfort. Krstic [7] proposed a method based on feedback control to compensate for the input delay of any length in a nonlinear control system. Mossolly et al. [8] proposed control strategies based on energy cost and thermal space transient model constraints and they used genetic algorithms to solve problems. This optimization plan/model is suitable for building floor case studies in Beirut. Powell et al. [9] modeled the characteristics of the complexity of large-scale energy systems and used recursive neural networks to accurately predict the hourly load capacity of regional energy systems 24 h in advance. Other researchers have studied the operation and simulation of central air-conditioning systems. Wei et al. [10] obtained the operating power consumption curve based on the mathematical.

Model of the established air conditioning system equipment to determine the optimal operating plan. Chan [11] proposed a solar heating and cooling (SHC) absorption chiller of central air-conditioning system design based on the TRNSYS simulation model for a hotel building. Different control strategies are loaded into the TRNSYS simulation model to evaluate the superiority of different control strategies. Li et al. [12] proposed a central air-conditioning solar heating and cooling (SHC) absorption chiller system based on TRNSYS and proposed three control schemes for the solar collector circuit to determine the preferred design strategy for these systems. Xue et al. [13] proposed a fast power demand response control strategy to investigate the performance of operational dynamics and energy systems in response to strategically controlled demand response events. From the analysis of research status at home and abroad, the current research on control strategies of central air-conditioning systems focuses on the combination of intelligent algorithms and hybrid models. However, the model has insufficient accuracy, inaccurate control, problems such as control errors and overshoot, and overly complex control models are not suitable for actual engineering control.

Judging from the progress of theoretical research at present, although there are many types of research on advanced HVAC system control technology [14,15], the control modes of building HVAC systems presently have a great limitation both in control methods and controlled parameters. From the control method point of view, PID control [16–18] is a kind of negative feedback control system, which is widely used in the control of HVAC systems of public buildings by using the proportional integral and differential method to calculate the control amount according to the system deviation. For controlled objects with inherent nonlinearity and hysteresis characteristics [19] such as the HVAC system, it is difficult to obtain an ideal PID control effect due to the uncertainty and time-varying nature of external environmental disturbances. From the controlled parameters point of view, constant pressure control [20] and constant temperature control [21,22] are widely applied. However, there are significant drawbacks to the constant pressure difference and constant temperature difference control of the air conditioning system. On one hand, for constant pressure difference control, there is no direct relationship between the load and pressure difference of the HVAC system. It is not possible to use the differential pressure as a controlled variable to ensure that the chilled water flow changes accurately following the load change [23]. Moreover, the return temperature of the chilled water is inconsistent with the water supply temperature due to the transmission delay of the HVAC system. It is unscientific to adjust the chilled water flow rate according to the temperature difference between the supply and return water detected at the same time as the controlled parameter. Therefore, the control mode based on water supply temperature commonly used for HVAC systems is only applicable to controlled objects or processes without time delay [24,25]. New control techniques and methods need to be adapted to meet the actual needs of the stability and rapid response of the central air conditioning system.

Aiming at the problems existing in HVAC system control technologies, a load forecast fuzzy control strategy was proposed. The predicted load obtained by the SVM method training is used as an input parameter to the controller in advance for feedforward fuzzy control, which can regulate the HVAC system in advance based on the forecast cooling load demand and overcome the shortcomings of controlled parameters. In this study, a simulation platform was established for the heat pump system in Tianjin University based on TRNSYS and MATLAB to confirm the advantages of the proposed LFF control strategy.

2. Case Description

The proposed strategy was validated by an air-conditioning system in Tianjin University Laboratory. The laboratory has two floors and a height of 9 m (see Figure 1). Temperature, humidity sensors, and wind sensors were used for testing. The room temperature measuring point is installed 1 m from the ground level. The central air conditioning system has an automatic platform for monitoring and recording the operating parameters of the air conditioning system. In this case, there are two heat units with a rated cooling capacity of 42 kW and two variable frequency water pumps with a rated power of 3 kW.

Figure 1. Lab Exterior View.

The actual operation strategy of the chilled water system is the supply water temperature control strategy, as shown in Figure 2. When the water supply temperature is below the minimum temperature (9 °C), units will be turned on one by one until the temperature rises to the maximum temperature (11 °C). Conversely, when the water temperature is higher than the maximum temperature, units will be turned off one by one until the water temperature is lowered to the minimum temperature. In addition, the unit is equipped with a supercooling protection device, which will shut down all units when the water supply temperature is lower than the minimum temperature for more than one minute. However, in the SWT control strategy, the pump is not controlled.

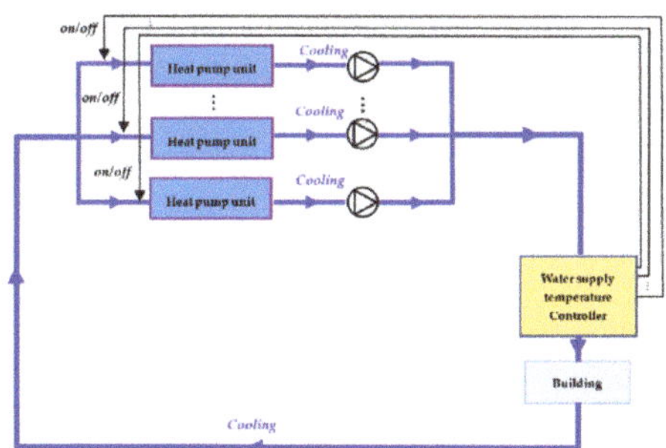

Figure 2. The logic diagram of the supply water temperature (SWT) control strategy.

Another common control strategy is the room temperature fuzzy control strategy. As a feedback control strategy, the control strategy blurs the indoor temperature and the time change rate of the indoor temperature to solve the operation of the control equipment to ensure that the room temperature is controlled at the set value [26]. The control logic diagram of RTF is shown in Figure 3.

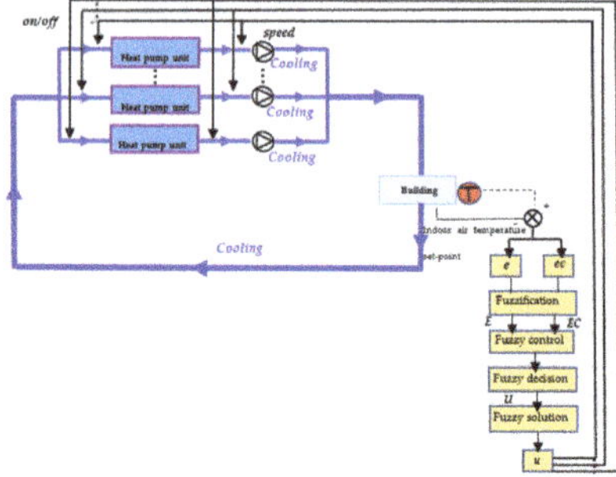

Figure 3. The logic diagram of the room temperature fuzzy (RTF) control strategy.

3. LFF Control Strategy

Control errors caused by system delay effects may cause adjustment effects to not be reflected in time, resulting in greater overshoot and oscillation. This is a detrimental effect on the air conditioning system control. The objective of the optimal control strategy is to solve the problem of time delay in control, which provides the HVAC system with the capability to operate at relatively high efficiency and save energy at various possible conditions in operation. The proposed LFF control strategy has the following characteristics. The predicted load was obtained based on the SVM [27] method and the predicted load was used as the input parameter of the feed-forward controller. The feed-forward controller can control the HVAC system in advance to eliminate the impact of system time delay on the control of the air conditioning system, and this advanced control time is the time delay of the chilled water transmission of the HVAC system. The predicted load and the time of the advance input are used as input parameters for fuzzy control to obtain control signals for the operation of the water pump and the unit. It is worth mentioning that compared with the SWT control strategy, the LFF control strategy controls the unit while controlling the operation mode of the pump. Compared with the feedback control strategy, the LFF control strategy is load-based feedforward control, which ensures the directness and accuracy of the control.

The control logic diagram of the LFF control strategy is shown in Figure 4. As shown in Figure 4, the cooling load was generated by the weather parameters acting on the building model. Based on the SVM method, the cooling load was trained to obtain the predicted cooling load. The predicted cooling load was applied to the fuzzy controller output control signal u to control the operation of the energy system equipment in advance T.

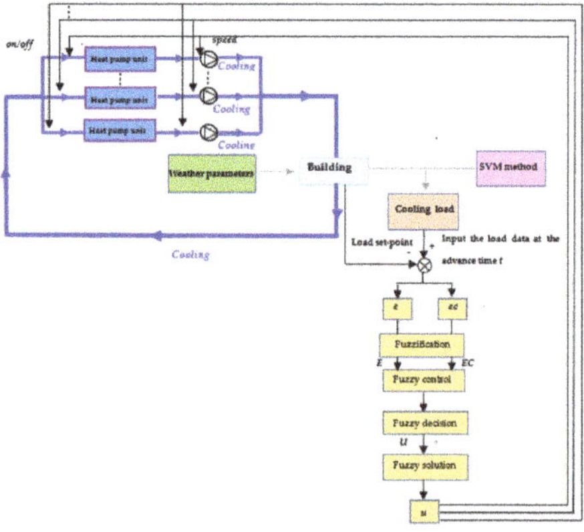

Figure 4. The logic diagram of the load forecast fuzzy (LFF) control strategy.

4. The Method of Control Optimizations

4.1. The Load Forecast Model Based on SVM

Key to the LFF control technology is the accurate input of the cooling load, which determines the effect and quality of control. This paper firstly performs the load simulation calculation on the building model based on TRNSYS, and then the load is predicted by using the support vector machines (SVM) method for the calculated load. SVM is the technique to solve the classification and regression problems [28]. Support vector regression (SVR) is a machine learning method based on statistical learning theory [29].

Support vector machines have strong generalization ability and can effectively solve practical problems such as nonlinearity and small samples. It mainly includes ε-SVR and υ-SVR models [30]. The ε-SVR method maps the input space from a low-dimensional feature space to a high-dimensional feature space based on a nonlinear mapping function $\varphi(x)$, and then uses Equation (1) to fit a linear function.

$$f(x) = \omega^T \times \varphi(x) + b \tag{1}$$

In the SVM method, the parameters ω and b are determined using the minimum structural risk. The insensitive loss function parameter ε is introduced to obtain optimization Equations (2)–(5) [31].

$$\min \frac{1}{2}\omega^T\omega + C\sum_{i=1}^{n}\left(\xi_i + \xi_i^*\right), \tag{2}$$

The constraints are as follows:

$$\text{s.t. } y_i - \omega^T \times \varphi(x_i) - b \leq \varepsilon + \xi_i, \tag{3}$$

$$\omega^T \times \varphi(x_i) + b - y_i \leq \varepsilon + \xi_i^*, \tag{4}$$

$$\xi_i \geq 0,\ \xi_i^* \geq 0,\ i = 1, 2, \ldots, n \tag{5}$$

The SVM includes two parameters: the penalty parameter "C" as the intrinsic parameter of SVM and the parameter γ in the kernel function. The penalty parameter "C" and the kernel function γ affect the correlation between the complexity, stability, and vector of the model, respectively. A Gaussian kernel function was introduced to represent the complex non-linear relationship between input and output [32]. The Gaussian kernel function is as follows:

$$K\left(x_i, x_j\right) = exp\left(-\gamma\|x_i - x_j\|^2\right), \gamma > 0, \tag{6}$$

The load of the building was simulated from 15th June to 15th September which is the actual operation period in summer. The meteorological data used in this paper are those of a typical meteorological year.

The flow chart of the SVM method is shown in Figure 5.

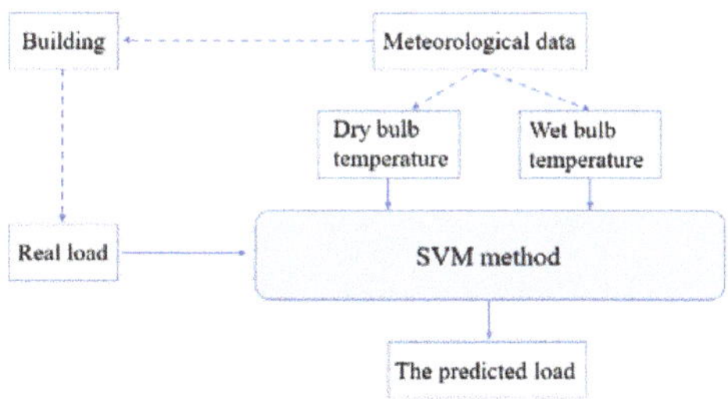

Figure 5. The flow chart of the support vector machines (SVM) method.

The specific steps of load prediction based on the SVM method are as follows.

- The objective function of the support vector machine is established from the training samples. The load calculated by the simulation and the corresponding dry bulb temperature (T, T_{h-1}, T_{h-2}, T_{h-3}) and wet bulb temperature (W, W_{h-1}) of the typical meteorological year are used as training samples to establish the support vector machine objective function, where T_{h-1}, T_{h-2} and T_{h-3} represent the dry bulb temperatures of 1 h, 2 h, and 3 h in advance and W_{h-1} represents the wet bulb temperature 1 h earlier.
- The optimal combination of key parameters of the SVM is determined via MATLAB program calculations. The best combination of key parameters of the SVM is Best c = 14.4149, g = 0.7136.
- The optimal combination parameters are substituted into the SVM model to obtain its decision regression model to obtain the predicted load. The annual (8760 h) maximum instantaneous cooling load obtained by the SVM method is 68.6 kW.

4.2. The Calculation of Time Delay

The time delay of the system includes the heat transfer delay caused by the terminal equipment of the air conditioning system and the time delay caused by the fluid flow of the air conditioning water system. The air conditioning system, in this case, is a fan coil system, and the heat transfer from the terminal to the air is accomplished by convection, which has a small time delay compared to the water system flow and can be ignored [33]. Therefore, only the time delay caused by the flow of the air conditioning water system is considered here.

The time delay of fluid flow, which is the time required for chilled water to flow from the outlet of the air conditioning system to the most unfavorable terminal, can be calculated from the hydraulics of the pipeline. Equation (7) is the ratio of branch pipe flow to total system flow under the assumption that the flow rates at the respective terminal are the same.

$$x_i = \frac{m_i}{M} \tag{7}$$

The system is assumed to have a total of n branches. The flow rate V_i of each section of the main pipe can be calculated by Equation (8).

$$V_i = \sum_{j=1}^{i} x_i \times \left(\frac{D_n}{D_i}\right)^2 \times V_n \tag{8}$$

The time delay of each main pipe is calculated according to the ratio of the length of the main pipe to the flow rate, which can be computed by Equation (9).

$$T_i = \frac{L_i}{V_i} = \frac{L_i}{\sum_{j=1}^{i} x_i \times \left(\frac{D_n}{D_i}\right)^2 \times V_n} \tag{9}$$

The sum of the time delay of the main sections is the time delay from the outlet of the chilled water to the most unfavorable terminal. The total time delay is calculated as Equation (10). The calculation results are shown in Table 1.

$$T = \sum_{i=1}^{n} T_i = \sum_{i=1}^{n} \frac{L_i}{\sum_{j=1}^{i} x_i \times \left(\frac{D_n}{D_i}\right)^2 \times V_n} \tag{10}$$

Table 1. The theoretical calculation of the time delay caused by the flow.

i	L_i(m)	D_i(mm)	x_i	V_i(m/s)	T_i(s)
1	5.30	32	1/12	0.12	44.24
2	7.24	40	1/12	0.15	47.22
3	4.41	50	1/12	0.15	29.96
4	7.32	50	1/12	0.20	37.30
5	8.01	50	1/12	0.25	32.65
6	5.45	65	1/12	0.17	31.29
7	6.80	65	1/12	0.20	33.46
8	4.69	65	1/12	0.23	20.19
9	8.52	65	1/12	0.26	32.61
10	7.21	80	1/12	0.19	37.62
11	4.69	80	1/12	0.21	22.25
12	16.4	80	1/12	0.23	71.30
		Total			440

4.3. Fuzzy Control Scheme

A two-dimensional fuzzy controller is established based on the fuzzy logic theory [34] and applied to the HVAC automatic control system. The signal obtained by the sensor is compared with the set value to obtain the deviation e and the deviation change rate ec, and then the deviation e and deviation change rate *ec* are taken as two inputs of the fuzzy controller. The fuzzy quantization process is performed to obtain the fuzzy variables *E* and *EC*. According to fuzzy rules, the fuzzy decision is made to obtain a fuzzy control quantity *U*. Finally, the actual control output is obtained through the defuzzification and the proportional transformation. For the LFF control strategy, the input of fuzzy control is the cooling load demand. The difference between the input load and the set load is the deviation *e*. The change rate of the load versus time is the deviation change rate *ec*. *e* and *ec* are the double inputs of the fuzzy control system and the output value *u* is the pump speed control value.

The sub-fuzzy system was selected to represent the control level. As shown in Equations (11)–(13).

$$E(e) \in \{NB, NM, NS, ZO, PS, PM, PB\} \tag{11}$$

$$EC(ec) \in \{NB, NM, NS, ZO, PS, PM, PB\} \tag{12}$$

$$U(u) \in \{NB, NM, NS, ZO, PS, PM, PB\} \tag{13}$$

E: The universe of *E* is {−6, −5, −4, −3, −2, −1, 0, 1, 2, 3, 4, 5, 6}. The minimum value of the load is 0 kW and the maximum value is 68.6 kW. In order to convert *e* into the domain of *E*, we need to multiply the coefficient ke. The value of ke is determined to be 0.175.

EC: The universe of *EC* is {−6, −5, −4, −3, −2, −1, 0, 1, 2, 3, 4, 5, 6}. The minimum value of load change rate is −27.2 and the maximum value is 27.2. The universe [−6,6] that converts *ec* to *EC* needs to be multiplied by the coefficient kec, the value of kec is determined to be 0.221.

U: The universe of *u* and *U* are both [0,1].

Fuzzy sets: Each input parameter is represented by a fuzzy set Ak with a membership function μ, see Equations (14) and (15). The most commonly used triangular membership function was used in this study [35].

$$Ak = \{(i, \mu(i)\} \tag{14}$$

$$\mu(i) \in [0,1] \tag{15}$$

4.4. Optimization of Pumps and Units Control

The control strategy of the pump is as follows: when the required flow of the system is less than or equal to half of the maximum flow, one pump is individually frequency-controlled and the other is not operated. If the required flow is greater than half of the maximum flow, one pump provides half the flow at full load and the other pump provides the remaining flow by frequency conversion. Specific control strategies for pumps is shown in Table 2. The fuzzy control output value u is between 0 and 1. The pump operates at a frequency and the operating frequency is proportional to the control signal.

Table 2. Pump control strategy.

u	Pump Control Method
0.0–0.2	Both pumps are off and the pump control signal is 0
0.2–0.5	One pump control signal is 0, the other pump control signal is $0.5 + u$
0.5–1.0	One pump control signal is 1 and the other control signal is $2u - 1$

In order to ensure the normal operation of the system, one unit is always running and the other unit is controlled by the set start-stop time. The start and stop time means that the unit is turned on within the set time, while the unit is turned off outside the set time. For example, a start-stop time of 0.5 means that the unit's on-time and off-time are each half of the total operating time. The specific operation control strategy of the unit is shown in Table 3.

Table 3. Heat units control strategy.

u	Heat Units Control Method
0.00–0.50	One unit start-stop time ratio is 1, another unit start-stop time ratio is 0
0.50–0.66	One unit start-stop time ratio is 1, another unit start-stop time ratio is 0.17
0.66–0.75	One unit start-stop time ratio is 1, another unit start-stop time ratio is 0.33
0.75–0.80	One unit start-stop time ratio is 1, another unit start-stop time ratio is 0.50
0.80–1.00	Start and stop time rate of both units is 1

5. TRNSYS Simulation Platform

5.1. Establishment of the Simulation Platform

The construction of the test platform based on TRNSYS is shown in Figure 6. In this case, the simulation time is from 15th June to 15th September of the typical meteorological year, for a total of 2231 h. The building is simulated by a single area building module (Type 12) in TRNSYS. A water-water heat pump system is used in this case. Type 668 was selected as the water-water heat pump unit module. The cooling transmission in the system is achieved by the variable speed pump module Type 110. In order to advance the predicted load in advance and consider the delay caused by the flow of the central air conditioning chilled water system, the delay module (Type 93) is added to the test platform. The model is closer to the actual system and lays the foundation for subsequent feedforward control. In the test platform, the switch differential controller Type 2 and the Type 155 read the chilled water supply temperature of the heat pump unit and collectively control the heat pump unit operation to make it the same as the actual operation logic.

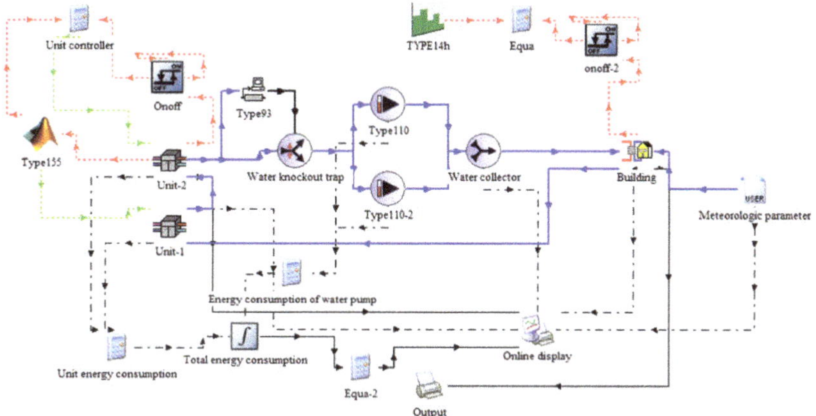

Figure 6. TRNSYS simulation platform.

5.2. Verification of Simulation Platform

5.2.1. Indoor Temperature Validation Test

Based on the actual monitoring data, the weather data measured from 6th August to 11th August, 2017 were selected to verify the system simulation model. In order to reduce the influence of the initial conditions on the simulation, the actual indoor temperature on 8th August when the system is on stable operation is compared with the indoor simulated temperature with and without considering the delay effect of the system respectively. The temperature comparison results are shown in Figure 7. It can be seen from Figure 7 that the indoor temperature curve obtained by the simulation platform with delay effect is basically consistent with the actual temperature curve during the period from 9 am to 8 pm. The maximum temperature difference at a single point is 0.8 °C, and the average relative temperature difference is 0.4 °C. However, there is a large deviation between the actual room temperature and the indoor simulation temperature without considering the system delay effect. The maximum single point temperature difference is 1.9 °C, and the average relative temperature difference is 1.1 °C. As a result of this, in the simulation of the room temperature, the model considering the delay effect of the system is more in line with the actual situation. In the period from 1 am to 9 am, the reason for the error in the simulation result of the model considering the system delay effect is that the unit was set to be completely closed when the chilled water outlet temperature is lower than 8 °C in the simulation model in order to reduce the frequent oscillation of the system parameters during the simulation. This will result in a lower temperature of the chilled water at night, which in turn will result in lower room temperature. The simulated temperature from 4 pm to 7 pm is also low because the simplified room model only considers the effect of outdoor temperature on the indoor load and does not consider the effects of radiation. The indoor load was influenced by solar radiation and the heat storage effect of the wall, which leads to increased room temperature.

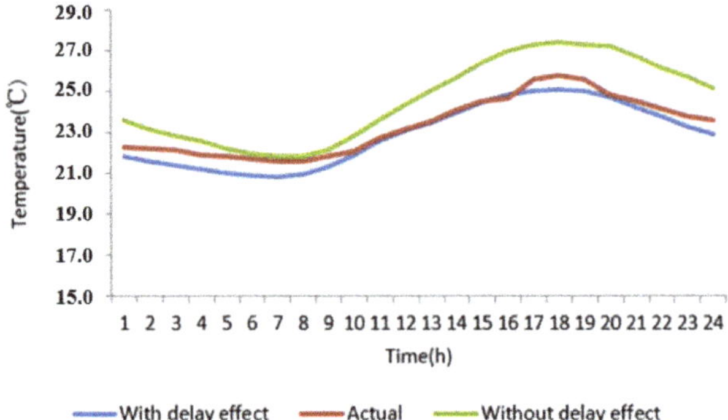

Figure 7. The temperature in three modes.

5.2.2. Heat Pump Unit and Pump Energy Consumption Validation Test

The energy consumption of the heat pump unit for the actual three hours based on the load rate and power of the heat pump unit during three time periods were monitored and calculated, from 9:22 to 10:22 on 7th August, 15:03 to 16:03 on 8th August, and from 13:17 to 14:17 on 11th August.

As shown in Figure 8, within three test days, the energy consumption of the heat pump unit under the consideration of time delay, without considering time delay, and actual control conditions have been marked, which shows the comparison of energy consumption within 3 h for the model unit considering the system delay effect, without considering the system delay effect, and the actual unit, respectively. It can be seen that the relative errors of the total energy consumption of the actual unit and the simulated total energy consumption of the unit with and without considering the system delay effect are 4.0% and 14.4%. This shows that a system with a delay effect is more energy-efficient than a system without a delay effect because the cooling load is input at an early time. From the above, the simulated energy consumption of the heat pump unit considering the system delay effect is more consistent with the actual energy consumption.

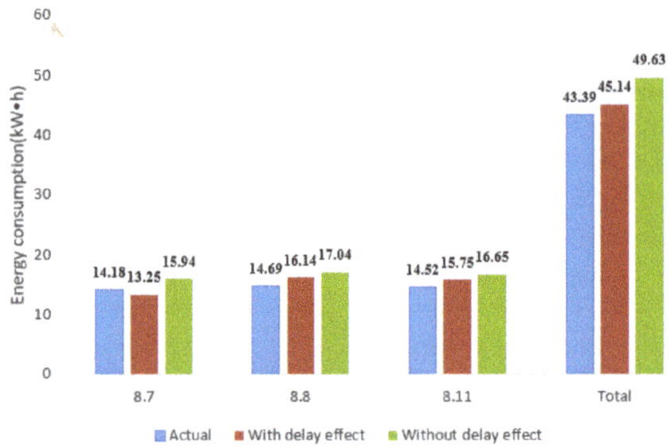

Figure 8. Comparison of simulated energy consumption and actual energy consumption of heat pump units.

The two pumps have been operating at full load, with a rated power of each pump at 3 kW. Therefore, the water pump energy consumption per hour is 6 kWh, and the actual total energy consumption of the three-hour pump is 18 kWh. Figure 9 shows the comparison of the system with and without the delay module, and the actual unit pumps energy consumption within three hours. The relative error between the actual total pump energy consumption and the simulated total energy consumption of the pump considering the delay effect of the system is 0.2%, without considering the delay effect of the system is 1.1%. The reason why the difference between both models is small is that the pump operates at a fixed frequency. Since the pump runs at a fixed frequency, the control signal cannot be controlled by the variable frequency like the control unit, and the energy consumption of the pump can only be reduced by changing the start and stop time of the pump. Therefore, the energy consumption of the system pump with or without delay effect is basically the same.

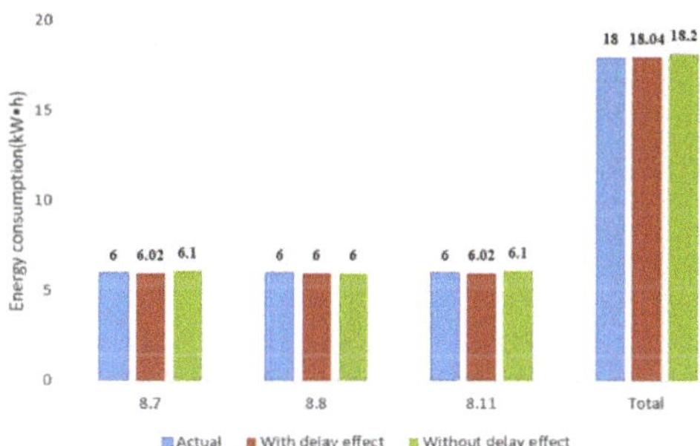

Figure 9. Comparison between simulated energy consumption and actual energy consumption of water pumps.

6. Results and Discussion

Based on the test platform, the indoor temperature and total system energy consumption under the three control strategies (LFF, RTF, and SWT) were simulated. The TRNSYS simulation platform of the LFF control strategy is shown in Figure 10.

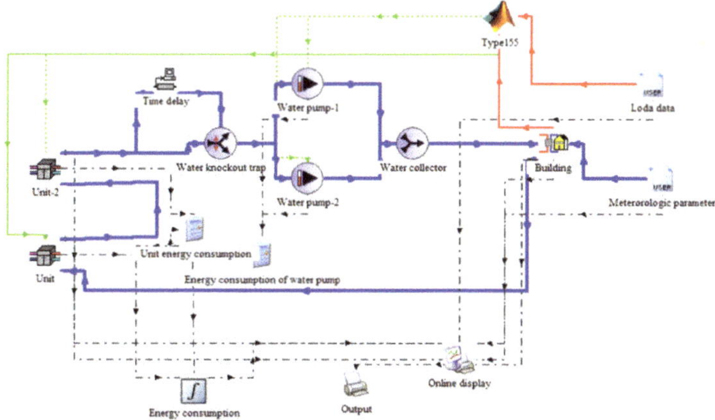

Figure 10. TRNSYS simulation platform of the LFF control strategy.

6.1. Simulation Results of Indoor Temperature

The room temperatures of the three control strategies were compared under both full-time simulation and hottest day simulation conditions. The temperature results of LFF, RTF, and SWT control strategies for the full-time simulation correspond to Figure 11a–c, respectively. As can be seen from Figure 11, three control strategies can control the room temperature between 25 °C and 27 °C when the air conditioning system is in operation. Indoor temperatures above 27 °C occur at night when the outdoor temperature is high.

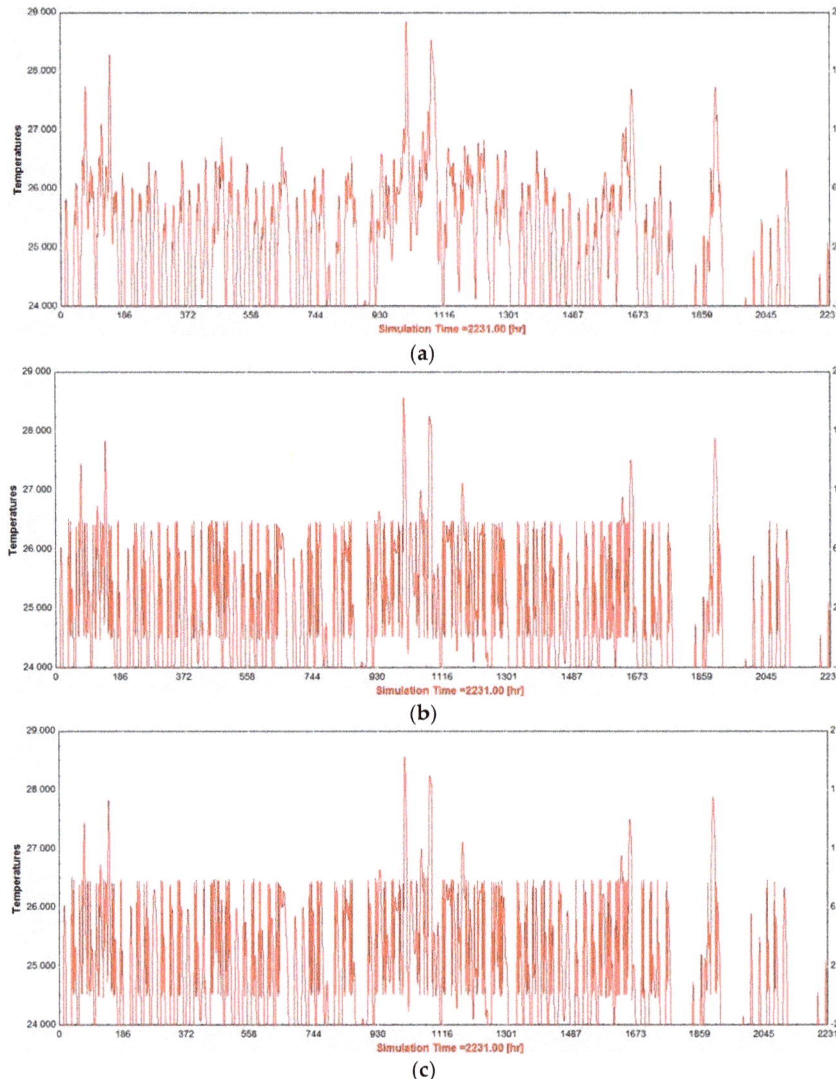

(a)

(b)

(c)

Figure 11. (**a**) Room temperature simulation results under LFF control; (**b**) Room temperature simulation results under RTF control; (**c**) Room temperature simulation results under SWT control.

During the entire simulation, the outdoor temperature was the highest on 19th June, so the simulation results of 19th June were selected to analyze the indoor temperature control of the HVAC system under the limit conditions. The room temperature comparison under different control strategies on 19th June is shown in Figure 12.

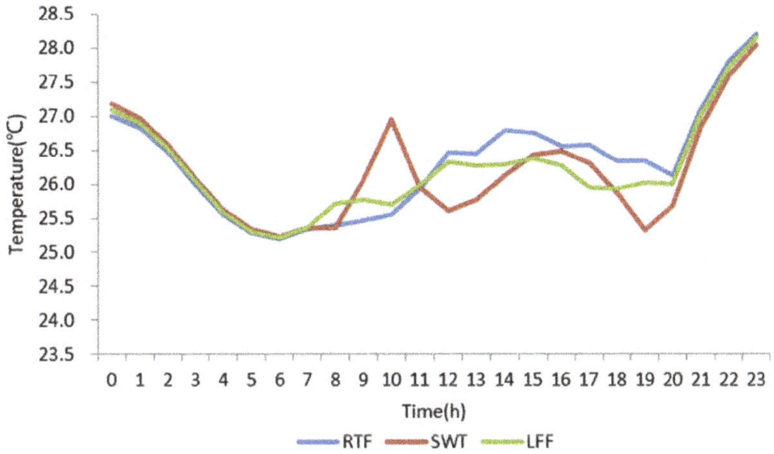

Figure 12. Comparison of room temperature under different control strategies on 19th June.

All three control strategies can control the room temperature within the operating period of the air conditioning system between 25 °C and 27 °C. Three kinds of control of the temperature range are acceptable. During the 0:00 to 7:00 and 20:00 to 23:00 periods which the air-conditioning chilled water system is not in operation, the indoor room temperature under the three control strategies is almost the same regardless of the trend or value. However, in the period of 7:00 to 20:00 when the air-conditioning chilled water system is in operation, there are obvious differences in the indoor room temperature control effect under the three control strategies. The room temperature under SWT control fluctuates the most. The room temperature reaches a maximum of 27 °C at 10:00. At this time, the temperature of the external wall will rise due to the increase of the outdoor temperature, thus making the indoor temperature rise. The room temperature reached a minimum of 25 °C at 19:00 because the chilled water temperature of the unit was turned off at 8 °C, which in turn caused the indoor temperature to drop. The room temperature under the RTF control fluctuated relatively less with a fluctuation range of 25.4 °C and 26.8 °C. The RTF control strategy is based on the indoor temperature as the control object, and the fuzzy temperature of the set temperature range is solved to ensure the thermal comfort of the indoor temperature. However, with the LFF control strategy of predictive load feed-forward control, the required load of the system is given from the demand-side response, which can ensure that the system operating temperature just matches the room temperature demand. Therefore, the room temperature fluctuation range under the LFF control strategy is between 25.4 °C and 26.4 °C, and among the three control strategies, the temperature fluctuation range is the smallest.

6.2. Simulation Results of Energy Consumption

The total hourly energy consumption of pumps and units under all three control strategies are shown in Figure 13.

The comparison of energy consumption under the three control strategies for the system is shown in Figure 14. It can be intuitively seen from Figure 14 that the LFF control strategy consumes the least amount of energy, whether it is the total energy consumption of the system or the energy consumption of the unit and the pump. The energy consumption of the system under the LFF control strategy is the least because the system is controlled from the demand- side, the system is feed-forward input according to the predicted load advance time delay, and the operation of the pumps and units is fuzzy controlled. Demand-side control can fundamentally solve problems such as the mismatch between system energy consumption and required cooling capacity, and avoid unnecessary startup and overload operation of the unit. Compared with the LFF control strategy, the RTF control strategy, as a feedback control, has a significantly better control effect than the actual control strategy based on the temperature difference between the supply and return water. However, due to problems such as time delay in feedback control, it is impossible to determine the operation of the system from the demand-side feedforward control, which will cause certain unnecessary energy loss in the system. Table 4 shows the energy consumption comparison between LFF and the other two control strategies (SWT, RTF), respectively. Compared to the SWT control strategy, the LFF control strategy has a heat pump unit energy-saving rate of 14.5%, a pump energy-saving rate of 10.2%, and a total energy-saving rate of 13.4%. Compared with the RTF control strategy, the LFF control strategy has a relatively low energy-saving rate of 9.2%, 4.1%, and 7.8%, respectively.

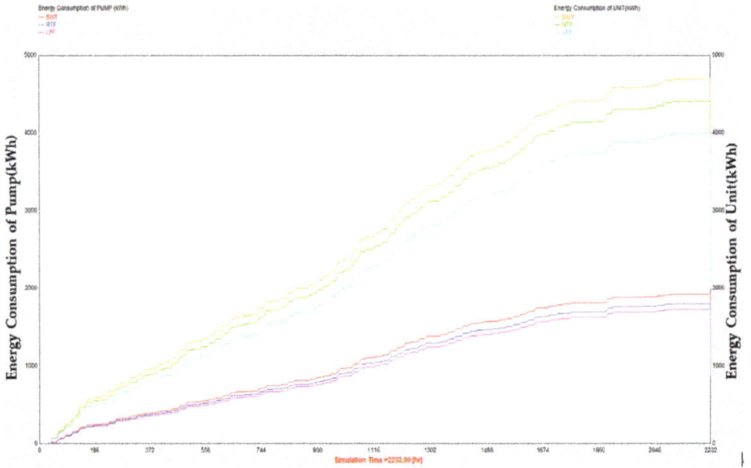

Figure 13. The simulation results of energy consumption under three different control strategy.

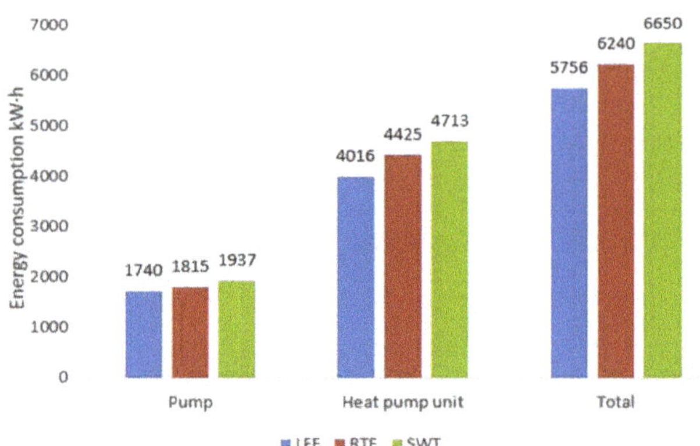

Figure 14. Comparison of energy consumption under three conditions.

Table 4. Comparison of energy-saving rates of LFF control strategies compared to the other two control strategies.

Contrast Item	Unit Energy Consumption (kW·h)/Energy Saving Rate (%)	Pump Energy Consumption (kW·h)/Energy Saving Rate (%)	Total Energy Consumption (kW·h)/Energy Saving Rate (%)
LFF	4016	1740	5756
SWT	4713/14.5%	1937/10.2%	6650/13.4%
RTF	4425/9.2%	1815/4.1%	6240/7.8%

7. Conclusions

In order to solve the problem of time delay in the control process of the air conditioning system, the LFF control strategy was developed. In this control strategy, the cooling load forecast is used as an input parameter to regulate and control the air conditioning system in advance and ignore the problem of parameter variation caused by time delay in the control process.

The energy performance under the LFF control strategy has been validated based on the actual case built TRNSYS simulation platform. The results show that compared with the commonly used feedback control strategy, the proposed effective control concept ensures indoor comfort and reduces system energy consumption. Under the LFF control, the indoor temperature fluctuations are minimal and the energy consumption under this control strategy is the lowest. Compared with the SWT and RTF controls, the total energy consumption of the LFF control at full-time simulation was reduced by 13.4% and 7.8%, respectively.

However, the simulation analysis results of the energy-saving rate have some limitations. By changing the simulation parameters, it can be found that the system energy-saving rate obtained by the TRNSYS simulation platform is mainly affected by outdoor meteorological parameters and thermal performance parameters of the building envelope. For different types of buildings and different outdoor meteorological parameters, the system energy-saving rates simulated by the LFF control strategy are often different.

Author Contributions: Methodology, J.Z.; software, R.G.; Y.S.; validation, Y.S. formal analysis, J.Z.; investigation, J.Z.; R.G.; resources, J.Z.; writing—original draft preparation, Y.S.; writing—review and editing, Y.S.; R.G. All authors have read and agreed to the published version of the manuscript.

Funding: This research was supported by the Natural Science Foundation of China, Project No. 51678398 and the Natural Science Foundation of Tianjin, Project No.18JCQNJC08400. "The funders had no role in the design of the study; in the collection, analyses, or interpretation of data; in the writing of the manuscript, or in the decision to publish the results".

Conflicts of Interest: The authors declare no conflict of interest.

Nomenclature

LFF	load forecast fuzzy
RTF	room temperature fuzzy
SWT	supply water temperature
SVM	Support Vector Machines
SVR	Support Vector Regression
C	error cost
γ	kernel parameter
m_i	the number of the terminal of the branch I of the air conditioning system
x_i	the ratio of the flow rate of each branch pipe to the total flow
D_n	the inner diameter of the chilled water outlet pipe
T_i	the delay time of each main pipe
NB	negative large
NS	negative small
PS	positive small
PB	positive large

Ak	fuzzy set
EC	the change rate of the load versus time
ω	weight vectors
$\varphi(x)$	nonlinear mapping function
b	threshold
(x_m, y_m)	a pair of input and output vectors
n	number of samples
ξ_i	upper training error
ξ_i^*	lower training error
M	the total number of the terminal of the air conditioning system
D_i	he inner diameter of each section of the main pipe
V_n	he real-time flow rate of the chilled water outlet.
L_i	the length of each main pipe
NM	medium
Z	zero
PM	medium
μ	a membership function
E	the difference between the input load and the set load
U	Pump speed control

References

1. Costa, A.; Keane, M.M.; Torrens, J.I.; Corry, E. Building operation and energy performance: Monitoring, analysis and optimisation toolkit. *Appl. Energy* **2013**, *101*, 310–316. [CrossRef]
2. Lam, J.C.; Wan, K.K.; Tsang, C.L.; Yang, L. Building energy efficiency in different cli-mates. *Energy Convers Manag.* **2008**, *49*, 2354–2366. [CrossRef]
3. Du, Z.M.; Jin, X.Q.; Fang, X.; Fan, B. A dual-benchmark based energy analysis method to evaluate control strategies for building HVAC systems. *Appl. Energy* **2016**, *183*, 700–714. [CrossRef]
4. Peng, Y.; Rysanek, A.; Nagy, Z. Using machine learning techniques for occupancy-prediction-based cooling control in office buildings. *Appl. Energy* **2018**, *211*, 1343–1358. [CrossRef]
5. Yordanova, S.; Merazchiev, D.; Jain, L. A Two-Variable Fuzzy Control Design with Application to an Air-Conditioning System. *Fuzzy Syst. IEEE Trans.* **2015**, *23*, 474–481. [CrossRef]
6. Wang, S.W.; Tang, R. Supply-based feedback control strategy of air-conditioning systems for direct load control of buildings responding to urgent requests of smart grids. *Appl. Energy* **2017**, *201*, 419–432. [CrossRef]
7. Krstic, M. Input Delay Compensation for Forward Complete and Strict-Feedforward Nonlinear Systems. *IEEE Trans. Autom. Control* **2010**, *55*, 287–303. [CrossRef]
8. Mossolly, M.; Ghali, K.; Ghaddar, N. Optimal control strategy for a multi-zone air conditioning system using a genetic algorithm. *Energy* **2009**, *34*, 58–66. [CrossRef]
9. Powell, K.M.; Sriprasad, A.; Cole, W.J.; Edgar, T.F. Heating, cooling, and electrical load forecasting for a large-scale district energy system. *Energy* **2014**, *74*, 877–885. [CrossRef]
10. Wei, B.; Tang, J.; Wei, Z. Dynamic simulation to the operation of central air conditioning system. *Ind. Electron. Appl.* **2010**, *100*, 861–867.
11. Chan, I.Y.T. Simulation and evaluation of control strategies for air conditioning system using TRNSYS. *Int. J. Prison. Health* **2012**, *8*, 92–99.
12. Li, Q.; Zheng, C.; Shirazi, A.; Mousa, O.B.; Moscia, F.; Scottc, J.A.; Taylor, R.A. Design and analysis of a medium-temperature, concentrated solar thermal collector for air-conditioning applications. *Appl. Energy* **2017**, *190*, 1159–1173. [CrossRef]
13. Xue, X.; Wang, S.; Yan, C.; Cui, B. A fast chiller power demand response control strategy for buildings connected to smart grid. *Appl. Energy* **2015**, *137*, 77–87. [CrossRef]
14. Shirazi, A.; Pintaldi, S.; White, S.D.; Morrison, G.L.; Rosengarten, G.; Tylor, R.A. Solar-assisted absorption air-conditioning systems in buildings: Control strategies and operational modes. *Appl. Therm. Eng.* **2016**, *92*, 246–260. [CrossRef]

15. Fiorentini, M.; Wall, J.; Ma, Z.; Braslavsky, J.H.; Cooper, P. Hybrid model predictive control of a residential HVAC system with on-site thermal energy generation and storage. *Appl. Energy* **2017**, *187*, 465–479. [CrossRef]
16. Attaran, S.M.; Yusof, R.; Selamat, H. A novel optimization algorithm based on epsilon constraint-RBF neural network for tuning PID controller in decoupled HVAC system. *Appl. Therm. Eng.* **2016**, *99*, 613–624. [CrossRef]
17. Lygouras, J.N.; Botsaris, P.N.; Vourvoulakis, J.; Kodogiannis, V. Fuzzy logic controller implementation for a solar air-conditioning system. *Appl. Energy* **2007**, *84*, 1305–1318. [CrossRef]
18. Wang, J.J.; An, D.W.; Jing, Y.Y. Genetic optimization algorithm of PID decoupling control for VAV air-conditioning system. *J. Tianjin Univ.* **2009**, *15*, 308–314. [CrossRef]
19. Xia, Y.; Deng, S.; Chen, M.Y. Inherent operational characteristics and operational stability of a variable speed direct expansion air conditioning system. *Appl. Therm. Eng.* **2016**, *73*, 268–277. [CrossRef]
20. Li, X.; Chen, J.; Chen, Z.; Liu, W.; Hu, W. A new method for controlling refrigerant flow in automobile air conditioning. *Appl. Therm. Eng.* **2004**, *24*, 1073–1085. [CrossRef]
21. Li, S. Optimum Control Strategy and Performance Simulation of a Constant Temperature and Humidity Air-conditioning System. *J. Refrig.* **2012**, *33*, 22–27.
22. Yu, X.; Wang, R.Z.; Zhai, X.Q. Year round experimental study on a constant temperature and humidity air-conditioning system driven by ground source heat pump. *Energy* **2011**, *36*, 1309–1318. [CrossRef]
23. Liu, X.; Liu, J.; Lu, Z.; Xing, K.; Mai, Y. Diversity of energy-saving control strategy for a parallel chilled water pump based on variable differential pressure control in an air-conditioning system. *Energy* **2015**, *88*, 718–733.
24. Liu, X.F.; Liu, J.P.; Lu, J.D.; Liu, L.; Zou, W. Research on operating characteristics of direct-return chilled water system controlled by variable temperature difference. *Energy* **2012**, *40*, 236–249. [CrossRef]
25. Wang, S.; Gao, D.C.; Sun, Y.; Xiao, F. An online adaptive optimal control strategy for complex building chilled water systems involving intermediate heat exchangers. *Appl. Therm. Eng.* **2013**, *50*, 614–628. [CrossRef]
26. Chiou, C.B.; Chiou, C.H.; Chu, C.M.; Lin, S.L. The application of fuzzy control on energy saving for multi-unit room air-conditioners. *Appl. Therm. Eng.* **2009**, *29*, 310–316. [CrossRef]
27. Cortes, C.; Vapnik, V.N. Support vector networks. *Mach. Learn.* **1995**, *20*, 273–295. [CrossRef]
28. Selakov, A.; Cvijetinović, D.; Milović, L. Hybrid PSO-SVM method for short-term load forecasting during periods with significant temperature variations in city of Burbank. *Appl. Soft Comput.* **2014**, *16*, 80–88. [CrossRef]
29. Awad, M.; Khanna, R. Support Vector Regression. *Neural Inf. Process. Lett. Rev.* **2007**, *11*, 203–224.
30. Fu, Y.; Li, Z.; Zhang, H. Using Support vector machine to predict next-day electricity load of public buildings with sub-metering devices. *Procedia Eng.* **2015**, *121*, 1016–1022. [CrossRef]
31. Shrivastava, N.A.; Khosravi, A.; Panigrahi, B.K. Prediction Interval Estimation of Electricity Prices Using PSO-Tuned Support Vector Machines. *IEEE Trans. Ind. Inform.* **2015**, *11*, 322–331. [CrossRef]
32. Ebtehaj, I.; Bonakdari, H.; Shamshirband, S. A combined support vector machine-wavelet transform model for prediction of sediment transport in sewer. *Flow Meas. Instrum.* **2016**, *47*, 19–27. [CrossRef]
33. Li, Y.M.; Wu, J.Y.; Shiochi, S. Experimental validation of the simulation module of the water-cooled variable refrigerant flow system under cooling operation. *Appl. Energy* **2010**, *87*, 1513–1521. [CrossRef]
34. Hien, L.V.; Trinh, H. Stability Analysis and Control of Two-Dimensional Fuzzy Systems with Directional Time-Varying Delays. *IEEE Trans. Fuzzy Syst.* **2018**, *26*, 1550–1564. [CrossRef]
35. Ozkop, E.; Altas, I.H.; Okumus, H.I.; Sharaf, A.M. A fuzzy logic sliding mode controlled electronic differential for a direct wheel drive EV. *Int. J. Electron.* **2015**, *102*, 1919–1942. [CrossRef]

Article

Steady-State Predictive Optimal Control of Integrated Building Energy Systems Using a Mixed Economic and Occupant Comfort Focused Objective Function

Christopher J. Bay [1,*,†], Rohit Chintala [1,†] and Bryan P. Rasmussen [2,*]

1 National Renewable Energy Laboratory, Golden, CO 80401, USA; rohit.chintala@nrel.gov
2 Department of Mechanical Engineering, Texas A&M University, College Station, TX 77843, USA
* Correspondence: christopher.bay@nrel.gov (C.J.B.); brasmussen@tamu.edu (B.P.R.)
† This work was completed while C.J. Bay and R. Chintala were doctoral students at Texas A&M University.

Received: 27 April 2020; Accepted: 1 June 2020; Published: 6 June 2020

Abstract: Control of energy systems in buildings is an area of expanding interest as the importance of energy efficiency, occupant health, and comfort increases. The objective of this study was to demonstrate the effectiveness of a novel predictive steady-state optimal control method in minimizing the economic costs associated with operating a building. Specifically, the cost of utility consumption and the cost of loss productivity due to occupant discomfort were minimized. This optimization was achieved through the use of steady-state predictions and component level economic objective functions. Specific objective functions were developed and linear models were identified from data collected from a building on Texas A&M University's campus. The building consists of multiple zones and is serviced by a variable air volume, chilled water air handling unit. The proposed control method was then co-simulated with MATLAB and EnergyPlus to capture effects across multiple time-scales. Simulation results show improved comfort performance and decreased economic cost over the currently implemented building control, minimizing productivity loss and utility consumption. The potential for more serious consideration of the economic cost of occupant discomfort in building control design is also discussed.

Keywords: energy optimization; steady-state control; building energy control system; comfort and engineering; buidling simulation (EnergyPlus and MATLAB)

1. Introduction

Energy use and consumption of natural resources has become a pertinent concern for current and future generations. In the U.S. alone, total energy consumption has tripled over the last 65 years from 34.6 quadrillions Btus (quads) in 1950 to 101 quads in 2019 [1]. Of the energy consumed in the U.S., non-renewable energies still represent nearly 90% of energy sources [2]. Many nations have put forth specific renewable energy targets which aim to reduce dependence on non-renewable energies and maintain a competitive edge in the global energy technology market. For example, the European Union's (EU) Renewable Energy Directive has established a goal of 20% final energy consumption from renewable sources by 2020 [3]. The U.S. Department of Energy has set a goal of having 20% electricity sourced from wind energy by the year 2030 [4]. While renewable sources are projected to grow, reductions in energy usage can work to achieve these goals as well.

Delving into the energy consumption practices in the U.S., approximately 40% of all energy goes to building operations in the commercial and residential sectors [5]. The data show that approximately 75% of the energy used in the building sector comes from fossil fuels. As a result, energy usage in buildings account for 40% of the total U.S. carbon emissions [6]. Additionally, the buildings' share of U.S. energy consumption has increased from 34% in 1980 to 40% as of 2010, and is projected to continue in growth [7]. The commercial buildings sector accounts for half of this energy usage and is a prime target for reduction.

Examining the top commercial site energy end uses reveals that space heating (27%), lighting (14%), space cooling (10%), water heating (7%), and ventilation (6%) are responsible for 64% of energy consumption [5]. These categories can be combined more generally to refer to services required mostly when a building is occupied (space conditioning and lighting). Economically, utilities cost businesses and building owners in the commercial sector $179.4 billion in 2010 [5]. The same five end uses listed above account for the top five cost areas, further motivating the need for energy reduction technologies.

Advanced building controls is a major area of research seeking to reduce energy usage by improving on the current practice of low-level controllers (proportional-integral or proportional-integral-derivative). In some buildings, there may be supervisory control, but frequently the components and systems operate independently and in a decentralized fashion. Because of the physically interconnected and complex nature of building systems, this uncoordinated control can often lead to inefficiencies where controllers compete with each other in achieving their desired outputs. To solve this issue, many advanced control strategies have been proposed, with Model Predictive Control (MPC) being a front-runner. A key part of MPC is the development of the objective function to be minimized. Strategies thus far have targeted energy and cost reduction, but often do not optimize occupant comfort and the associated economic impact of discomfort, accounting for occupant comfort using limits of thresholds. Thus, the energy optimal solution results in the maximum tolerated occupant discomfort. One exception included the cost of occupant discomfort as a portion of an employee's lost salary, as described in [8]. Such an objective function would enable the identification of possible energy and cost savings, serving to guide building managers and researchers as to where their efforts for increases in performance and efficiency should be focused.

This paper contributes a novel, scalable steady-state economic controller that accounts for both the cost of utilities as well as the loss of productivity due to occupant discomfort. Detailed are the development of a steady-state optimal control method based on data collected from a building on Texas A&M University's campus, the comparison of a simulated standard control implementation versus the proposed supervisory controller, and a discussion of the impact of including the cost of occupant discomfort in the control strategy. While the decision was made to focus on one building for this work, the basic structure of the algorithm is scalable and enables the optimization of larger systems/buildings that contain multiple chillers, AHUs, dozens of VAVs and hundreds of zones. Additionally, this work tackles several practical challenges that are not often addressed in current literature, including accounting for humidity as well as temperature issues as they relate to energy/comfort, having a time-varying objective function that is updated based on current operating conditions, empirical fits for component models to reflect the systems in an actual building, and switching models to capture changes in system behavior due to mechanical limits. First, a background on recent efforts in control of building energy systems is given, focusing on economic optimization. Then, information about the building and development of the steady-state control method are given, followed by the simulation methods. Simulation results are then presented, followed by a discussion of the importance of occupant comfort in building control strategies. The paper ends with a discussion of the study's outcome as well as future work.

2. Background

Control of buildings presents several unique challenges, many of which arise form the interconnected nature of building systems. For example, in a large building there may be multiple chillers that are used to chill a secondary fluid, such as water. This chilled water is then pumped to various systems and areas of buildings where heat exchangers in air handling units (AHUs) use the chilled water to cool and dehumidify air streams. A network of fans and ducts then deliver the cooled air to the desired locations. The flow of this cooled air into the zones can be controlled by variable air volume (VAV) units, in which there may be another heat exchanger that utilizes heated water to warm the air, if necessary. The heated water for this process is provided by a different set of centralized pumps and heat exchangers. The zones themselves are connected to one another, either by conduction through barriers or convection through shared doorways/open spaces. All these interconnections and couplings result in coordinated control problems.

Another difficulty comes from nonlinear dynamics evolving over multiple time scales across these various building systems. While changes in damper position of a VAV or fan speed in an AHU are relatively fast (on the order of seconds), changes in desired chilled water temperatures can take longer (minutes), or changes in room air or wall temperature can be even slower (hours). There are also slow, overarching changes to take into consideration such as the shift in solar loads as the sun moves throughout the day, gradual changes in outdoor air temperature due to change in weather or seasons, and the slow deterioration of equipment through use over time. These all contribute to shifting system behaviors and disturbances that can cause undesired performance. There are also discrete changes to take into account, such as whether an area/room is occupied, how many people are in the room, opening and closing of windows or doors, or changes in real-time pricing of utilities. Furthermore, the system sensors are distributed (not always equally), monitoring is performed centrally, and devices are driven by localized controls. All of this occurs within constraints, either due to hardware limitations, limited resources, or issues of health and comfort.

While the challenges of building control are numerous, one control method that has emerged as a capable solution is model predictive control (MPC), also known as receding horizon control. MPC has been chosen as the most appropriate control method in several building thermal control research projects [9–11]. MPC predicts the change in dependent variables of a modeled system by optimizing independent variables. Using the current state information, dynamic models of the system, and an objective function, MPC will determine the changes in the independent variables that will minimize the user-defined objective function while honoring given constraints on both dependent and independent variables. Once this series of changes is determined, the controller will apply the first determined control action, and then repeat the calculations for the next time step. Figure 1 displays how a typical reference tracking MPC implementation would behave. It can be seen at time k that the controller determines what the predicted output would be along with its optimal control trajectory. After completing the computations, the control would be applied and the system would move on to time $k + 1$, repeating the predictions and optimization with the new measurements. More details about MPC can be found in [12].

MPC applications for buildings have been studied mostly in simulation [13–22], with a few experimental efforts [9–11,23–25]. In simulation, MPC has been adapted for controlling building systems such as floor heating [26], water heating [27], cooling [11], and ventilation [28], among others. There is a large body of work on MPC in buildings, but what is not as clear is if full MPC is needed to solve the challenges presented by controlling buildings. Thus, the research presented here proposes a steady-state method that is inspired by previous MPC efforts.

For most reported applications of MPC for buildings, the cost takes the form of economic MPC (E-MPC), where the objective function is a linear combination of the monetary cost of building energy consumption [17]. In these applications when E-MPC is used, the amount of energy consumed is being minimized, while authors occasionally account for occupant comfort limits through constraints on the variables [13,15,17,18,20,23,24]. These proposed control strategies are able to respond to real-time changes in utility pricing and optimize energy usage; however, occupant comfort becomes a second priority with limits on the ranges of parameters such as temperature and lighting levels. These limits ensure the controller does not drift too far from the user-defined comfort zone, but does not actually optimize occupant comfort or the associated cost of discomfort. In fact, the energy optimal solution generally maximizes discomfort within prescribed limits.

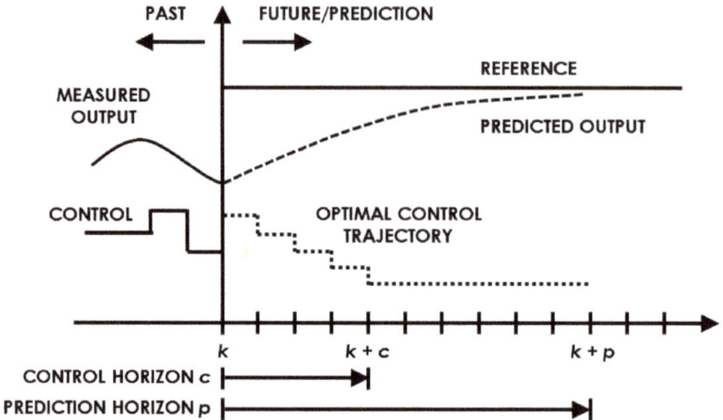

Figure 1. How MPC, or receding horizon control, typically functions.

Some researchers have included comfort in the optimization beyond limits on parameters, e.g., Corbin et al. [19] minimized the sum of electricity used by all HVAC equipment and a comfort penalty. This comfort penalty was defined as an area-weighted sum of the number of zone occupied hours outside of a predicted mean vote (PMV) threshold of ± 0.5. PMV is an index that determines the thermal comfort of an average individual dependent upon a variety of factors, including air temperature, relative humidity, relative air velocity, metabolic rate, clothing insulation level, work output, and several other variables [29]. While the researchers of [19] worked to optimize occupant comfort, they did so with the focus of reducing the cost of energy usage, neglecting the economic cost associated with occupant comfort and productivity.

The authors of [21] also included a discomfort cost with their monetary energy cost that was based on different lower and upper thermal limits; however, the physical meaning of this discomfort cost is arbitrary as the cost increases to unity until the temperature limits are exceeded and then becomes significantly large, not following any physical or measured relationship. The cost function in [22] included regulation of occupant comfort based on PMV, though it took the quadratic cost form, with the first term being the difference between the predicted PMV of the zone and the PMV setpoint for the zone, quantity squared, multiplied by a weighting factor. The second term consisted of the square of the change in control action, or increment, multiplied by a weighting factor. With this form, the MPC will balance maintaining the desired zone PMV while limiting large control action rates, specifically changes in water flow and air velocity. This will help keep occupants comfortable but the economic cost of the control actions is not accounted for. Morosan et al. [16] used a linear objective function that penalized the error between the predicted room temperature and the future room temperature reference as well as the energy usage to condition the room. In their efforts, comfort is accounted for as a comfort index that acts as a penalty when the room temperature does not meet its setpoint. However, in the presented simulations, the temperature setpoints are arbitrarily chosen and not dependent on any comfort information. Additionally, this method, similar to the previous ones mentioned, does not account for the economic aspect of occupant comfort.

One objective function from the literature that appears more unique than others appeared in [25]. This objective function consisted of three different linear terms: (1) a weighting coefficient multiplied by the predicted percentage of dissatisfied (PPD) people, which PPD can be calculated from PMV; (2) a weighting coefficient multiplied by the summation of cost of energy consumed by the heating and cooling devices; and (3) a weighting coefficient multiplied by the summation of the green house gas intensities of the various energy sources (electricity and natural gas). This objective function displays the power of MPC to determine optimal control actions with respect to a user's desired metrics, in this case occupant comfort, monetary cost of energy, and environmental impact of energy sources. While providing great flexibility in allowing the building operator to prioritize the three metrics with the weighting factors, the respective economic impact of the three areas is not represented in the objective function due to differing units and arbitrary weights.

Overall, previous methods have accounted for the economic cost of energy usage and/or attempted to maintain occupant comfort through optimization constraints or optimizing comfort itself, but none have accounted for the economic aspect of comfort on occupant productivity alongside utility costs. Additionally, it is unknown whether comparable results can be achieved with less computational burden by using a steady-state predictive model as opposed to a full dynamic MPC solution. This research effort aimed to develop a novel supervisory control method that optimizes a building's economic cost, due to both the consumption of utilities and the economic cost of loss of productivity due to occupant discomfort. This control method emulates MPC in that it uses models to predict the system's behavior at steady-state, such as was done by Elliott and Rasmussen [30]. The focus of this paper is on the development of the economic costs of the system, while a full MPC implementation will be completed in future work. Through analysis of the performance of this control method, the authors intend to identify areas having the most potential for savings and guide the priorities of building managers and researchers for future work.

3. Development of Economic Objective Function for Advanced Building Systems Control

Considering the literature and previous works, a general component-level objective function of the form shown in Equation (1) was chosen. The quadratic terms (first and third) were included to allow for standard convex optimization when desired, where e represents the error for the system, Q the weighting placed upon the error, u the control action, and S the weighting placed on control action. The linear terms (second and fourth) were included to facilitate calculating cost purely in economic terms (dollars), as opposed to nonsensical units such as dollars squared. R and T can be formulated in such a manner to transform e and u into economic cost. This is demonstrated in subsequent sections using a specific building at Texas A&M's campus as a case study.

$$J_{component} = e^T Q e + e^T R + u^T S u + u^T T \quad (1)$$

3.1. Building Configuration

For operating a typical building, there are economic costs associated with electrical cost, thermal costs, and comfort/productivity costs. The electrical costs can stem from running equipment such as electric motors in AHU fans, or from reheating air at local VAVs before it enters a zone. Thermal costs can occur from site-wide chillers that produce chilled water used to cool the air locally in buildings. Of course, comfort/productivity costs come from the thermal discomfort of the occupants. These costs exists across a wide selection of buildings/situations and can be accommodated by the general cost function above. To provide a more detailed understanding of how to derive the costs for a specific building, as well as how to construct/apply the proposed steady-state method, a case study is presented focused on a specific building, as detailed below.

Working with the Utilities Energy Management (UEM) Office at Texas A&M University, limited access was granted to the Utilities Business Office (UBO) for the purpose of collecting data and future implementation of advanced controllers. Individual component objective functions were developed for specific equipment in the UBO; however, the UBO represents a typical office building and thus the work can be generalized to other commercial buildings. What follows is a description of the UBO and its behavior to inform the development of the component objective functions.

The UBO is a rectangular, single-story building consisting of 11 zones, 10 of which are actively controlled. The general layout can be seen in Figure 2, with the thermostats shown as white boxes and the approximate locations of the VAVs shown as blue boxes. In this initial development, the decision was made to focus solely on the cooling aspects of the system because: (1) the majority of the year is spent cooling due to the location and climate of the building; and (2) simplified operating conditions will help to validate the developed control strategy in its initial implementation. The process flow and current control structure is displayed in Figure 3. The rest of this section details each of the subsystems and currently implemented controls.

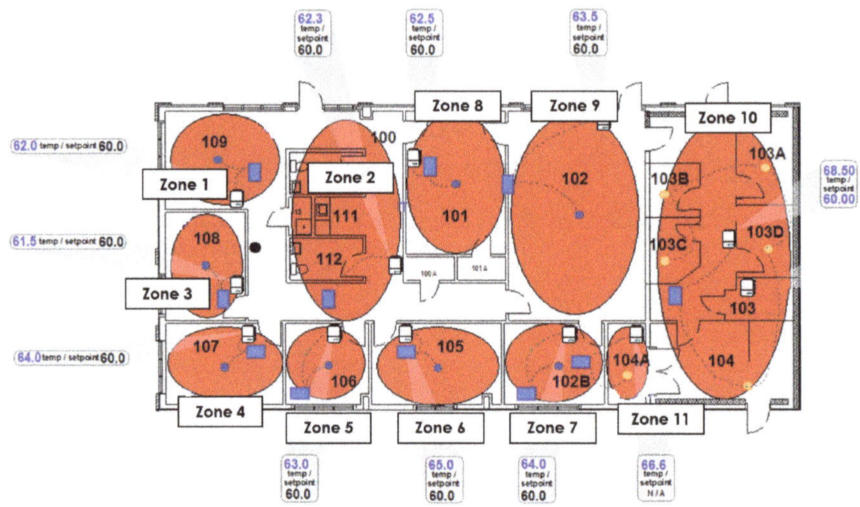

Figure 2. Zone layout for the Utilities Business Office (UBO) at Texas A&M University.

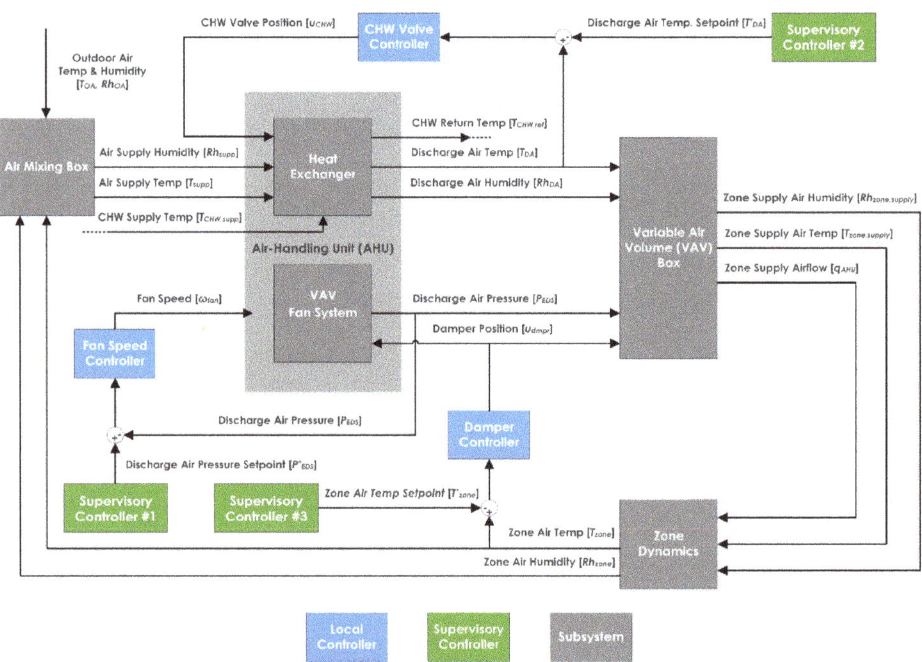

Figure 3. Process flow and current control implementation for the UBO.

3.1.1. Rooftop Air Handling Units

Most commercial buildings are serviced by rooftop air handling units (AHUs). These AHUs serve to condition a combination of indoor air and outdoor air to meet the comfort/health requirements in one or multiple zones. In the presented case study, the UBO is serviced by a single rooftop AHU. The AHU consists of a variable air volume (VAV) fan, a chilled water coil, an outdoor air damper, a return air damper, a discharge air temperature sensor, and an end static pressure sensor. The organization of these components can be seen in Figure 4. During normal operation, the VAV fan is driven by a Proportional-Integral-Derivative (PID) loop to maintain an end static pressure setpoint given by a supervisory Proportional-Integral (PI) controller (see Figure 3 and Table 1). This PI controller's feedback signal is a pressure demand calculation dependent upon the individual damper positions of the zone terminal boxes. This calculation is given by Equation (2).

$$D_{EDS} = \frac{1}{n}\sum_{i=1}^{n} u_{dmpr,i} \cdot 0.4 + \max(u_{dmpr,i}) \cdot 0.6 \tag{2}$$

where D_{EDS} is the end-static pressure demand, $n = 10$ is the number of individual dampers, and u_{dmpr} is the damper position. The end-static pressure demand setpoint for this supervisory PI controller stays constant at 60 (unitless) and was set during tuning by the building technicians.

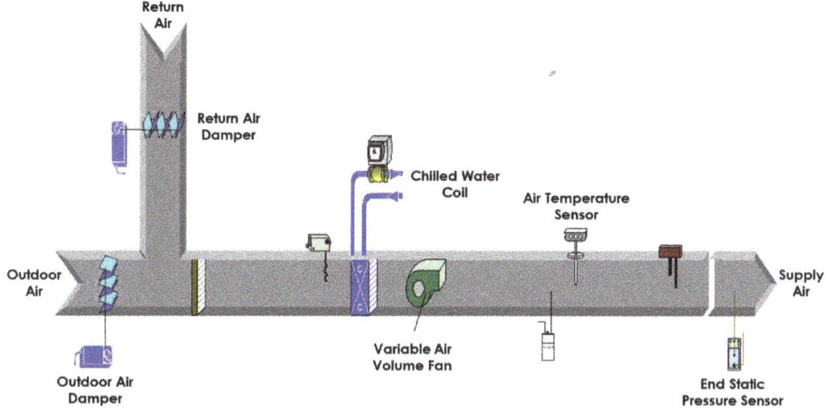

Figure 4. Rooftop Air Handling Unit (AHU) for the Utilities Business Office (UBO) at Texas A&M University.

Table 1. Specifications for control loops in the UBO.

Controller	Reference	Feedback	Output	K_P	K_I	K_D
Supervisory Controller #1	D^*_{EDS}	D_{EDS}	P^*_{EDS}	12	0.1	0
Fan Speed Controller	P^*_{EDS}	P_{EDS}	ω_{fan}	0.5	0.15	0.1
Supervisory Controller #2	D^*_{CLG}	D_{CLG}	T^*_{DA}	12	0.01	2000
Chilled Water Valve Controller	T^*_{DA}	T_{DA}	u_{CHW}	20	1	10
Supervisory Controller #3	Deadband Controller			N/A		
Damper Controller	T^*_{zone}	T_{zone}	u_{dmpr}	Set by Manufacturer		

3.1.2. Chilled Water Production

In many sites, chilled water is produced at a central location and used for conditioning the air in local buildings. There is a cost associated with the production of this chilled water, and it is important to capture in an economic control method. In the presented case study, chilled water provided by the central plant is passed through the UBO's AHU chilled water coil to lower the temperature of the moving air as well as reduce the air's humidity. The amount of chilled water passing through the coil is controlled by a valve which is actuated by a PID control loop. This control loop is driven by a difference in a discharge air temperature setpoint and the discharge air temperature. The discharge air temperature setpoint is the output of a supervisory PID control loop (see Figure 3 and Table 1) that is trying to maintain a cooling demand setpoint. A cooling demand calculation supplies the feedback signal for this supervisory controller that is a weighted combination of the average and maximum cooling loopout values from the individual zone control loops.

Figure 5 details the chilled water loop. A chilled water pump works to maintain a specific supply pressure to provide the required chilled water flow for the AHU. The chilled water supply and return temperatures are available to measure within the energy management system.

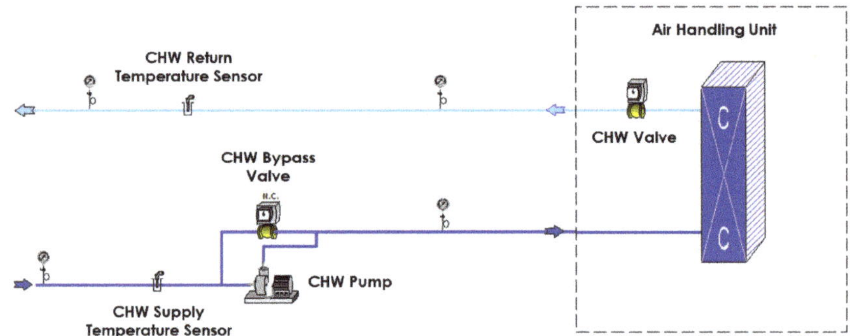

Figure 5. Chilled water loop for the Utilities Business Office (UBO) at Texas A&M University.

3.1.3. Conditioned Zones in Buildings

Zones within buildings can consist of one or multiple rooms, and can be serviced by one or multiple pieces of equipment. Frequently, zones in commercial buildings are assigned a VAV terminal box or damper that regulates the amount of conditioned air that is introduced to that space. There is typically a thermostat placed in the zone to provide feedback for the zone controller to regulate temperature.

As described above, the UBO consists of 11 zones, 10 of which are actively controlled. Each controllable zone is serviced by a VAV terminal box equipped with hot water reheat capabilities, an example of which is shown in Figure 6. The flow of conditioned air into the room is regulated by a damper in the terminal box whose position is determined by a PID control loop. The error signal for the control loop is the difference between the respective room temperature setpoint and the measured room temperature while the output is the damper position (see Table 1). The room temperature setpoint is determined by whether the room is occupied or unoccupied and if the zone is in heating or cooling mode. A supervisory deadband control method (see Figure 3) is employed such that if the room is occupied, the zone VAV will heat the room to 70 °F or cool the room to 74 °F. If the room is unoccupied, the VAV will heat the room to 60 °F or cool the room to 85 °F. If the occupancy sensors in all the rooms read unoccupied, then the main AHU will turn off and the room temperatures will freely fluctuate.

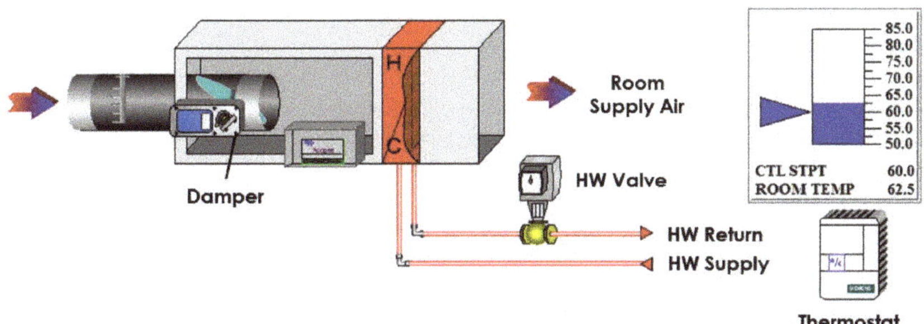

Figure 6. Variable Air Volume (VAV) box in the Utilities Business Office (UBO) at Texas A&M University.

3.2. Economic Cost of Operating Equipment

Having described the existing equipment and controls, the following sections will outline the economic cost functions for said equipment, particularly an AHU and individual zones. Examining a typical AHU, two sources of economic cost become apparent: (1) the cost of electricity used by the AHU fan to move the air; and (2) the cost associated with the production of chilled water for cooling of the air. At the zone level, there is usually no building equipment that consumes a significant amount of power or resources (when considering cooling only applications; VAVs with local reheat would be a different situation). For example, the damper in a VAV terminal box is the only actuated component and the power required to move it is negligible.

For the presented case study, an economic objective function that minimizes the cost of occupant discomfort as a measure of the loss of productivity at the zone level was developed. As mentioned in the Background Section, select previous building energy optimizations have included some form of occupant comfort, whether simply as constraints on the optimization or as a measure to be minimized; however, to the knowledge of the authors, no previously simulated or implemented building energy optimization has considered the economic cost of occupant comfort. The following sections detail the development of these objective functions.

3.2.1. AHU Fan Economic Objective Function Development

As described above, the fan in an AHU works to maintain an end static pressure in the duct to move the required amount of air to condition the individual zones. To accomplish this, the fan motor requires electricity and there is a cost associated with the power used. A simple way to measure the power consumed by a fan motor would be to use a power meter on the electrical lines to the fan and measure the power consumed; however, power meters are not often included on individual fans within existing building energy systems. Therefore, for the presented case study of the UBO, the power consumed by the fan was instead determined by other available data. Specifically, the change in pressure across the fan and the volume flow rate of air were used. With these two values, the work being performed by the fan on the air can be estimated and converted into a measure of power. The equation for work performed by a fan is given in Equation (3):

$$P_{fan} = 0.1175 \cdot \frac{q_{AHU} \cdot \Delta P}{\mu_f \cdot \mu_b \cdot \mu_m} \tag{3}$$

where P_{fan} is the power consumed by the fan [W], 0.1175 is a conversion factor for imperial units, q_{AHU} is the air flow rate through the AHU [cfm], ΔP is the change in pressure across the fan [in. H_2O], and μ_f, μ_b, and μ_m are efficiencies for the fan blade, the fan belt, and the fan motor, respectively. The efficiencies were all assumed to be 0.9 based on comparable values to reflect some degradation from the ideal efficiencies of new/perfectly tuned equipment. The change in pressure across the fan was taken from the end static pressure sensor in the AHU. As for the volume flow rate, the flow rate meters included in each of the VAVs were used in summation to calculate the total air flow rate through the AHU, assuming minimal duct losses and leaks.

The most appropriate control action u for the AHU fan objective function would be the end static pressure setpoint. Assuming the local PID controllers have zero steady state error and sample significantly faster than the new supervisory controller, a reasonable assumption can be made that the end static pressure will equal the end static pressure setpoint. As such, assuming an electric utility rate of \$0.12 per kWh, the fan power cost can be calculated as:

$$J_{fan} = \frac{0.1175}{1000} \cdot \frac{C_{elec} \cdot q_{AHU} \cdot P^*_{EDS} \cdot t_s}{\mu_f \cdot \mu_b \cdot \mu_m} \tag{4}$$

where C_{elec} is the rate of electricity cost [\$/kWh], P^*_{EDS} [in. H_2O] is the end static pressure setpoint, and t_s [hr] is the sampling time of the supervisory controller. Reformulating Equation (4) to fit Equation (1) gives:

$$\begin{aligned} J_{fan} &= e^T Qe + e^T R + u^T Su + u^T T \\ Q = 0 \quad R &= 0 \quad S = 0 \\ u &= P^*_{EDS} \\ T &= \frac{0.1175 \cdot C_{elec} \cdot q_{AHU} \cdot t_s}{1000 \cdot \mu_f \cdot \mu_b \cdot \mu_m}, \end{aligned} \tag{5}$$

thus establishing an objective function that can minimize the cost of electricity used by the AHU's VAV fan by optimizing the end static pressure setpoint.

3.2.2. AHU Chilled Water Economic Objective Function Development

As described in the building operation details, chilled water is used in the AHU to condition the zone supply air. If specific details about the chiller are known, the consumption of power can be calculated and then used to determine the cost of producing said chilled water. In the case study, utility usage data were leveraged to determine this cost. The UEM Office at Texas A&M University maintains utility usage data, specifically the cost per unit of energy of the respective utility. For chilled water, this is the dollar cost associated with producing one mmBtu of chilled water at the campus wide chilled water temperature of 45 °F. In the AHU, the discharge air temperature setpoint is tracked by a PID loop that actuates the valve metering how much chilled water flows through the chilled water coil. The energy associated with chilled water usage can be determined by Equation (6):

$$\dot{Q} = \dot{m} \cdot c \cdot \Delta T \tag{6}$$

where $\dot{Q}$ is the rate of change of heat, or energy, $\dot{m}$ is the mass flow rate, c is the specific heat of the respective fluid, and ΔT is the change in temperature of the fluid. If a mass flow rate sensor is available on the AHU for the chilled water, then this measurement can be used to calculate the rate of change of energy of the chilled water and thus the overall cost; however, mass flow rate sensors are not usually installed on chilled water lines at the AHU level. If this is the case, then a volume flow rate can be used such that:

$$\dot{m} = \rho \cdot q \tag{7}$$

where ρ is the density of water and q is the volume flow rate. With the UBO, the maximum flow rate through the chilled water valve was determined to be $3.41e^{-3}$ m^3/s (54.05 gal/min). Assuming a linear valve/flow relationship, the chilled water valve position multiplied by the maximum possible flow will give the current volume flow rate of the chilled water. Using data values for the discharge air temperature setpoint and the chilled water valve position, a fit was generated to transform the setpoint to a valve position [31]. Combining the relationships gives Equation (8)

$$\begin{aligned} J_{CHW} &= \alpha \cdot q_{max} \cdot \rho \cdot c \cdot \Delta T_{h_{2}o} \cdot t_s \cdot 0.00341214 \cdot \frac{3682}{241} \\ \alpha &= \left(3.021 T_{DA}^{*2} - 109.4 T_{DA}^{*} + 1002\right) \end{aligned} \tag{8}$$

where α is the fitted relationship between the discharge air temperature setpoint (T_{DA}^{*} [°C]) and the chilled water valve position [%], q_{max} [m^3/s] is the maximum flow through the valve (at 100% opening), ρ [kg/m^3] is the density of water, c [kJ/kg °C] is the specific heat of water, $\Delta T_{h_{2}o}$ [°C] is the change in temperature of the supply and return chilled water, t_s [h] is the sample time of the supervisory controller, 0.00341214 [mmBtu/kWh] is a conversion factor from kWh to mmBtu, and the last term is the economic cost for the UBO of the consumed chilled water. Reformulating Equation (8) to fit Equation (1), where u equals the discharge air temperature setpoint, gives:

$$\begin{aligned} J_{CHW} &= e^T Q e + e^T R + u^T S u + u^T T \\ Q &= 0 \quad R = 0 \quad S = 0 \\ u = \alpha &= \left(3.021 \cdot T_{DA}^{*2} - 109.4 \cdot T_{DA}^{*} + 1002\right) \\ T &= q_{max} \cdot \rho \cdot c \cdot \Delta T_{h_{2}o} \cdot t_s \cdot 0.00341214 \cdot \frac{3682}{241} \end{aligned} \tag{9}$$

where the linear cost term $u^T T$ is equal to Equation (8). This result actually lends itself well to traditional convex optimization of the setpoint T_{DA}^{*}, due to the quadratic nature of the fit.

3.2.3. Zone Occupant Comfort Economic Objective Function Development

To measure an occupants level of discomfort, many have relied on the use of predicted mean vote (PMV), developed by Fanger in the 1970s [29]. PMV is a measure on the American Society of Heating, Refrigeration, and Air-Conditioning Engineers (ASHRAE) thermal sensation scale of −3 to 3, where negative numbers represent being too cold, positive numbers represents being too warm, and a value of zero represents being comfortable. Performing a large study of people, Fanger collected data regarding occupants votes and created an equation to determine the PMV across a variety of environmental factors. These factors include air temperature, air relative humidity, relative air speed, mean radiant temperature, an occupant's insulation level due to clothing, an occupant's metabolic rate, and an occupant's work output, among several other variables. The details of the equation can be found in [29]. From PMV, Fanger determined the Predicted Percentage of Dissatisfied (PPD) of people. The relationship of PPD to PMV can be seen in Figure 7a.

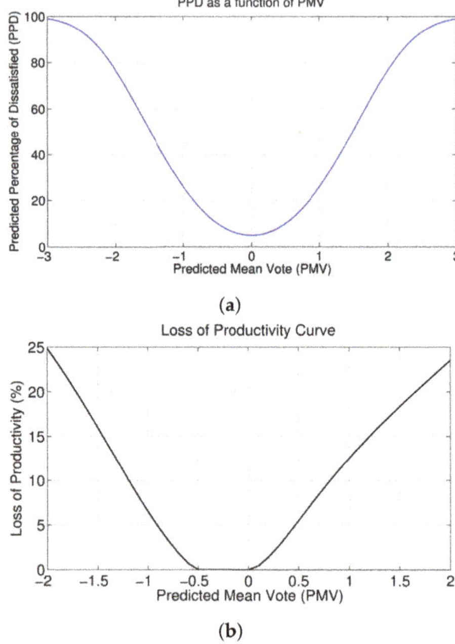

(b)

Figure 7. Curves used in the zone occupant comfort economic objective function development. These equations/relationships and their values are defined in [29,32]. (**a**) Relationship between PPD and PMV as calculated in Fanger [29]. (**b**) Shape of Loss Of Productivity function as calculated in [32].

One can see that at −3 and 3 PMV approximately 100% of the population would be dissatisfied. Of note is that, even at 0 PMV, 5% of the population, on average, will still be dissatisfied, attesting to the fact that each individual has specific preferences. While PMV determines how many people will be dissatisfied, a relationship to tie this to an economic cost is still needed.

Fortunately, researchers have investigated the effect of occupant comfort on worker productivity, as mentioned in the literature review. One effort in particular tied PMV to a measure of Loss of Productivity (LOP) [%]. By using regression analysis, a direct relation can be calculated between a worker's loss of performance and the PMV of an indoor climate by including the calculations of equivalent thermal situations from Gagge's two-layer human model [33] and Fanger's comfort equation [29]. For a detailed explanation of the relationship, see [32]. The results of Roelofsen's work [32] are two sets of coefficients for a regression fit for the cold side of the PMV comfort zone and for the warm side of the PMV comfort zone. The regression is the sixth-order fit shown in Equation (10):

$$LOP = c_0 + c_1 PMV + c_2 PMV^2 + c_3 PMV^3 + c_4 PMV^4 + c_5 PMV^5 + c_6 PMV^6 \quad (10)$$

where LOP is the loss of productivity and $c_0, \ldots, c_6$ are the regression coefficients. The value of the Roelofsen's coefficients are repeated in Table 2, for reference. Roelofsen constructed the regression on the cold side to be zero at −0.5 PMV and for the warm side to be zero at 0 PMV, leaving a region between −0.5 and 0 PMV where LOP is zero. This bias is because several studies found a region of conditions near and below 0 PMV that had a negligible effect on productivity. Figure 7b shows the change in LOP as PMV varies.

Table 2. Regressions for loss of productivity fit of PMV; values presented in [32].

Regression Coefficients	Cold Side of PMV Comfort Zone	Warm Side of PMV Comfort Zone
c_0	1.2802070	−0.15397397
c_1	15.995451	3.8820297
c_2	31.507402	25.176447
c_3	11.754937	−26.641366
c_4	1.4737526	13.110120
c_5	0.0000000	−3.1296854
c_6	0.0000000	0.29260920

To connect LOP to an economic cost, a multiplication of the lost productivity percentage by the amount of salary an employee earns over the sample period can be used, as shown in Equation (11):

$$\beta = LOP \cdot \left(\frac{p_{year} \cdot t_s}{52 \cdot 40} \right) \tag{11}$$

where β [\$] is the lost productivity in wages, p_{year} [\$] is the annual salary for the zone, and t_s [h] is the sampling time. A 40-h work week for the entire year was used as a conservative assumption. In reality, there will be some variation due to holidays, vacation, and overtime.

Considering the individual zones in the presented case study with respect to Equation (1), it is noted that, while there is no control action u at the zone level that consumes energy, there is an error signal present. That is the error e in Equation (1), which is given by:

$$e = T_{zone} - T^*_{zone} \tag{12}$$

where T_{zone} and T^*_{zone} are the zone temperature and zone temperature setpoint, respectively. The error is defined in this manner as this publication focuses on the systems when in cooling mode; thus, the error will be mostly positive during cooling mode. If the zone were in heating mode, the sign of the error term would need to reversed. Recognizing that the zone temperature setpoint can be optimized by the supervisory controller to minimize the LOP by optimizing the zone's PMV, and that the error signal is a difference of temperatures, the sensitivities of the above relationships can be determined and combined to give an economic objective function.

In determining PMV, only the zone air temperature and zone air relative humidity are changing. As such, a fit of the PMV equation was created and used to reduce computation time and complexity. A metabolic rate of 70 [W/m^2], a clothing insulation factor of 0.75 [m^{2}K/W], and a relative air velocity of 0.2 [m/s] were assumed. Additionally, the mean radiant temperature was assumed to be equal to the zone air temperature. The fit is defined as:

$$PMV = 0.5542 \cdot Rh_{zone} + 0.23 \cdot T_{zone} - 5.44 \tag{13}$$

where Rh_{zone} [%] is the relative humidity of the zone and T_{zone} [°C] is the zone air temperature. Equation (13) is used to determine the sensitivity of PMV to changes in air temperature. The resulting combination of sensitivities is shown in Equation (14):

$$\begin{gathered} J_{room} = e^T Q e + e^T R + u^T S u + u^T T \\ Q = 0 \quad S = 0 \quad T = 0 \\ R = \left[\frac{\partial PMV}{\partial T_a} \right] \cdot \left[\frac{\partial LOP}{\partial PMV} \right] \cdot \left[\frac{\partial \beta}{\partial LOP} \right] \end{gathered} \tag{14}$$

where the sensitivities are determined to be:

$$\left[\frac{\partial PMV}{\partial T_a}\right] = 0.23 \tag{15}$$

$$\left[\frac{\partial LOP}{\partial PMV}\right] = c_1 + 2c_2PMV + 3c_3PMV^2 + 4c_4PMV^3 + 5c_5PMV^4 + 6c_6PMV^5 \tag{16}$$

$$\left[\frac{\partial \beta}{\partial LOP}\right] = \left(\frac{p_{year} \cdot t_s}{52\text{wks} \cdot 40\text{hrs}}\right). \tag{17}$$

The cost calculation assumes that, if more than one occupant is in a zone, the sum of the occupant's salaries is used for p_{year}. For Equation (16), the coefficients are as described in Table 2. If the zone's PMV falls within the range of -0.5 to 0, then Equation (16) is equal to zero. The overall trend of the objective function as the zone air temperature varies is shown in Figure 8. The abrupt changes occur at the PMV values of -0.5 and 0, due to the fit of the LOP equation.

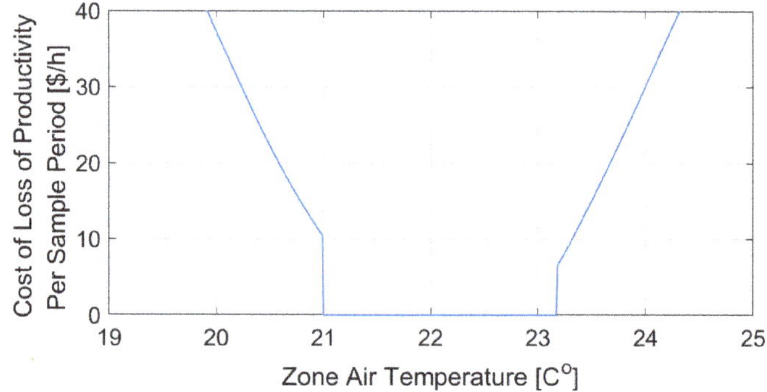

Figure 8. The change in objective cost of the loss of productivity due to discomfort due to changes in zone air temperature during a 15 min period.

This curve will shift depending on other variables in the objective function that are varied, such as annual salary and relative humidity of the zone. In addition, depending on the value of the user-defined setpoint, the objective cost can become negative. This does not mean that the zone is earning money, but occurs because of the structure of the objective function. An optimization competition can occur between the error term and the coefficient R. As the error is defined as the difference in the zone temperature and zone temperature setpoint, the objective cost will be zero when the zone reaches this defined setpoint; however, if this setpoint is not equal to the optimal comfort temperature, as determined by the PMV function, then the objective cost will also be zero if the zone air temperature is equal to the optimal comfort temperature.

This behavior presents an interesting control question. To operate at the most cost effective point, as defined by the objective function, would require that the zone temperature setpoint be set to the PMV optimal temperature, but, in doing so, the ability for occupants or building managers to provide the system feedback about their comfort is removed. Thus, which is more important: the ability for occupants to choose their zone temperatures or allowing the system to determine what is best for the occupants? This question deserves additional investigation but is beyond the scope of this publication. Fortunately, with this objective function, the system will propose a compromise: a temperature between the user-defined setpoint and the PMV optimal comfort temperature.

For a centralized implementation, the objective function then becomes:

$$J_{total} = J_{fan} + J_{CHW} + \sum_{i=1}^{n} J_{zone,i} \tag{18}$$

where n is the number of zones in the system. Equation (18) is the system objective function that will be minimized in the optimization.

4. Simulation Design

A model of the UBO was first created with the aid of SketchUp [34], a 3D modeling software. Using dimensions taken from the building, the single story layout was replicated. Utilizing the plug-in from OpenStudio, thermal zones, boundaries, and interactions were defined. This model was then exported as an input file (IDF) for EnergyPlus [35], an open-source energy simulation program that has been developed by the Building Technologies Office (BTO) under the U.S. Department of Energy (DOE). Using EnergyPlus, HVAC equipment was added to the model based on the equipment present in the UBO. While EnergyPlus excels at modeling and simulation, it is not immediately accessible for controller development and implementation. As such, controllers for the AHU and the zone level VAVs were created in MATLAB. To enable co-simulation between EnergyPlus and MATLAB, two programs were used. The first was MLE+ [36], an open-source MATLAB toolbox for creating the necessary configuration files and providing functions to connect EnergyPlus and MATLAB with an easy-to-use graphical interface. MLE+ utilizes the Building Controls Virtual Test Bed (BCVTB) [37] as the communication backend between EnergyPlus and MATLAB, providing the co-simulation functionality.

The individual zones of the model were setup as defined in the building layout and exterior doors and windows were placed as accurately as possible. The AHU and zone VAVs were added using EnergyPlus' HVACTemplate objects. A central electric chiller was also added to supply chilled water to the AHU. The chilled water output temperature is regulated to 45 °F, the same as the supply chilled water temperature for the UBO.

Currently, EnergyPlus does not offer modeling of pressure with variable air volume systems, thus another solution was necessary to simulate the UBO's control and physical limitations of end static pressure and flow in the AHU. To accomplish this, the authors assumed that the dynamics of the fan speed and end static pressure were fast enough compared to the simulation timestep (1 min) to be considered instantaneous. Additionally, the assumption that the fan would supply the requested end static pressure, constrained by the physical limitations of the AHU and ducting, was made. To determine this constraint, data from the real UBO building were analyzed and a maximum possible work performed by the fan was calculated (910 W). During the optimizations, the constraint is calculated by using Equation (19).

To determine the total volume flow through the AHU, individual models of the zone VAVs were generated from data based on VAV damper position and end static pressure. The effect of outdoor air temperature was also considered on the VAVs as the AHU draws in outdoor air, but was shown to be minimal. This is most likely due to the fact that AHU is able to meet its discharge air temperature setpoint, even at varying flows, effectively isolating the VAV supply air from the outdoor air conditions. The flows from the VAV models are summed to obtain the total flow through the AHU, assuming minimal duct losses. The constraint can be written as:

$$910 \geq 0.1175 \cdot \frac{q_{AHU} \cdot P_{EDS}}{\mu_f \cdot \mu_b \cdot \mu_m} \tag{19}$$

Thus, the optimization will only ever choose an end static pressure setpoint that is physically achievable by the system. This end static pressure is then passed to the VAV models which, along with the commanded damper positions, produce individual zone air volume flows.

The overall control hierarchy can be seen in Figure 9. The supervisory controller supplies the zone temperature setpoints (T^*_{ZONES}) to the respective PID reference inputs and the end static pressure setpoint (P^*_{EDS}) to the VAV models. The discharge air temperature setpoint (T^*_{AHU}) is supplied directly to EnergyPlus as the control for the chilled water valve is implemented using an appropriate setpoint manager within EnergyPlus. The zone temperature error is fed to the cooling PID controller which outputs a desired percentage of maximum flow setting (0–100%). If the minimum flow setting for a zone is zero, then the signal remains unchanged. The desired flow percentage is then passed to the flow PID controller which then produces a desired damper position. This damper position is used in the VAV models as previously described. The outputs from the VAV models, desired zone air volume flows, are converted to air flow fractions and then sent to zone VAVs in EnergyPlus where the room dynamics are simulated for one timestep.

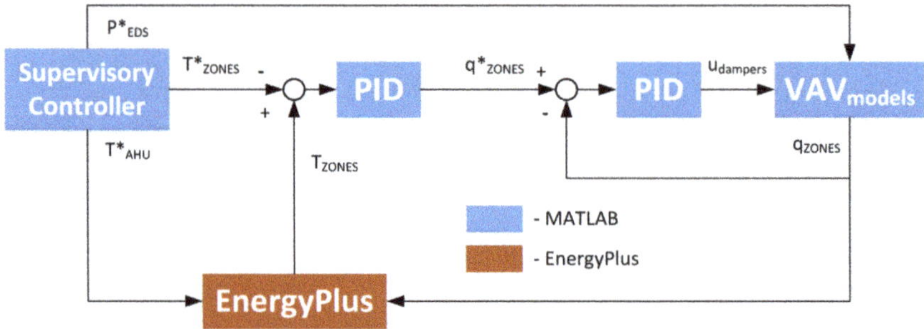

Figure 9. Control hierarchy used in simulation of the UBO.

For the supervisory controllers, a prediction model is necessary to determine the future zone temperatures as the setpoints are optimized. A previously developed modeling algorithm was employed to generate models for the individual zones [38]. One of the main reasons this algorithm was chosen is because of its ease in producing models, requiring the user to only select input and output data. The full details of this process are beyond the scope of this publication, but more information can be found in [38]. Briefly, data from the EnergyPlus simulation are analyzed to develop discrete, linear models of the selected parameters. These models take the form of:

$$\begin{aligned} x_i(k+1) &= A_i x_i(k) + B_i u_i(k) + K_i e_i(k) \\ y_i(k) &= C_i x_i(k) \end{aligned} \tag{20}$$

where A, B, C, and K are the identified system matrices, u is the model input vector, y is the predicted output, and e is the error defined as the difference between the current value and the previous predicted value. The algorithm automatically identifies significant coupling interactions between zones and includes the respective zone temperatures ($T_{zone\ dist.,j}$) as measured disturbances. In addition to the temperature of these disturbance zones, other parameters such as outdoor air temperature (T_{OA}), outdoor air relative humidity (Rh_{OA}), AHU discharge air temperature (T_{AHU}), end static pressure (P_{EDS}), and zone temperature setpoints ($T^*_{zone,i}$) are used as inputs to ARX, ARMAX, Output Error, and Box-Jenkins modeling methods with the output being the respective zone temperatures ($T_{zone,i}$). The best fitting model is selected and then the individual models are combined into a centralized model of the entire system. Steady-state predictions from this centralized model are then used by the supervisory controller to optimize the UBO's 12 setpoints (discharge air temperature setpoint, end static pressure setpoint, and 10 zone temperature setpoints) to minimize the economic cost functions previously described. The optimization occurs every 15 min with the R, S, and T matrices updating based on current operating conditions to reflect the time varying nature of the system dynamics.

Steady-state relationships were chosen over dynamic relationships for the initial implementation and validation of the proposed economic objective function strategy. This choice takes advantage of the fact that in the building energy systems, the control variable's dynamics (AHU discharge air temperature, end static pressure, VAV damper position) change quickly versus the other system variables (outdoor air temperature, outdoor relative humidity, room temperature), which change relatively slowly over the optimized timestep. The exact model inputs and outputs are shown in Table 3.

Table 3. Inputs and outputs for the generated zone models.

Model Input	Model Output
T_{OA}	
Rh_{OA}	
$T^*_{zone,i}$	$T_{zone,i}$
T^*_{AHU}	
P^*_{EDS}	
$T_{zone\ dist.,j} - T_{zone,i}$	

Models were identified for two operational cases: (1) where the zone VAV damper still had actuator range (i.e. the damper was not fully open); and (2) where the zone VAV damper is fully open. The separate models were necessary as the effect of the model inputs varies greatly between the two cases. In Case 1, the effect of T^*_{AHU} and P^*_{EDS} are minimal compared to $T^*_{zone,i}$. This is due to the fact that the VAV damper is actuated by a PID controller with $T^*_{zone,i}$ as the reference. The PID controller is able to reject changes in T^*_{AHU} and P^*_{EDS} by changing the damper position to achieve $T^*_{zone,i}$. However, in Case 2, when the damper is fully open, T^*_{AHU} and P^*_{EDS} become the inputs of significance as they directly effect the zone temperature, determining the amount and the temperature of the incoming conditioned air.

A decision process was necessary to determine when each model should be used. This decision was made based off of two conditions. The first determined if the predicted zone temperature using the Case 1 models was greater than the prediction from the Case 2 models. This condition served to verify whether the predicted temperature from the Case 1 models was currently achievable with the state of the AHU. If the Case 1 predicted temperature was lower than the Case 2 predicted temperature, then the VAV would not be able to achieve the Case 1 predicted temperature with the current T^*_{AHU} and P^*_{EDS} values. The second checked if the current VAV damper position was less than 95%, or, in other words, if the VAV still had actuator range of the damper. The value of 95% was used as opposed to 100% to serve as a threshold and help prevent the system from oscillating between cases. If both these conditions were true, then the Case 1 models were used; otherwise, the Case 2 models were used. To help ensure smooth transfer between the two models and more accurate predictions, errors were calculated between the previous predictions the measured temperatures and included in the current prediction.

5. Results

Simulations were completed to determine the steady-state optimal control method's performance compared to the current control strategies in place in the UBO, using Equation (18) as the objective function. An additional simulation was completed to demonstrate the current control method's ability to track LOP optimal (PMV = −0.25) zone temperature setpoints. By using LOP optimal temperature setpoints, a more direct performance difference can be determined between the current control method and the proposed steady-state optimal control method. Lastly, a simulation showcasing the proposed optimal control method's ability to prioritize certain zones over other zones was completed.

5.1. UBO Simulation with Current Building Controls

The current control method was simulated on the UBO building under two operational cases: (1) with the building operator defined zone temperature setpoints (23 °C); and (2) with LOP optimal zone temperature setpoints (the air temperature at which PMV = −0.25). The numerical results of the first case are used as a baseline to compare against, while this section details the performance of the second case. Figure 10 shows one day of the zone temperatures for the UBO building using the current control method and PMV optimal temperature setpoints. The outdoor air temperature (dashed line) is included in the plot for reference.

The current control method with demand calculations and local PID control shows the ability to track the LOP setpoints fairly accurately. Worth noting is that beginning around 2 pm, Zone 1's temperature starts to drift upwards away from the optimal temperature. This is due to Zone 1's damper being fully open combined with the fact that the AHU fan has reached its power limit and the discharge air temperature is not decreasing fast enough to provide the additional required cooling. Figure 11 shows the system's end static pressure (dashed green line) and the total air flow (solid blue line) in the AHU. Shortly after 1 PM, the end static pressure begins to decrease as the total air flow continues to increase. This is the point where the AHU fan has reached its maximum power capabilities. As the zone dampers continue to open, there is less obstruction to the passage of air, decreasing the pressure and increasing the flow.

Figure 12 shows the chilled water flow and discharge air temperature of the AHU. The discharge air temperature gradually decreases throughout the day (dashed green line), responding to the increase in cooling demand. As the temperature drops, more chilled water is required to cool the air, displayed by the increase in the mass flow rate of the chilled water (solid blue line).

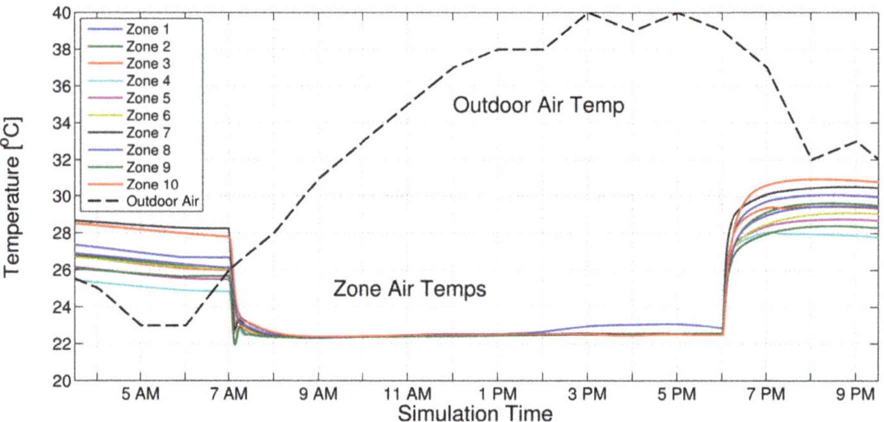

Figure 10. Zone and outdoor air temperatures for the UBO using the currently implemented methods with PMV optimal temperature setpoints. None of the rooms violated the prescribed PMV thresholds.

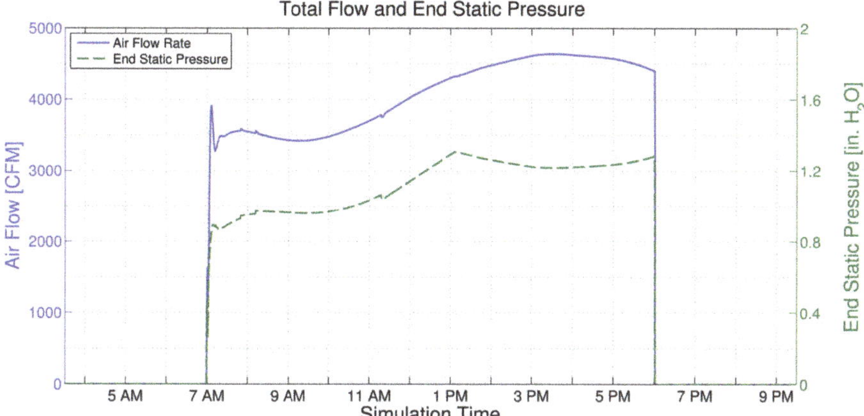

Figure 11. Total air flow and end static pressure in the AHU for the UBO using the currently implemented control methods with PMV optimal temperature setpoints.

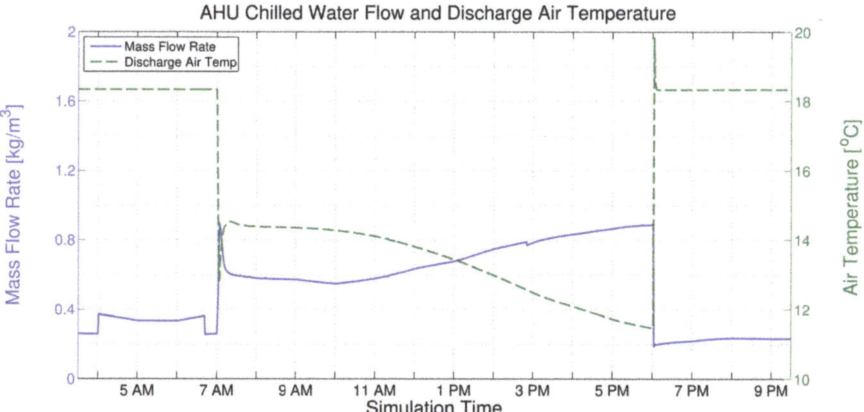

Figure 12. Chilled water flow and discharge air temperature for the UBO using the currently implemented control methods with PMV optimal temperature setpoints.

5.2. *Steady-State Optimal Control Simulation*

The proposed steady-state optimal control method was simulated on the UBO building with the user-defined temperature setpoints equal to the PMV optimal temperature. All the zone temperatures can be seen in Figure 13. Compared to Figure 10, the zone temperatures appear to vary slightly more through out the day. This is not because the zones temperatures are not optimal, but because of the range of PMV (-0.5 to 0) for zero loss of productivity. This range of PMV's allows the optimization a band in the individual zone temperatures while minimizing the utility cost of the chilled water and electricity and leveraging the coupling that exists between zones.

Figure 14 shows the end static pressure and the total air flow through the AHU. Comparing to the current control method simulation, the air flows follow relatively similar paths, with the pressure in the steady-state method simulation taking a higher value but remaining more constant throughout the day. Figure 15 shows a lower discharge air temperature for the steady-state case. While this results in increased flow rates of the chilled water, the cost may not necessarily be higher as the return chilled water temperature may be lower, meaning the chiller has to cool the water over a smaller difference in temperatures. This lower discharge air temperature helps the steady-state optimal control method to achieve more reduction in the cost of lost productivity due to discomfort, enabling lower temperatures in the zones.

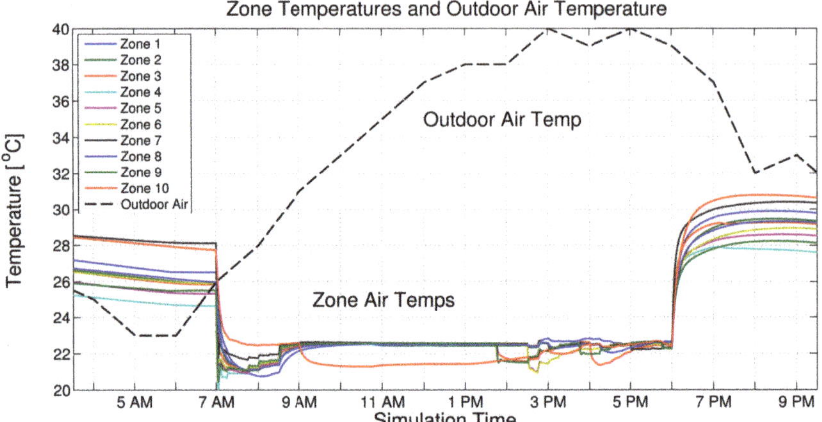

Figure 13. Zone and outdoor air temperatures for the UBO using the steady-state control method with PMV optimal temperature setpoints. None of the rooms violated the prescribed PMV thresholds.

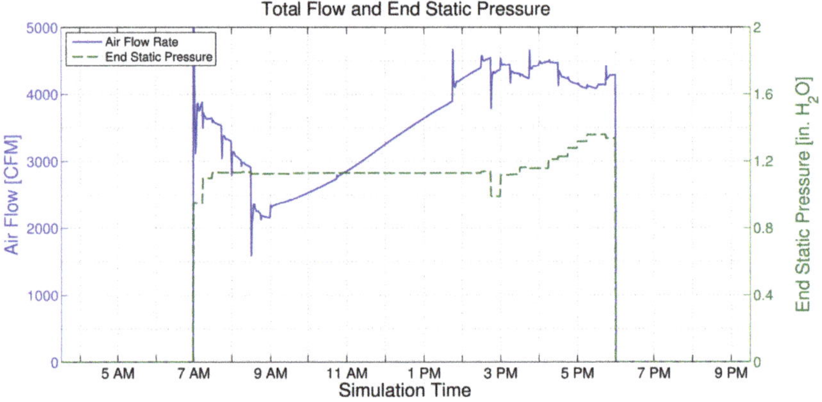

Figure 14. Total air flow and end static pressure in the AHU for the UBO using the steady-state control method with PMV optimal temperature setpoints.

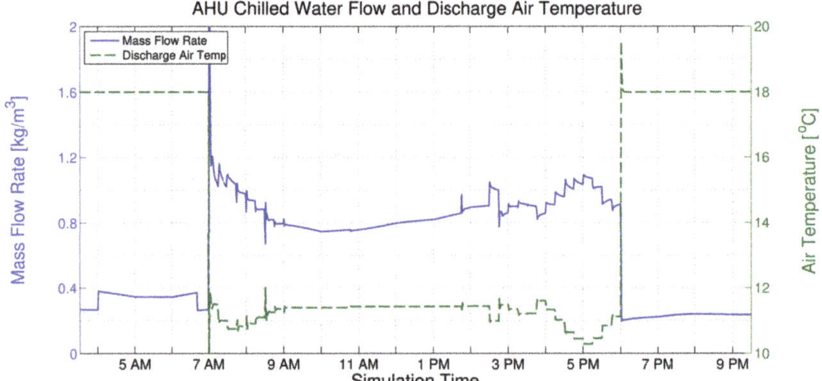

Figure 15. Chilled water flow and discharge air temperature for the UBO using the steady-state control method with PMV optimal temperature setpoints.

5.3. Very Important Person Simulation

To demonstrate one of the proposed steady-state algorithms capabilities, a simulation in which the comfort conditions of one zone was valued significantly more over the other zones in the building was completed. This can occur in the situation where there is a very important person (VIP) that requires comfortable conditions to be maintained, or in the case where other rooms are less important to maintain at a specific comfort level and can be warmer to reduce utility usage. In this simulation, Zone 5 was chosen as the VIP zone. Figure 16 shows all of the zone temperatures. The other zones are higher in temperature throughout the day, while Zone 5 is maintained at a lower temperature. Several zone temperatures can be seen rising above the 0 PMV threshold after 1 PM, when the cooling demand for the day is the greatest. This departure from the optimal LOP range between −0.5 and 0 PMV is due to the optimization balancing the cost of discomfort in the zones with the cost of the utilities.

Figure 17 provides further insight into the maintaining of comfort in Zone 5. The zone temperature is shown with the solid blue line and two thresholds are displayed: (1) the dashed red represents the 0 PMV threshold; and (2) the dashed green represents the −0.5 PMV threshold. After the building is initially occupied, the zone temperature is maintained between the two thresholds resulting in zero loss of productivity for Zone 5.

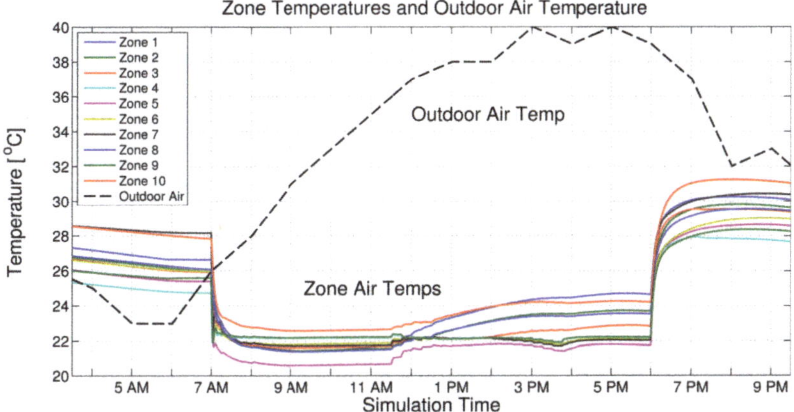

Figure 16. Zone and outdoor air temperatures for the UBO using the steady-state optimal control method with a VIP zone.

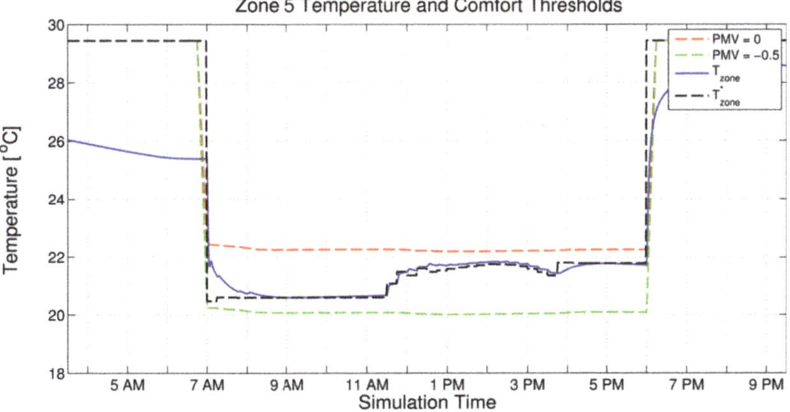

Figure 17. Zone and outdoor air temperatures for the UBO using the steady-state optimal control method with a VIP zone.

6. Discussion

Table 4 shows the annualized costs from the two simulations performed with the current control method and the simulation performed with the proposed steady-state optimal control method. The costs from the simulations were annualized using cooling degree days for the College Station, TX area. The first simulation in the table is the UBO as it is currently operated. The building technician defined temperature of 23 °C was used as the zone temperature setpoint. While this simulation used the least amount in utilities, it also had the greatest cost in terms of loss productivity due to discomfort. The technician defined setpoints is above the PMV optimal range for loss productivity, thus this is to be expected.

Table 4. Annualized economic costs of the different simulation scenarios.

Control Method	Zone Comfort Cost (Productivity)	CHW Cost	Fan Cost	Total Utility Cost	Total Cost
Current Control with User Setpoints	$6869.38	$4093.61	$1255.67	$7046.03	$13,915.41
Current Control with PMV Setpoints	$476.37	$4562.48	$1677.62	$8131.19	$8607.56
Optimal Predicted Steady-State Setpoints with PMV Setpoints	$299.21	$4383.28	$1226.63	$7426.72	$7725.93

The second simulation in the table is the UBO and its current control method but with PMV optimal temperature setpoints (setpoints that give 0 PMV). Worth noting is the significant decrease of 93.1% in the cost of loss productivity just by changing the user-defined setpoints. This of course comes at an increase (approximately 15.4%) in utility cost; however, the total cost was reduced by $5307.85, or 38.1%. The third simulation is the UBO with the proposed steady-state optimal control method. This method resulted in the greatest decrease of the cost of lost productivity of 95.6% with a slightly higher cost in utilities of 5.4%. The steady-state optimal control method also gave the greatest decrease in overall cost, saving $6189.48, or 44.5%, of the original cost. This translates to utility savings of $704.47, or 8.7%, and total cost savings of $881.63, or 10.2%, over the current control method with PMV optimal setpoints.

A significant observation is that, just by changing the current zone temperature setpoints, the UBO building operators could have immediate savings in terms of increased productivity for a slight increase in utility cost with no change in control methods. Furthermore, additional savings can be had through the use of the advanced steady-state optimal control method. Other benefits of using the advanced controller are that, over time, the modeling identification algorithm used will update and improve the steady-state prediction models automatically, providing for the potential for further savings over time. In addition, the models will adapt as seasonal climate shifts occur and equipment efficiency changes while the current control method would require manual tuning as the system parameters change to maintain the same level of performance.

In the UBO, user overrides of the current control system and setpoints are common. While these overrides may reduce in frequency with a change in zone temperature setpoints to PMV optimal values, the impact of overrides would still be greater with the current control system compared to the proposed steady-state optimal controller. The steady-state optimal controller balances the optimal economic zone setpoint with the user-defined setpoint, theoretically reducing the number of overrides. The advanced controller also has the added benefit of allowing the building operator to easily prioritize zones to maximize comfort by adjusting the weight of the annual salary of respective zones.

The authors acknowledge that not all building operational situations call for maximum comfort and productivity, but propose that the importance of occupant comfort and its significant economic impact on businesses and organizations merits further investigation. As building design and control move forward, optimizing occupant comfort should be considered a priority as opposed to maintaining comfort within a predefined range.

Future Directions

The presented work has raised some interesting questions, such as the importance of occupant comfort and its associated economic cost of loss productivity versus utility cost. Further study into this relationship is necessary and can include additional economic and psychological measures, such as the impact of productivity in specific working environments, as well as investigating if integrating user feedback into the control would be beneficial or not. In addition, while the simulations showed the proposed steady-state controller to be successful in reducing the overall cost, verification on the actual building is needed. It would also be of interest to investigate the performance of steady-state predictions versus dynamic predictions and determine exactly how much benefit there is from implementing the more computationally difficult dynamic models of fully implemented MPC.

7. Conclusions

This paper presents a novel economic steady-state optimal control method for control of energy systems in buildings. The control method used economic objective functions that were based on real data from systems found in the Utilities Business Office at Texas A&M University to minimize the economic cost associated with operating the building. Specifically, the cost of utilities (electricity and chilled water) were optimized alongside the cost of loss productivity due to occupant discomfort. In these optimizations, the effect of practical realities that are under represented in the literature were taken into account, including humidity and its effect on comfort, the switching behavior of systems due to mechanical limits, and the effect of current operating conditions on the optimization. Co-simulations of the steady-state optimization controller applied to the Utilities Business Office building were performed with EnergyPlus and MATLAB. The simulation results show improved comfort performance and economic savings with the use of the steady-state optimization controller over the current control method implemented. While the proposed algorithm was deliberately applied to one AHU, the basic structure allows for the optimization of larger building systems that could include several chillers, AHUs, dozens of VAVs, and hundreds of zones.

Author Contributions: Conceptualization, C.J.B. and B.P.R.; methodology, C.J.B., R.C., and B.P.R.; software, C.J.B. and R.C.; validation, C.J.B. and B.P.R.; formal analysis, C.J.B.; investigation, C.J.B.; data curation, C.J.B.; writing—original draft preparation, C.J.B.; writing—review and editing, C.J.B., R.C., and B.P.R.; visualization, C.J.B.; supervision, B.P.R.; project administration, B.P.R.; and funding acquisition, B.P.R. All authors have read and agreed to the published version of the manuscript.

Funding: This material was based on work supported by the National Science Foundation under Grant No. NSF CMMI-1563361. Any opinions, findings, and conclusions or recommendations expressed in this material are those of the author(s) and do not necessarily reflect the views of the National Science Foundation.

Acknowledgments: The authors would like to recognize Texas A&M University's Utilities Energy Management Office, specifically Chris Dieckert, for their contribution of building data and access.

Conflicts of Interest: The authors declare no conflict of interest.

Abbreviations

The following abbreviations are used in this manuscript:

MPC	Model Predictive Control
AHU	Air Handling Unit
VAV	Variable Air Volume
E-MPC	Economic Model Predictive Control
HVAC	Heating, Ventilation, and Cooling
PMV	Predicted Mean Vote
PPD	Predicted Percentage of Dissatisfied
e	Error
Q	Weighting for quadratic error
R	Weighting for linear error

u	Control action
S	Weighting for quadratic control action
T	Weighting for linear control action
UEM	Utilities Energy Management
UBO	Utilities Business Office
PID	Proportional-Integral-Derivative
CHW	Chilled water
PI	Proportional-Integral
K_P	Proportional gain
K_I	Integral gain
K_D	Derivative gain
D^*_{EDS}	End-static pressure demand setpoint
D_{EDS}	End-static pressure demand
P^*_{EDS}	End-static pressure setpoint
P_{EDS}	End-static pressure
ω_{fan}	Fan speed
D^*_{CLG}	Cooling demand setpoint
D_{CLG}	Cooling demand
T^*_{DA}	Discharge air temperature setpoint
T_{DA}	Discharge air temperature
u_{CHW}	Chilled water valve position
T^*_{zone}	Zone temperature setpoint
T_{zone}	Zone temperature
u_{dmpr}	Damper position
P_{fan}	Fan power
q_{AHU}	Air flow rate of the air handling unit
ΔP	Change in pressure across the fan
μ_f	Fan blade efficiency
μ_b	Fan belt efficiency
μ_m	Fan motor efficiency
kWh	KiloWatt-hour
J_{fan}	Fan power cost
C_{elec}	Rate of electricity cost
t_s	Sampling time
$\dot{Q}$	Rate of change of heat
$\dot{m}$	Mass flow rate
c	Specific heat
ΔT	Difference in temperature
ρ	Density
q	Volume flow rate
J_{CHW}	Chilled water cost
q_{max}	Maximum flow rate through the chilled water valve
$mmBtu$	One million British thermal units
ASHRAE	American Society of Heating, Refrigeration, and Air-Conditioning Engineers
LOP	Loss of productivity
β	Lost productivity in wages
p_{year}	Annual salary
Rh_{zone}	Relative humidity of a zone
T_a	Air temperature
J_{total}	Total cost
IDF	Energy Plus input file
DOE	Department of Energy
BCVTB	Building Controls Virtual Test Bed

References and Note

1. U.S. Energy Information Administration (EIA). *Total Energy—Data*; EIA: Washington, DC, USA, 2020.
2. U.S. Energy Information Administration (EIA). *Total Energy—Data*; EIA: Washington, DC, USA, 2016.
3. European Commission. *Renewable Energy—Moving towards a Low Carbon Economy*; European Commission: Brussels, Belgium, 2009.
4. Department of Energy. *20% Wind Energy by 2030: Increasing Wind Energy's Contribution to U.S. Electricity Supply*; Department of Energy: Washington, DC, USA, 2008.
5. U.S. Energy Information Administration. *Annual Energy Review 2011*; Technical Report; U.S. Department of Energy: Washington, DC, USA, 2012.
6. U.S. Energy Information Administration. *Emissions of Greenhouse Gases in the United States 2008*; Technical Report; U.S. Department of Energy: Washington, DC, USA, 2009.
7. U.S. Department of Energy. Building Share of U.S. Primary Energy Consumption (Percent). 2012. Available online: http://buildingsdatabook.eren.doe.gov (accessed on 1 January 2020).
8. Putta, V.; Zhu, G.; Kim, D.; Hu, J.; Braun, J. Comparative evaluation of model predictive control strategies for a building HVAC system. In Proceedings of the 2013 American Control Conference, Washington, DC, USA, 17–19 June 2013; pp. 3455–3460.
9. Gyalistras, D.; The OptiControl Team. *Final Report: Use of Weather and Occupancy Forecasts for Optimal Building Climate Control (OptiControl)*; Technical Report; Terrestial Systems Ecology ETH: Zurich, Switzerland, 2010.
10. Ghiaus, C.; Hazyuk, I. Calculation of optimal thermal load of intermittently heated buildings. *Energy Build.* **2010**, *42*, 1248–1258. [CrossRef]
11. Ma, Y.; Borrelli, F.; Hencey, B.; Packard, A.; Bortoff, S. Model Predictive Control of thermal energy storage in building cooling systems. In Proceedings of the 48h IEEE Conference on Decision and Control (CDC) Held Jointly with 2009 28th Chinese Control Conference, Shanghai, China, 15–18 December 2009; pp. 392–397. [CrossRef]
12. Rawlings, J.B.; Mayne, D.Q. *Model Predictive Control Theory and Design*; Nob Hill Pub: Madison, WI, USA, 2009.
13. Hazyuk, I.; Ghiaus, C.; Penhouet, D. Optimal temperature control of intermittently heated buildings using Model Predictive Control: Part II—Control algorithm. *Build. Environ.* **2012**, *51*, 388–394. [CrossRef]
14. Oldewurtel, F.; Parisio, A.; Jones, C.N.; Gyalistras, D.; Gwerder, M.; Stauch, V.; Lehmann, B.; Morari, M. Use of model predictive control and weather forecasts for energy efficient building climate control. *Energy Build.* **2012**, *45*, 15–27. [CrossRef]
15. Halvgaard, R.; Poulsen, N.K.; Madsen, H.; Jorgensen, J.B. Economic Model Predictive Control for building climate control in a Smart Grid. In Proceedings of the 2012 IEEE PES Innovative Smart Grid Technologies (ISGT), Washington, DC, USA, 16–20 January 2012; pp. 1–6. [CrossRef]
16. Morosan, P.D.; Bourdais, R.; Dumur, D.; Buisson, J. Building temperature regulation using a distributed model predictive control. *Energy Build.* **2010**, *42*, 1445–1452. [CrossRef]
17. Touretzky, C.R.; Baldea, M. Integrating scheduling and control for economic MPC of buildings with energy storage. *J. Process Control* **2014**, *24*, 1292–1300. [CrossRef]
18. Oldewurtel, F.; Parisio, A.; Jones, C.N.; Morari, M.; Gyalistras, D.; Gwerder, M.; Stauch, V.; Lehmann, B.; Wirth, K. Energy efficient building climate control using Stochastic Model Predictive Control and weather predictions. In Proceedings of the 2010 American Control Conference, Baltimore, MD, USA, 30 June–2 July 2010; pp. 5100–5105. [CrossRef]
19. Corbin, C.D.; Henze, G.P.; May-Ostendorp, P. A model predictive control optimization environment for real-time commercial building application. *J. Build. Perform. Simul.* **2013**, *6*, 159–174. [CrossRef]
20. Oldewurtel, F.; Ulbig, A.; Parisio, A.; Andersson, G.; Morari, M. Reducing peak electricity demand in building climate control using real-time pricing and model predictive control. In Proceedings of the 49th IEEE Conference on Decision and Control (CDC), Atlanta, GA, USA, 15–17 December 2010; pp. 1927–1932. [CrossRef]
21. Molina, D.; Lu, C.; Sherman, V.; Harley, R.G. Model Predictive and Genetic Algorithm-Based Optimization of Residential Temperature Control in the Presence of Time-Varying Electricity Prices. *IEEE Trans. Ind. Appl.* **2013**, *49*, 1137–1145. [CrossRef]

22. Alvarez, J.D.; Redondo, J.L.; Camponogara, E.; Normey-Rico, J.; Berenguel, M.; Ortigosa, P.M. Optimizing building comfort temperature regulation via model predictive control. *Energy Build.* **2013**, *57*, 361–372. [CrossRef]
23. Ma, Y.; Borrelli, F.; Hencey, B.; Coffey, B.; Bengea, S.; Haves, P. Model Predictive Control for the Operation of Building Cooling Systems. *IEEE Trans. Control Syst. Technol.* **2012**, *20*, 796–803. [CrossRef]
24. Bengea, S.C.; Kelman, A.D.; Borrelli, F.; Taylor, R.; Narayanan, S. Implementation of model predictive control for an HVAC system in a mid-size commercial building. *HVAC&R Res.* **2014**, *20*, 121–135. [CrossRef]
25. West, S.R.; Ward, J.K.; Wall, J. Trial results from a model predictive control and optimisation system for commercial building HVAC. *Energy Build.* **2014**, *72*, 271–279. [CrossRef]
26. Karlsson, H.; Hagentoft, C.E. Application of model based predictive control for water-based floor heating in low energy residential buildings. *Build. Environ.* **2011**, *46*, 556–569. [CrossRef]
27. Duburcq, G.S. Advanced control for intermittent heating. In Proceedings of the Clima 2000 Conference, Brussels, Belgium, 30 August to 2 September 1997.
28. Yuan, S.; Perez, R. Multiple-zone ventilation and temperature control of a single-duct VAV system using model predictive strategy. *Energy Build.* **2006**, *38*, 1248–1261. [CrossRef]
29. Fanger, P. *Thermal Comfort: Analysis and Applications in Environmental Engineering*; McGraw-Hill: New York, NY, USA, 1972.
30. Elliott, M.; Rasmussen, B.P. Optimal Setpoints for HVAC Systems via Iterative Cooperative Neighbor Communication. *J. Dyn. Syst. Meas. Control* **2014**, *137*, 011006. [CrossRef]
31. Bay, C.J. Advancing Embedded and Extrinsic Solutions for Optimal Control and Efficiency of Energy Systems in Buildings. Ph.D. Thesis, Texas A&M University, Uvalde, TX, USA, 2017.
32. Roelofsen, P. The impact of office environments on employee performance: The design of the workplace as a strategy for productivity enhancement. *J. Facil. Manag.* **2002**, *1*, 247–264. [CrossRef]
33. Gagge, A.P.; Fobelets, A.P.; Berglund, L.G. A Standard Predictive Index of Human Response to the Thermal Environment. *ASHRAE Trans.* **1986**, *92*, 709–731.
34. Trimble. 3D modeling for everyone. 2017.
35. Crawley, D.B.; Pedersen, C.O.; Lawrie, L.K.; Winkelmann, F.C. EnergyPlus: Energy Simulation Program. *ASHRAE J.* **2000**, *42*, 49–56.
36. Bernal, W.; Behl, M.; Nghiem, T.X.; Mangharam, R. MLE+: A Tool for Integrated Design and Deployment of Energy Efficient Building Controls. In Proceedings of the 4th ACM Workshop On Embedded Sensing Systems For Energy-Efficiency In Buildings, (BuildSys'12), Toronto, ON, Canada, 6–9 November 2012.
37. Wetter, M. Co-simulation of building energy and control systems with the Building Controls Virtual Test Bed. *J. Build. Perform. Simul.* **2011**, *4*, 185–203. [CrossRef]
38. Chintala, R.H. A Methodology for Automating the Implementation of Advanced Control Algorithms Such As Model Predictive Control on Large Scale Building HVAC Systems. Ph.D. Thesis, Texas A&M University, Uvalde, TX, USA, 2018.

Article

Solar+ Optimizer: A Model Predictive Control Optimization Platform for Grid Responsive Building Microgrids

Anand Krishnan Prakash [1,†], Kun Zhang [1,†], Pranav Gupta [1], David Blum [1], Marc Marshall [2], Gabe Fierro [3], Peter Alstone [2], James Zoellick [2], Richard Brown [1] and Marco Pritoni [1,*]

1 Lawrence Berkeley National Laboratory, Berkeley, CA 94720, USA; akprakash@lbl.gov (A.K.P.); kunzhang@lbl.gov (K.Z.); pranavhgupta@lbl.gov (P.G.); dhblum@lbl.gov (D.B.); rebrown@lbl.gov (R.B.)
2 Schatz Energy Research Center, Humboldt State University, Arcata, CA 95521, USA; mwm1@humboldt.edu (M.M.); peter.alstone@humboldt.edu (P.A.); james.zoellick@humboldt.edu (J.Z.)
3 Department of Electrical Engineering and Computer Sciences, University of California Berkeley, Berkeley, CA 94720, USA; gtfierro@cs.berkeley.edu
* Correspondence: mpritoni@lbl.gov; Tel.: +1-530-220-4394
† These authors contributed equally to this work.

Received: 18 May 2020; Accepted: 9 June 2020; Published: 15 June 2020

Abstract: With the falling costs of solar arrays and battery storage and reduced reliability of the grid due to natural disasters, small-scale local generation and storage resources are beginning to proliferate. However, very few software options exist for integrated control of building loads, batteries and other distributed energy resources. The available software solutions on the market can force customers to adopt one particular ecosystem of products, thus limiting consumer choice, and are often incapable of operating independently of the grid during blackouts. In this paper, we present the "Solar+ Optimizer" (SPO), a control platform that provides demand flexibility, resiliency and reduced utility bills, built using open-source software. SPO employs Model Predictive Control (MPC) to produce real time optimal control strategies for the building loads and the distributed energy resources on site. SPO is designed to be vendor-agnostic, protocol-independent and resilient to loss of wide-area network connectivity. The software was evaluated in a real convenience store in northern California with on-site solar generation, battery storage and control of HVAC and commercial refrigeration loads. Preliminary tests showed price responsiveness of the building and cost savings of more than 10% in energy costs alone.

Keywords: demand flexibility; control system; optimization; resiliency; smart buildings; distributed energy resources; model predictive control

---

1. Introduction

The United States electrical grids face a range of new challenges to safe and reliable operation: aging infrastructure, increased penetration of less predictable renewable generators to mitigate climate change and the increasing occurrence of extreme weather events all place stress on the grid [1]. To address these issues, more decentralized grid architectures have been proposed [2] based on distributed energy resources (DERs) and microgrids [3–5]. Collections of buildings with local DER and energy storage could operate in grid-disconnected (islanded) mode in case of outages, improving system resiliency [6,7]. Buildings that participate as DERs could also provide additional opportunities for energy storage using their thermal systems (Heating, Ventilation and Air Conditioning (HVAC) and Refrigeration) [8]. With the emergence of low-cost solar and battery storage, small-size microgrids are now a commercially viable option at sites with a high value for resilience and several products to control them have appeared in the market [9–11]. However, microgrid software typically focuses on site protection and battery control and does not coordinate with control of building systems, especially HVAC [12].

HVAC systems represent the largest fraction of the energy use and demand in commercial buildings [13]. Traditional HVAC control strategies deploy rather simple control algorithms [14] that might reside in zone-level thermostats (in small and medium commercial buildings) or in a centralized building automation system (in larger commercial buildings). These algorithms rely on static schedules, irrespective of the actual occupancy of the building or grid needs. While more modern HVAC control strategies exist, they do not typically incorporate DER and are largely proprietary in their implementation, making them hard to extend.

The lack of interoperability between these new control systems limits the realization of the full potential for microgrid systems [15]. For instance, a microgrid controller that has very little information about the status of the building and its projected load cannot operate the battery optimally. Furthermore, there can be missed opportunities for HVAC and refrigeration controllers to take full advantage of periods of high solar generation to pre-cool or pre-heat, utilizing innate thermal storage in the mass of buildings and refrigerated goods.

To address this gap, this paper presents a software platform called "Solar+ Optimizer" (SPO), which was developed, deployed and tested in a pilot application at a fueling station with convenience store in Blue Lake, California, United States. The software provides optimal control of the building loads and DERs and has been built exclusively on open-source libraries. The controller takes into account the time-varying costs of energy and demand and the status of the grid connection to reduce the overall operational cost for the building owner, and it provides demand-response services to the grid. The platform is designed to be scalable, as well as vendor and protocol agnostic. This allows building managers to take advantage of a larger market of connected devices (both sensors and actuators) without being tied down to any particular manufacturer ecosystem, and it could enable less costly adaptation and modification of the system over the expected multi-decade lifetime of microgrid hardware.

The paper is organized as follows. Section 2 reviews the existing literature of advanced control studies with a focus on experimental and field studies. Section 3 introduces the overall software architecture and details the principal components within. Section 4 describes the deployment of the hardware and software on a case study and presents experimental results and analysis. Section 5 discusses the results and technical challenges encountered during the demonstration. The paper ends with conclusions and future work in Section 6.

2. Literature Review and Contribution

Research on advanced control strategies and algorithms (e.g., MPC and reinforcement learning) with application to building systems and DERs has grown significantly in the last decade [16], as the potential of these advanced controls to provide flexibility to the electrical grid has become more evident. Previous studies have investigated MPC applications in a range of HVAC and thermal energy storage systems in buildings: These include MPC utilizing an ice storage tank and building thermal mass [17,18]); MPC for Air Handlers (AHU) and Variable Air Volume (VAV) systems [19–21]; and MPC applied to window operation for mixed natural and mechanical ventilation in an office building [22,23]. Studies have also demonstrated the coordination of HVAC, energy storage and PV generation using MPC based controls in simulated commercial buildings [24–27]. However, most of these studies developed customized solutions, closely tied to the specific building and equipment setup; the gap that this work fills is an MPC application that is built to be extensible and scalable.

There are several control algorithms that have been used for optimizing building systems and DER. Linear programming was employed to minimize the conditional value at risk in the objective function while providing resilience and cost minimization in commercial buildings through local energy generation and storage [7,28]. Genetic algorithms have also been used to optimize the building thermal loads [29,30]. Reduced energy costs and improved occupant thermal comfort within buildings were achieved by using particle swarm optimization [31,32]. Dynamic programming was employed by Benjamin Heymann and Jiménez-Estévez [33] to reduce the energy costs while meeting the building load requirements. A neural network model was trained using the Levenberg–Marquardt algorithm to predict the optimal boiler operation period in commercial buildings [34]. Another emerging type of advanced control is Deep Reinforcement Learning (DRL) [35–40]. While DRL is a model-free approach, most DRL methods use detailed models to generate synthetic data for training purposes. EnergyPlus models of the buildings have been used during the training period [36–40], along with the Building Controls Virtual Test-Bed (BCVTB) [41], which has been leveraged for controls based on the co-simulation framework [37,38]. The DRL algorithms were used to minimize energy costs while maintaining thermal comfort by controlling room temperature setpoints [37], air flowrate in the VAV boxes [38], supply water temperature [39] or outdoor air damper positions [40]. This robust area of new research is promising, with many approaches to optimal control still in their infancy. There is not yet an agreed-on standard approach for these problems.

While the majority of research on advanced control algorithms is conducted on simulations, some studies have deployed those advanced controls on real systems, which allows researchers to test their software in a live environment with unforeseen system behavior that is difficult or impossible to include in simulation. MPC control of multiple air handling units (AHUs) or rooftop units (RTUs) in multi-zone real buildings are demonstrated in [42–46]. Experiments using MPC to control building systems, behind-the-meter energy storage and DERs have also been demonstrated [47–49]. MPC based controllers were deployed in office buildings to determine the setpoints for the supply air temperature, fan speed and the zone air temperature for each AHU [42,43], and West et al. [42] additionally controlled the chilled water and hot water valve position. MPC has also been used to control the air conditioning in large spaces that were served by multiple RTUs by turning on/off different stages of operation of each RTU. These controls were demonstrated in a gym space of a university campus [44] and in a restaurant [45]. Carli et al. [46] used MPC to control the fan speed of a fan coil unit that supplies conditioned air to a single office space in a university building. Frequency regulation as a grid service (by varying the air flow rate setpoints, which resulted in modifying the fan speed) was implemented in the FLEXLAB® testbed [50] at Lawrence Berkeley National Laboratory [47,48]. A public school building in southern Italy was chosen as a site of demonstration in [49], where the authors performed their optimizations in the cloud and send the control signals to the Internet of Things (IoT) devices and controllers that were retrofitted in the school building to control the battery and various loads through an intermediary gateway. DRL strategies have also been implemented in real buildings, although in small experiments. Chen et al. [51] used DRL to control the damper position of a VAV box in a single conference room in a building and Zhang and Lam [52] used it to control the supply water setpoint to the HVAC system in a experimental test-bed office in a university building.

A core barrier in progressing from simulations to real-world deployment of advanced controls is the lack of a robust and reliable software infrastructure to implement those controls in real-world building systems, which often have an eclectic mix of various controllers and systems in place. There has been work on middle-ware software platforms that collect data from various connected sensors and actuators across different systems within a building [42,43,46,48,49,53–56]. The ability to retrieve data from various IoT sensors and devices, store these data centrally and use them for simulations of better control algorithms have been demonstrated in [46,49,53–56]. Interfacing with the existing proprietary control software for gathering data and publishing control signals is another common solution, but this requires site specific implementations as seen in [42,43,48]. Bruno et al. [49] and Carli et al. [46] introduced a software stack and also demonstrate actual controls capabilities on real buildings based on decisions determined by the optimization engine. However, the architecture seems rather case specific with no mention of expansion capabilities to other types of buildings or systems. The VOLTTRON [57] platform has also been used in research studies for data collection and optimization of flexible building loads and grid integration, but these works have been in simulation [58–60], in laboratory environments [61], or are still in progress [62,63]. Most of the platforms reviewed here employ a publish-subscribe method of communication (usually Message Queue Telemetry Transport, MQTT [64]) for messaging, which relies upon a central broker, usually residing in the cloud. This is in line with broader trends for information technology and software services moving to a cloud-based software architecture, but the buildings industry is justifiably reluctant to adopt this model. Cybersecurity vulnerabilities and the possible loss of data and control signals during a network outage are major issues that concern building managers, particularly in commercial buildings. Furthermore, in a microgrid application there can be loss of network connectivity during a blackout, when the system is expected to continue working and provide resilience. Hence, having controls that are able to operate in a mode with local communication and computation may be required to enable widespread adoption of smart control platforms.

Among the field demonstrations of advanced controls, the scope of controls were limited to a very small subset of buildings, mostly office spaces and buildings on university campuses, laboratories or experimental test-beds. Experiments were conducted in well-instrumented environments and thus presented more measured data for advanced controls deployment than general buildings do. Most existing studies rely on proprietary software and do not describe the effort required to deploy both software and hardware, which is critical for replication. While some papers discuss software implementation, the focus was generally on communication and software architecture, with less emphasis on the integration of systems and controls with those software. Those that did demonstrate integration between the controller and the building, deployed a cloud based solution that is susceptible to network outages and more vulnerable to security risks. The SPO solution presented in this paper represents an advance in the field studies of advanced building controls by narrowing the gap between MPC based simulation studies and field demonstrations. The main contributions of this paper are:

- An integrated software architecture that supports a vendor- and protocol-agnostic data acquisition and control framework that enables both local and/or cloud based controls. The architecture is extensible to other building types, equipment and DERs.
- A field demonstration of this software controlling HVAC, refrigeration and DER using MPC. The software is deployed in the local area network in a real-world small commercial building.
- A proof-of-concept demonstration of a building controller that is responsive to various grid signals (time varying energy costs) and that supports demand response events.
- A description of the implementation challenges experienced during deployment and operations, intended to accelerate the effort of future researchers and practitioners who could avoid these barriers that have been identified .

3. Controller Architecture and Components

The Solar+ Optimizer (SPO) is a software solution that has been developed to integrate sensors and controllers for building systems and DERs and to identify real-time optimal control actions for the connected systems such as building loads and batteries. It supports integration across multiple devices and protocols, as illustrated in the architecture diagram in Figure 1. Through support for several communication protocols and APIs, SPO allows integration of systems that are typical for small and medium commercial buildings. It can operate completely within a local network, but can also be configured to operate in tandem with cloud-based resources. This section describes the different software components of SPO that enable these capabilities.

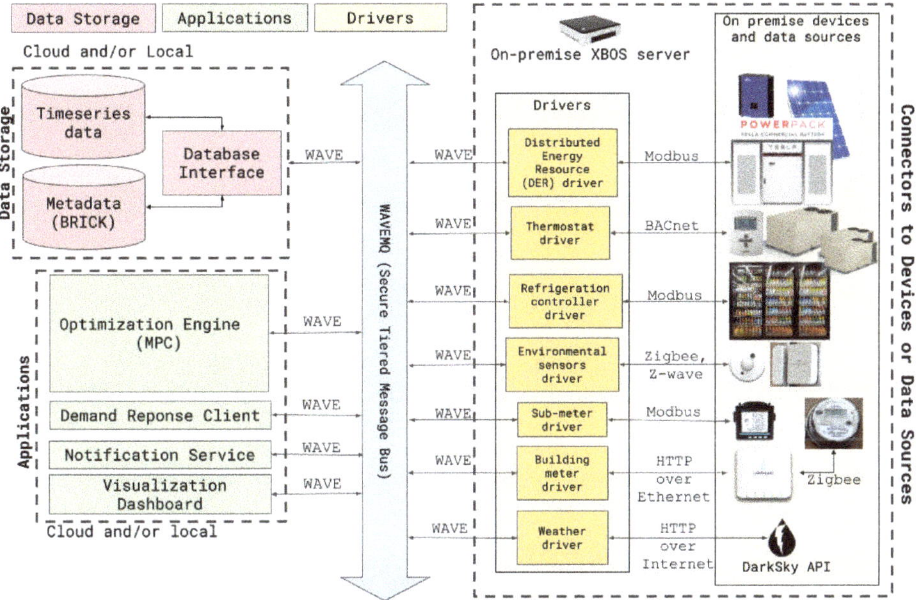

Figure 1. Software architecture diagram of the Solar+ Optimizer system.

3.1. eXtensible Building Operating System (XBOS)

To coordinate the numerous heterogeneous connected devices and controllers within a building, robust network communication is a key requirement. Most commercial and academic solutions use middleware, i.e., software that resides between the hardware devices and other data sources that produce data and the applications that use these data. SPO uses XBOS, an open-source building operating system developed for real-time data acquisition from sensors and control of building actuators [65]. XBOS consists of the following components:

- **WAVE and WAVEMQ**: WAVE is an authentication engine that handles permissions and access control. WAVEMQ is a multi-tier publish-subscribe message bus that allows exchange of data and control signals.
- **Drivers**: Drivers are connectors to real devices and other data sources (e.g., web based services, emulated devices, etc.). A driver is responsible for gathering data from a device and for controlling the device in response to requests from an external controller. With the required permissions, a driver can publish and subscribe to messages on WAVEMQ.
- **Data Storage**: Both operational and configuration data are stored on dedicated databases. There are separate data stores for the building metadata represented using the Brick schema [66] and for the continuous real-time data that are being collected by the drivers.
- **Applications**: Developers can write applications on the XBOS platform using real-time data that is being published on the message bus (e.g., notification service and visualization dashboard) or using historical data that have been stored in the database (e.g., MPC based optimization engine and fault detection tools). Applications can publish control signals for the devices on WAVEMQ and can trigger a change in their mode of operation.

3.2. WAVE and WAVEMQ

WAVE is "an authorization framework offering decentralized trust: no central services can modify or see permissions and any participant can delegate a portion of their permissions autonomously" [67].

WAVEMQ is a tiered publish-subscribe message bus that all drivers and applications on XBOS use to transmit messages [68]. WAVE-based keys and permissions authorize drivers and applications to communicate with one another. The tiered nature of WAVEMQ is implemented in the form of a single "designated router" that is deployed on reliable hardware, typically located in the cloud, and additional message routers at each site, called "site routers", as illustrated in Figure 2. The services, which can be drivers or applications that have been granted the required WAVE permissions, can publish (i.e., write) on or subscribe (i.e., read) to a particular topic (i.e., sensor measurements or actuation commands) on the bus.

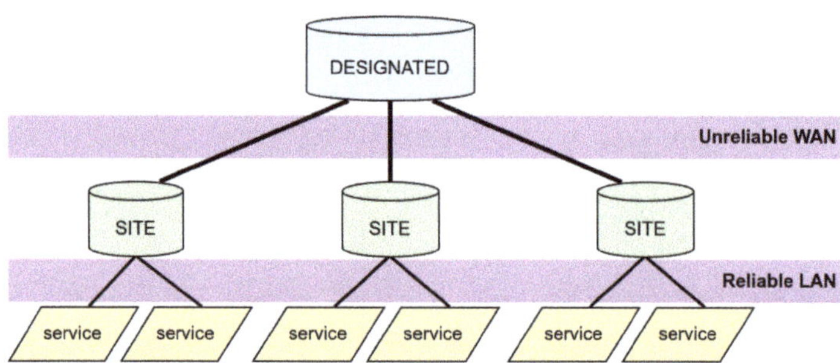

Figure 2. WAVEMQ: A tiered publish-subscribe message bus [68].

During network outages and loss of connectivity between the site routers and the designated router, this tiered architecture of WAVEMQ enables the site router to continue message delivery across the services hosted locally. By buffering messages that have not yet been delivered to the designated router since the loss of Internet connectivity, and by re-attempting to publish these messages once the connectivity has been restored, WAVEMQ also mitigates the risk of data loss. Messages in XBOS are defined using Google's Protocol Buffers or protobufs [69] and the native functions and services are implemented as remote procedural calls using the gRPC framework [70]. These technologies support platform and language independent development, which means drivers and applications can be developed in any programming language and they can use the auto-generated language specific bindings for functions calls and message transmission.

3.3. Drivers

Drivers are the components of XBOS responsible for presenting a uniform communication interface, to local hardware devices and external software applications, that is agnostic to the particular protocols and networks used by those devices and applications. Figure 1 shows some examples of hardware devices that can be integrated using XBOS drivers: environmental sensors, electric meters, HVAC controllers, battery controllers, solar inverters, etc. Internet weather services and utility APIs for power prices are few examples of external software applications from which drivers gather data. These devices and services communicate over a variety of protocols, including older, legacy equipment such as refrigeration or HVAC controllers that communicate over wired, non-Internet Protocol (non-IP) based protocols such as Modbus serial [71] or BACnet MS/TP [72]. In such cases, it is essential to set up the drivers locally within the building's physical network. The drivers also translate the data from the device to the necessary protobuf format required to publish it on WAVEMQ and also interpret the control signals from XBOS applications to equivalent commands for the devices (e.g., changes to the HVAC setpoints or battery charge rate, etc.).

3.4. Data Storage

Data storage to the database on each WAVEMQ router is handled by an XBOS service called the "data ingester." This service, along with the tiered WAVEMQ message bus, allows SPO to access multiple data stores for the same data at different designated and/or site routers, as needed, without configuring additional data replication or mirroring scripts. SPO takes advantage of this feature by providing the capability of deploying the whole system—the drivers, data storage and optimization engine—locally with a building's local area network, provided that all the required data can be obtained from local devices and services.

For buildings and sites that allow both cloud and local communication, the preferred approach is to deploy the essential and critical drivers and applications such as sensor and actuator drivers and optimization engine locally and deploy drivers that communicate with web services (e.g., utility demand response (DR) servers, weather APIs, etc.) and read-only applications (e.g., benchmarking tools and visualization dashboards) on the cloud. Such a strategy limits the drivers that have access to local actuators to be located within the local network, and also reduces the chance that a controller remains in a suspended state due to loss of connectivity to cloud-based intelligence. With this network configuration, multiple data ingester services can be configured across the local site router and the cloud designated router. While the local ingester ensures that the most recent data are always available locally for the optimization engine and other critical applications, the ingester on the cloud can store the messages to a more persistent cloud data store for permanent storage and analytical applications. This preferred architecture is illustrated in Figure 3.

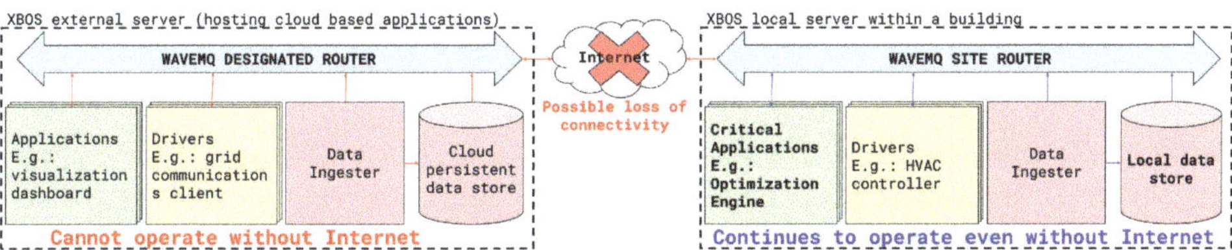

Figure 3. The services on the site router continue to run even during network outages.

The caveat in Figure 3 is that the critical applications on the site router should not require any data from the drivers on the designated router. If applications require data from external web services, it is recommended to host the drivers that are querying them directly on the site router to minimize points of failure.

3.5. Applications: Optimization Engine

Existing solutions often use proprietary platforms and site-specific specifications for generating and sending optimal control signals to devices. Similarly to how XBOS is used as middleware platform by SPO, the open-source package MPCPy [73] is used to implement MPC-based optimization in an extensible and open-source framework. SPO integrates MPCPy as an XBOS application, extending its capabilities to interact with real-time systems. This application, labeled as the "Optimization Engine" queries historical data and future forecasts from the data store, solves the optimization problem using the MPCPy framework and publishes control signals to devices through WAVEMQ. Figure 4 depicts this interaction between the optimization engine, the data store and the device drivers. Through this implementation, SPO provides a scalable, protocol- and manufacturer-independent solution for implementing advanced building controls. This section details the components of the optimization engine.

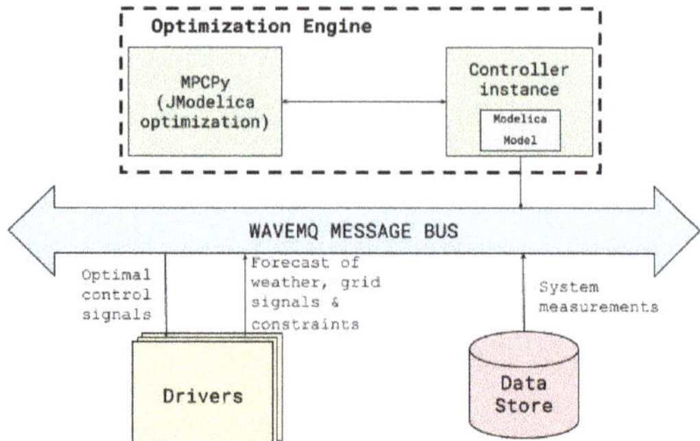

Figure 4. The optimization engine component and its relationship to other parts of SPO.

3.5.1. MPCPy

The MPCPy package [73] is designed to facilitate the simulation, testing and implementation of MPC for building systems. It includes several generic capabilities such as solving optimal control problems with constraints, parameter estimation and validation, interaction with real and/or emulated building systems, and data processing including weather, internal loads, grid signals, etc.

3.5.2. Model Formulation

MPCPy utilizes models defined in the Modelica language [74], an equation-based multi-domain language to model complex physical systems including mechanical, electrical and deterministic control systems. These models can be used to predict the future system behavior and to support linked simulation-optimization problems. For use in optimization, MPCPy focuses on the use of simplified physical models, often known as "grey-box" models. These grey-box models include sufficient detail about the building systems to simulate overall responses but not the specific details of typical expert-defined building energy models. The goal is to balance model specificity with computational efficiency for use with optimization solvers. MPCPy utilizes JModelica.org [75] to generate and solve a control optimization problem based on the user-provided Modelica model, objective and constraint information and input data (e.g., weather or electricity prices). JModelica.org uses CasADi [76] to compute function derivatives and the optimization algorithm IPOPT [77] to solve the resulting nonlinear problem. IPOPT, short for interior point optimizer, is a state-of-the-art nonlinear optimization library to solve large-scale continuous system optimization problems.

3.5.3. Optimization Configuration for Grid Interactions

Using the MPCPy framework, the SPO optimization engine is designed to minimize the cost of building operations subject to various DR scenarios or grid price signals. SPO can optimize for different types of peak-load reduction DR events (e.g., through utility programs such as Peak Day Pricing [78] or Critical Peak Pricing [79]), as well as dynamic prices [80,81]. At its most basic level, SPO minimizes electricity bills of buildings that are subject to Time-Of-Use (TOU) tariffs that contain both energy and demand charges. Other modes of operation for responding to signals from the grid are shown in Table 1, including real-time pricing, demand limiting, load shedding, load shifting and load tracking.

The optimization engine is structured in a flexible way so that the various responses to grid signals can be easily configured and swapped. This is achieved by formulating the objective function in a generic way, translating the grid signals into components of this parameterized function and into constraints of the optimization problem. The various options are stored in variables of a configuration file to easily switch between modes. Table 1 summarizes the five grid signals and the corresponding configuration in the optimization engine. The constraints presented here only relate to the grid-responsive modes, excluding other constraints related to system operation (e.g., indoor temperature boundaries).

Table 1. MPC configuration for handling different grid signals.

Mode	Objective	Information in the Signal	Constraints
Price based	Minimize cost of energy and/or power	Energy price and/or demand charges, estimated peak power	None
Demand limiting	Minimize cost of energy and/or power	Energy price and/or demand charges, estimated peak power, maximum power P_{max}	$P \leq P_{max}$
Load shedding	Minimize cost of energy and/or power	Energy price and/or demand charges, estimated peak power, baseline power $P_{baseline}$, P_{shed}, start time, final time	$P \leq P_{baseline} - P_{shed}$ for $[t_{start}, t_{final}]$
Load shifting	Minimize cost of energy and/or power	Energy price and/or demand charges, estimated peak power, baseline power $P_{baseline}$, $P_{decrease}$, $P_{increase}$, start increase time, final increase time, start decrease time, final decrease time	$P \geq P_{baseline} + P_{increase}$ for $[t_{start,increase}, t_{final,increase}]$ and $P \leq P_{baseline} - P_{decrease}$ for $[t_{start,decrease}, t_{final,decrease}]$
Load tracking	Minimize error against reference power profile	Reference power profile $P_{reference}$	$\mid P - P_{reference} \mid < \epsilon$

3.5.4. Supervisory Control Scheme

MPC is a receding horizon control process, which can be briefly described as follows: at each sampling interval, system states are measured or estimated and fed back to the controller model to update the model states. With the updated information, the controller predicts the system behavior based on the built-in model within the control horizon (e.g., the next 6 h) along with disturbance forecast such as weather and occupancy. The optimization algorithm then tries to find an optimal solution by minimizing the objective function subject to the latest system constraints. The first control action is implemented and then the MPC engine relaunches a new optimization at the next control interval. The SPO optimization engine follows this approach, but it is designed to be a supervisory controller, i.e., it does not replace but rather interacts with local controllers such as thermostats. The optimizer periodically (e.g., every 5 min) sends optimal setpoints to the local controllers, and they control the equipment through traditional control loops at a finer time scale (e.g., 1 s).

3.5.5. Weather Forecast

The optimization engine requires weather forecast data to operate. The SPO weather forecast module predicts solar radiation using data from external weather forecast services. Hourly forecasts of outdoor dry bulb temperature, relative humidity, wind speed and cloud cover for the duration of the control horizon (e.g., 6 h into the future) are required by the controller. A solar forecast model has also been implemented in this module to calculate the global solar radiation [82]. The direct and diffuse solar radiation are further computed based on the predicted global solar radiation [83]. Based on the predicted solar radiation, the plane-of-array solar radiation can be easily calculated and used as inputs for building and photovoltaic (PV) system models.

4. Evaluation

In this section, the performance of the SPO solution in a real commercial building is evaluated. It begins with the description of the site where SPO has been deployed, the hardware and software details, followed by the specifications of the optimization engine and concludes with the results of how SPO controls the building's loads and its distributed energy resources in response to different grid signals.

4.1. Site Description

The SPO has been deployed in a convenience store/gas station in Blue Lake, California, United States at the Blue Lake Rancheria. Being located at the edge of the utility service territory and in a generation-constrained area of Humboldt County in California, this site is at a strategic point for supporting grid reliability. The refueling and cold storage services provided by the store are recognized as critical community needs by the operators, particularly during blackouts or natural disasters, which drives their interest in energy resiliency at the site. Strongly tied to this motivation is the requirement for network resilience, that is, being able to operate in an Internet-disconnected mode as well. A unique aspect about this particular building is that it is attached to a casino complex and hence the store also contains gaming machines. The presence of these machines within the convenience store means there are regulatory requirements for network and firewall restrictions that prevent any control signals from originating outside their local network. The site has a unique combination of a real need for resilience and a site with professional IT network staff who require robust cybersecurity.

The store is typically open 24 h a day, 7 days a week and serves frequent customers throughout a day. The space inside the store is conditioned by two separate two-stage roof top units (RTUs), virtually (but not physically) dividing the store into an "east zone" and a "west zone". These RTUs are controlled by individual thermostats that have been installed in the corresponding zones. The store also has a walk-in refrigerator for storing and displaying beverages through a set of glass reach-through doors and a walk-in freezer for storing and displaying ice-cream and frozen food. The site has 60 kW of local solar power generation capacity along with a battery with 174 kWh of energy storage (and power capacity of 109 kW). As seen in Figure 5, the solar panels are installed on the gas station canopy and the energy storage unit is a Tesla Powerpack.

(**a**) 60 kW local solar power capacity

(**b**) 174 kWh local energy storage capacity with a peak output of 109 kW

Figure 5. Distributed Energy Resources installed on site.

The typical daily energy consumption is about 800 kWh (or 33 kW average demand), with an hourly average load profile illustrated in Figure 6. However, the gaming machines in the store contribute a significant portion of the total electrical load, and only about 15 kW of the electrical demand is for the thermal systems (i.e., the two RTUs, the refrigerator and the freezer), which are controllable by SPO. The remaining uncontrolled demand is primarily from gaming machines that are located in the store. The building exhibits peak demand in the early morning and early evening periods—both prime targets for thermal load shifting to alleviate system stress around sunrise and sunset. It currently operates on the E19S electricity tariff offered by Redwood Coast Energy Authority, a community-choice aggregator that provides retail options to customers in Humboldt County who could otherwise be served by the distribution utility, Pacific Gas and Electric Company [84]. The site soon plans on moving to the B19 tariff [85]. Both these tariffs contain time-of-use energy charges ($/kWh) and time-of-use demand charges ($/kW), with the key difference being the timing of the peak prices. In B19, the peak prices occur later in the day, between 16:00 and 21:00 daily to account for the growing duck-curve problem in California [86], while E19S is a legacy tariff with the peak prices during the 12:00–18:00 period that historically included coincident peak demand. The building operators are motivated to reduce their overall energy costs by reducing energy consumption and by effective management of the peak loads on the grid.

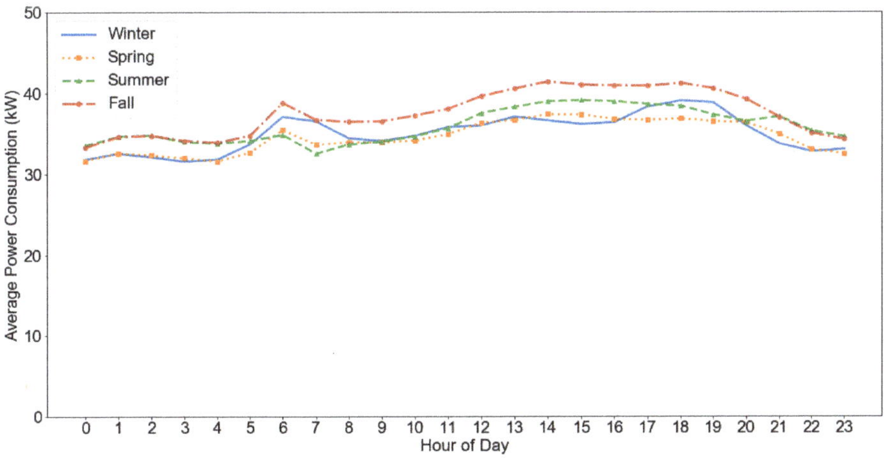

Figure 6. Daily load profile of the convenience store across different seasons.

4.2. Hardware and Software Set Up

Given the cybersecurity requirements at the site, the SPO system uses local drivers with local storage and an MPC-based optimization engine. The local SPO server with a WAVEMQ site router has been hosted on an Intel NUC computer with an Intel i7 processor and 2TB of internal disk capacity [87] at the site. The designated WAVEMQ router is hosted on a cloud server, which is also attached to a persistent data store, also in the cloud. InfluxDB [88] is utilized as the timeseries database, both in the cloud and on the local server. The local server has been configured with a retention policy to only store the most recent two weeks of data that might be relevant for the optimization engine. The local server also hosts the drivers that communicate with each of the devices and/or services (over the protocols mentioned) listed below, reading data once every minute and publishing to the WAVEMQ message bus. There are data ingester services set up both at the site router and the designated router to store these published messages in the local and the cloud-based InfluxDB databases, respectively. The drivers for the controllers also subscribe to the 'optimal setpoints' being published by the optimization engine so that they can change the setpoints of the controllers.

The list of data sources and/or controllers, with their respective communication protocols are given below:

- **Electricity Meters:** The building has six Continental Controls Wattnode power meters [89] that measure the power consumption of: whole building, west zone RTU compressor, east zone RTU compressor, refrigerator compressor, refrigerator evaporator fan and freezer compressor and evaporator fan, respectively. These devices measure instantaneous power parameters and communicate over Modbus serial.
- **HVAC:** The two RTUs are controlled by two separate KMC Flexstat 120063CEW [90] thermostats, which communicate over BACnet/IP. The preferred temperatures in the west and east zone are 20.56 and 21.67 °C (or 69 and 71 °F, respectively), respectively. The heating and cooling setpoints of the thermostats are controlled by the optimizer.
- **Refrigeration:** Sporlan Parker PSK214 Modbus (serial) [91] refrigeration controllers are used to control the large refrigeration systems. By default the freezer is set to an indoor temperature of −21.67 °C (−7 °F) and the refrigerator to 0.56 °C (33 °F). During the experiments, the SPO controlled the indoor cabinet temperature setpoint of the equipment.
- **PV and Battery:** The inverters of the PV panels and the battery are interfaced with a Schweitzer Engineering Laboratories SEL-3505 Real Time Automation Controller (RTAC) [92]. The RTAC is the short-timescale microgrid controller for this system and it handles power flows, circuit switching and safety aspects during the grid-islanded system operation. However, due to certain restrictions at this site, the RTAC only allows 'read' operations (over Modbus(TCP)) to be performed by the SPO.
- **Emulated Battery:** As the Tesla battery on site only allows 'read' operations, a software-based emulated battery is used for the experiments. This battery has been scaled down to a size that makes sense for conventional small and medium convenient stores (without power-hungry gaming machines, but HVAC and refrigeration). The emulated battery has a total capacity of 27 kWh, with a peak output of 14 kW (equivalent to two Tesla Powerwalls [93]).
- **Weather:** The current outdoor temperature, cloud cover, relative humidity and wind speed data, along with their 48-h forecasts, are collected from the DarkSky weather service's REST API [94].
- **Grid Signals:** Provides information about the prices based on tariffs or dynamic prices and/or could also publish information about scheduled demand response events. While this is currently implemented as library function that retrieves the grid signals from a static database, retrieving the real-time or day-ahead Independent System Operator (ISO) prices or dynamic prices from a utility using protocols such as IEEE 2030.5 [95] or OpenADR [96] are planned future work.

Table 2 provides a list of data and their respective sources that are required for the optimization engine. A summary of the controllers and the variables that can be controlled, with the default values and lower and upper bounds, is given in Table 3. These limits were provided by the convenience store operators after considering the requirements for indoor air conditioning and food storage. The limits have been encoded into the respective drivers so that even if the optimization engine produces a setpoint outside the limits, the drivers can ensure that those values are not set on the actual device. It is to be noted that the list of variables in Tables 2 and 3 are a subset of variables that can be read from or written to each of the devices.

Table 2. List of data sources and relevant variables (Inputs to the optimization engine).

Source	Protocol	Variables
Wattnode Meter	Modbus (serial)	active power (W)
KMC Thermostat	BACnet/IP	indoor temperature (°C)
Refrigeration Controller	Modbus (serial)	cabinet temperature (°C)
RTAC	Modbus (TCP)	PV production (W) battery charge/discharge rate (W) battery state of charge (%)
Battery (emulated)	-	battery charge/discharge rate (W) battery State of Charge (%)
Dark Sky API	HTTP	outdoor temperature (°C) cloud cover (%) relative humidity (%) wind speed (mph) and 48-hour forecasts of these
Grid Signals (tariffs, dynamic prices, Demand response events)	-	price of energy ($/kWh) demand charge ($/kW) amount of load limit/shed/shift (W)

Table 3. List of Controllers and their relevant variables (Outputs of the optimization engine).

Controller	Protocol	Control Variables	Default	Lower Limit	Upper Limit
Thermostat East	BACnet/IP	heating setpoint cooling setpoint	19.44 °C 21.67 °C	17.22 °C 19.44 °C	22.22 °C 23.33 °C
Thermostat West	BACnet/IP	heating setpoint cooling setpoint	18.33 °C 20.56 °C	17.22 °C 19.44 °C	22.22 °C 23.33 °C
Freezer	Modbus (serial)	cabinet temp. setpoint	−21.67 °C	−34.44 °C	−18.89 °C
Refrigerator	Modbus (serial)	cabinet temp. setpoint	0.56 °C	0.56 °C	3.33 °C
Battery (emulated)	-	charge/discharge rate	0W	−14 kW	14 kW

4.3. Optimization Engine Set Up

4.3.1. Modeling

As mentioned in Section 4.1, the main occupied zone in the store is conditioned by two RTU systems. One of the two food storage rooms is conditioned by a custom-made refrigeration system; the other by a freezer. The layout of these spaces is captured by the building model, which includes four thermal zones: two RTU zones, one refrigerator zone and one freezer zone. The building zones are modeled using the lumped resistance and capacitance approach: each zone is represented by one resistance and one capacitance and connected with each other thermally. The overall building thermal model is therefore a linear fourth-order model. The battery system is modeled based on the bucket model approach by considering the battery as a repository for energy [97]. The state variable is the battery SOC and the input is the real power that should be stored in or extracted from the battery. The PV system is modeled with a constant efficiency and the input is the predicted plane-of-array solar radiation. The linearity of these models allows efficient computation for the optimization algorithm, which takes about one minute to converge to an optimal solution in these experiments. Figure 7 shows the MPC model structure and its interaction with the local controllers.

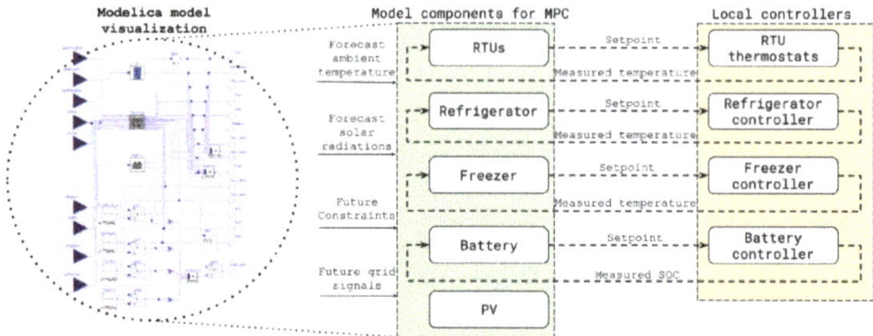

Figure 7. MPC model structure.

The weather parameters considered for the building are the dry bulb temperature and the solar radiation. The dry bulb temperature forecast is obtained from the Dark Sky API based on the site latitude and longitude. The solar radiation forecasts for the building and PV system are predicted using the built-in model as described in Section 3.5.5. The prediction horizon used in the experiments is 6 h. The store zones are occupied by staff for 24 h each day. The occupancy disturbance is therefore not considered. The major internal load disturbance in the store is from the gaming machines, which emit heat directly to the zone. These machines are in operation 24 h per day. This internal thermal load is therefore assumed to be constant in the model. The system constraints for MPC used in the experiments are defined by the facility staff and are summarized in Table 3. The thermal comfort setpoints for the HVAC system are constrained to values between 19.44 °C (67 °F) and 23.33 °C (74 °F). The constraints for the freezer cabinet temperature setpoint are from −34.44 °C to −18.89 °C (−30 °F to −2 °F). The constraints for the refrigerator cabinet temperature are from 0.56 °C to 3.33 °C (33 °F to 38 °F). The battery cannot be charged more than 95% and discharged below 25% of its total capacity. The maximum charge or discharge power is 14 kW. The upper and lower bounds for the power consumption are calculated for each grid signal, as shown in Table 1.

The system states of the model are the space temperatures of the RTU systems, cabinet temperatures of the refrigerator and freezer system and the battery SOC. These values are all measured and thus there is no need for state estimations in the MPC formulation. The measured states are updated in the model at each control interval (5 min). The control inputs for the model are the heating or cooling rate for the RTUs, the cooling rates for the refrigerator and the freezer, and the battery charging and discharging rate. The control outputs are the supervisory setpoints for each individual system: zone setpoints for the RTUs, refrigerator setpoint, freezer setpoint and charge/discharge power setpoint for the battery. When the setpoints are sent to each controller, the controller decides its operation mode based on its internal control loop implemented by the manufacturer. For instance, the thermostat receives the optimal setpoint and determines whether to switch the RTU heating or cooling state ON or OFF.

4.3.2. Controller Start-Up

The controller needs to be instantiated once the system model has been specified. During this process, the software loads and initializes all the components including the MPC model, the optimization problem, the system states (measured temperatures and SOC), the inputs (outdoor dry bulb temperature, solar radiation, grid signals and constraints) and the outputs (setpoints), based on the configuration. The objective function and associated constraints to grid signals, as summarized in Table 1, are automatically updated according to the goal of each experiment. At every control interval, the software calls the instantiated controller to solve the optimization problem and pushes the solutions to the respective devices over WAVEMQ. A single instance of the controller can handle different types of grid signals and only requires re-instantiation when configuration parameters (e.g., new input sources and modified outputs) are changed.

4.4. Experiments

To evaluate the performance of the SPO, the system was subjected to different grid signals (as shown in Table 1) and the optimal setpoints that were generated were pushed to the equipment in the building. This paper presents the results of three tests: (1) a dynamic pricing signal; (2) a Time-Of-Use (TOU) tariff; and (3) a demand limiting signal. While most commercial buildings in the US are enrolled in a TOU tariff rate, some utilities are moving towards use of dynamic pricing for grid integration [80,81]. These prices are also considered as the replacement for event-based demand response programs. These use cases are a representative collection of legacy (i.e., TOU tariffs and event-based demand response programs) and emerging (i.e., dynamic pricing) interaction mechanisms between grid and building-level microgrids.

By specifying a linear regression model based on the weather, solar production and building load data from ten days prior to each experiment (excluding days when SPO was being run), a weather-normalized baseline has been calculated for each of the following experiments. The battery was excluded from the baseline formulation because it was not operated during the period of baseline data collection. For all the experiments, the preferred temperatures for the HVAC zones are 20.56 °C (69 °F) and 21.67 °C (71 °F) for the west and east zones, respectively. The key metric used to evaluate the performance of the SPO system is the total cost of electricity (including both energy and demand charges) since the primary objective of SPO's optimization engine is to minimize this cost. This is the cost incurred due to the net load consumption supplied by the grid. All variables in the following equation are assumed to be non-negative.

$$netload = totalbuildingload - powerproducedbyPV + batterychargingrate - batterydischargingrate.$$

4.4.1. Dynamic Prices

This test, conducted on 17 May 2020 evaluated the behavior of the SPO in response to dynamic prices. A grid signal containing a 24 hour price forecast for the event day was generated and published on WAVEMQ. The prices were based on the wholesale market prices obtained from the California Independent System Operator (CAISO), the entity responsible for the California energy markets. These prices reflect the duck-curve dynamics in California, which occur when the net electricity demand drops during mid-day due to large amounts of solar generation during that time. The demand ramps up rapidly in evening hours as the sun sets, but air conditioning loads continue to be present due to the thermal lags in buildings, and many people return home and begin evening activities.

Figure 8 shows the results of the dynamic price test. Section (a) of the figure shows the driving variables: dynamic prices and solar irradiance. Section (b) shows the battery state of charge, and the resulting net load controlled by the SPO, compared to the baseline power profile. Section (c) shows the SPO generated setpoints for the HVAC zones and their corresponding temperature profiles.

The expected behavior of the system is that the battery, the HVAC and the refrigeration loads will be controlled to minimize consumption during high price times and shift consumption to low price times, subject to the constraints on comfort and other factors. In the operational test, this behavior was observed in broad terms. Figure 8b shows how the net building load was lower at times of high prices and higher during the beginning and the middle parts of the day when prices were low. This was achieved through a combination of battery dispatch and modification of thermal setpoints. Deviations from this baseline are attributed to the behavior of the control system.

The baseline load was much lower the first 6 h of the day as the battery was charged to full capacity during these hours. This was in preparation of the increasing prices, starting from 06:00, when the battery discharges completely to support the building load (Figure 8b). The battery was controlled similarly during the high price periods that occurred later in the day as well. From the slope of the battery state of charge, it is evident that the rate of charging and discharging change also vary according to the the price fluctuations. In Figure 8c, it can be seen that the responses of HVAC systems were mainly for the second peak in the afternoon. From around 12:00 to 17:00, the battery was in charging mode and the cooling power usage was kept minimum as the indoor temperatures were increasing. Once the battery began to discharge, the HVAC systems began to bring the indoor temperatures back to the preferred temperatures slowly. Even though the HVAC systems used more power during the high price duration, the net energy use decreased as battery was discharging. Due to very strict constraints on the temperatures and also owing to undersized equipment, there were hardly any changes in the operation of refrigeration systems. However, the temperatures of both the freezer and refrigerator were maintained within the bound set by the convenience store operators.

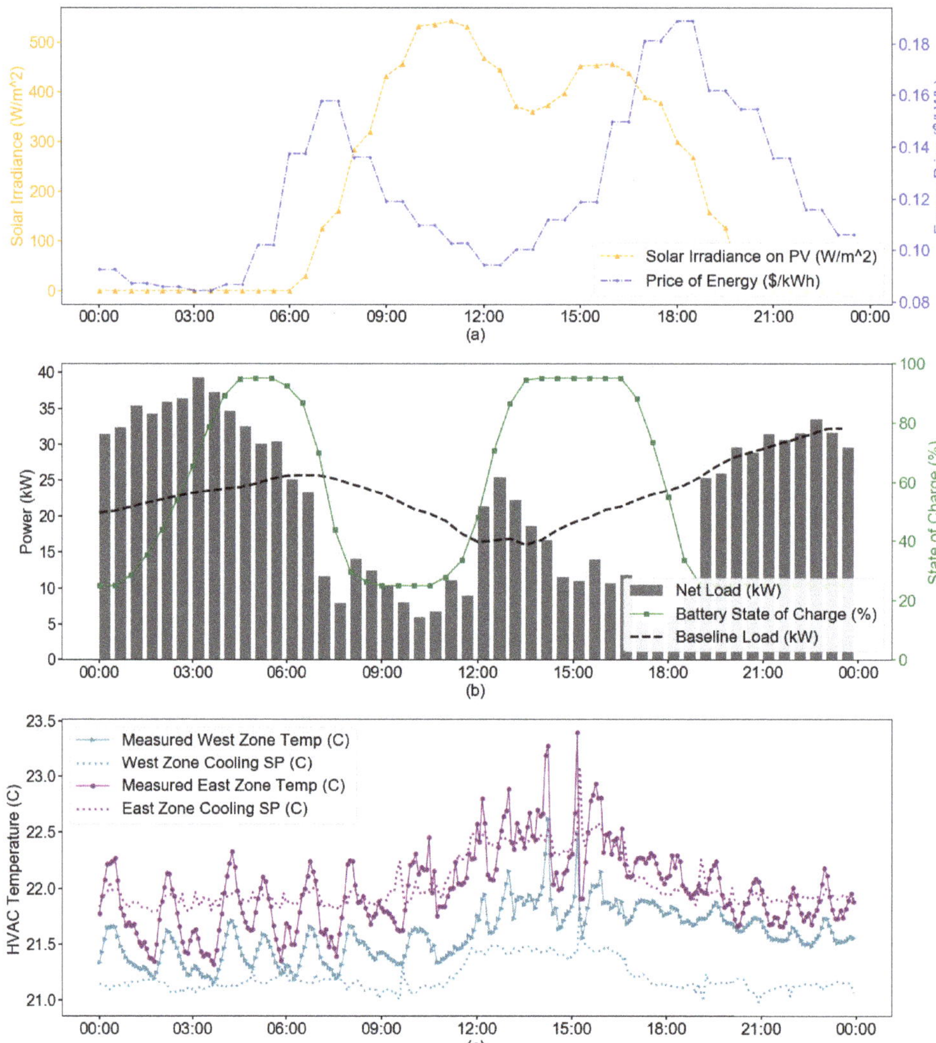

Figure 8. (**a**) SPO's response to hourly dynamic prices and varying solar irradiance. (**b**) Through battery discharge and reduction in building load, where the net load is minimum during high price times and the battery charges during the low price times anticipating the high price period. (**c**) SPO changes the thermostat setpoints to vary the zone temperature. Preferred temperatures: 20.56 °C (69 °F) for the West Zone and 21.67 °C (71 °F) for the East Zone.

As there were no demand charges in this signal, the total cost of electricity constitutes only the energy cost, which is calculated by multiplying the hourly energy charge with the hourly energy consumption. The total cost for this day was $59.14 for the SPO optimized actions as compared to the $68.72 for the baseline load, generating a savings of 13.94%.

4.4.2. Time-Of-Use Prices

This experiment, conducted on 15 May 2020, tested the response of SPO to the winter TOU rates of the B19 tariff [85]. This tariff contains time of use based energy prices and demand charges. Figure 9a shows the solar irradiance, demand and energy charges. The peak price window for both demand and energy charges are from 16:00 to 21:00 Figure 9b shows the battery and net load responses, controlled by the SPO. Figure 9c shows the HVAC zone setpoints that were generated by SPO and the actual temperatures for these two zones (gaps in the chart are due to missing data, where the thermostats temporarily lost network connection).

In response to the peak prices, the battery began to charge quite late until the PV generation began to get online and charged up to its maximum of 95% of its capacity right before 16:00 and started to discharge until 21:00, the end of the peak period. The HVAC systems started pre-cooling very early at around 07:00 Then, it started to increase the indoor temperatures once the battery began to charge. During the beginning hours of the peak period, the temperatures were maintained as high to reduce the HVAC load. Then, the HVAC west began to cool down the space as the accumulated penalty of deviating from the setpoint began to get high. With regards to the refrigeration systems, there were still hardly any periods when the temperatures were allowed to rise, but they always remained within the limits.

Table 4 compares total electricity costs incurred in these 24 hour for the SPO optimized building against the building's baseline load for this day (shown in Figure 9b). The energy cost is calculated as described in the previous experiment and it can be seen that SPO was able to save 10.33% of the total energy cost. Calculating the total demand cost and the subsequent savings is slightly more complicated. The total demand cost is the sum of the demand costs across all the different TOU rate periods (e.g., 16:00–21:00). The maximum load (load refers to the 15-min average power consumption) for each rate period across the whole billing cycle (typically monthly), or, in this case, the whole day, multiplied by the corresponding demand charge is the demand cost for that period. This introduces the caveat that the demand cost shown in Table 4 might not be the final demand cost for the full billing cycle. Hence, for the purposes of evaluating this experiment, the billing cycle is assumed to be one day. Under this assumption, SPO reduced the total demand cost by $21.13. For a realistic billing cycle of a month, through continuous peak demand management by SPO, much higher demand cost savings can be accrued.

Table 4. Comparison of estimated costs for energy and demand between SPO and Baseline controls.

Costs	SPO Optimized Load	Baseline Load
Energy Cost	$61.42	$68.63
Demand Cost	$793.86	$814.99
Total Cost	$855.28	$883.62

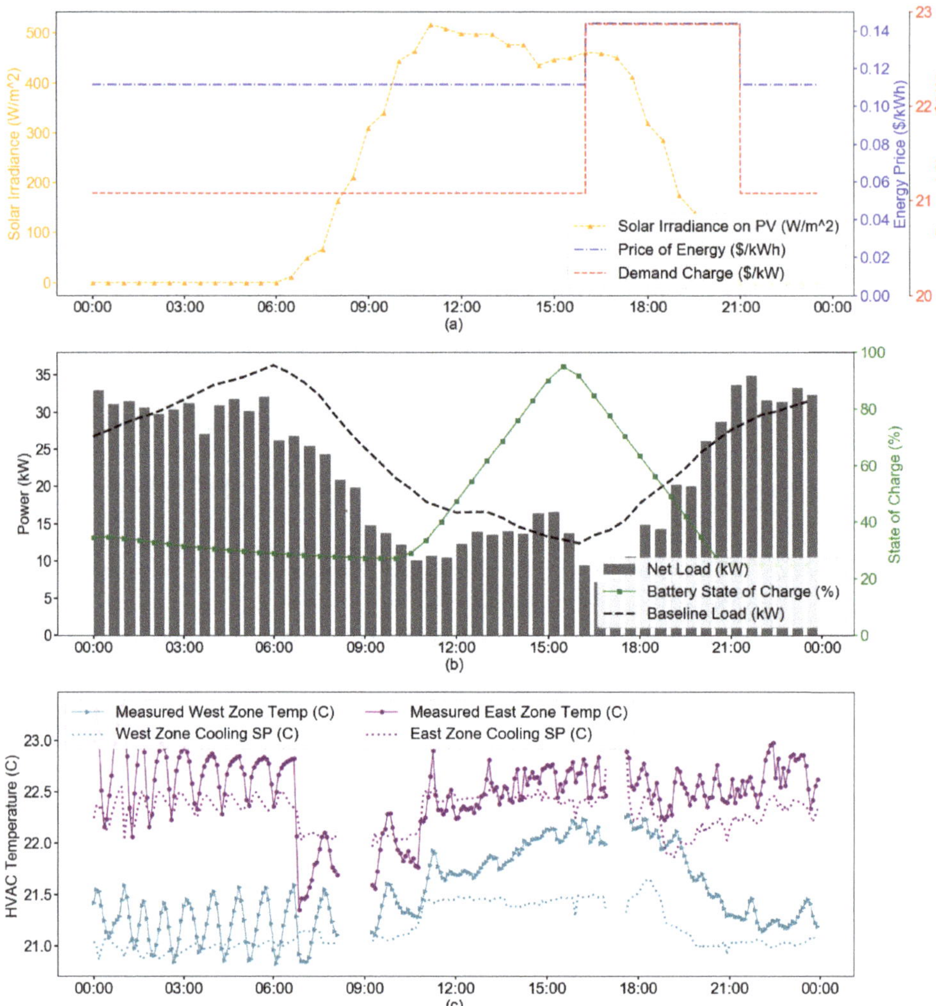

Figure 9. (**a**) SPO's response to a TOU tariff with both demand charges and energy costs and varying solar irradiance. (**b**) The battery is used effectively for reducing net load during peak price hours. (**c**) SPO changes the thermostat setpoints to vary the zone temperature. Preferred temperatures: 20.56 °C (69 °F) for the West Zone and 21.67 °C (71 °F) for the East Zone.

4.4.3. Demand Limiting Event

Demand limiting "refers to shedding loads when pre-determined peak demand limits are about to be exceeded ... and this is typically done to flatten the load shape when the pre-determined peak is the monthly peak demand" [98]. Coordinated load limiting efforts across multiple buildings helps to reduce the stress on the utility during peak hours.

An experiment was conducted in May 2020 to test the response of SPO to a demand-limiting signal. The signal constrained demand to 26 kW from 06:00 to 08:00., given that minimum baseline load during this period is 32.9 kW. Figure 10a depicts the response of SPO: the building reduced its average power consumption at 06:00 to 26.30 kW from 32 kW at 05:30. Figure 10b shows that this was achieved by reducing in the power consumption of the two HVAC units from 06:00 (in yellow and orange) and a drop in power consumption of the freezer unit at 07:30. However, the battery state of charge remained nearly flat throughout the event as SPO decided not to employ the battery during it. This is evidence of SPO's intelligent load control capabilities as it is able to coordinate the controllable load without depending on the battery to handle grid signals.

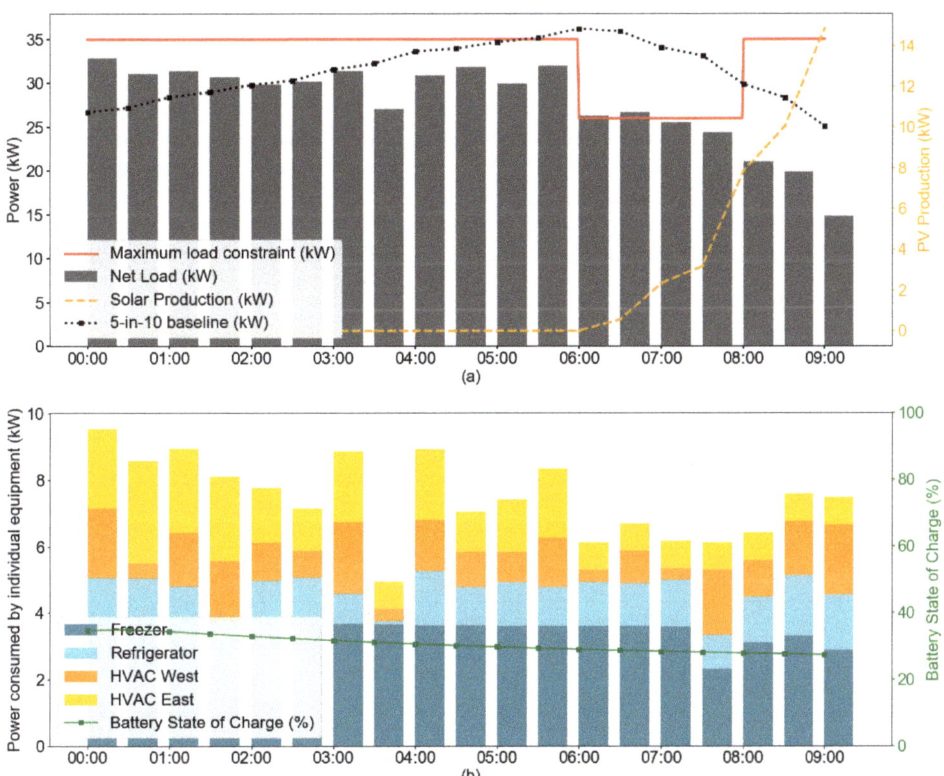

Figure 10. (**a**) Reduction in net load during a 26 kW demand limiting grid signal from 06:00 to 08:00 (**b**) Breakdown of controlled loads; HVAC and freezer loads cause the decrease in net load consumption; the battery was not used during this event.

5. Discussion

Overall, this work represents a proof of concept for an open-source, extensible, cyber-secure software system that can optimally control energy use in a microgrid.

5.1. Benefits to Developers

A core design intent of SPO was the ability to be modular and extensible, unlike solutions presented in previous studies [43,46]. For this reason, SPO was developed on top of modern open-source projects such as XBOS and MPCPy. Its core communication infrastructure, based on a distributed and secure message bus [68] and gRPC [70], allows developers to seamlessly replace components such as databases, optimization engines and drivers. For instance, the native XBOS data store was replaced by another time-series database [88], due to the cybersecurity requirements of the demonstration site. The use of gRPC is an innovative feature compared to other open-source projects [57,99], and it allows developers to use languages they are more familiar with or are more advantageous for a particular application. Hence, XBOS drivers have been developed in both Go and Python as part of this project. Further, native integration with BRICK metadata schema [66] is a distinctive feature of SPO. BRICK is an emerging "open-source effort to standardize semantic descriptions of the physical, logical and virtual assets in buildings and the relationships between them" [100]. BRICK avoids hard-coding sensor names in the software, making the application more portable between buildings. For instance, instead of representing a sensor as "supply_temp_VAV002_AHU01", BRICK queries can search for the ID of the supply air temperature sensor in any VAV belonging to any AHU.

During the model construction and calibration phase, the use of MPCPy was crucial to make the software more flexible. MPCPy affords the construction of multiple systems that can be quickly replaced with others, as long as the inputs and outputs stay the same. This allows rapid switching between emulated components and real components (e.g., a virtual battery with a real battery) and vice versa. Furthermore, it provides the possibility to run the optimization engine in "shadow mode" (i.e., the optimal controller computes setpoints, but it does not send them to the actual devices) and, if necessary, to quickly swap to controlling the real systems. This setup allowed debugging the system with real-time data but avoided loss of comfort or disruption of business operation.

5.2. Challenges of the Real World Deployment

The real-world deployment of SPO was a useful step to test the robustness of the implementation and understand the challenges with real systems. The lessons learned during this process are summarized in Table 5.

Table 5. A summary of challenges faced during the implementation and deployment of SPO.

#	Category	Description of Challenge
1	Technical	Limited choice of secure connected devices with local communication interfaces
2	Technical	Uncertain service and support for connected products, web services and underlying libraries (e.g., discontinued services, APIs changes).
3	Technical	Complex interaction of advanced supervisory control with local control in each connected device (e.g., thermostat hysteresis and defrost control).
4	Technical	Overconstrained systems (e.g., undersized refrigerator with tight temperature control bands)
5	Technical	Faulty equipment and sensors that make modeling harder due to unexpected behavior and incorrect representation of the system state
6	Technical	Unmodeled and unmeasured effects in the systems (e.g., unknown occupancy, door opening, internal gains due to uncontrolled equipment)
7	Organizational	Conflicting objectives and different risk tolerance between occupants/managers and researchers (e.g., thermal comfort, refrigerator temperature swings)
8	Organizational	Strict site/organization procedures and requirements (e.g., cybersecurity procedures)
9	Logistic	Delays in equipment deliveries (e.g., components in high demand)
10	Logistic	Faulty new equipment that needs to be replaced
11	Regulatory	Long lead times to work with highly-regulated, risk-adverse entities (e.g., utilities to sign off on the interconnect agreement, receiving approval before battery and PV commissioning)
12	Exceptional	Unfortunate and unforeseen natural disasters (e.g., power shutoffs due to threats of wildfires, COVID-19 pandemic)

Selecting sensors and controllers for the project proved to be the first hurdle. While there is a proliferation of new IoT and smart home devices on the market, very few met the cybersecurity and local control requirement of the project. In particular, to access the API of most smart thermostats, an Internet connection is required. This was undesirable for a microgrid that needs to work during network outages. Eventually, a BACnet thermostat was selected for installation, although its cost was significantly higher than other alternatives and it did not provide native data encryption. There is a clear need for modern connected devices that meet the requirements of commercial buildings and microgrids. The deployment process also revealed the effect of the fast rate of change of the online services and IoT market. In the middle of the project, the weather service used to gather weather forecasts for the site had to be replaced, because its free service was terminated after a corporate acquisition. As this paper is written, the new weather forecast service was also acquired by a different company and there is uncertainty about the future of the service and API.

Precise control of refrigerator and RTU operation through these networked controllers represented another challenge. Often, the details about the local control loop (e.g., deadbands, hysteresis and other safeguard mechanisms) were not provided by the manufacturer and they had to be reverse-engineered by the research team. While these features are useful to protect the equipment, they added uncertainty to the results of each control action determined by the supervisory controller. Additional issues related to the ability to control the system emerged when it became clear that the refrigeration system was undersized and its temperatures tightly prescribed, allowing for little flexibility in its operation. Unlike typical commercial buildings, the 24-h operation of the store also added challenges and unique problems for the testing of the system deployment. Conflicts between local temperature adjustments by the staff and the setpoints suggested by the system also emerged. In the system monitoring process, unexpected data were often observed. In one instance, the store temperature kept dropping even though the cooling systems had not been running . Analysis revealed that the door openings had a large impact on the thermal conditions of the store. This type of event was very difficult to characterize in the model. The placement of the sensors in the store also impacted the controls. As additional sensors were not installed in the store, the measurements from the temperature sensors located within the thermostats and the refrigeration controllers were used to determine the state of the system. Unfortunately, these devices were installed away from the areas where staff and customers typically orbit, and hence did not capture the user comfort.

As it is typical of deployment projects, the research team had to work with multiple departments within the same organization, each having their own set of requirements. Building occupants expressed concerns about temperature oscillations during experiments. The IT department defined strict cybersecurity requirements and access control procedures. These issues were resolved through clear and responsive communication with all parties involved at all times, providing frequent and regular updates and expectations. Providing the local staff with a real-time dashboard and alarm/notification system was a very effective way of keeping them engaged and updated. However, these needs also impacted the number of experiments allowed and their boundaries.

Further, the overall project schedule was significantly delayed because of logistical issues (i.e., unforeseen delays in equipment delivery and faulty equipment) regulatory requirements (i.e., working with utilities to sign off on the relatively novel microgrid interconnect agreement and receiving approval before battery and PV commissioning) and unfortunate and unforeseen natural disasters (i.e., power shutoffs due to wildfires and the COVID-19 pandemic). Although this combination of circumstances was unique, field deployments of advanced technologies should account for delays in their timeline. The delays that were experienced due to the wildfires and the pandemic only underscore the value of research that is focused on accelerating deployment of these systems. While it is unreasonable to expect that every microgrid would face similar challenges, a core goal of deployment-focused research must be to identify solutions to these challenges and improve the ability of developers to deploy advanced energy systems at scale.

5.3. Limitations and Future Work

During field tests, the SPO system demonstrated the ability to respond to both price- and and demand- based grid signals, using all the controllable loads: the battery, the two RTUs, the freezer and the refrigerator. The system has been collecting data from the local controllers and sensors for almost a year and has been controlling the equipment for seven months at the time of writing this paper. However, thus far the experiments have lasted approximately one week at a time and the results presented in the paper covered only a single day of operations. Thus, there needs to be further prolonged testing to evaluate the performance and the robustness of the system in the long run. Additionally, the baseline load for each of the experiments has been determined using a linear regression model based on historical environmental and building load data. In addition, the focus of this paper is to demonstrate an integrated software solution in a real building rather than precisely investigating the performance of the optimal control algorithm . The next steps in the research project involve determining more accurate baselines (statistical techniques such as Random Forest Regression, Autoregressive integrated moving average (ARIMA), etc.) and improving and evaluating the performance of the optimization algorithm by comparing it against other advanced control solutions.

Currently, the measurements recorded by the temperature sensors that are embedded into the thermostats and the refrigeration controllers have been used by SPO's optimization engine. While this characteristic allows easy deployment of SPO, there has been work planned to improve the building model by collecting data from temporarily installed temperature loggers across the store. Installing additional sensors to record occupancy related information would also be greatly beneficial. For example, with a door sensor, SPO would have been able to track the door open/close events and the model would have been able to take into account the effect of outside air in the store.

While this pilot site has unique characteristics, this test successfully demonstrates the feasibility of the SPO system for similar types of buildings. Many convenience stores in operation today (around 12,000 in California alone) have similar HVAC and refrigeration systems and can be upgraded to become a building-scale microgrid. Quick serve restaurants, hotels and grocery stores are other possible candidates for deployment of these systems. Deploying SPO in such buildings would provide data points regarding the portability and the scalability of the whole system and identifying another deployment site is also a part of the future work.

6. Conclusions

This paper presents the design, implementation and preliminary test of Solar+ Optimizer (SPO), a control software that provides demand flexibility to building-scale microgrids. The software uses Model Predictive Control (MPC) to optimally coordinate the operation of building loads and the distributed energy resources on site. SPO is designed to be vendor-agnostic and protocol-independent and is built on open-source software. The software has been tested in a convenience store in northern California with on-site solar generation, battery storage and control of HVAC and commercial refrigeration loads and preliminary results show the ability to shed load in response to price signals and to curtail demand, generating more than 10% savings in energy costs alone. Future work includes more extensive testing and publishing the project as an open-source library as well as sharing the data obtained during the project.

Author Contributions: Conceptualization, K.Z., D.B., and M.P.; Methodology, K.Z., D.B., and M.P.; Software, A.K.P., K.Z., P.G., D.B., M.M. and G.F.; Validation, K.Z., D.B., P.A., J.Z., R.B. and M.P.; Formal analysis, A.K.P. and K.Z.; Investigation, A.K.P., M.M. and M.P.; Resources, A.K.P., M.M., J.Z. and R.B.; Data curation, A.K.P.; Writing—original draft preparation, A.K.P., K.Z. and M.P.; Writing—review and editing, A.K.P., P.G., D.B., P.A., R.B. and M.P.; Visualization, A.K.P. and K.Z.; Supervision, D.B., R.B. and M.P.; Project administration, P.A., J.Z. and R.B.; Funding acquisition, P.A., R.B. and M.P. All authors have read and agreed to the published version of the manuscript.

Funding: This research was funded by California Energy Commission grants EPC-17-002 and EPC-15-057 and by the U.S.–China Clean Energy Research Center (CERC) 2.0 on Building Energy Efficiency (BEE), supported by the U.S. Department of Energy, Building Technologies Office. Lawrence Berkeley National Laboratory is a multiprogram science lab in the national laboratory system supported by the U.S. Department of Energy through its Office of Science under Contract No. DE-AC02-05CH11231.

Acknowledgments: The authors wish to acknowledge the Blue Lake Rancheria for supporting this research by providing cost share and providing their convenience store as a pilot site to deploy the Solar+ Optimizer software.

Conflicts of Interest: The authors declare no conflict of interest.

References

1. Li, Z.; Guo, J. Wisdom about age [aging electricity infrastructure]. *IEEE Power Energy Mag.* **2006**, *4*, 44–51.
2. Mariam, L.; Basu, M.; Conlon, M.F. Microgrid: Architecture, policy and future trends. *Renew. Sustain. Energy Rev.* **2016**, *64*, 477–489. [CrossRef]
3. Hirsch, A.; Parag, Y.; Guerrero, J. Microgrids: A review of technologies, key drivers, and outstanding issues. *Renew. Sustain. Energy Rev.* **2018**, *90*, 402–411. [CrossRef]
4. Yoldaş, Y.; Önen, A.; Muyeen, S.; Vasilakos, A.V.; Alan, İ. Enhancing smart grid with microgrids: Challenges and opportunities. *Renew. Sustain. Energy Rev.* **2017**, *72*, 205–214. [CrossRef]
5. Parhizi, S.; Lotfi, H.; Khodaei, A.; Bahramirad, S. State of the art in research on microgrids: A review. *IEEE Access* **2015**, *3*, 890–925. [CrossRef]
6. Bahramirad, S.; Khodaei, A.; Svachula, J.; Aguero, J.R. Building Resilient Integrated Grids: One neighborhood at a time. *IEEE Electrif. Mag.* **2015**, *3*, 48–55. [CrossRef]
7. Tavakoli, M.; Shokridehaki, F.; Akorede, M.F.; Marzband, M.; Vechiu, I.; Pouresmaeil, E. CVaR-based energy management scheme for optimal resilience and operational cost in commercial building microgrids. *Int. J. Electr. Power Energy Syst.* **2018**, *100*, 1–9. [CrossRef]
8. Lizana, J.; Chacartegui, R.; Barrios-Padura, A.; Valverde, J.M. Advances in thermal energy storage materials and their applications towards zero energy buildings: A critical review. *Appl. Energy* **2017**, *203*, 219–239. [CrossRef]
9. Eaton. Microgrid and Distributed Energy Resources. Available online: https://www.eaton.com/us/en-us/catalog/services/microgrid-and-distributed-energy-resources.html (accessed on 14 May 2020).
10. Company, S.E. Don't Panic. Make Your Complex Microgrid Easy with S&C. Available online: https://www.sandc.com/en/solutions/microgrids/ (accessed on 14 May 2020).
11. Electric, S. Microgrid Solutions. Available online: https://www.se.com/us/en/work/solutions/microgrids/ (accessed on 14 May 2020).
12. Reynolds, J.; Rezgui, Y.; Hippolyte, J.L. Upscaling energy control from building to districts: Current limitations and future perspectives. *Sustain. Cities Soc.* **2017**, *35*, 816–829. [CrossRef]
13. EIA. Annual Energy Outlook 2019: Commercial Sector Key Indicators and Consumption Reference Case. Available online: https://www.eia.gov/outlooks/aeo/data/browser/#/?id=5-AEO2019&sourcekey=0. (accessed on 7 May 2020).
14. Hydeman, M.; Taylor, S.T.; Eubanks, B. Control sequences & controller programming. *ASHRAE J.* **2015**, *57*, 58–62.
15. Joos, G.; Reilly, J.; Bower, W.; Neal, R. The need for standardization: The benefits to the core functions of the microgrid control system. *IEEE Power Energy Mag.* **2017**, *15*, 32–40. [CrossRef]
16. Henze, G.P. Model predictive control for buildings: A quantum leap? *J. Build. Perform. Simul.* **2013**, *6*, 157–158. [CrossRef]
17. Henze, G.P.; Kalz, D.E.; Liu, S.; Felsmann, C. Experimental Analysis of Model-Based Predictive Optimal Control for Active and Passive Building Thermal Storage Inventory. *HVAC R Res.* **2005**, *11*, 189–213. [CrossRef]
18. Henze, G.P.; Felsmann, C.; Knabe, G. Evaluation of optimal control for active and passive building thermal storage. *Int. J. Therm. Sci.* **2004**, *43*, 173–183. [CrossRef]
19. Nabil Nassif, S.K.; Sabourin, R. Optimization of HVAC Control System Strategy Using Two-Objective Genetic Algorithm. *HVAC R Res.* **2005**, *11*, 459–486. [CrossRef]
20. Nassif, N.; Kajl, S.; Sabourin, R. Simplified Model-based Optimal Control of VAV Air-conditioning System. In Proceedings of the Ninth International IBPSA Conference, Montreal, QC, Canada, 15–18 August 2005; pp. 823–830.
21. Wang, S.; Jin, X. Model-based optimal control of VAV air-conditioning system using genetic algorithm. *Build. Environ.* **2000**, *35*, 471–487. [CrossRef]
22. May-Ostendorp, P.; Henze, G.P.; Corbin, C.D.; Rajagopalan, B.; Felsmann, C. Model-predictive control of mixed-mode buildings with rule extraction. *Build. Environ.* **2011**, *46*, 428–437. [CrossRef]
23. Corbin, C.D.; Henze, G.P.; May-Ostendorp, P. A model predictive control optimization environment for real-time commercial building application. *J. Build. Perform. Simul.* **2013**, *6*, 159–174. [CrossRef]

24. Lee, Y.M.; Horesh, R.; Liberti, L. Optimal HVAC Control as Demand Response with On-site Energy Storage and Generation System. *Energy Procedia* **2015**, *78*, 2106–2111. [CrossRef]
25. Biyik, E.; Kahraman, A. A predictive control strategy for optimal management of peak load, thermal comfort, energy storage and renewables in multi-zone buildings. *J. Build. Eng.* **2019**, *25*, 100826. [CrossRef]
26. Marušić, D.; Lešić, V.; Capuder, T.; Vasak, M. Price-Optimal Energy Flow Control of a Building Microgrid Connected to a Smart Grid. In Proceedings of the 2018 26th Mediterranean Conference on Control and Automation (MED), Zadar, Croatia, 19–22 June 2018; pp. 1–9.
27. Xue, Y.; Todd, M.; Ula, S.; Barth, M.J.; Martinez-Morales, A.A. A comparison between two MPC algorithms for demand charge reduction in a real-world microgrid system. In Proceedings of the 2016 IEEE 43rd Photovoltaic Specialists Conference (PVSC), Portland, OR, USA, 5–10 June 2016; pp. 1875–1880.
28. Tavakoli, M.; Shokridehaki, F.; Marzband, M.; Godina, R.; Pouresmaeil, E. A two stage hierarchical control approach for the optimal energy management in commercial building microgrids based on local wind power and PEVs. *Sustain. Cities Soc.* **2018**, *41*, 332–340. [CrossRef]
29. Reynolds, J.; Rezgui, Y.; Kwan, A.; Piriou, S. A zone-level, building energy optimisation combining an artificial neural network, a genetic algorithm, and model predictive control. *Energy* **2018**, *151*, 729–739. [CrossRef]
30. Magnier, L.; Haghighat, F. Multiobjective optimization of building design using TRNSYS simulations, genetic algorithm, and Artificial Neural Network. *Build. Environ.* **2010**, *45*, 739–746. [CrossRef]
31. Bonthu, R.K.; Pham, H.; Aguilera, R.P.; Ha, Q.P. Minimization of building energy cost by optimally managing PV and battery energy storage systems. In Proceedings of the 2017 20th International Conference on Electrical Machines and Systems (ICEMS), Sydney, NSW, Australia, 11–14 August 2017; pp. 1–6.
32. Wang, Z.; Yang, R.; Wang, L. Intelligent multi-agent control for integrated building and micro-grid systems. In Proceedings of the ISGT 2011, Anaheim, CA, USA, 17–19 January 2011; pp. 1–7.
33. Benjamin Heymann, J. Frédéric Bonnans, P.M.F.J.S.F.L.; Jiménez-Estévez, G. Continuous optimal control approaches to microgrid energy management. *Energy Syst.* **2018**, *9*, 59–77. [CrossRef]
34. Macarulla, M.; Casals, M.; Forcada, N.; Gangolells, M. Implementation of predictive control in a commercial building energy management system using neural networks. *Energy Build.* **2017**, *151*, 511–519. [CrossRef]
35. Vázquez-Canteli, J.R.; Nagy, Z. Reinforcement learning for demand response: A review of algorithms and modeling techniques. *Appl. Energy* **2019**, *235*, 1072–1089. [CrossRef]
36. Jia, R.; Jin, M.; Sun, K.; Hong, T.; Spanos, C. Advanced Building Control via Deep Reinforcement Learning. *Energy Procedia* **2019**, *158*, 6158–6163. [CrossRef]
37. Wang, Y.; Velswamy, K.; Huang, B. A Long-Short Term Memory Recurrent Neural Network Based Reinforcement Learning Controller for Office Heating Ventilation and Air Conditioning Systems. *Processes* **2017**, *5*, 46. [CrossRef]
38. Wei, T.; Wang, Y.; Zhu, Q. Deep reinforcement learning for building HVAC control. In Proceedings of the 2017 54th ACM/EDAC/IEEE Design Automation Conference (DAC), Austin, TX, USA, 18–22 June 2017; pp. 1–6.
39. Zhang, Z.; Chong, A.; Pan, Y.; Zhang, C.; Lu, S.; Lam, K. A Deep Reinforcement Learning Approach to Using Whole Building Energy Model For HVAC Optimal Control. In Proceedings of the 2018 ASHRAE/IBPSA-USA Building Performance Analysis Conference and SimBuild, Chicago, IL, USA, 26–28 September 2018.
40. Ahn, K.U.; Park, C.S. Application of deep Q-networks for model-free optimal control balancing between different HVAC systems. *Sci. Technol. Built Environ.* **2020**, *26*, 61–74. [CrossRef]
41. Wetter, M. Co-simulation of building energy and control systems with the Building Controls Virtual Test Bed. *J. Build. Perform. Simul.* **2011**, *4*, 185–203. [CrossRef]
42. West, S.R.; Ward, J.K.; Wall, J. Trial results from a model predictive control and optimisation system for commercial building HVAC. *Energy Build.* **2014**, *72*, 271–279. [CrossRef]
43. Bengea, S.C.; Kelman, A.D.; Borrelli, F.; Taylor, R.; Narayanan, S. Implementation of model predictive control for an HVAC system in a mid-size commercial building. *HVAC R Res.* **2014**, *20*, 121–135. [CrossRef]
44. Kim, D.; Braun, J.; Cai, J.; Fugate, D. Development and experimental demonstration of a plug-and-play multiple RTU coordination control algorithm for small/medium commercial buildings. *Energy Build.* **2015**, *107*, 279–293. [CrossRef]

45. Kim, D.; Braun, J.E. Development, implementation and performance of a model predictive controller for packaged air conditioners in small and medium-sized commercial building applications. *Energy Build.* **2018**, *178*, 49–60. [CrossRef]
46. Carli, R.; Cavone, G.; Ben Othman, S.; Dotoli, M. IoT Based Architecture for Model Predictive Control of HVAC Systems in Smart Buildings. *Sensors* **2020**, *20*, 781. [CrossRef] [PubMed]
47. Vrettos, E.; Kara, E.C.; MacDonald, J.; Andersson, G.; Callaway, D.S. Experimental Demonstration of Frequency Regulation by Commercial Buildings—Part I: Modeling and Hierarchical Control Design. *IEEE Trans. Smart Grid* **2018**, *9*, 3213–3223. [CrossRef]
48. Vrettos, E.; Kara, E.C.; MacDonald, J.; Andersson, G.; Callaway, D.S. Experimental Demonstration of Frequency Regulation by Commercial Buildings—Part II: Results and Performance Evaluation. *IEEE Trans. Smart Grid* **2018**, *9*, 3224–3234. [CrossRef]
49. Bruno, S.; Giannoccaro, G.; La Scala, M. A Demand Response Implementation in Tertiary Buildings Through Model Predictive Control. *IEEE Trans. Ind. Appl.* **2019**, *55*, 7052–7061. [CrossRef]
50. FLEXlab. Advanced Integrated Building & Grid Technologies Testbed. Available online: https://flexlab.lbl.gov/ (accessed on 4 May 2020).
51. Chen, B.; Cai, Z.; Bergés, M. Gnu-RL: A Precocial Reinforcement Learning Solution for Building HVAC Control Using a Differentiable MPC Policy. In Proceedings of the 6th ACM International Conference on Systems for Energy-Efficient Buildings, Cities, and Transportation, New York, NY, USA, 13–14 November 2019; Association for Computing Machinery: New York, NY, USA, 2019; pp. 316–325. [CrossRef]
52. Zhang, Z.; Lam, K.P. Practical Implementation and Evaluation of Deep Reinforcement Learning Control for a Radiant Heating System. In Proceedings of the 5th Conference on Systems for Built Environments, Shenzhen, China, 7–8 November 2018; Association for Computing Machinery: New York, NY, USA, 2018; pp. 148–157. [CrossRef]
53. Brundu, F.G.; Patti, E.; Osello, A.; Giudice, M.D.; Rapetti, N.; Krylovskiy, A.; Jahn, M.; Verda, V.; Guelpa, E.; Rietto, L.; et al. IoT Software Infrastructure for Energy Management and Simulation in Smart Cities. *IEEE Trans. Ind. Inform.* **2017**, *13*, 832–840. [CrossRef]
54. Fagiani, M.; Severini, M.; Valenti, M.; Ferracuti, F.; Ciabattoni, L.; Squartini, S. rEMpy: A comprehensive software framework for residential energy management. *Energy Build.* **2018**, *171*, 131–143. [CrossRef]
55. Bottaccioli, L.; Aliberti, A.; Ugliotti, F.; Patti, E.; Osello, A.; Macii, E.; Acquaviva, A. Building Energy Modelling and Monitoring by Integration of IoT Devices and Building Information Models. In Proceedings of the 2017 IEEE 41st Annual Computer Software and Applications Conference (COMPSAC), Turin, Italy, 4–8 July 2017; Volume 1, pp. 914–922.
56. Berbakov, L.; Batić, M.; Tomašević, N. Smart Energy Manager for Energy Efficient Buildings. In Proceedings of the IEEE EUROCON 2019 -18th International Conference on Smart Technologies, Novi Sad, Serbia, Serbia, 1–4 July 2019; pp. 1–4.
57. Akyol B.A.; Haack, J.; Ciraci, S.; Carpenter, B.; Vlachopoulou, M.; Tews, C. VOLTTRON: An Agent Execution Platform for the Electric Power System. In Proceedings of the Third International Workshop on Agent Technologies for Energy Systems (ATES 2012), Valencia, Spain, 5 June 2012.
58. Evins, R.; David, N. Using Simple Predictive Models to Improve Control of Complex Building Systems. In Proceedings of the 4th ACM International Conference on Systems for Energy-Efficient Built Environments, Delft, Netherlands, 8–9 November 2017; Association for Computing Machinery: New York, NY, USA, 2017. [CrossRef]
59. Mirakhorli, A.; Dong, B. An Open Source Smart Building Energy Management Platform through VOLTTRON. In Proceedings of the 2017 International Building Performance and Simulation Association Conference, San Francisco, CA, USA, 7–9 August 2017.
60. Hao, H.; Liu, G.; Huang, S.; Katipamula, S. *Coordination and Control of Flexible Building Loads for Renewable Integration; Demonstrations using VOLTTRON*; U.S. Department of Energy Office of Scientific and Technical Information: Oak Ridge, TN, USA, 2016. [CrossRef]
61. Luo, J.; Pourbabak, H.; Su, W. The application of distributed control algorithms using VOLTTRON-based software platform. In Proceedings of the 2017 8th International Renewable Energy Congress (IREC), Amman, Jordan, 21–23 March 2017; pp. 1–6.

62. Raker, D.; Sellers, A.; Kini, R.; Green, M.; Stuart, T.; Ellingson, R.; Khanna, R.; Heben, M. Grid Integration of Building Systems and 1 MW Photovoltaic Array using VOLTTRON. In Proceedings of the 2017 IEEE 44th Photovoltaic Specialist Conference (PVSC), Washington D.C., USA, 25–30 June 2017; pp. 2926–2930.
63. Raker, D.; Kini, R.; Huntsman, R.; Green, M.; Sapci, O.; Stuart, T.; Ellingson, R.; Khanna, R.; Heben, M. Transactive Mitigation Of Variability In The Output Of 1 MW Photovoltaic Array Using VolttronTM. In Proceedings of the 2018 IEEE 7th World Conference on Photovoltaic Energy Conversion (WCPEC) (A Joint Conference of 45th IEEE PVSC, 28th PVSEC 34th EU PVSEC), Waikoloa Village, HI, USA, 10–15 June 2018; pp. 1462–1467.
64. For the Advancement of Structured Information Standards (OASIS), T.O. MQTT. Available online: http://mqtt.org/ (accessed on 14 May 2020).
65. Fierro, G. XBOS Overview. Available online: https://docs.xbos.io (accessed on 6 May 2020).
66. Balaji, B.; Bhattacharya, A.; Fierro, G.; Gao, J.; Gluck, J.; Hong, D.; Johansen, A.; Koh, J.; Ploennigs, J.; Agarwal, Y.; et al. Brick: Towards a Unified Metadata Schema For Buildings. In Proceedings of the 3rd ACM International Conference on Systems for Energy-Efficient Built Environments, Stanford, CA, USA, 15–17 November 2016; Association for Computing Machinery: New York, NY, USA, 2016; pp. 41–50. [CrossRef]
67. Andersen, M.P.; Kumar, S.; AbdelBaky, M.; Fierro, G.; Kolb, J.; Kim, H.S.; Culler, D.E.; Popa, R.A. WAVE: A Decentralized Authorization Framework with Transitive Delegation. In Proceedings of the 28th USENIX Security Symposium (USENIX Security 19), Santa Clara, CA, United States, 14—16 August 2019; Association: Santa Clara, CA, USA, 2019; pp. 1375–1392.
68. Andersen, M.P. WAVEMQ—Tiered message bus for WAVE 3. Available online: https://github.com/immesys/wavemq (accessed on 6 May 2020).
69. Developers, G. Protocol Buffers. Available online: https://developers.google.com/protocol-buffers (accessed on 7 May 2020).
70. gRPC Authors 2020. gRPC: A High-Performance, Open Source Universal RPC Framework. Available online: https://grpc.io (accessed on 7 May 2020).
71. Organization, T.M. Modbus. Available online: http://www.modbus.org/specs.php (accessed on 14 May 2020).
72. American Society of Heating, R.; (ASHRAE), A.C.E. BACnet. Available online: http://www.bacnet.org/ (accessed on 14 May 2020).
73. Blum, D.H.; Wetter, M. MPCPy: An Open-Source Software Platform for Model Predictive Control in Buildings. In Proceedings of the 15th Conference of International Building Performance Simulation, San Francisco, CA, USA, 7–9 August 2017.
74. Mattsson, S.E.; Elmqvist, H. Modelica—An international effort to design the next generation modeling language. In Proceedings of the 7th IFAC Symposium on Computer Aided Control Systems Design, Gent, Belgium, 28–30 April 1997.
75. Åkesson, J.; Gäfvert, M.; Tummescheit, H. JModelica—An Open Source Platform for Optimization of Modelica Models. In Proceedings of the 6th Vienna International Conference on Mathematical Modelling, Vienna, Austria, 11–13 February 2009.
76. Andersson, J.; Åkesson, J.R.; Diehl, M. CasADi: A Symbolic Package for Automatic Differentiation and Optimal Control. *Recent Adv. Algorithm. Differ.* **2012**, *87*, 297–307.
77. Wächter, A.; Biegler, L.T. On the Implementation of a Primal-Dual Interior Point Filter Line Search Algorithm for Large-Scale Nonlinear Programming. *Math. Program.* **2006**, *106*, 25–57. [CrossRef]
78. Gas, P.; Company, E. Find out if Peak Day Pricing Is Right for Your Business. Available online: https://www.pge.com/en_US/small-medium-business/your-account/rates-and-rate-options/peak-day-pricing.page. (accessed on 15 May 2020).
79. Edison, S.C. Get to Know Critical Peak Pricing (CPP). Available online: www.sce.com/business/rates/cpp (accessed on 18 May 2020).
80. International, P.G. SMUD Sees Early Success with Dynamic Pricing Pilot. Available online: https://www.power-grid.com/2013/07/10/smud-sees-early-success-with-dynamic-pricing-pilot/ (accessed on 15 May 2020).
81. Fund, E.D. The Costs and Benefits of Real-Time Pricing. Available online: https://citizensutilityboard.org/wp-content/uploads/2017/11/FinalRealTimePricingWhitepaper.pdf. (accessed on 15 May 2020).

82. Zhang, Q.; Huang, J.; Lang, S. Development of typical year weather files for Chinese locations. *ASHRAE Trans.* **2002**, *108*, 1063–1086.
83. Perez, R.; Ineichen, P.; Seals, R.; Michalsky, J.; Stewart, R. Modeling daylight availability and irradiance components from direct and global irradiance. *Sol. Energy* **1990**, *44*, 271–289. [CrossRef]
84. Authority, R.C.E. RCEA Community Choice Energy Program: Commercial, Industrial & Agricultural Generation Rates. Available online: https://redwoodenergy.org/wp-content/uploads/2020/03/March-2020-Com-Rates-for-Website.pdf (accessed on 16 May 2020).
85. Gas, P.; Company, E. ELECTRIC SCHEDULE B-19. Available online: https://www.pge.com/tariffs/assets/pdf/tariffbook/ELEC_SCHEDS_B-19.pdf (accessed on 16 May 2020).
86. of Energy Efficiency & Renewable Energy. Confronting the Duck Curve: How to Address Over-Generation of Solar Energy. Available online: https://www.energy.gov/eere/articles/confronting-duck-curve-how-address-over-generation-solar-energy. (accessed on 16 May 2020).
87. Intel. Intel NUC. Available online: https://www.intel.com/content/www/us/en/products/boards-kits/nuc.html (accessed on 14 May 2020).
88. InfluxData. InfluxDB. Available online: https://www.influxdata.com/ (accessed on 15 May 2020).
89. Systems, C.C. Wattnode Modbus. Available online: https://ctlsys.com/product/wattnode-modbus/ (accessed on 14 May 2020).
90. Controls, K. BAC-120063CEW. Available online: https://www.kmccontrols.com/product/bac-120063cew/ (accessed on 14 May 2020).
91. Corporation, P.H. Sporlan Electronic Refrigeration Controllers: PSK214N9EXBS. Available online: https://ph.parker.com/us/en/sporlan-electronic-refrigeration-controllers/psk214n9exbs (accessed on 16 May 2020).
92. Laboratories, S.E. SEL-3505/3505-3: Real-Time Automation Controller (RTAC). Available online: https://selinc.com/products/3505/ (accessed on 14 May 2020).
93. Tesla. Meet Powerwall, Your Home Battery. Available online: https://www.tesla.com/powerwall (accessed on 16 May 2020).
94. Company, T.D.S. Dark Sky API. Available online: https://darksky.net/dev (accessed on 14 May 2020).
95. Association, I.S. 2030.5-2018-IEEE Standard for Smart Energy Profile Application Protocol. Available online: https://standards.ieee.org/standard/2030_5-2018.html (accessed on 16 May 2020).
96. Alliance, O. OpenADR. Available online: https://www.openadr.org/ (accessed on 14 May 2020).
97. Reniersa, J.M.; Mulder, G.; Ober-Blöbaum, S.; Howey, D.A. Improving optimal control of grid-connected lithium-ion batteries through more accurate battery and degradation modelling. *J. Power Sources* **2018**, *379*, 91–102. [CrossRef]
98. Kiliccote, S.; Piette, M.A.; Hansen, D. Advanced Controls and Communications for Demand Response and Energy Efficiency in Commercial Buildings. In Proceedings of the Second Carnegie Mellon Conference in Electric Power Systems: Monitoring, Sensing, Software and Its Valuation for the Changing Electric Power Industry, Pittsburgh, PA, USA, 11–12 January 2006.
99. Pipattanasomporn, M.; Kuzlu, M.; Khamphanchai, W.; Saha, A.; Rathinavel, K.; Rahman, S. BEMOSS: An agent platform to facilitate grid-interactive building operation with IoT devices. In Proceedings of the 2015 IEEE Innovative Smart Grid Technologies-Asia (ISGT ASIA), Bangkok, Thailand, 3–6 November 2015; pp. 1–6.
100. Consortium, B. Brick: A Uniform Metadata Schema for Buildings. Available online: https://brickschema.org/ (accessed on 15 May 2020).

Article

Performance Assessment of Data-Driven and Physical-Based Models to Predict Building Energy Demand in Model Predictive Controls

Alice Mugnini [1], Gianluca Coccia [1,*], Fabio Polonara [1,2] and Alessia Arteconi [1,3]

1 Dipartimento di Ingegneria Industriale e Scienze Matematiche, Università Politecnica delle Marche, Via Brecce Bianche 12, 60131 Ancona, Italy; a.mugnini@pm.univpm.it (A.M.); f.polonara@univpm.it (F.P.); a.arteconi@univpm.it (A.A.)
2 Consiglio Nazionale delle Ricerche, Istituto per le Tecnologie della Costruzione, Viale Lombardia 49, 20098 San Giuliano Milanese (MI), Italy
3 Department of Mechanical Engineering, KU Leuven, B-3000 Leuven, Belgium
* Correspondence: g.coccia@univpm.it; Tel.: +39-071-2204770

Received: 11 May 2020; Accepted: 15 June 2020; Published: 16 June 2020

Abstract: The implementation of model predictive controls (MPCs) in buildings represents an important opportunity to reduce energy consumption and to apply demand side management strategies. In order to be effective, the MPC should be provided with an accurate model that is able to forecast the actual building energy demand. To this aim, in this paper, a data-driven model realized with an artificial neural network is compared to a physical-based resistance–capacitance (RC) network in an operative MPC. The MPC was designed to minimize the total cost for the thermal demand requirements by unlocking the energy flexibility in the building envelope, on the basis of price signals. Although both models allow energy cost savings (about 16% compared to a standard set-point control), a deterioration in the prediction performance is observed when the models actually operate in the controller (the root mean square error, *RMSE*, for the air zone prediction is about 1 °C). However, a difference in the on-time control actions is noted when the two models are compared. With a maximum deviation of 0.5 °C from the indoor set-point temperature, the physical-based model shows better performance in following the system dynamics, while the value rises to 1.8 °C in presence of the data-driven model for the analyzed case study. This result is mainly related to difficulties in properly training data-driven models for applications involving energy flexibility exploitation.

Keywords: model predictive control; data-driven model; artificial neural network; physical building model; energy flexibility

---

1. Introduction

Advanced control methods for energy management in buildings are required if the goal is obtaining an optimized operational performance [1]. Model predictive control (MPC) represents one of the most investigated controls in academic literature [2,3] given its ability to easily merge the principles of feedback control and numerical optimization [4]. The basic concept of MPC is to use a dynamic model to forecast a system behavior and to optimize the actuations in order to operate under the best sequence of decisions [5]. A key feature of MPCs consists in selecting future control actions, taking into account both predictions of future disturbances and system constraints [4], while the goal is pursued.

In buildings, MPCs can be applied for many purposes: (i) to exploit the energy storage capability in high-massive buildings, (ii) to maximize the use of renewable energy sources (RES), or (iii) to implement demand side management (DSM) such as demand response (DR). However, in order to be truly effective, an MPC must be based on a reliable model of the system under study [6].

Buildings are complex systems consisting of smaller systems which interact with the occupants [7]. In order to accurately predict their thermal dynamics, different aspects need to be considered. The IEA (International Energy Agency) EBC (Energy and Buildings Communities) annex 53 [8] defined six main factors that determine the energy consumption in buildings: (a) climate, (b) envelope, (c) systems and equipment, (d) operation and maintenance, (e) user behavior, and (f) indoor environmental quality. As highlighted by Geraldi and Ghisi [9], the six factors can be grouped into two categories: the first three factors, (a), (b), and (c), account for the building dimension, while the remaining, (d), (e), and (f), are related to the human dimension. The latter category can have a great impact on the assessment of the energy demand of buildings [10]. However, it is not always easy to predict users' behavior [7]; thus, a certain level of uncertainty is always present in building models which are not able to exactly predict the occupancy profiles.

In general, for short-time predictions, three categories of building energy modeling and forecasting are available: physical-based, data-driven, and hybrid models [11].

Physical-based systems are white box models [12] that need a detailed description of the building's physical and thermal properties in order to describe the building's dynamics with mathematical equations. Typically, they solve energy conservation equations based on heat transfer phenomena. No training data are required, and the parameters of the model are usually obtained from design plans, manufacture catalogues, or on-site measurements [11]. Most of the popular software, such as Energy Plus [13], TRNSYS [14], DOE-2 [15], or ESP-r [16], is based on a physical-based approach [17].

On the other hand, data-driven (or black box) models do not require a physical knowledge of the system, but they need a large amount of training data to be collected over an exhaustive period [11], i.e., both the data and the considered period should be statistically representative of the system operation. Statistical models have been directly applied in order to capture the correlation between building energy consumption and available measurement data [12]. The most common black box models are [18] support vector machines (SVM) [19], statistical regression (e.g., linear auto regressive models with exogenous inputs, ARX [20]), and artificial neural networks (ANNs) [21]. Unlike white box models, in which all the model parameters have a physical meaning, the parameters involved in a black box model cannot be interpreted in such terms.

A compromise between the two approaches is represented by hybrid (or grey box) models. Grey box models are a combination of physical-based and data-driven prediction models; thereby, some internal parameters and equations are physically interpretable, while others are estimated with a data-driven approach. Grey box models are widespread in building energy modeling [22], although they require both the system structure and training data.

Many works are available in the literature concerning the performance evaluation of the different hybrid and data-driven building models to be used in an MPC. Hietaharju et al. [23] introduced a generalizable grey box model based on heat transfer laws to predict temperature inside buildings. Testing the model structure on five buildings, of which real data were available, they found an average modeling error constantly below 5% during the 28-h prediction horizon. Ferracuti et al. [24] compared the performance of three different data-driven models for short-term thermal prediction in a real building: a lumped element grey box model, an ARX, and a nonlinear ARX. This work demonstrates that all the data-driven models investigated can be used to predict the short-term flexibility of the building for DR applications. In fact, for a prediction horizon of one hour, all the models showed a maximum root mean square error, *RMSE*, less than 0.5 °C in the tested period (among the grey box models, the third order one showed the best performance). Relying on a real neighborhood, Walker et al. [25] tested the use of machine learning algorithms (boosted-tree, random forest, SVM-linear, quadratic, cubic, and fine-Gaussian, as well as ANNs) to predict the electricity demand at individual and aggregated building levels, using data from 47 commercial buildings. Their results showed that boosted-tree, random forest, and ANN provided the best performance in predicting the hourly energy demand when computational time and error accuracy were compared. Touretzky and Patil [26] developed an ARX model to forecast the building power demand, also adopting physics-based modeling approaches for building energy management. They investigated different configurations of options for inputs and outputs in relation to the available measurements, highlighting the importance of an appropriate selection of exogenous inputs in order to capture the effect of common demand management practices. In order to evaluate the user behavior impact on overheating in a domestic environment, Baborska–Narozny and Grudzinska [27] developed a grey box model to simulate different scenarios in relation to fabric, occupant ventilation, and shading practices. The results showed that overheating could be entirely avoided if blinds were deployed to prevent excessive solar heat gains and mechanical extract ventilation was installed in the building.

Besides the above, other studies focus on the energy performance improvement that is obtained when a predictive control is used in a building with respect to a classic ruled-based control. Drgoňa et al. [28], for example, obtained energy use savings equal to 53.5% and a thermal comfort improvement of 36.9% for an office building in Belgium when a white box MPC based on first-principle physical equations was adopted. Moreover, Ferreira et al. [29] found similar energy savings (greater than 50%) when an MPC was adopted in the building sector. In this case, they proposed a discrete MPC that used radial-basis-function ANNs as predictive models and demonstrated the feasibility of the model with experimental results obtained in a building of the University of Algarve. Joe and Karava [30] introduced a smart operation strategy based on an MPC in order to optimize the performance of hydronic radiant floor systems in office buildings. They obtained a 34% cost saving compared to the baseline feedback control during the cooling season and a 16% energy use reduction during the heating season.

However, all the mentioned works focus either on evaluating the best model configuration to be adopted in an MPC (e.g., parameters identification, selection of inputs and outputs) or on the energy benefits that can be obtained through the adoption of such controls in buildings.

The purpose of this work is to combine these two types of analysis when the energy flexibility provided by the thermostatically controlled load can be also exploited. The two opposite modeling approaches (physical-based and data-driven) to predict the building energy demand are compared from two different points of view: (i) the capability of the models to reproduce the building energy behavior of a reference case and (ii) the practical implementation of a simple MPC designed to minimize the energy supply cost. In particular, the relationship between the model structure and its effectiveness in predicting the energy flexibility behavior will be explored.

With a dynamic cost tariff and the possibility of activating the building energy's flexibility [31] by allowing the indoor temperature to vary in a wider comfort range, the MPC can apply load shifting strategies to reach the goal. For the physical-based model, a lumped-capacitance model based on thermal–electrical analogy was used, while an ANN was chosen as data-driven model. The reference building, from which training data were extrapolated and in which the MPC was tested, was designed in a TRNSYS [14] simulation environment. The goal was to highlight the advantages and disadvantages of the two approaches when they were implemented in an MPC.

After this introductive section, Section 2 describes the methodology used to design the two models and the optimization process carried out by the MPC in the two cases. The case study is reported in Section 3, while the results of the study are provided in Section 4. The conclusions of the paper can be found in Section 5.

2. Methodology

In this study, the goal of the MPC was to minimize the total energy cost for the building thermal demand satisfaction. The typical structure of an MPC is shown in Figure 1. It is mainly composed of two parts: the building predictive model and the optimizer. The building predictive model should be able to dynamically forecast the building's energy response in a certain period (prediction horizon, *ph*), while its inputs can vary both in a controlled (manipulated variables) and in an uncontrolled (disturbances) way. To solve the optimization problem, it is important to define a proper objective function and to respect the system constraints; in this way, the optimizer has the possibility to select the best control actions to maximize the performance. Following a "receding horizon" logic, the MPC updates the best control action at each timestep, moving the prediction horizon forward and repeating the optimization [5].

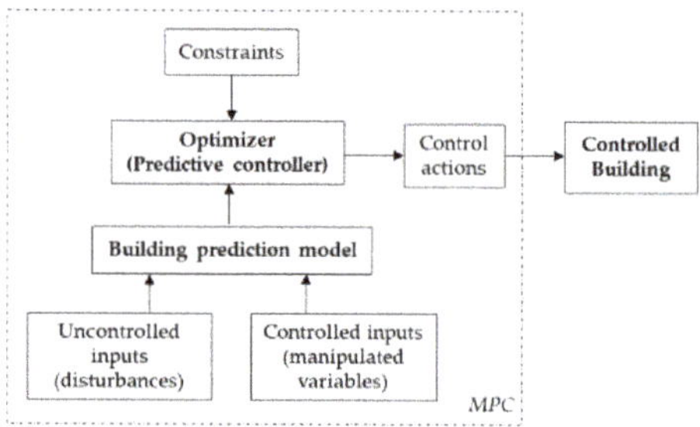

Figure 1. Architecture of a typical building's model predictive control (MPC).

As anticipated, two different approaches for the building prediction model are used in this work: a data-driven model based on an ANN and a physical-based model with a thermal–electrical analogy structure. In order to obtain training data for the ANN and to check the MPC effectiveness, a detailed building model designed in TRNSYS [14] was adopted as a control building.

The MPC routine was written in MATLAB [32], and for each time step the controlled building started the MATLAB engine to run the controller. The uncontrolled inputs of the models were weather conditions and heat gains, while the manipulated variable was the hourly building energy demand. Since the objective of the work was to focus on the comparison between the two approaches for building modeling in an operative MPC, an ideal HVAC system was considered in the building, i.e., the thermal energy demand was treated as a control action (Figure 1).

The optimizer solved the optimization problem in the prediction horizon, *ph*, in order to minimize the total energy cost with constraints on the internal comfort conditions. As an incentive for the exploitation of the energy flexibility, a dynamic energy cost tariff was considered [33]. In addition, to amplify the cost variations, a penalty signal was used in the MPC optimizer. It was obtained with a statistical method based on mean and standard deviation:

$$p(t) = \frac{c(t) - \mu_c}{\sigma_c} \tag{1}$$

where p is the penalty signal at each time t, c is the energy cost, while μ_c and σ_c are the cost signal mean value and standard deviation, respectively. Additional details about the penalty signal will be discussed in Section 3.

The performance of each model was assessed by evaluating the root mean square error, *RMSE*, which is defined as:

$$RMSE = \sqrt{\frac{1}{n}\sum_{j=1}^{n}\left(y_{\text{model,j}} - y_{\text{data,j}}\right)^2} \tag{2}$$

where y is the variable being evaluated and n is the number of points considered. Another index that will be used in the results is the root square error, *RSE*, defined as:

$$RSE_i = \left|y_{\text{model,j}} - y_{\text{data,j}}\right| \tag{3}$$

The first analysis consisted of evaluating the deviation of the models with respect to the building reference data. Then, in order to assess the accuracy of the models in the operative MPC, the *RMSE* was calculated between the prediction of the MPC building model and the actual thermal behavior of the building. Due to the different mathematical formulations of the two models, the optimization problem was different for the two cases. The following Sections 2.1 and 2.2 discuss the mathematical formulations of the two approaches.

2.1. MPC with Physical-Based System Model

A lumped-parameter model based on the thermal–electrical analogy was chosen as the building prediction model in the physical-based MPC. The building thermal dynamics is represented by an equivalent circuit of thermal resistances and capacitances [34]. A third order model was selected since it represented a good compromise between network complexity and capability of predicting the short-term dynamics of the building [35].

As shown in Figure 2, three thermal nodes were used. Each node is described by a temperature (T) and a thermal capacitance (C). Then, four thermal resistances (R) were used to model the heat transfer between the nodes.

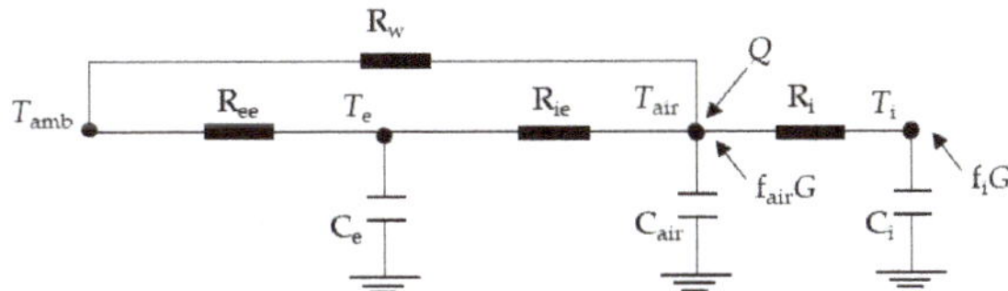

Figure 2. Thermal network for building model.

All the numerical values of the parameters (R and C) were deducted by the knowledge of the thermal and geometrical building features. Specifically, the first node (T_e, C_e) represents the external building thermal mass, the second node (T_{air}, C_{air}) is the indoor air node, while the last node (T_i, C_i) is the internal building thermal mass. As suggested by EN ISO 13790 [36], C_e and C_i are calculated by summing the heat capacitances of all the building elements (up to the thermal insulation) in direct thermal contact with the internal air zone.

As concerns the thermal resistance, R_w is the thermal resistance from the indoor air node temperature to the ambient air temperature (T_{amb}), due to air changes and windows. R_{ee} and R_{ie} are the thermal resistances between the external building thermal mass node and T_{amb} and T_{air}, respectively. They are calculated as equivalent thermal resistances due to the conductive heat transfer of all the building envelope layers, from outdoor to the thermal insulation for R_{ee}, and from thermal insulation to indoor for R_{ie}. These thermal resistances also take into account the convective heat transfer phenomena between the external surface and ambient temperature (R_{ee}) and between the internal building envelope surface and indoor air temperature (R_{ie}). In the same fashion, R_i considers the thermal resistance between the indoor air node and the internal thermal mass T_i. The heat fluxes, directly applied to the internal thermal nodes T_{air} and T_i, are the cooling power derived by an ideal HVAC system (Q) and the total heat gains (G). The latter includes both solar and internal contributions, which are provided with a scalar factor (f) for both the internal air and the internal mass node.

The dynamics of the resistance–capacitance (RC) model can be represented by the following equations:

$$C_{air}\frac{dT_{air}}{dt} = \frac{(T_e - T_{air})}{R_{ie}} + \frac{(T_{amb} - T_{air})}{R_w} + \frac{(T_i - T_{air})}{R_i} + f_{air}G + Q \tag{4}$$

$$C_e\frac{dT_e}{dt} = \frac{(T_{amb} - T_e)}{R_{ee}} + \frac{(T_{air} - T_e)}{R_{ie}} \tag{5}$$

$$C_i\frac{dT_i}{dt} = \frac{(T_{air} - T_i)}{R_i} + f_iG \tag{6}$$

Using these relations, a discrete time invariant state–space formulation can be set up:

$$[X(k+1)] = [A][X(k)] + [B][U(k)] \tag{7}$$

where $[X] = [T_{air}\ T_e\ T_i]^T$ represents the system state at each discrete time k, $[U] = [T_{amb}\ G\ Q]^T$ is the vector of the inputs, and [A] and [B] are coefficient matrices. A discrete time (k) of 1 hour is adopted, as well as for the simulation time step (t).

Along with the building prediction model, the MPC must include an optimizer. In the present work, the objective of the MPC was to minimize the total energy cost for the building thermal demand satisfaction while the indoor temperature remains in an allowed comfort range. For the physical-based MPC, a simple linear programming (LP) problem was formulated. In this optimization problem, both the objective function and constraints must be linear functions of the decision variables. The objective function (OF) is calculated as the sum of the building thermal energy consumption (Q_k) multiplied by the penalty signal p_k at each time step k in the whole prediction horizon (ph):

$$\mathrm{OF} = \sum_{k=1}^{ph} p_k Q_k \tag{8}$$

Therefore, the minimization problem can be written as:

$$\min \mathrm{OF} \tag{9}$$

subject to the following comfort and ideal HVAC constraints:

$$\forall\, k = 1, \ldots, ph \qquad T_{\min} \le T_{air,k} \le T_{\max} \tag{10}$$

$$\forall\, k = 1, \ldots, ph \qquad 0 \le Q_k \le Q_{\max} \tag{11}$$

The LP optimization problem is solved at each time step within a MATLAB script, according to a "dual-simplex" algorithm. The actual air zone temperature, $T_{air}(t)$, is passed to the MPC as the starting condition for the optimization. Based on the receding horizon principle, the control action at the controlled building time, t, is the first value of Q_k in the optimal decision variable sequence. At the following time step, t+1, the optimization problem is re-solved, moving forward the dataset of the disturbances by a time step. Figure 3 shows the scheme of the operation.

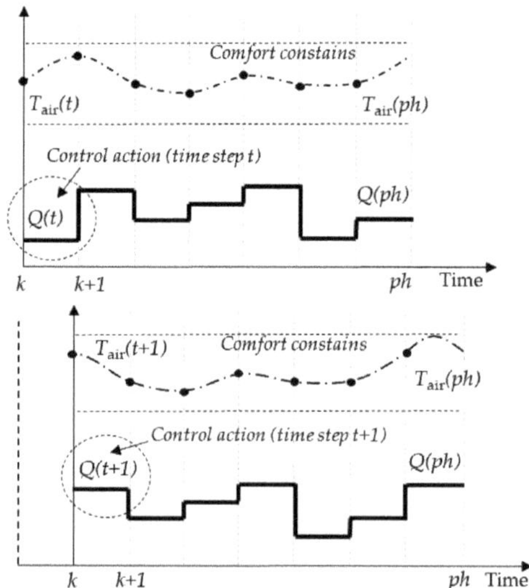

Figure 3. MPC receding horizon scheme.

2.2. MPC with Data-Driven System Model

The data-driven system used in the present study is based on an artificial neural network (ANN), a mathematical model that reflects the functioning of a biological brain [37]. In an ANN, the inputs (x) have the same role of biological dendrites, while the outputs (y) can be regarded as biological axons. The processing of the data takes place in the neurons, which, in an ANN, apply a nonlinear activation function, g, on the input data. Being a pure mathematical model without physical meaning, an ANN needs to be trained with existing data. The purpose of the training, which can be carried out with different error minimization techniques, is to determine the coefficient weights and biases of the network, i.e., the parameters that fully describe an ANN. Taking into account a feedforward ANN consisting of only one layer of neurons (also called a hidden layer because its activation values are not directly accessible from outside the network) and a linear activation function in the output layer (with just one output), the mapping carried out by the ANN on the input can be expressed as follows:

$$y = \sum_{j=1}^{\mathrm{m}} \left(\hat{w}_j \mathrm{g} \left(\sum_{i=1}^{\mathrm{d}} \mathrm{w}_{ji} x_i + \mathrm{b} \right) + \hat{\mathrm{b}} \right) \tag{12}$$

where d is the number of inputs, w_{ji} is the weights matrix of the inputs, b is the bias vector of the inputs, m is the number of neurons in the hidden layer, $\hat{w}_j$ is the weights matrix of the hidden layer, and $\hat{\mathrm{b}}$ is the bias vector of the hidden layer.

Generally, training data are divided into inputs and targets, and a well-trained ANN is expected to determine its outputs with a low deviation in respect to the provided targets. The available data of the system under study need to be examined carefully in order to train the ANN only with the inputs that most influence the objective targets. If the physics of the system under study is complex, a method to individuate the most effective inputs involves the use of statistical approaches such as factor analysis. Instead, if the physics of the system is not entirely unknown, the operator can try to select the input variables that mostly influence the desired target. In the present work, since the thermal behavior of the building is known, the four input variables (outdoor temperature, solar gains, internal gains, and indoor temperature) that most significantly influence the target variable (the thermal power required by the building) were selected.

The ANN was trained with 168 hourly-based input/target data referred to as a typical week. These data, provided by the building simulation environment designed in TRNSYS, were not obtained with a fixed set-point condition. In this case, in fact, the ANN's performance would have been insignificant, as there would have been no correlation between the output and the controlled variable (the indoor temperature). Indeed, the control unlocks the energy flexibility of the building and allows indoor temperature variations in a given comfort range, as explained previously. Therefore, to improve the ANN's prediction capability, the building simulation environment was allowed to work with multiple indoor set-point temperatures, varying within a reasonable comfort range. To avoid the overfitting of the data, i.e., an exaggerated interpolating behavior of the ANN, only a fraction of the dataset was actually used to train the network. Specifically, the initial dataset of 168 points was randomly divided into three subsets: a training set (60%), a validation set (20%), and a test set (20%). While the training data were used by the ANN to complete its training, the validation set was used as an internal interrupt criterium to end the training if overfitting occurred. The test set, instead, was used to evaluate the ANN's performance after training, in order to check its prediction capability with new data.

ANNs are available in different architectures, based on the physical–mathematical problem that is being studied. In the present work, since the goal was to estimate the thermal power required by a building, a fitting ANN was chosen. In MATLAB, fitting ANNs have a feedforward architecture and are trained according to a Levenberg–Marquardt backpropagation algorithm, which uses regression analysis and *RMSE* to evaluate the performance. As it is well-known that even one layer of neurons is sufficient to represent complex, nonlinear problems [37], in this study, one hidden layer with five neurons was used. The neurons used a hyperbolic tangent sigmoid as activation function. The ANN therefore had the architecture as represented in Figure 4.

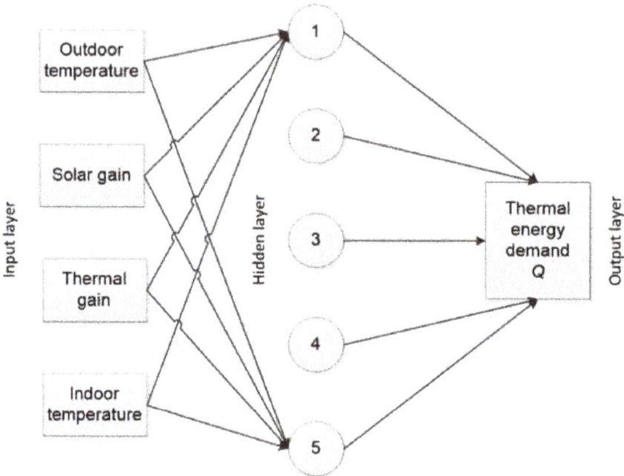

Figure 4. Artificial neural network (ANN) architecture of building prediction model.

After training, the ANN model was used in the MPC to predict the building's thermal demand for a given prediction horizon, *ph*. To this purpose, the ANN inputs were divided into uncontrolled (outdoor temperature, solar gains, and internal gains) and controlled (indoor temperature) variables (Figure 1). In this way, it was possible to manipulate the ANN as a function of the indoor temperature and to carry out an optimization process in order to minimize the overall energy cost in the time period *ph*. The objective function of the optimization algorithm can be therefore written as in Equation (8), subject to the constraints defined in Equations (10) and (11). Since the ANN function for the thermal demand was nonlinear, the optimization algorithm chosen in MATLAB was a programming solver based on the gradient method that uses an initial value for the indoor temperature as first attempt of solution. In the same fashion as the LP optimization problem defined for the physical-based model, the control action of the ANN-based MPC works according to the receding horizon principle (Figure 3).

3. Case Study

For the case study, a detailed building model was implemented in TRNSYS using Type 56. The model was composed of a single thermal zone and the envelope characteristics were extrapolated by the Tabula Project [38] for detached houses. The building was north-facing, while all the walls faced outwards and the floor was placed on the ground (considered at a constant temperature of 15 °C). Table 1 reports the main geometrical and thermal properties of the building that was considered (a single family house), which were extrapolated by UNI-TR 11552:2014 [39].

Table 1. Case study envelope properties: thermal transmittances and surfaces.

Property	External Walls	Roof	Floor	Windows
Thermal transmittance (W m^{-2} K^{-1})	0.34	0.28	0.33	2.20
Surface (m^2)	223.3	96.4	96.4	24.1

The structure was characterized by high levels of thermal insulation and double-glazed windows, which were air-filled, were selected (g-value of 0.7). An air change per hour (ACH) equal to 0.2 h^{-1} was selected. Internal gains included occupancy and artificial lighting [40]. The former was 120 W per person (occupancy density of 24 m^2 per person), while an artificial light density of 5 W m^{-2} was considered (artificial light turns on if total horizontal radiation is less than 120 W m^{-2} and turns off when the value exceeds 200 W m^{-2}). The building was located in Rome, Italy (41°55′ N, 12°31′ E), and a Meteonorm [41] weather file was used as a typical meteorological year.

In this work, the cooling season was chosen for the MPC test. Analogous results, however, could have been obtained for the heating season. As mentioned in Section 2, no specific HVAC system was modeled; instead, an ideal cooling control was used in Type 56 to extrapolate the training data [14]. In this way, the MPC control actions were applied as convective heat gains to the air nodes (positive for heating and negative for cooling). An indoor air temperature range of 25–27 °C was chosen as the comfort condition (i.e., T_{min} and T_{max} in Equation (10)) and a maximum cooling load power of 7 kW was fixed (i.e., Q_{max} in Equation (11)) [42]. Since the cooling power was directly applied to the air-zone, the ideal HVAC can be compared to a traditional heat pump split system. Assuming an average COP of 2.5, the thermal energy requirement can be converted into electricity consumption, and the penalty signal can be obtained consequently (Equation (1)).

For the reference case, a fixed set-point of 26 °C was simulated in order to provide a comparison between the building as controlled with the MPC and without the MPC.

4. Results

As discussed in Section 2, the evaluation of the forecast performance of the two building prediction models was realized according to two points of view: (i) the ability of the models to match the behavior of a known reference building and (ii) their dynamic operation when applied within the controller of the same building. Since short-term dynamics are involved in an MPC, a representative summer week was selected for the analysis (from July 30 to August 6). Figure 5 shows the uncontrollable inputs in the selected period (i.e., ambient temperature and total gains).

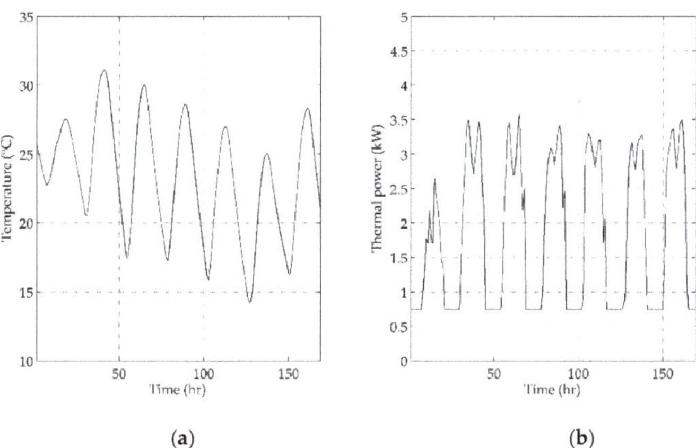

Figure 5. MPC uncontrollable inputs (disturbances) for the selected summer week: (**a**) ambient temperature, T_{amb}; (**b**) total gains, *G*.

As concerns the performance analysis, the ANN training data were selected as comparison terms to test the two building models. As mentioned in Section 3, the ANN training data were obtained with daily random set-points, which could range in the allowed comfort band. Figures 6 and 7 show the results in the entire 168-point dataset for the ANN-based model and the RC network, respectively. Since the output of the models was different in the two approaches, for the ANN the hourly cooling power forecasting was evaluated (Figure 6a), while for the RC network the internal air node temperature was considered (Figure 7a). As can be seen, both the prediction models were able to reply to the dynamic variations of the training data. In the first case, the *RMSE* was 0.26 kW, while the value found for the RC network was 0.34 °C. As highlighted by the *RSE* profile in Figure 6b, for the ANN the deviation was mainly due to the inability of the network to simulate the cases with reduced or zero cooling demand. For the physical-based model, instead (Figure 7b), there seemed to be a regular prediction error rather than specific peaks. It is worth noting that the *RSE* assumes a maximum value of 0.9 °C in the RC model and a value of 0.8 kW in the ANN model.

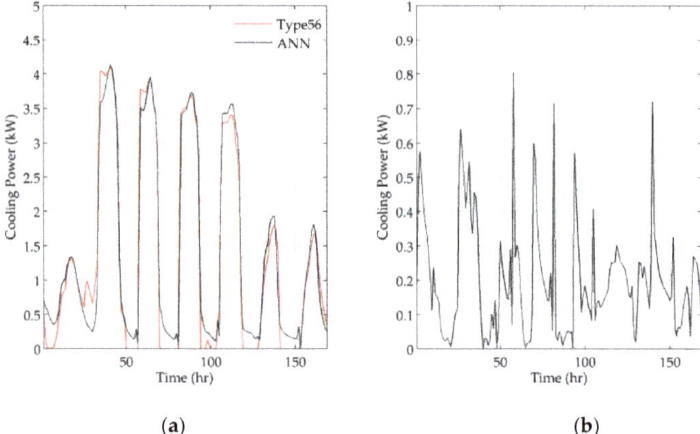

Figure 6. ANN model prediction results compared to training data: (**a**) cooling power demand; (**b**) root square error (*RSE*).

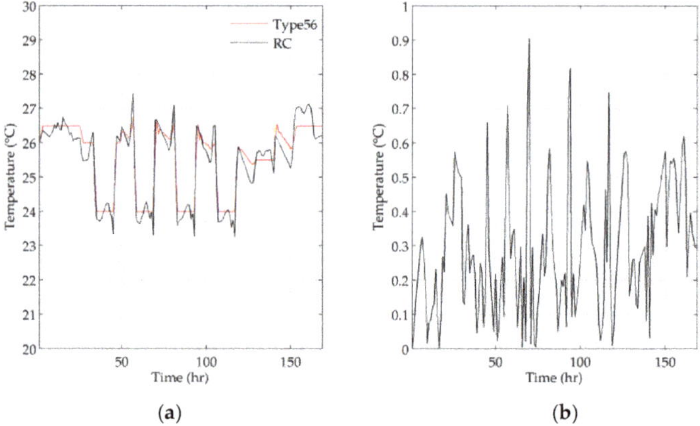

Figure 7. Resistance–capacitance (RC) model prediction results compared to training data: (**a**) indoor air temperature; (**b**) *RSE*.

When the building was allowed to be controlled by the MPC, at each time step the cooling power demand that was selected was the one that minimized the total energy cost. The original energy cost profile and its corresponding penalty signal in the representative summer week are shown in Figure 8a,b, respectively. The use of the penalty signal, instead of the actual energy cost, allowed us to amplify the cost variation and, thus, to incentivize the unlocking of the building's energy flexibility.

Figures 9 and 10 show the MPC results for both the prediction models. The results are presented with a prediction horizon, *ph*, of 6 hours. Specifically, Figures 9a and 10a show the comparison between the building's actual internal air temperature (Type 56) and the MPC's predicted value for the same time step (*t*). Instead, in Figures 9b and 10b, the control actions ($Q(t)$ in Figure 3) selected by the controller at each time step are represented. Looking at the black curves in Figures 9a and 10a, it is possible to note that both the prediction models were able to activate the building's energy flexibility, exploiting the whole comfort temperature range. Low temperature values are preferred when the energy cost is low and subsequent increases are expected. Conversely, the temperature is maintained close to the higher comfort range when high energy costs are detected.

The application of the MPC with both the prediction models produced a total cost reduction of about 16% if compared to the reference building with a fixed set-point of 26 °C. The *RMSE* between the actual air temperature and that predicted by the MPC at each time step also shows similar error values for the two models: 1.1 °C for the controller with the ANN and 0.99 °C for the RC model. However, comparing these values with the *RMSE*s found in the first part of the analysis, a degradation in the prediction performance can be noted for both the approaches. This is due to the fact that the building operated in variable dynamic conditions when the energy flexibility was activated. Thus, the predictions depend on constantly updated factors (such as the starting temperature conditions, the charge and discharge level of thermal inertia, etc.) which clearly amplify the prediction error.

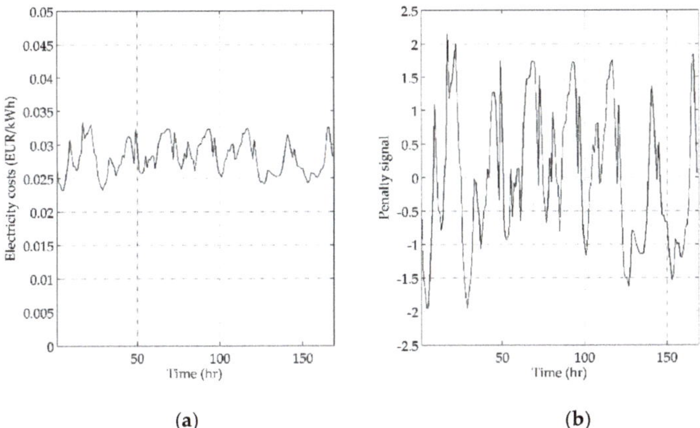

Figure 8. Electricity cost signal for the selected summer week: (**a**) hourly energy cost, *c*; (**b**): penalty signal, *p*.

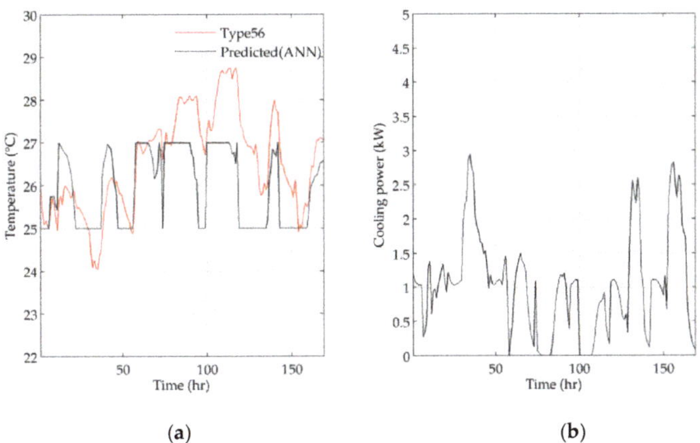

Figure 9. MPC with ANN as building prediction model: (**a**) internal air temperature, comparison between the actual Type 56 air zone temperature and ANN prediction at each timestep; (**b**) cooling power profile (control action sequences).

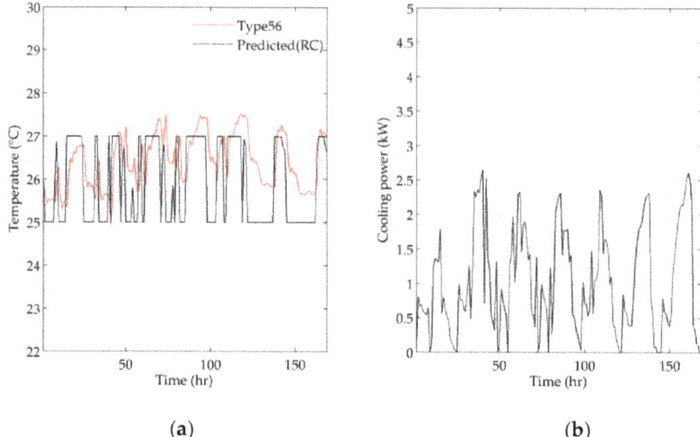

(a) (b)

Figure 10. MPC with RC as building prediction model: (**a**) internal air temperature, comparison between the actual Type 56 air zone temperature, and RC prediction at each timestep; (**b**) cooling power profile (control action sequences).

Although the models seem to have similar performances, different on-time trends for the two models can be expected if the actual building air temperature is observed (the red curves in Figures 9a and 10a). In particular, the MPC with the RC network seemed to follow the system dynamics more accurately than the controller with the ANN. When the ANN was operatively used in the controller, it appeared to perform less effectively than the RC network. In both cases, in the second half of the tested period, the prediction error started to grow but, in the case of the ANN, the actual air zone temperature exceeded the upper comfort limit by more than one degree (28.8 °C was the maximum value reached with the ANN in the controller, against 27.5 °C in the case of the RC model). This behavior is also confirmed by the duration curves reported in Figure 11. In the building regulated by the ANN-based MPC, the indoor temperature was found to be above the upper control limit for 36% of the simulation time. This percentage dropped to 24% when the RC network was used. This behavior was due to the difficulty of the control to maintain the comfort level when the temperature was too close to the upper comfort boundary; a small error in prediction can also cause temperature violations.

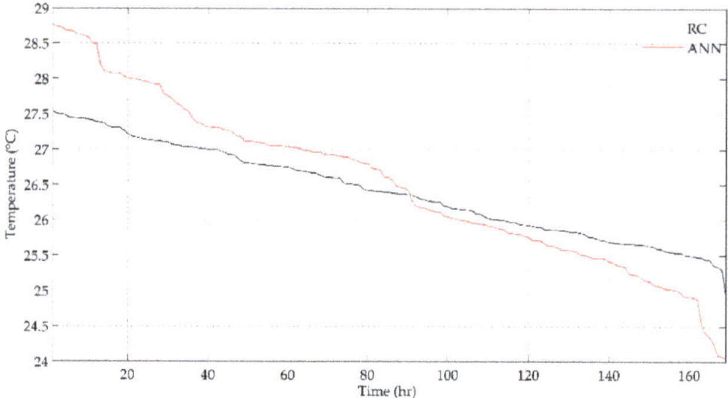

Figure 11. Indoor air temperature duration curves.

In summary, a reversal of performance between the two models can be found when the evaluation is carried out for the application in a realistic controller. The main reason for this behavior is related to the difficulty in selecting the proper dataset for the ANN training. In fact, the model must not only be able to replicate the response of the building in the same input conditions (and this is done well in the present study), but it must also be able to predict the system's responses in different scenarios, taking into account the contribution of energy flexibility. When the latter is introduced, it becomes difficult to identify a dataset that can train the black box model adequately, since the problem becomes dynamically affected by the operation of the system. Although the flexibility contribution seems to be better represented by the white box model based on the RC network, even in this case, a degradation in the prediction performance is detected when the controller is applied to the building. When using a white box model, however, a relevant amount of detailed data relating to the design and construction characteristics of the building should be known in order to implement an accurate model. Moreover, it is not always obvious which is the best network structure to use in the physical-based approach. For very complex buildings, it may be exceedingly difficult to identify an appropriate model and the corresponding parameters, even if detailed knowledge of the building is available. Another aspect that should be taken into account when using a purely physical-based model is that some dynamics (e.g., occupancy) cannot be considered in any way unless dedicated models are added. When, instead, measured data are used for the model training, such information may be intrinsically provided to the model. For these reasons, it could be convenient to monitor data and use hybrid models.

To conclude, Table 2 reports a comparison summary between the two building model approaches, subdivided into the main steps of configuration and implementation in an MPC.

Table 2. Comparison summary between a physical-based and a data-driven approach in an operative MPC.

Step	Physical-Based Approach	Data-Driven Approach
Preparation of the building model	• No measured data are required • An accurate knowledge of the geometrical and thermal characteristics of the building is needed	• Measured data are required • The knowledge of the geometrical and thermal characteristics of the building is not needed
Identification of the model configuration	• Difficulty in selecting the proper RC network and the numerical values of the parameters for complex buildings • Necessary to provide an accurate occupancy model if this aspect needs to be considered	• There is no systematic procedure to choose the best network architecture, and the optimal number of neurons is the result of a trial-and-error process • Difficulty in selecting proper input and output quantities when representing energy flexibility
Model development	• Linear model that can be represented with a state–space formulation	• Nonlinear model, one hidden layer of neurons is generally sufficient
Comparison with the reference building	• Good ability to replicate the reference case behavior at the same inputs (*RMSE* = 0.34 °C)	• Good ability to replicate the reference case behavior at the same inputs (*RMSE* = 0.26 kW)
Implementation in an MPC	• Improved performance in terms of objective functions compared to the reference operation • Good ability to represent flexibility and follow thermal dynamics, with occasional comfort constraint violations • An amplification of the prediction error occurs when real implementation is tested (comfort violation of 24%, maximum temperature deviation of 0.5 °C)	• Improved performance in terms of objective functions compared to the reference operation • Relevant errors in demand prediction when energy flexibility is managed • Comfort constraint violations can occur during real implementation • An amplification of the prediction error occurs when real implementation is tested (comfort violation of 36%, maximum temperature deviation of 1.8 °C)

5. Conclusions

In this paper, a comparison between a data-driven model implemented with an artificial neural network (ANN) and a physical-based model realized with an RC network is provided in an operative model predictive control (MPC). The controller was designed to provide a minimization of the total energy cost for the thermal demand satisfaction.

Focusing on the evaluation of the cooling season, a 16% reduction in the weekly cost with respect to the reference case was obtained, with an *RMSE* of about 1 °C in both cases (1.1 °C with the data-driven model and 0.9 °C with the physical-based approach). Although the data-driven model shows a good performance in replicating the building's thermal power profile, this trend is not confirmed when it works operatively in the controller. In fact, the comfort constraints are not respected for the 36% of the simulation time, with a maximum temperature deviation from the upper comfort limit of 1.8 °C. The physical-based model, instead, shows a discomfort percentage of 24% but a maximum deviation of only 0.5 °C.

The main conclusions of the presented work can be summarized as follows:

- Both the controllers show a good ability to replicate the reference case behavior at equal inputs.
- In the case of real implementation, for the ANN-based MPC, comfort constraint violations can occur more frequently. Indeed, the use of a data-driven model poses an issue for the training dataset to operatively represent the energy flexibility contribution in buildings.
- The physical-based model seems to better reproduce the system dynamics. However, an accurate knowledge of the building, which is not always possible, is required.
- Both the approaches show an amplification of the prediction error when their dynamic operation in the MPC is considered, even if the physical-based MPC limits this issue.

This latter point highlights the difficulty in implementing MPCs in real controls for energy flexible systems and suggests the need for further investigation. Indeed, it is important to consider that the results presented in this paper are related to a relatively simple building and no dedicated occupancy models were considered. When such advanced controls must be applied to more complex buildings where the role of users is not negligible, a purely physical-based approach can be computationally heavy, and it could be more convenient to use hybrid models. As a future development, it would be interesting to test the use of a hybrid model with an RC network structure and to provide a sensitivity analysis of the model according to the users' behavior in order to assess whether improvements in operational performance are also evident in the case of a simplified building.

Author Contributions: Conceptualization, A.M. and G.C.; methodology, A.M. and G.C.; validation, A.M. and G.C.; writing—original draft preparation, A.M.; writing—review and editing, G.C.; supervision, A.A.; project administration, F.P. and A.A.; funding acquisition, F.P. and A.A. All authors have read and agreed to the published version of the manuscript.

Funding: This research was funded by the Italian Ministero dell'Istruzione, dell'Università e della Ricerca (MIUR) within the framework of PRIN2015 project "Clean Heating and Cooling Technologies for an Energy Efficient Smart Grid", Prot. 2015M8S2PA.

Conflicts of Interest: The authors declare no conflict of interest.

Nomenclature

A	State space model coefficient matrices for state vector
ACH	Air changes for hours
ANN	Artificial neural network
B	State space model coefficient matrices for input
b	Bias vector
c	Hourly thermal energy cost (EUR kWh^{-1})
C	Thermal capacitance (J kg^{-1} K^{-1})
COP	Coefficient of performance
DR	Demand response
DSM	Demand side management
d	Number of inputs
f	Scalar factor
G	Gains (W)
g	Activation function
HVAC	Heating, ventilation, and air conditioning
k	Discrete timestep (hr)
LP	Linear programming
MPC	Model predictive control

m	Number of neurons
n	Samples number
OF	Objective function
p	Penalty signal
ph	Prediction horizon
Q	Thermal power (W)
R	Thermal resistance (K W^{-1})
RC	Resistance–capacitance
RES	Renewable energy source
RSE	Root square error
RSME	Root mean square error
SVM	Support vector machine
T	Temperature (°C)
t	Simulation time step (hr)
U	Input vector
w	Weights matrix
X	State vector
x	Input variable
y	Output variable
μ	Mean value
σ	Standard deviation

Subscripts

air	Internal air
amb	Ambient
c	Cost
e	External
ee	External Envelope
i	Internal
ie	Internal Envelope
k	Discrete time step
max	Maximum value
min	Minimum value
n	sample
w	Windows and air changes

References

1. Treado, S.; Chen, Y. Saving building energy through advanced control strategies. *Energies* **2013**, *6*, 4769–4785. [CrossRef]
2. Afram, A.; Janabi-Sharifi, F. Theory and applications of HVAC control systems—A review of model predictive control (MPC). *Build. Environ.* **2014**, *72*, 343–355. [CrossRef]
3. Thieblemont, H.; Haghighat, F.; Ooka, R.; Moreau, A. Predictive control strategies based on weather forecast in buildings with energy storage system: A review of the state-of-the art. *Energy Build.* **2017**, *153*, 485–500. [CrossRef]
4. Serale, G.; Fiorentini, M.; Capozzoli, A.; Bernardini, D.; Bemporad, A. Model predictive control (MPC) for enhancing building and HVAC system energy efficiency: Problem formulation, applications and opportunities. *Energies* **2018**, *11*, 631. [CrossRef]
5. Rawlings, J.; Manye, D.Q. *Model Predictive Control: Theory and Design*; Nob Hill Publishing: Madison, WI, USA, 2012; ISBN 9780975937709.
6. Prívara, S.; Cigler, J.; Váňa, Z.; Oldewurtel, F.; Sagerschnig, C.; Žáčeková, E. Building modeling as a crucial part for building predictive control. *Energy Build.* **2013**, *56*, 8–22. [CrossRef]
7. D'Oca, S.; Hong, T.; Langevin, J. The human dimensions of energy use in buildings: A review. *Renew. Sustain. Energy Rev.* **2018**, *81*, 731–742. [CrossRef]

8. IEA EBC annex 53: Total energy use in buildings. 2013. Available online: https://www.iea-ebc.org/Data/publications/EBC_PSR_Annex53.pdf (accessed on 15 June 2020).
9. Geraldi, M.S.; Ghisi, E. Building-level and stock-level in contrast: A literature review of the energy performance of buildings during the operational stage. *Energy Build.* **2020**, *211*, 109810. [CrossRef]
10. Palmer, J.; Terry, N.; Armitage, P. Building Performance Evaluation Programme: Findings from Non-domestic Projects (Getting the Best from Buildings). 2016. Available online: https://assets.publishing.service.gov.uk/government/uploads/system/uploads/attachment_data/file/497761/Non-Domestic_Building_performance_full_report_2016.pdf (accessed on 15 June 2020).
11. Foucquier, A.; Robert, S.; Suard, F.; Stéphan, L.; Jay, A. State of the art in building modelling and energy performances prediction: A review. *Renew. Sustain. Energy Rev.* **2013**, *23*, 272–288. [CrossRef]
12. Li, X.; Wen, J. Review of building energy modeling for control and operation. *Renew. Sustain. Energy Rev.* **2014**, *37*, 517–537. [CrossRef]
13. EnergyPlus. Available online: https://energyplus.net (accessed on 15 June 2020).
14. TRNSYS. Available online: http://www.trnsys.com (accessed on 15 June 2020).
15. DOE-2. Available online: http://doe2.com (accessed on 15 June 2020).
16. ESP-r. Available online: http://www.esru.strath.ac.uk/applications/esp-r (accessed on 15 June 2020).
17. Harish, V.S.K.V.; Kumar, A. A review on modeling and simulation of building energy systems. *Renew. Sustain. Energy Rev.* **2016**, *56*, 1272–1292. [CrossRef]
18. Wei, Y.; Zhang, X.; Shi, Y.; Xia, L.; Pan, S.; Wu, J.; Han, M.; Zhao, X. A review of data-driven approaches for prediction and classification of building energy consumption. *Renew. Sustain. Energy Rev.* **2018**, *82*, 1027–1047. [CrossRef]
19. Li, Q.; Meng, Q.; Cai, J.; Yoshino, H.; Mochida, A. Predicting hourly cooling load in the building: A comparison of support vector machine and different artificial neural networks. *Energy Convers. Manag.* **2009**, *50*, 90–96. [CrossRef]
20. Guo, Y.; Nazarian, E.; Ko, J.; Rajurkar, K. Hourly cooling load forecasting using time-indexed ARX models with two-stage weighted least squares regression. *Energy Convers. Manag.* **2014**, *80*, 46–53. [CrossRef]
21. Khosravani, H.R.; Castilla, M.D.M.; Berenguel, M.; Ruano, A.E.; Ferreira, P.M. A comparison of energy consumption prediction models based on neural networks of a bioclimatic building. *Energies* **2016**, *9*, 57. [CrossRef]
22. Brastein, O.M.; Perera, D.W.U.; Pfeifer, C.; Skeie, N.O. Parameter estimation for grey-box models of building thermal behaviour. *Energy Build.* **2018**, *169*, 58–68. [CrossRef]
23. Hietaharju, P.; Ruusunen, M.; Leiviskä, K. A Dynamic Model for Indoor Temperature Prediction. *Energies* **2018**, *11*, 1477. [CrossRef]
24. Ferracuti, F.; Fonti, A.; Ciabattoni, L.; Pizzuti, S.; Arteconi, A.; Helsen, L.; Comodi, G. Data-driven models for short-term thermal behaviour prediction in real buildings. *Appl. Energy* **2017**, *204*, 1375–1387. [CrossRef]
25. Walker, S.; Khan, W.; Katic, K.; Maassen, W.; Zeiler, W. Accuracy of different machine learning algorithms and added-value of predicting aggregated-level energy performance of commercial buildings. *Energy Build.* **2020**, *209*, 109705. [CrossRef]
26. Touretzky, C.R.; Patil, R. Building-level power demand forecasting framework using building specific inputs: Development and applications. *Appl. Energy* **2015**. [CrossRef]
27. Baborska-Narozny, M.; Grudzinska, M. Overheating in a UK high-rise retrofit apartment block—ranking of measures available to case study occupants based on modelling. *Energy Procedia* **2017**, *111*, 568–577. [CrossRef]
28. Drgoňa, J.; Picard, D.; Helsen, L. Cloud-based implementation of white-box model predictive control for a GEOTABS office building: A field test demonstration. *J. Process Control* **2020**, *88*, 63–77. [CrossRef]
29. Ferreira, P.M.; Ruano, A.E.; Silva, S.; Conceição, E.Z.E. Neural networks based predictive control for thermal comfort and energy savings in public buildings. *Energy Build.* **2012**, *55*, 238–251. [CrossRef]
30. Joe, J.; Karava, P. A model predictive control strategy to optimize the performance of radiant floor heating and cooling systems in office buildings. *Appl. Energy* **2019**. [CrossRef]
31. Jensen, S.Ø.; Marszal-Pomianowska, A.; Lollini, R.; Pasut, W.; Knotzer, A.; Engelmann, P.; Stafford, A.; Reynders, G. IEA EBC annex 67 energy flexible buildings. *Energy Build.* **2017**, *155*, 25–34. [CrossRef]
32. MATLAB. Available online: https://www.mathworks.com/products/matlab.html (accessed on 15 June 2020).
33. GSE (Gestore Servizi Energetici). Available online: https://www.gse.it (accessed on 15 June 2020).

34. Bagheri, A.; Feldheim, V.; Ioakimidis, C.S. On the Evolution and Application of the Thermal Network Method for Energy Assessments in Buildings. *Energies* **2018**, *11*, 890. [CrossRef]
35. Reynders, G.; Diriken, J.; Saelens, D. Quality of grey-box models and identified parameters as function of the accuracy of input and observation signals. *Energy Build.* **2014**. [CrossRef]
36. ISO—International Organization for Standardization. *ISO 13790:2008 Energy Performance of Buildings. Calculation of Energy Use for Space Heating and Cooling*; British Standards Institution: London, UK, 2008; pp. 1–24.
37. Bishop, C.M. Neural networks and their applications. *Rev. Sci. Instrum.* **1994**, *65*, 1803–1832. [CrossRef]
38. Corrado, V.; Ballarini, I.; Corgnati, S.P. *Typology Approach for Building Stock: D6.2 National Scientific Report on the TABULA Activities in Italy*; Dipartimento di Energetica, Gruppo di Ricerca TEBE, Politecnico di Torino: Torino, Italy, 2012; ISBN 9788882020392.
39. *UNI/TR 11552 Opaque Envelope Components of Buildings—Thermo-Physical Parameters*; UNI: Milan, Italy, 2014; pp. 1–44.
40. *UNI/TS 11300-1 Energy Performance of Buildings Part 1: Evaluation of Energy Need for Space Heating and Cooling*; UNI: Milan, Italy, 2014.
41. Meteonorm. Available online: https://meteonorm.com (accessed on 15 June 2020).
42. ASHRAE. *ASHRAE Handbook Fundamentals*; ASHRAE: Atlanta, GA, USA, 2005.

Article

Energy Performance Investigation of a Direct Expansion Ventilation Cooling System with a Heat Wheel

Miklos Kassai

Department of Building Service Engineering and Process Engineering, Faculty of Mechanical Engineering, Budapest University of Technology and Economics, Muegyetem rkp. 3., H-1111 Budapest, Hungary; kassai@epgep.bme.hu; Tel.: +36-20-362-8452

Received: 11 October 2019; Accepted: 7 November 2019; Published: 8 November 2019

Abstract: Climate change is continuously bringing hotter summers and because of this fact, the use of air-conditioning systems is also extending in European countries. To reduce the energy demand and consumption of these systems, it is particularly significant to identify further technical solutions for direct cooling. In this research work, a field study is carried out on the cooling energy performance of an existing, operating ventilation system placed on the flat roof of a shopping center, located in the city of Eger in Hungary. The running system supplies cooled air to the back office and storage area of a shop and includes an air-to-air rotary heat wheel, a mixing box element, and a direct expansion cooling coil connected to a variable refrigerant volume outdoor unit. The objective of the study was to investigate the thermal behavior of each component separately, in order to make clear scientific conclusions from the point of view of energy consumption. Moreover, the carbon dioxide cross-contamination in the heat wheel was also analyzed, which is the major drawback of this type heat recovery unit. To achieve this, an electricity energy meter was installed in the outdoor unit and temperature, humidity, air velocity, and carbon dioxide sensors were placed in the inlet and outlet section of each element that has an effect on the cooling process. To provide continuous data recording and remote monitoring of air handling parameters and energy consumption of the system, a network monitor interface was developed by building management system-based software. The energy impact of the heat wheel resulted in a 624 kWh energy saving and 25.1% energy saving rate for the electric energy consumption of the outdoor unit during the whole cooling period, compared to the system without heat wheel operation. The scale of CO_2 cross-contamination in the heat wheel was evaluated as an average value of 16.4%, considering the whole cooling season.

Keywords: building energy efficiency; heat wheel; direct expansion cooling; ventilation system; energy consumption

---

1. Introduction

The use of environmental control systems has significantly increased in the building sector in order to reduce the energy consumption of heating, ventilation, and air-conditioning (HVAC) systems [1]. Air handling units (AHUs) are one of the most complex building service systems [2], and can include heating, cooling, humidifier, mixing element, and heat recovery units, in order to provide the required indoor air quality and thermal comfort in conditioned spaces [3].

In a typical AHU, chilled water in the cooling coils cools the air, and hot water (or steam) in the heating coils heats the air, in order to maintain the desired temperature of the supply [4]. The supply and return fans assist in moving the air for heat exchange, as well as circulating it in the HVAC system at the required flow rate [5]. Several components are part of a typical system, i.e., the chiller, the boiler, the supply and return fans, and the water pump that consumes a lot of energy [6].

Direct expansion ventilation units are becoming more commonly used central air-conditioning technical solutions, in which a refrigerant is directly delivered to the cooling (and heating) coil [7]. These systems have the potential to save cooling and heating energy use, since they do not require any water pumps for their operation, compared to water-based central air conditioning systems [8,9].

Developers are working really hard to minimalize the energy consumption of their developed devices; however, there are many imperfections in the actual available product catalogues, technical data, and technical support service systems, especially for the annual energy designing provided by the ventilation producers for building service and energy design engineers [10,11]. Therefore, it would be particularly significant to have measured and recorded data obtained from field studies [12,13], which may be utilized in the course of design work, and which would allow a proper estimation of the expected realizable annual energy consumption of air handling elements in the function of the temperature and relative humidity of ambient and indoor air and operating parameters [14].

Stamatescu et. al [15] presented the implementation and evaluation of a data mining methodology based on collected data from a more than one-year operation. The case study was carried out on four AHUs of a modern campus building for preliminary decision support for facility managers. The results are useful for deriving the behavior of each piece of equipment in various mode of operation and can be built upon for fault detection or energy efficiency applications. The imperfection of their work is the missing data for air condition parameters (temperature and humidity) between the coils and mixing box; before and after the fans, which cannot be neglected, since the electrical motor of the fans increases the air temperature and decreases the relative humidity; and the air volume flow rate, which changes during the operation. All these missing parameters have a significant effect on the energy efficiency of the ventilation system.

Hong et. al [16] conducted a case study on a running AHU for data-driven predictive model development. In order to develop the optimal model, input variables, the number of neurons and hidden layers, and the period of the training data set were considered. The results and conclusions presented for the one-year field study could have much better reflected the reality from the view point of energy performance, if further temperature and relative humidity sensors had been placed between the coils and humidifier element, before and after the fans. Only focusing on energy performance data recording is not enough, since the desired indoor air quality and thermal comfort are also significant parameters that need to be considered. To draw a more exact conclusion from this point of view, the CO_2 parameter should also have been monitored and recorded in the outdoor air inlet (OA) and supply air outlet (SA) sections in the investigated AHU.

Based on a literature review of the field, there are some case studies in which the heat recovery unit has also been considered in the ventilation system. Noussan et. al [17] presented results obtained from an operation data analysis of an AHU serving a large university classroom. The main drivers of energy consumption are highlighted, and the classroom occupancy is found to have a significant importance in the energy balance of the system. The availability of historical operation data allowed a comparison of the actual operation of the AHU and the expected performance from nominal parameters to be performed. Calculations were made considering the operation analysis of the heat recovery unit over different years; however, the existing system does not include any heat or energy recovery devices, so there are no exact measured data from this point of view.

Bareschino et. al [18] compared three alternative hygroscopic materials for desiccant wheels considering the operation of the air handling unit they are installed in. Their results demonstrated that a primary energy saving of about 20%, 29%, and 15% can be reached with silica-gel, milgo, and zeolite-rich tuff desiccant wheel-based air handling units, respectively. The results were given based on a simulation and there is no exact measured data, which would be significant for making precise and clear energetic conclusions.

In this work, a field study is carried out on an existing, operating ventilation system that includes an air-to-air rotary heat wheel, a mixing box element, and a direct expansion cooling coil connected to a variable refrigerant volume outdoor unit. One of the main objectives of the present paper is to investigate the cooling energy performance and thermal behavior of each air handling component separately. To achieve this, an advanced data recording and remote monitoring system was considerately developed by building management system-based software. The system includes an electricity energy meter installed in the outdoor unit, as well as temperature, humidity, air velocity, and CO_2 sensors placed in the inlet and outlet section of all the air handling elements that have an effect on the cooling process. The purpose of the CO_2 measurements was to investigate the CO_2 cross-contamination, which occurs from the exhaust air flow to the supply air flow in the air-to-air rotary heat wheel, resulting in indoor air quality degradation. The novelty of this research is the accurate determination of the seasonal effectiveness and the energy saving impact of the heat wheel on the electric energy consumption of the outdoor unit. Moreover, the relative average and maximum value of CO_2 cross-contamination in the rotary heat recovery using the developed measurement system in the cooling period are presented. A further innovation in this study is the analytical evaluation method developed, which shows a good agreement between the calculated and measured energy consumption.

2. Materials and Methods

The selected air handling unit (AHU) is located on the flat roof of a shopping center, located in the city of Eger in Hungary, which has supplied fresh air to the back-office and storage area of a shop since 2017.

2.1. Description of the Investigated Central Ventilation System

The main air handling components of the system are an air-to-air rotary heat wheel, a mixing box element, and direct expansion cooling/heating (DX) coil connected to a variable refrigerant volume outdoor unit. Figure 1 shows the elements of the investigated AHU.

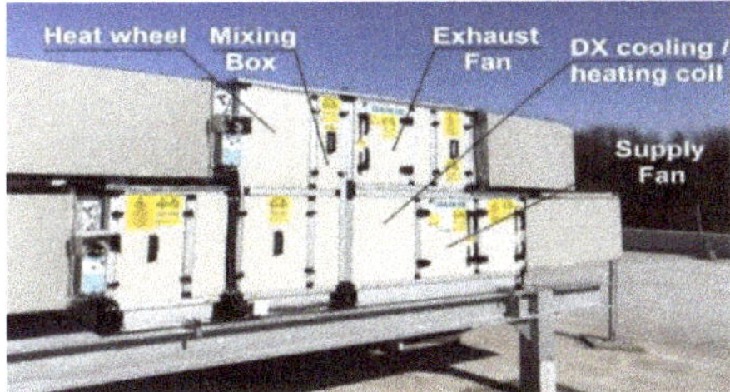

Figure 1. Photo from the investigated air handling units (AHU).

The specification of the AHU can be seen in Table 1.

Table 1. Specification of the investigated AHU [19].

Parameter	Value	Unit
Width × Height × Length	1450 × 1340 × 2897	mm
Air flow	1060	m^3/h
External Pressure Drop	280	Pa
Weight	595	kg

Figure 2 shows the outdoor unit which is connected with refrigerant pipes to the DX coil and is located in the AHU.

Figure 2. Photo from the outdoor unit.

Technical data of the unit can be seen in Table 2.

Table 2. Technical data of the outdoor unit [19].

Parameter	Value	Unit
Total Cooling Capacity	10.9	kW
Refrigerant	R410a	-
EER	3.99	-
Fin Material	Aluminium	-
Tube Material	Copper	-

Table 3 shows the specification of the air-to-air recovery heat wheel in the cooling period.

Table 3. Specification of the investigated heat wheel [19].

Parameter	Value	Unit
Heat recovered	2	kW
Effectiveness	74.9	%
Diameter	600	mm

2.2. Description of the Developed Measurement System

In total, six temperature and relative humidity sensors, three CO_2 sensors and three air velocity sensors were placed in the inlet and outlet section of each air handling element and an electricity energy meter was installed in the outdoor unit. The placement of the measurement points can be seen in Figure 3. The technical data of the installed sensors and instrument can be read in Table 4.

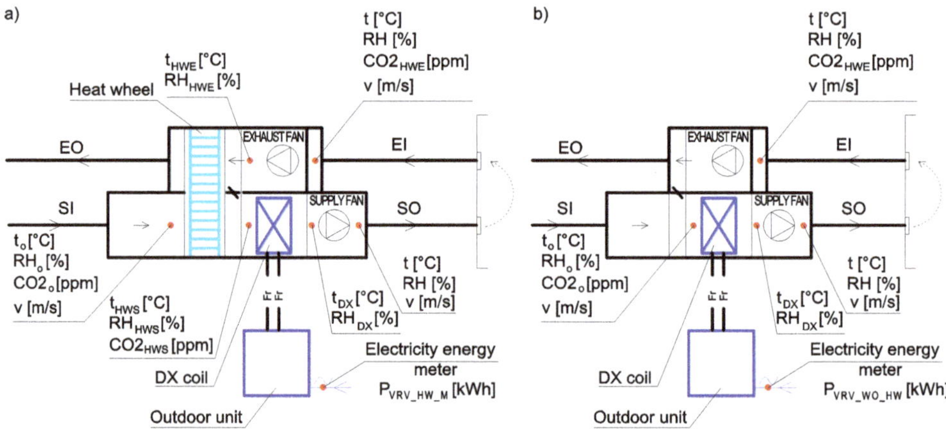

Figure 3. Schematic diagram (**a**) for the placement of the measurement points on the real operating AHU and (**b**) the AHU without a heat recovery operation assumption.

Table 4. Specification of the sensors and instrument.

Model	Device	Working Range	Accuracy
Honeywell VF20-3B65NW	Temperature sensor	−40–150 °C	±0.4 °C
Honeywell LFH20-2B65	Humidity sensor	10–90%	±3%
Honeywell AQS-KAM-20	CO_2 sensor	0–2000 ppm	±50 ppm
Honeywell AV-D-10	Air velocity sensor	2–20 m/s	±0.2 m/s
Inepro Metering Pro 380	Electricity energy meter	5–100 A	±1%

The recording of the measured data took place in an hourly period. With regards to the measurement accuracy, temperature sensors are normally used with a ±0.4 °C accuracy, humidity sensors with a ±3% accuracy, an air velocity sensor with a ±0.2 m/s accuracy, carbon dioxide sensors with a ±50 ppm accuracy, and an electric energy meter with a 1% of full scale accuracy. Among the monitoring air handling data, the air temperature and relative humidity data of the inlet and outlet sections of the DX cooling coil, energy recovery unit, and outdoor were used to investigate the energy performance and thermal behaviour of these air handling elements in the AHU in the cooling season.

The specification of the sensors and electricity energy meter used for monitoring of the investigated AHU can be seen in Table 4.

For the monitoring and recording of the various air condition parameters and the electrical energy consumption of the outdoor unit, the CentraLine Building Management System (BMS) software (version 2019) solution from Honeywell was implemented on a central server. Figure 4 shows a picture of the target building, along with a representative BMS screen for the investigated air handling unit.

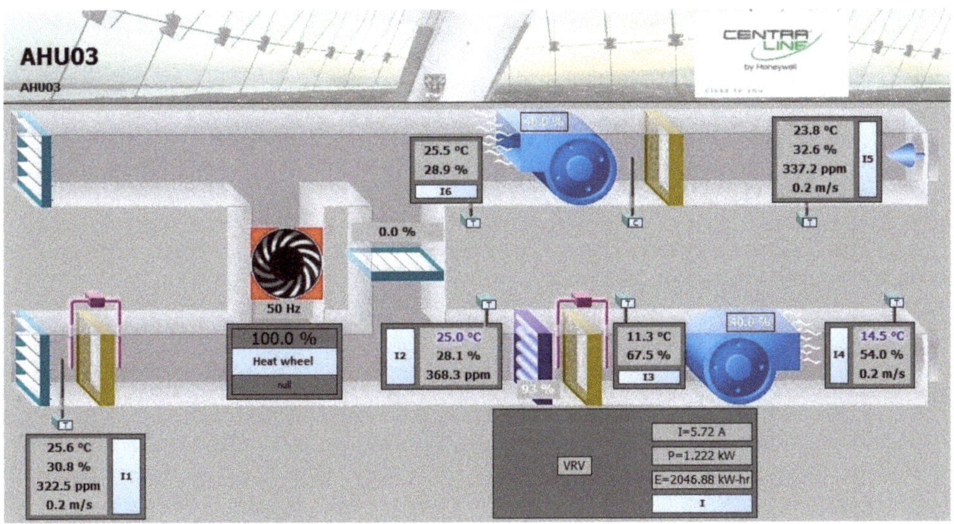

Figure 4. A screenshot of the investigated AHU in the Building Management System (BMS).

Access to the BMS software was remotely enabled. Within this technical context, the necessary data were collected for this field study. Data were collected online at hourly intervals, saved, and stored on a computer from a distance.

3. Evaluation of the Data Recorded

To investigate the energy performance of the AHU, using the measurements, the following mathematical approaches were implemented.

3.1. Calculation Formulas for Measured Data Evaluation

Using the measured air temperature and relative humidity data, the specific humidity could be calculated to obtain the enthalpy of the air. To achieve this, the water vapor saturation pressure (P_{ws}) was first calculated with Equation (1) [20]:

$$P_{ws} = A \cdot 10^{\left(\frac{mt}{t+t_n}\right)} \cdot 100, \tag{1}$$

where P_{ws} is the saturation pressure of the water vapor in Pa; t is the air temperature in °C; and A, m, and t_n are constant values in -. Since the temperature range during the measurements was between −20 and +50 °C, the constant values (in 0.083% maximum error) were as follows: $A = 6.116441$; $m = 7.591386$; $t_n = 240.7263$ [20]. The constant value of 100 in Equation (1) represents the conversion of the saturation pressure of water vapor from hPa to Pa.

To obtain the moisture content, the partial pressure of water vapor in the air at a given relative humidity was also calculated with Equation (2) [20]:

$$P_w = P_{ws} \cdot \frac{RH}{100}, \tag{2}$$

where P_w is the partial pressure of water vapor in Pa and RH is the relative humidity of the air in %. The constant value of 100 in Equation (2) represents the conversion of relative humidity from % to -.

The moisture content was calculated with Equation (3) [21]:

$$x = 0.622 \cdot \frac{P_w}{P_o - P_w}, \tag{3}$$

where x is the moisture content of the air in kg/kg, P_o is the barometric pressure in Pa, and the 0.622 constant value is the molecular weight ratio of water vapor to dry air.

The enthalpy was calculated with Equation (4) [21]:

$$h = c_{pa} \cdot t + x \cdot (c_{pw} \cdot t + 2500), \tag{4}$$

where h is the enthalpy of the air in kJ/kg, c_{pa} is the specific heat of air at constant pressure in kJ/(kg·°C), c_{pw} is the specific heat of water vapor at constant pressure in kJ/(kg·°C), and the constant value of 2500 represents the evaporation heat in kJ/(kg·°C).

3.2. Formulas for Energy Calculations

Considering the fact that there is balanced ventilation, the effectiveness values of the heat wheel were determined from the air temperature measured values using Equation (5) [22,23]:

$$\varepsilon_S = \frac{(t_{HWS} - t_o)}{(t_{HWE} - t_o)}, \tag{5}$$

where ε_S is the real sensible effectiveness of the heat wheel given by the measured data in -, t_{HWS} is the air temperature in the supply outlet section of the heat wheel in °C, t_{HWE} is the air temperature in the exhaust inlet section of the heat wheel in °C, and t_o is the ambient air temperature which is equal to the air temperature in the supply inlet section of the heat wheel in °C.

To get information about the seasonal energy performance of the heat recovery during the cooling period, the average of the sensible effectiveness was calculated with Equation (6):

$$\overline{\varepsilon}_{S\_AV} = \frac{\sum_{i=1}^{n} \varepsilon_{S\_i}}{n}, \tag{6}$$

where $\overline{\varepsilon}_{s\_AV}$ is the average of the sensible effectiveness of the heat wheel given by the measured data in the cooling season in - and n is the number of measurements.

The maximum value of the sensible effectiveness was also analyzed during the whole cooling season, which was calculated with Equation (7):

$$\varepsilon_{s\_MAX} = MAX\ (\varepsilon_{s\_i} \ldots \varepsilon_{s\_n}), \tag{7}$$

where $\overline{\varepsilon}_{s\_MAX}$ is the maximum value of the sensible effectiveness of the heat wheel given by the measured data in the cooling season in -.

To calculate the energy saving of the heat wheel in the cooling season, Equation (8) was used:

$$Q_{HW\_saved} = \dot{m}_s \cdot (h_o - h_{HWS}) \cdot \tau, \tag{8}$$

where Q_{HW_saved} is the energy saving of the heat wheel in kWh; $\dot{m}_s$ is the air mass flow rate delivered by the fans in kg/s, which is calculated by the multiplication of the measured air velocity in m/s and the internal cross-section of air duct in 0.7398 m^2 and approached a 1.2 kg/m^3 constant air density; h_o is the ambient air enthalpy which is equal to the air enthalpy in the supply inlet section of the heat wheel in kJ/kg; h_{HWS} is the air enthalpy in the supply outlet section of the heat wheel in kJ/kg; and τ is the time in hours. The average air volume flow rate was evaluated as 1060 m^3/h during the cooling season.

To calculate the cooling energy consumption of the DX coil, Equation (9) was used:

$$Q_{DX\_HW} = \dot{m}_s \cdot (h_{HWS} - h_{DX}) \cdot \tau, \tag{9}$$

where Q_{DX_HW} is the cooling energy consumption of the DX coil in kWh, and h_{DX} is the air enthalpy in the supply outlet section of the DX coil in kJ/kg, which is equal to the supply air condition.

In order to investigate more the energy saving impact of the heat wheel on the DX coil, the cooling energy consumption of DX coil was also determined by Equation (10), neglecting the air-to-air rotary heat wheel operation, when the DX coil directly cools the hot ambient air to the supply air conditions.

$$Q_{DX\_WO\_HW} = \dot{m}_s \cdot (h_o - h_{DX}) \cdot \tau, \tag{10}$$

where $Q_{DX\_WO\_HW}$ is the cooling energy consumption of the DX coil without the heat wheel operation in kWh.

The calculated electric energy consumption of the outdoor unit was calculated with Equation (11):

$$P_{VRV\_HW} = \frac{Q_{DX\_HW}}{EER}, \tag{11}$$

where P_{VRV_HW} is the calculated electric energy consumption of the outdoor unit with the heat wheel operation in kWh, and EER is the energy efficiency ratio, given by the producer in -.

Moreover, the real electric energy consumption of the outdoor unit ($P_{VRV\_HW\_M}$) was also measured during the cooling season, in order to see the agreement between values of the measured data and calculations using the recorded air condition parameters (P_{VRV_HW}). The difference between the measured and calculated electric energy consumption was determined with Equation (12):

$$\Delta P_{VRV\_HW} = P_{VRV\_HW\_M} - P_{VRV\_HW}, \tag{12}$$

where ΔP_{VRV_HW} is the difference between the measured and calculated electric energy consumption of the outdoor unit with the heat wheel operation in kWh.

The rate of deviation of the measured and calculated electric energy consumption of the outdoor unit related to the measured data was calculated with Equation (13):

$$\Delta P_{VRV\_HW\_REL} = \frac{\Delta P_{VRV\_HW}}{P_{VRV\_HW}} \cdot 100, \quad (13)$$

where $\Delta P_{VRV\_HW\_REL}$ is the rate of deviation of the measured and calculated electric energy consumption of the outdoor unit in %.

The electric energy consumption of the outdoor unit without the heat wheel operation was calculated with Equation (14):

$$P_{VRV\_WO\_HW} = \frac{Q_{DX\_WO\_HW}}{EER}, \quad (14)$$

where $P_{VRV\_WO\_HW}$ is the electrical energy consumption of the outdoor unit without the heat wheel operation in kWh when it directly cools the hot ambient air to the supply air conditions via the DX coil during the cooling season.

The energy saving of the heat wheel in terms of the electric energy consumption of the outdoor unit was calculated with Equation (15):

$$\Delta P_{VRV\_HW\_saved} = P_{VRV\_WO\_HW} - P_{VRV\_HW}, \quad (15)$$

where $\Delta P_{VRV\_HW\_saved}$ is the amount of energy saved by the heat wheel in terms of the calculated electric energy consumption of the outdoor unit compared to that without the heat recovery operation in kWh.

The energy saving impact of the heat wheel on the electric energy consumption of the outdoor unit, compared to the system without the heat wheel operation, was calculated with Equation (16):

$$\Delta P_{VRV\_HW\_saved\_REL} = \frac{\Delta P_{VRV\_HW\_saved}}{P_{VRV\_WO\_HW}} \cdot 100, \quad (16)$$

where $\Delta P_{VRV\_HW\_saved\_REL}$ is the energy saving rate of the heat wheel for the electric energy consumption of the outdoor unit, compared to the system without heat the wheel operation, in %.

The value of the actual energy efficiency ratio of the outdoor unit given obtained the field study was determined with Equation (17) for the investigated cooling season to compare the data provided by the producer:

$$EER_M = \frac{Q_{DX\_HW}}{P_{VRV\_HW\_M}} \cdot 100, \quad (17)$$

where EER_M is the evaluated energy efficiency ratio (-) based on the measurement during the whole investigated cooling season.

3.3. Formulas for Carbon Dioxode Cross-Contamination in the Heat Wheel

The scale of carbon dioxide (CO_2) cross-contamination in the air-to-air rotary heat recovery wheel was also investigated by measurements in the heat wheel during the operation of the air handling unit in the cooling period. To achieve this, the CO_2 concentration difference between the supply inlet and outlet sections of the heat wheel was first determined with Equation (18):

$$\Delta C_{CO2\_cross} = C_{CO2\_HWS} - C_{CO2\_o}, \quad (18)$$

where ΔC_{CO2_cross} is the scale of the CO_2 cross-contamination in the heat wheel in a given hour in ppm; C_{CO2_HWS} is the CO_2 concentration in the supply outlet section of the heat wheel in ppm; and C_{CO2_o} is the CO_2 concentration of the ambient air in ppm, which is equal to the CO_2 concentration in the supply inlet section of the heat wheel in ppm.

Having completed the measurements, the average of the CO_2 cross-contamination values was taken, and the ratio of the result and the supplied average CO_2 concentration was calculated by Equation (19):

$$\overline{\Delta C}_{CO2\_cross\_AV} = \frac{\sum_{i=1}^{n} \Delta C_{CO2\_cross\_i}}{n}, \quad (19)$$

where $\Delta C_{CO2\_cross\_AV}$ is the average of the CO_2 cross-contamination values in ppm and n is the number of measurements.

Since CO_2 cross-contamination occurs from the exhaust section to the supply section in the heat wheel, the average of the measured CO_2 values in the exhaust inlet section of the heat wheel was also calculated with Equation (20) using the data measured:

$$\overline{C}_{CO2\_HWE\_AV} = \frac{\sum_{i=1}^{n} C_{CO2\_HWE\_i}}{n}, \quad (20)$$

where $C_{CO2\_HWE\_AV}$ is the average value of the CO_2 concentration in the exhaust inlet section of the heat wheel in ppm and n is the number of measurements.

Equation (21) was used to obtain the relative average of differences:

$$\overline{\Delta C}_{CO2\_REL} = \frac{\overline{\Delta C}_{CO2\_cross\_AV}}{\overline{C}_{CO2\_HWE\_AV}} \cdot 100, \quad (21)$$

where ΔC_{CO2_REL} is the relative average of CO_2 cross-contamination in %, considering the CO_2 concentration content in the exhaust inlet section of the heat wheel in ppm.

The maximum value of CO_2 cross-contamination was also analyzed during the whole cooling season, which was calculated with Equation (22):

$$C_{CO2\_REL\_MAX} = \left[MAX \left(\frac{C_{CO2\_cross\_i}}{C_{CO2\_HWE\_i}} \cdots \frac{C_{CO2\_cross\_n}}{C_{CO2\_HWE\_n}} \right) \right] \cdot 100, \quad (22)$$

where $C_{CO2\_REL\_MAX}$ is the maximum value of CO_2 cross-contamination in the heat wheel in the cooling season given by the measured data in %.

4. Results and Discussion

The reference period of the study is the year 2019, more specifically, the cooling period from June 1st to August 31st for a total of 92 days and 25,296 data samples for each of the used measurement points. The AHU is intermittently operated 12 h/day from 8:00 till 20:00 7 days/week. Since this research work focused on the ventilation energy saving of the heat recovery unit's DX cooling coil, the mixing box was shut off during the data recording.

The air handling parameters obtained from the field study for the investigated AHU are illustrated in Figures 5–7 with a monthly timescale. Since the ambient air temperature was the highest in June during the whole cooling season, this relevant month was selected to present the measured data resulting from the data collection.

Figure 5 shows the temperature of the outdoor air (t_o), the air in the supply outlet sections of the heat wheel (t_{HWS}) and DX coil (t_{DX}), and the exhaust inlet section of the heat wheel (t_{HWE}) over time at hourly intervals in June.

Considering the hottest periods in the cooling season, the ambient air temperature decreased by about 4–5 °C due to the pre-cooling effect of the heat wheel, and by an additional 18–20 °C, provided by air cooling of the DX coil.

Figure 6 shows the measured relative humidity of the outdoor air (RH_o), the air in the supply outlet sections of the heat wheel (RH_{HWS}) and DX coil (RH_{DX}), and the exhaust inlet section of the heat wheel (RH_{HWE}) over time at hourly intervals in June.

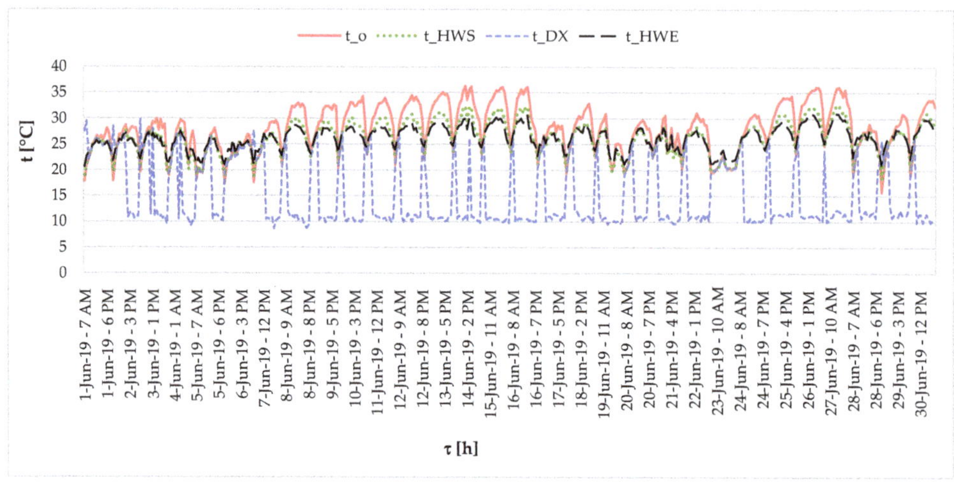

Figure 5. The air temperature values in the air handling processes.

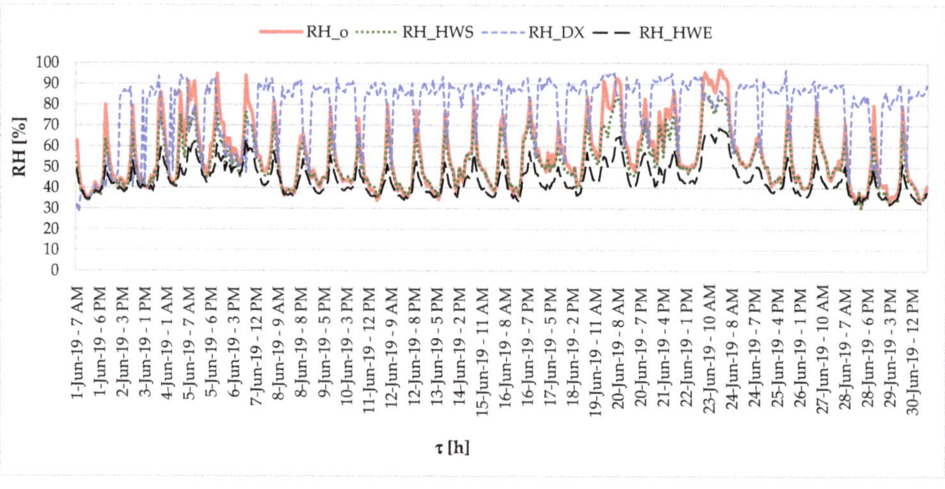

Figure 6. The air relative humidity values in the air handling processes.

The ambient air relative humidity decreased by about 60% due to the air cooling process. In this way, the supplied air relative humidity was around 90%.

Figure 7 shows the enthalpy of the outdoor air (h_o), the air in the supply outlet sections of the heat wheel (h_{HWS}) and DX coil (h_{DX}), and the exhaust inlet section of the heat wheel (h_{HWE}) over time at hourly intervals in June.

Considering the hottest periods in the cooling season, the ambient air enthalpy decreased by about 8–10 kJ/kg due to the pre-cooling effect of the heat wheel, and by an additional 30–35 kJ/kg, provided by air cooling of the DX coil.

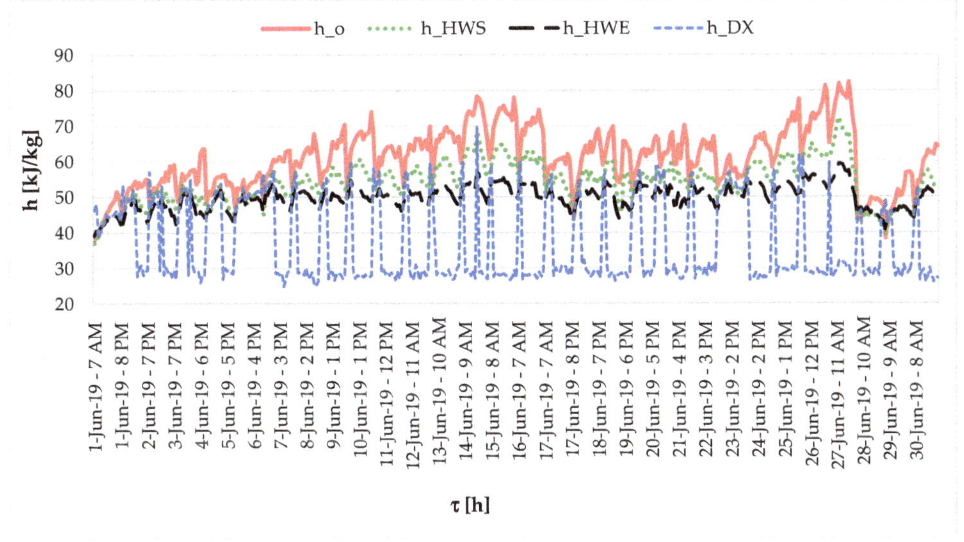

Figure 7. The air enthalpy values in the air handling processes.

Figure 8 shows the sensible effectiveness data (ε_s) for the outdoor air temperature in June.

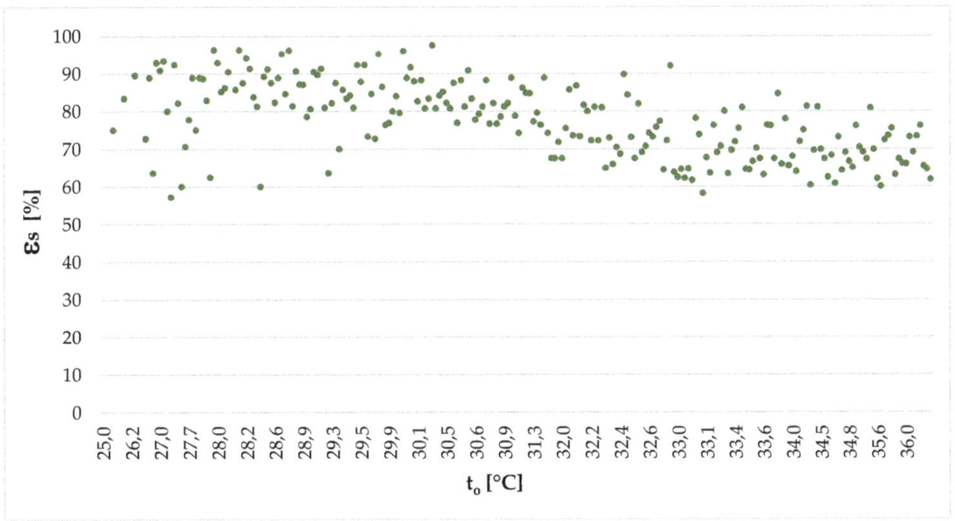

Figure 8. The sensible effectiveness values as a function of outdoor air temperature in June.

Based on the results, the average sensible effectiveness of the heat wheel was 79.6% during the whole cooling season and the maximum value of 97.6% was recorded in June.

Figure 9 shows the energy saving of the air-to-air rotary heat wheel (Q_{HW_saved}) in terms of the energy consumption of the DX coil, and the cooling energy consumption of the DX coil with the heat wheel operation (Q_{DX_HW}) and without the heat wheel operation ($Q_{DX\_WO\_HW}$), when the DX coil directly cools the hot ambient outdoor air to the supply air conditions during the cooling season.

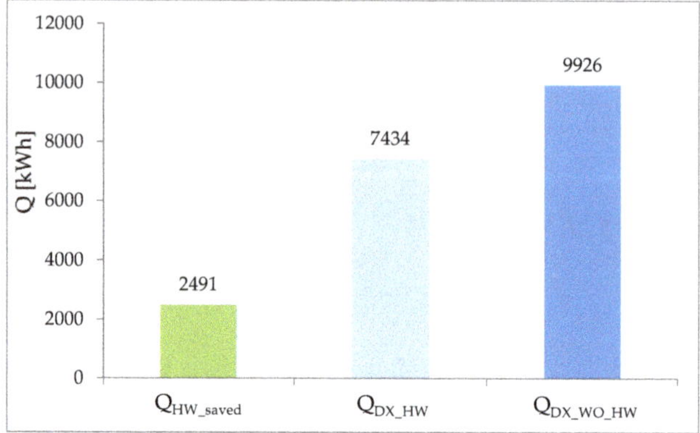

Figure 9. The energy recovery and auxiliary cooling energy consumption for ventilation.

Based on the results, the energy saving of the heat wheel was 2491 kWh in terms of the energy consumption of the DX coil, the cooling energy consumption of the DX coil with the heat wheel operation was 7434 kWh, and that without the heat wheel operation was 9926 kWh.

Figure 10 shows the electric energy consumption of the outdoor unit based on the direct real electric energy consumption measurements ($P_{VRV\_HW\_M}$) and the calculations made using the recorded air condition parameters with (P_{VRV_HW}) and without the heat wheel operation ($P_{VRV\_WO\_HW}$) for the whole cooling period.

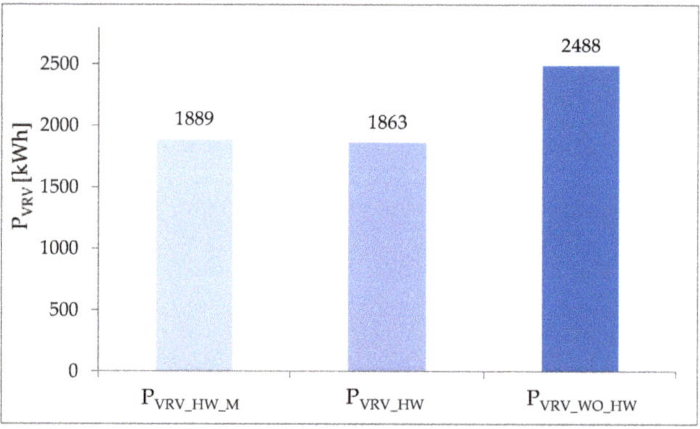

Figure 10. The electric energy consumption of the outdoor unit.

The real electric energy consumption of the outdoor unit based on the measurements was 1889 kWh and the calculations resulted in 1863 kWh consumption with and 2488 kWh consumption without the heat wheel operation for the whole cooling period.

Since the difference (ΔP_{VRV_HW}) is only 26 kWh and rate of deviation ($\Delta P_{VRV\_HW\_REL}$) is 1.36% between the values of the measured and calculated electric energy consumption of the variable refrigerant volume (VRV) outdoor unit with the heat wheel operation, Figure 10 shows very good agreement between the experimental and numerical results. The evaluated energy efficiency ratio is 3.94 based on the measurements (EER_M) conducted for the whole investigated cooling season, which is only 0.05 less than the value of 3.99 given by the producer. The energy impact of the heat wheel results in 624 kWh energy being saved ($\Delta P_{VRV\_HW\_saved}$), which is equivalent to a 25.1% energy saving rate ($\Delta P_{VRV\_HW\_saved\_REL}$) in terms of the electric energy consumption of the outdoor unit for the whole cooling period, compared to the system without the heat wheel operation.

Figure 11 shows the measured CO_2 concentration of the outdoor air (C_{CO2_o}), the air in the supply outlet section of the heat wheel (C_{CO2_HWS}), and the exhaust inlet section of the heat wheel (C_{CO2_HWE}) over time at hourly intervals in June. There are a few hours in Figure 11 when the recorded CO_2 values of the air were lower in the supply outlet section than in the exhaust inlet section of the heat wheel, probably due to the uncertainties and transient response characteristics of the CO_2 sensors.

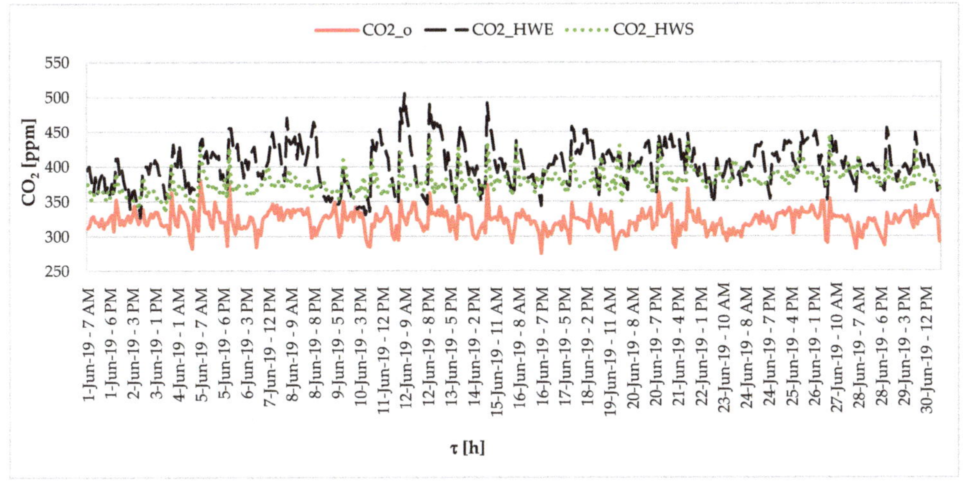

Figure 11. The carbon dioxide values in the investigated supply and exhaust sections of the heat wheel.

Having completed the measurements of the whole cooling period, the average CO_2 cross-contamination value ($\overline{\Delta C}_{CO2\_cross\_AV}$) was 63.9 ppm. The average value of the CO_2 concentration in the exhaust inlet section of the heat wheel ($\overline{C}_{CO2\_HWE\_AV}$) was 390.1 ppm. Based on the results, the relative average of CO_2 cross-contamination ($\overline{\Delta C}_{CO2\_REL}$) was 16.4% and the maximum value ($C_{CO2\_REL\_MAX}$) was recorded as 30.1%, considering the whole cooling season. To determine how the obtained values influence the indoor air quality inside of the conditioned spaces, further indoor air quality measurements are necessary (with the use of further measurement devices and questionnaires), which can act as a continuation of this research work, but exceed the limitation of this recent ongoing research project.

5. Conclusions

In this research work, a field study was carried out on the cooling energy performance of an existing, operating ventilation system under the operation of an air-to-air rotary heat wheel and direct expansion cooling coil, connected to a variable refrigerant volume outdoor unit. The major findings obtained from the study can be summarized as follows:

1. The operation of the heat wheel has a significant cooling energy saving impact on the electric energy consumption of the outdoor unit. Comparing the measured ventilation system with an air handling unit without a heat wheel operation, the cooling energy consumption is 25.1% higher;

2. Based on the measurements, the real sensible effectiveness and the CO_2 cross-contamination of the heat wheel are not in accordance with the design assumptions for the cooling period;

3. The sensible effectiveness of the heat wheel performed 4.7% higher than the data (74.9%) given in the technical data book of the producer;

4. Having completed the measurements for the whole cooling period, the amount of CO_2 cross-contamination in the heat wheel was much higher (with 16.4% relative average and 30.1% maximum values) than predicted during the designing phase.

Future work will focus on heating and annual energy performance investigations by conducting further field studies on the system. Moreover, simulation model development will also be considered for an annual energy consumption investigation of the existing ventilation system and model validation is planned based on data given by an annual field study. The long-term goal is to develop a simulation model which is suitable for determination of the energy consumption of ventilation systems in the design phase with a high accuracy.

Funding: The sensors, instruments, and transducers were financially supported and installed, and BMS software for data recording was made for this research work, by the national Intherm Ltd. In addition, the research project was financially supported by the National Research, Development and Innovation Office from the NRDI Fund [grant number: NKFIH PD_18 127907], the János Bolyai Research Scholarship of the Hungarian Academy of Sciences, Budapest, Hungary. Moreover, the research reported in this paper has been supported by the National Research, Development and Innovation Fund (TUDFO/51757/2019-ITM, Thematic Excellence Program), as well as the Higher Education Excellence Program of the Ministry of Human Capacities in the frame of Artificial Intelligence research area of Budapest University of Technology and Economics (BME FIKP-MI).

Acknowledgments: The author wishes to thank Árpád Nagy and Gyula Szabó from the group of Intherm Ltd., as well as Balázs Zuggó and Noémi Bálint from the group of Daikin Hungary Ltd., for providing their professional technical and practical knowledge for the research background of this research. Moreover, special thanks go to Laith Al-Hyari for his contribution in simple data saving.

Conflicts of Interest: The author declares no conflicts of interest.

Nomenclature

Abbreviations

A	Constant value (-)
AHU	Air Handling Unit (-)
CO_2	Carbon dioxide (ppm)
c_{pa}	Specific heat of air at constant pressure (kJ/(kg·°C))
c_{pw}	Specific heat of water vapor at constant pressure (kJ/(kg·°C))
EER	Energy efficiency ratio (-)
h	Enthalpy (kJ/kg)
m	Constant value (-)
$\dot{m}$	Air mass flow rate (kg/h)
P	Pressure (Pa); electric energy consumption of the outdoor unit (kWh)
RH	Relative humidity (%)
t	Temperature (°C)
t_n	Constant value (-)
VRV	Variable refrigerant volume (-)
$\dot{V}$	Air volume flow rate (m^3/h)
x	Absolute humidity ($g_{water}/kg_{dry\ air}$)
Greek Letters	
ϵ_s	Sensible effectiveness (-)
τ	Time (hr)
Subscripts	
DX	Supply outlet section direct expansion evaporator
EI	Exhaust air inlet
HWE	Exhaust inlet section of the heat wheel
HWS	Supply outlet section of the heat wheel
O	Outdoor
s	Saturation
SI	Supply air inlet
SO	Supply air outlet
w	Water vapor

References

1. Zhong, C.; Yan, K.; Dai, Y.; Jin, N.; Lou, B. Energy Efficiency Solutions for Buildings: Automated Fault Diagnosis of Air Handling Units Using Generative Adversarial Networks. *Energies* **2019**, *12*, 527. [CrossRef]
2. Misevičiūtė, V.; Valančius, K.; Motuzienė, V.; Rynkun, G. Analysis of exergy demand for air heating of an air handling unit. *Energy Effic.* **2017**, *10*, 989–998.
3. Cui, X.; Mohan, B.; Islam, M.R.; Chua, K.J. Investigating the energy performance of an air treatment incorporated cooling system for hot and humid climate. *Energy Build.* **2017**, *151*, 217–227. [CrossRef]
4. Ukai, M.; Tanaka, H.; Tanaka, H.; Okumiya, M. Performance analysis and evaluation of desiccant air-handling unit under various operation condition through measurement and simulation in hot and humid climate. *Energy Build.* **2018**, *172*, 478–492. [CrossRef]
5. Shea, R.P.; Kissock, K.; Selvacanabady, A. Reducing university air handling unit energy usage through controls-based energy efficiency measures. *Energy Build.* **2019**, *194*, 105–112. [CrossRef]
6. Kusiak, A.; Zeng, Y.; Xu, G. Minimizing energy consumption of an air handling unit with a computational intelligence approach. *Energy Build.* **2013**, *60*, 355–363. [CrossRef]
7. Guo, Y.; Zhang, G.; Zhou, J.; Wu, J.; Shen, W. A techno-economic comparison of a direct expansion ground-source and a secondary loop ground-coupled heat pump system for cooling in a residential building. *Appl. Therm. Eng.* **2012**, *35*, 29–39. [CrossRef]
8. Jorissen, F.; Boydens, W.; Helsen, L. Validated air handling unit model using indirect evaporative cooling. *J. Build. Perform. Simul.* **2018**, *11*, 48–64. [CrossRef]
9. Wallin, J.; Claesson, J. Improving heat recovery using retrofitted heat pump in air handling unit with energy wheel. *Appl. Therm. Eng.* **2014**, *62*, 823–829. [CrossRef]
10. Yun, G.Y.; Choi, J.; Kim, J.T. Energy performance of direct expansion air handling unit in office buildings. *Energy Build.* **2014**, *77*, 425–431. [CrossRef]
11. Ukai, M.; Okumiya, M. Comparison of Behaviour and Energy Performance of Desiccant Air Handling Unit under Various Control Method. In *IOP Conference Series: Earth and Environmental Science, Proceedings of the 4th Asia Conference of International Building Performance Simulation Association (ASIM2018), Hong Kong, China, 3–5 December 2018*; IOP Publishing: Bristol, UK, 2019; Volume 238, p. 012027.
12. Angrisani, G.; Roselli, C.; Sasso, M.; Tariello, F. Dynamic performance assessment of a solar-assisted desiccant-based air handling unit in two Italian cities. *Energy Convers. Manag.* **2016**, *113*, 331–345. [CrossRef]
13. Kim, S.H.; Moon, H.J. Case study of an advanced integrated comfort control algorithm with cooling, ventilation, and humidification systems based on occupancy status. *Build. Environ.* **2018**, *133*, 246–264. [CrossRef]
14. Li, N.; Xia, L.; Shiming, D.; Xu, X.; Chan, M. Dynamic modeling and control of a direct expansion air conditioning system using artificial neural network. *Appl. Energy* **2012**, *91*, 290–300. [CrossRef]
15. Stamatescu, G.; Stamatescu, I.; Arghira, N.; Fagarasan, I. Data-Driven Modelling of Smart Building Ventilation Subsystem. *J. Sens.* **2019**, *2019*, 3572019. [CrossRef]
16. Hong, G.; Kim, B.S. Development of a Data-Driven Predictive Model of Supply Air Temperature in an Air-Handling Unit for Conserving Energy. *Energies* **2018**, *11*, 407. [CrossRef]
17. Noussan, M.; Carioni, G.; Degiorgis, L.; Jarre, M.; Tronville, P. Operational performance of an Air Handling Unit: Insights from a data analysis. *Energy Procedia* **2017**, *134*, 386–393. [CrossRef]
18. Bareschino, P.; Pepe, F.; Roselli, C.; Sasso, M.; Tariello, F. Desiccant-Based Air Handling Unit Alternatively Equipped with Three Hygroscopic Materials and Driven by Solar Energy. *Energies* **2019**, *12*, 1543. [CrossRef]
19. Daikin Ventilation Catalogue. Daikin Europe Naamloze Vennootschap, Zandvoordestraat 300, 8400 Oostende, Belgium, RPR Oostende. 2019. Available online: https://www.daikin.eu/content/dam/document-library/catalogues/Ventilation%20catalogue%20-%20ECPEN13-203_Catalogues_English.pdf (accessed on 8 August 2019).
20. Oyj, V. Vaisala, Humidity Conversion Formulas. Helsinki, Finland. 2013. Available online: https://www.hatchability.com/Vaisala.pdf (accessed on 8 November 2019).
21. Rajput, R.K. *Thermal Engineering*, 6th ed.; Firewall/Laxmi Publications Ltd.: New Delhi, India, 2006; ISBN 81-7008-834-8.

22. European Committee for Standardization. *Ventilation for Buildings—Performance Testing of Components/Products for Residential Ventilation—Part 7: Performance Testing of Components/Products of Mechanical Supply and Exhaust Ventilation Units (Including Heat Recovery) for Mechanical Ventilation Systems Intended for Single Family Dwellings*; EN 13141-7:2010; CEN: Brussels, Belgium, 2010.
23. American Society of Heating. *Method of Testing Air-To-Air Heat Exchangers, ASHRAE Standard 84-1991*; American Society of Heating, Refrigerating and Air Conditioning Engineers Inc.: Atlanta, GA, USA, 1991.

Article

Study of the Improvement on Energy Efficiency for a Building in the Mediterranean Area by the Installation of a Green Roof System

Elisa Peñalvo-López [1], Javier Cárcel-Carrasco [1,*], David Alfonso-Solar [1], Iván Valencia-Salazar [2] and Elias Hurtado-Pérez [1]

1 Universitat Politècnica de València, Camino de Vera 14, 46022 València, Spain; elpealpe@upvnet.upv.es (E.P.-L.); daalso@iie.upv.es (D.A.-S.); ejhurtado@die.upv.es (E.H.-P.)
2 Instituto Tecnológico de Veracruz, Miguel Ángel de Quevedo 2779, Veracruz 91897, Mexico; ivvasa@itver.edu.mx
* Correspondence: fracarc1@csa.upv.es; Tel.: +34-963-877-000

Received: 2 February 2020; Accepted: 3 March 2020; Published: 7 March 2020

Abstract: Rooftop gardens ona building have proved to be a good way to improve its storm water management, but many other benefits can be obtained from the installation of these systems, such as reduction of energy consumption, decrease of the heat stress, abatement on CO_2 emissions, etc. In this paper, the effect from the presence of these rooftop gardens on abuilding's energy consumption has been investigated by experimental campaigns using a green roof ona public building in a Mediterranean location in Spain. The obtained results demonstrate a substantial improvement by the installation of the green roof onthe building's cooling energy demand for a standard summer day, in the order of 30%, and a reduction, about 15%, in the heating energy demand for a winter day. Thus, given the longer duration of the summer conditions along the year, a noticeable reduction on energy demand could be obtained. Simulation analysis, using commercial software TRNSYS code, previously calibrated using experimental data for typical summer and winter days, allows for the extrapolation to the entire year of these results deducing noticeable improvement in energy efficiency, in the order of 19%, but with an increase of 6% in the peak power during the winter period.

Keywords: green roofs; buildings; air conditioning; energy efficiency; mediterranean area

---

1. Introduction

The application of rooftop gardens onbuildings, or green roofs [1], which introduces a layer of vegetation, growing media and an additional drainage/auxiliary layers, has evidenced to improve storm water management [2,3], but this is not the only positive outcome resulting from these systems. They also produce positive impacts in many other aspects [4], such as reducing the heat island effect by decreasing the temperature in main city centers [5,6], ameliorating air pollution [7] and reducing energy consumption of buildings [8–10]. In relation to this last aspect, roofs are a critical part of the building envelopes, since they are highly susceptible to solar radiation and other environmental changes. Thereby, they have a significant influence on the indoor comfort conditions of the occupants. Roofs account for large amounts of heat gains and losses, especially onone-floor buildings with large roof area. In these cases, green roofs improve the performance of the building's energy behavior by either decreasing the heat load during the winter period [11] or the cooling requirements during summer time [12]. Green roofs also reduce the temperature fluctuation of the roof membrane along the year [13] and, consequently, increase the efficiency of photovoltaic (PV) systems installed on the roof [14]. In summary, a green roof is a good alternative to improve sustainability in urban areas by reducing energy consumption, heat stress, air pollution and CO_2 emissions.

All these possibilities, and the fact that thermal behavior of a building and thus, the impact of green roof installation on the building energy consumption is not an easy subject, explain the important effort developed during the last decade for research on these systems, both from the theoretical simulation and the experimental point of view [15].Thermal conductivity of employed materials is an important factor, but other factors, such as internal loads (lights, computers, people etc.) or roof reflectance to solar radiation, can play a very important role, especially in the summer period. Therefore, the contribution of the green roof to the improvement of the energy efficiency in the building will be highly dependent on local conditions and studies should be addressed to model and experimentally quantify that contribution for different climate areas. Thusfar, published works have focusedon cold [11,16] and hot climates [12,17] applications. In this last case, special emphasis has been given to the Mediterranean area [18]. Other research studies in the area indicate the benefits of integrating green roofs onbuildings, contributing to reduce a building's energy use while mitigating greenhouse gases (GHG) in urban areas [19–21].

This paper summarizes a long-term study using a green roof designed, built and installed ona public building located in the Mediterranean coast of Spain. The main emphasis of the study was to deduce its impact on the energy consumption of the building's air conditioning system by monitoring key energy and environmental variables, covering winter and summer periods. This approach allows evaluating the energy consumption of the building and address a complete comparison for similar periods before and after the installation of the green roof. Simulation studies using commercial software TRNSYS 17 allow for the extrapolation of the results to the entire year. Obtained results are representative for buildings in theMediterranean climate area. Section 2 introduces the experimental setup used for this study, while the main experimental results are presented in Section 3. Section 4 includes an extrapolation of these experimental results to the entire year period using commercial software TRNSYS 17.

2. Experimental Setup

A green roof was installed ona building located on Benaguasil, a small town in the Mediterranean coast of Spain. As a reference, aclimograph of Valencia was included, which is the nearest city (at a distance of about 18 km and with similar altitude) with available weather data (Valencia weather station of Viveros). In Valencia, the average annual temperature is 23.0 °C during the day and 13.8 °C at night. In January (the coldest month), the temperature typically ranges from 14 °C to 20 °C during the day and 4 °C to 12 °C at night. In August (the warmest month), the average temperature registered over the last 80 years in Valencia was around 25 °C (Figure 1). Furthermore, specific temperature data of the Benaguasil area ranged from 28 °C to 34 °C during the day and about 22 °C at night.

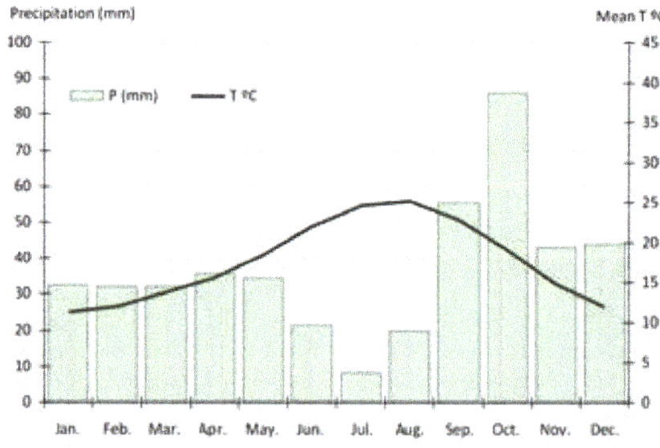

Figure 1. Climograph of the 1938–2018 period for Valencia city (Viveros).

Building Description

The building is a multipurpose social center with 1160 m^2 anda single floor, located in the southeast of the town (Figure 2). Since it was initially designed as a day care center for senior citizens, the building has some common spaces and facilities such as a dining room, dressing rooms, kitchen etc.

Figure 2. Building main facade (north).

The building has a flat roof with an "inverted roof" typology, characterized by having the thermal insulation (extruded polystyrene, XPS) over the waterproofing membrane, and over this thermal insulation liesa geotextile filter and a layer of gravels.

In our case, the detailed structure (Figure 3) incorporates the following elements: gravel layer (gravel diameter in the range of 20 to 50 mm and 1700 kg/m^3 bulk density, 50 mm of layer thickness), geotextile filter layer (2 mm), extruded polystyrene (XPS) insulation (40 mm), waterproofing membrane (5 mm) and concrete hollow block (300 mm).

INITIAL ROOF

Gravel
Geotextile
XPS insulation
Waterproof membrane
Concrete compression layer
Concrete hollow block

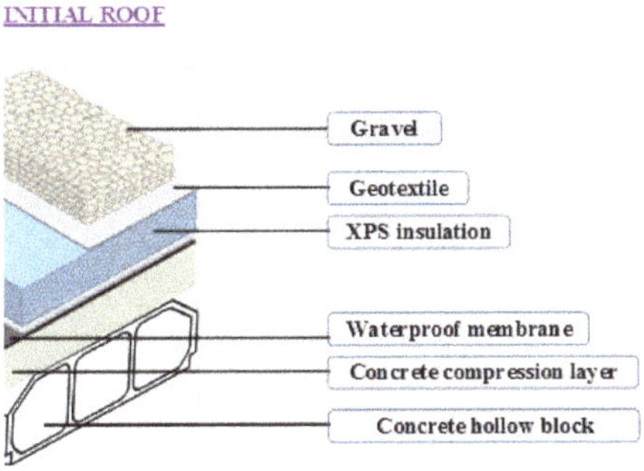

Figure 3. Initial roof structure.

The green roof was built over the present "inverted roof". It was decided to remove the layer of gravel and a water retention layer was added below the growing medium (separated with a filter fabric layer). This storage layer increases the capacity of the roof for retaining water after a rain episode and significantly reduces the amount of runoff generated. Figure 4 displays the green roof structure, which includes the following layers: growth medium (80 mm thickness), permeable textile layer (2 mm), drainage layer (water storage layer, 30 mm), geotextile layer/root barrier layer (3 mm), XPS insulation (40 mm), waterproofing membrane (5 mm) and a concrete hollow block (300 mm).

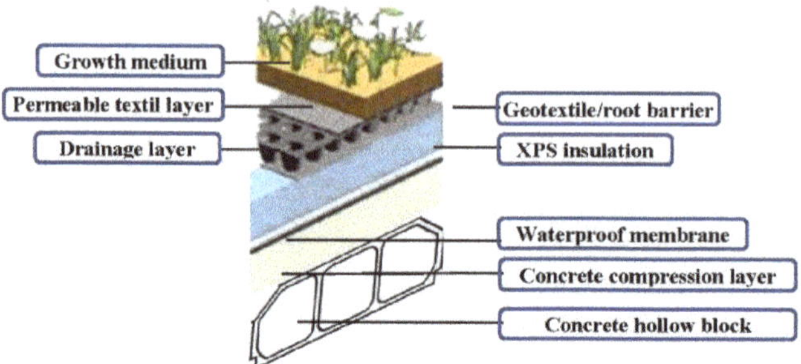

Figure 4. Green roof structure.

The growth medium is a mixture of conventional gardening organic substrate (40%), volcanic lava rocks (40%) and silica sand (20%). In the upper part of the green roof, there are plants covering almost the entire area with a height in the range of 50 to 150 mm. These plants are genus sedum (a mixture of sedum album AH, sedum floriferum, sedum sediform, sedum reflexum, sedum spurium, sedum moranense and sedum acre).

Figure 5 displays the plan view of the building, denoting the roof area where the green roof was installed by the dotted line. The building area under controlled conditions with green and conventional rooftop was 280 m^2. Figure 6 shows the green roof already installed.

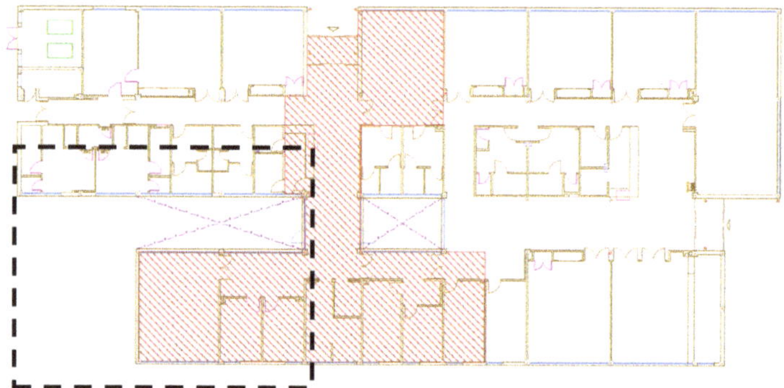

Figure 5. General view of building roof (dotted line indicating green roof affected area).

Figure 6. General view of greenroof.

The monitoring system used for these experiments is presented in Figure 7.

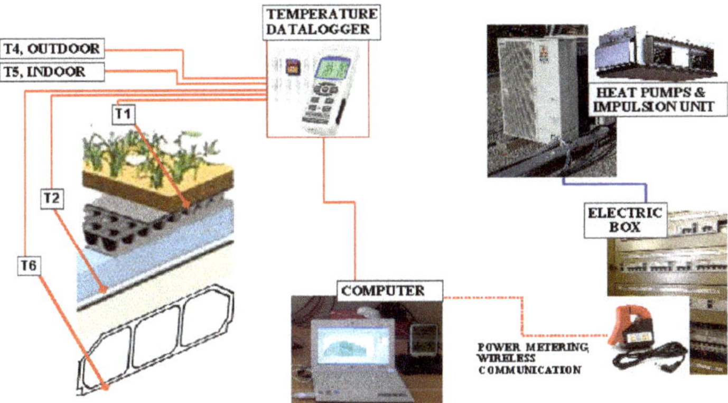

Figure 7. Monitoring system scheme.

A set of six type T thermocouples, whose positions and missions are detailed in Table 1, enables to determine the evolution of the temperature at different layers of the roof. These temperature measurements allow identifying similar ambient temperature conditions in the registered data sets and provide the data required for the simulation tools.

Table 1. Thermocouple positions.

Thermocouple #	Position
T1	Below first level layer (gravel or growth medium)
T2	Under XPS insulation layer
T3	Under insulated layer in the area not covered by the green roof. (Used as a reference)
T4	Outdoors of the building
T5	Indoors of the building
T6	Internal side of the roof

Specifications of the thermocouple sensors were: Probe PT100 RS PRO M16, PT100, +100 to +450 °C, diameter 6 mm, Connection head, Class B 4 Stainless Steel.

All these thermocouple signals were stored every minute in a data logger, together with the electricity consumption of the two heat pumps and the impulsion unit of the air conditioning system of the area covered by the green roof. In addition, wind velocities and solar radiation were provided by a nearby meteorological station.

The area affected by the green roof is about 1/4 of the total building surface and it has an independent air conditioning system. There are two heat pumps (Mitsubishi Electric PEA-RP250GA, with an input power of 8.455 kW each). Additionally, there is a common air impulsion unit of 3.05 kW of input power. Thus, maximum total input power, for both cooling and heating models, is about 19.96 kW.

This monitored building area is about 280 m^2 and it was closed during the testing periods to guarantee the control on internal loads and other factors that could affect the energy consumption. In this way, the test conditions are the same along the experimental campaigns with the conventional and the green roof. Only the external variables (temperature, solar radiation, wind etc.), which are experimentally monitored, change. These campaigns cover different indoor comfort temperatures, in arange from 22 °C to 28 °C.

3. Experimental Results

The experimental campaigns check the green roof effects on the building energy efficiency during 20 months (July 2017 to March 2019). During the first period, from July 2017 to February 2018, measurements correspond to the "conventional" roof. In February 2018, the external gravel layer of the conventional roof was replaced by the green roof. Thus, from March 2018 to March 2019, registered data corresponded to the green roof effects. Figure 8 displays typical traces for the temperatures and solar radiation obtained for the conventional and green roof campaigns, respectively, along an entire week during the summer period.

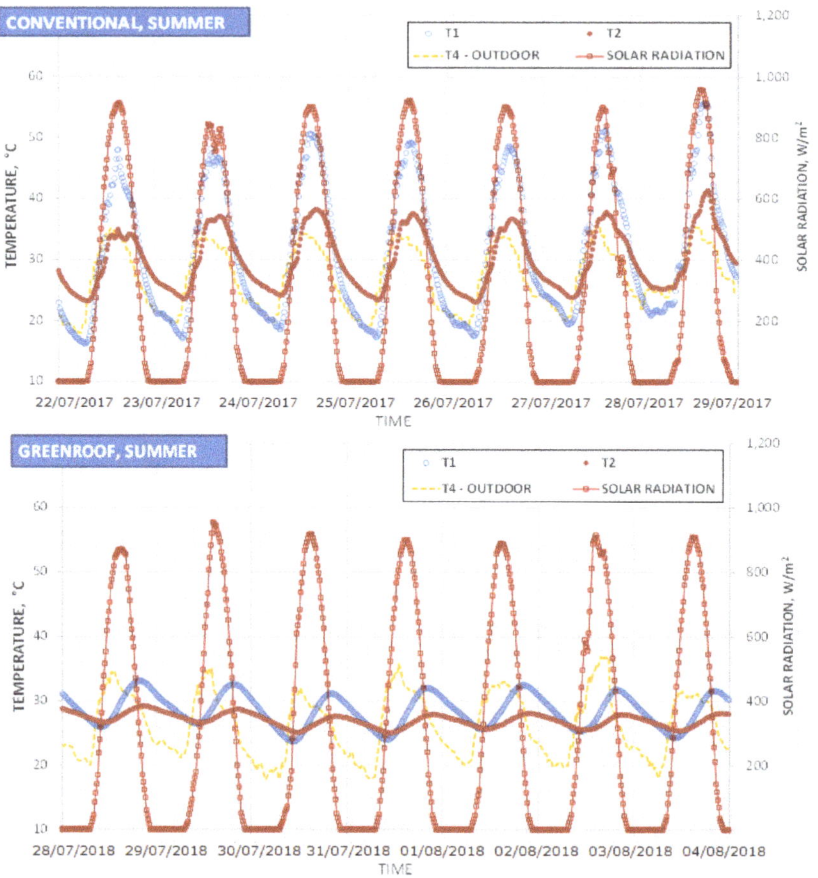

Figure 8. Typical traces for different temperatures and solar radiation in summer period.

The obtained results in both situations are summarized in Table 2, where the consumption from the air conditioning system (peak demand, energy consumption range during the entire campaign and its average daily range) and the external roof temperature are detailed for both types of roofs. Because the operation time of the offices located in the building was restricted to the morning hours, the operating time of the Heating, Ventilating and Air Conditioning (HVAC) system was only from 8:00 to 15:00, thus, the time span for comparison purposes was fixed for the period 9:00–13:00.

Table 2. Energy demand for conventional and green roof.

Dates	Conventional	Roof	Green	Roof
	Winter	Summer	Winter	Summer
	November 2017 to February 2018	July to October 2017	November 2018 to February 2019	July to October 2018
Maximum power demand (kW)	13.7	19.4	16.6	16.4
Daily energy consumption range (kWh)	20–25	30–35	24–28	20–25
Average daily consumption (kWh)	22.6	31.3	26.6	21.9
Range of Top roof temperature (°C)	20–30	45–55	5–15	22–35

We can deduct from these measurements that, during summer time, energy consumption was higher in the conventional roof in comparison to the winter one. However, energy consumption during winter is lower in the conventional roof. This behavior could be partly explained due to the presence of gravel in external layer of the roof, which acts as a heat storage system with high temperature. This fact introduces an additional load to be compensated by the air conditioning system during the summer. Given that the summer period is much longer than the winter one in the Mediterranean area, any effort to increase energy efficiency of the building should be concentrated onthe summer months.

Once the green roof is installed, the differences in energy consumption between the summer and winter periods reduced significantly to less than 15%, higher for the winter period in this case, whichcould be explained by the absence of the gravel layer as aheat source. This interpretation is supported by the data presented in Figures 9 and 10, where the evolution of the temperature along a similar day during the summer and winter periods, respectively, is presented for the conventional and green roof situations. Similar days were selected in terms of similar ambient outdoor temperature, humidity and solar radiation. During the summer, ambient outdoor temperature was 35 °C and the gravel reaches 50 °C, while with the presence of the green roof, this effect is smoothed and the temperature in the roof does not exceed the ambient outdoor temperature; in fact, it is below that value, namely, 32 °C.

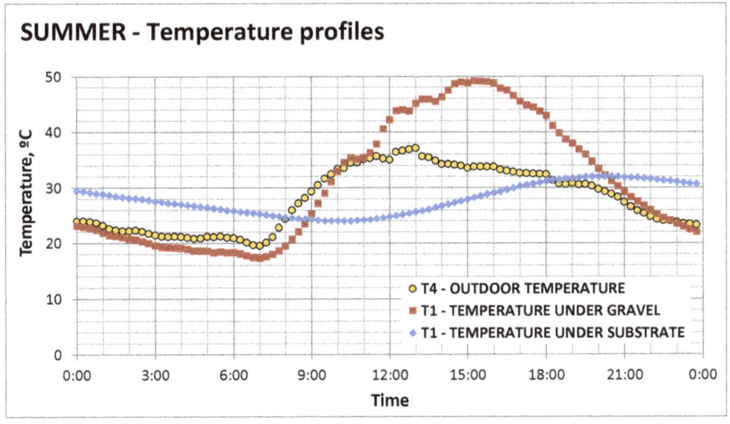

Figure 9. Temperature evolution for a typical summer day with and w/o green roof.

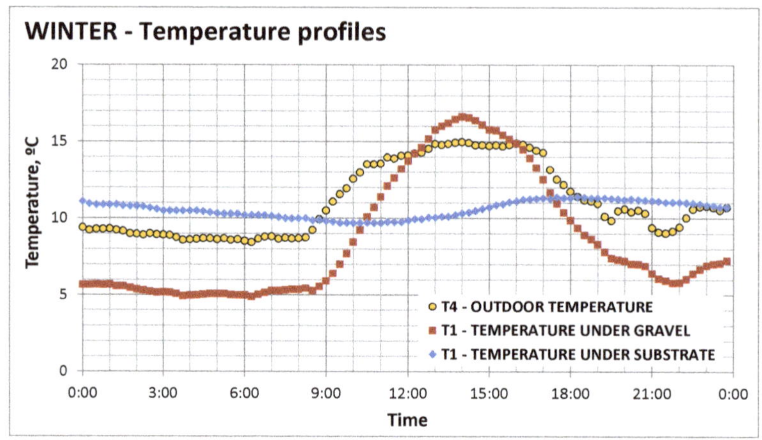

Figure 10. Temperatureevolution for a typical winter day with and w/o green roof.

Similar behaviorwasobtained for the winter period, as data plotted in Figure 10 shows. In this case, the smoothing of the temperature fluctuation due to the green roof reduces the heat input to the building and forces to higher energy consumption for the same level of indoor comfort.

Analysis of the data presented at Table 2 enables to evaluate the impact of the green roof on the energy requirements of the building. It can be deduced that in the winter period there is an increase for the total daily energy consumption and the requirement in electric power to be used by an order of 15%. This fact can be explained by the increase in the roof external layer temperature produced by the solar radiation, which is 50% higher in the case of the conventional roof than with the green roof due to the isolation produced by the former one. This heat source helps to heat up the building, reducing the energy requirements in winter, while this is not available when the green roof is installed. On the contrary, during the summer period, the energy saving increases up to 30%, with a reduction of the required peak power of about 15%. Given the higher percentage of savings and the longer duration of the summer period, it can be concluded that the global energy savings for the entire year is going to be highlysignificant. Table 3 summarizes the percentages in energy requirement variation due to installing the green roof for the winter and summer periods.

Table 3. Energy and power demand variation due to the green roof presence.

Energy and Power Demand	Winter	Summer
Maximum peak power demand (kW)	+17%	−15%
Average daily consumption (kWh)	+15%	−30%

Data comparingtwo similar days during summer time with both types of roofs areshown in Figure 11. An initial peak power was observed to start building conditioning, as is the case for the conventional roof; once a stable situation is reached, however, the power demand with the green roof is 20% less than with the conventional one.

A similar comparison is presented in Figure 12 for two similar days in the winter period. Power demand is very similar for both types of roof, but slightly lower in the case of the conventional one.

Obtained experimental results are comparable with other experimental studies conducted in Mediterranean climate conditions [22], in which a 15% to 17% lower energy consumption was observed during warm periods, whilehigher energy consumption (10% to 12%) was observed during cold times.

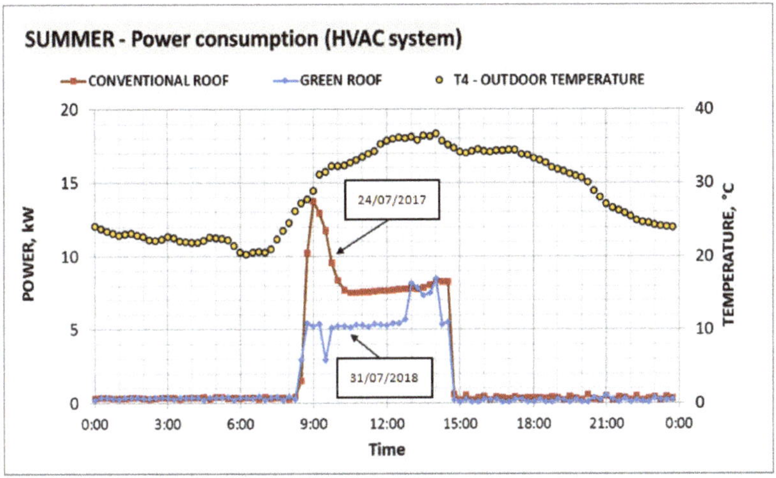

Figure 11. Comparison of similar days with and without a green roof in the summer period.

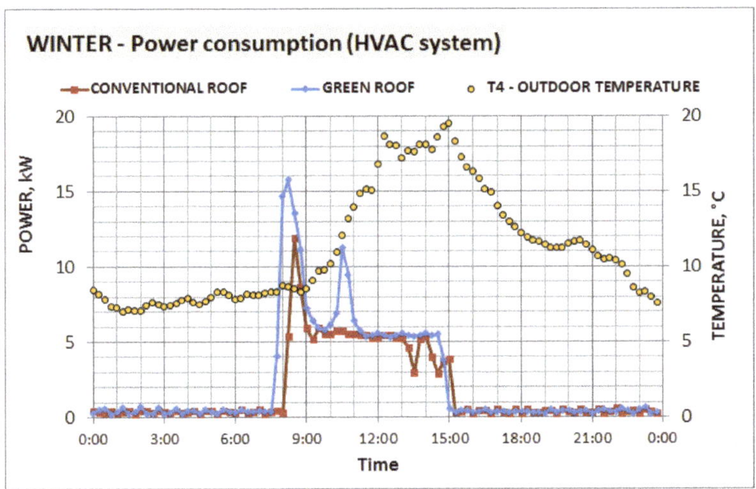

Figure 12. Comparison of similar days with and without a green roof in the winter period.

4. Simulation Results

In order to evaluate the benefits of the green roof for a longer period of time, commercial software TRNSYS 17 was used as theenergy simulation tool of the experimental setup. During the simulation, conventional and green roof were modeled based on their constructive characteristics and energy relationships. TRNSYS model was defined with 6 layers with different thickness: gravel or green roof (0.07 m), polystyrene insulation (0.05 m), membrane (0.001 m), concrete (0.075 m), building slab (0.17 m) and air chamber (0.5 m). The estimated U-value for the traditional roof was considered 0.518 W/m^2·°K, while for the green roof, estimation was 0.409 W/m^2·°K.Solar absorption of the roof was modelled with 0.8 for the gravel and 0.2 for the green roof.

Figure 13 shows the input data for the simulation, external conditions, building characteristics and temperature set points, and the deduced outputs, temperatures in the different layers of the roof and cooling and heating demand along the year.

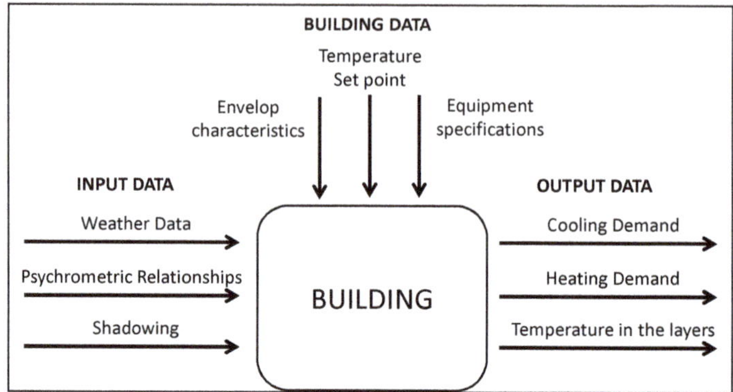

Figure 13. Parameters of TRNSYS Model.

Initially, rooftop models developed in TRNSYS were validated using collected experimental data. Validation was carried out in two stages, using a down-top approach: first, it was checked that modelled temperatures were similar to the experimentally measured at the rooftop layers along a week period. Then, the model was calibrated using the data from experimental daily profiles in the summer period.

In the case of the week validation, temperatures of the different layers of the roof were determined with the model and results (Figure 14) were compared with the corresponding experimental data. Maximum simulated temperatures of 59 °C were obtained during summertime in the gravel, while registered data during the same time period revealed a temperature range of 45 °C to 55 °C.

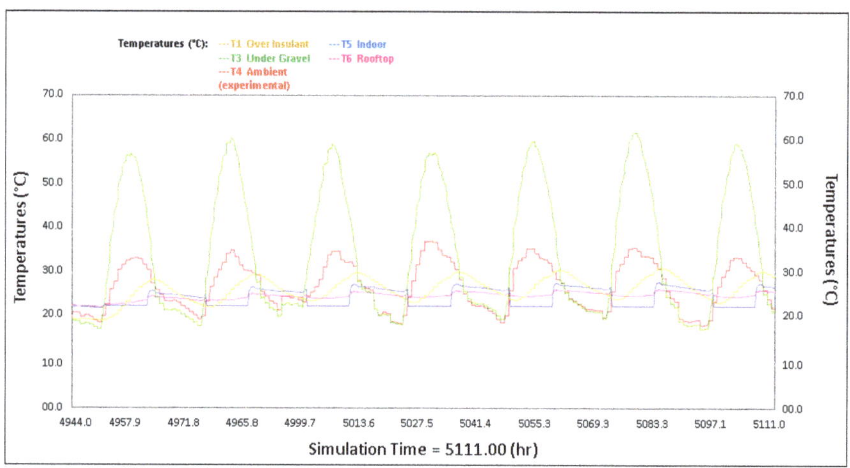

Figure 14. Simulated temperatures of the initial rooftop.

One-day validation was performed using as a reference two similar days in terms of environmental parameters (temperature, humidity etc.) and use-of-space (workday) during the experimental campaign, one for each of the two different types of roof considered, on 24th July 2017 (conventional roof) and 31st July 2018 (green roof), in order to deduce the parameters used in the simulation. Comparison of the simulation results, detailed in Figure 15, with the experimental data plotted in Figure 11, shows a good enough agreement, thus, the simulation for the entire year could be addressed using these parameters. In order to complete the simulation task, a selected time window during the day with stable conditions was identified. This time span was from 11:00 to 13:00. During this time span, TRNSYS simulations were addressed considering a set-point of 22 °C in heating demand (October to March) and 26 °C for cooling needs (April to September).

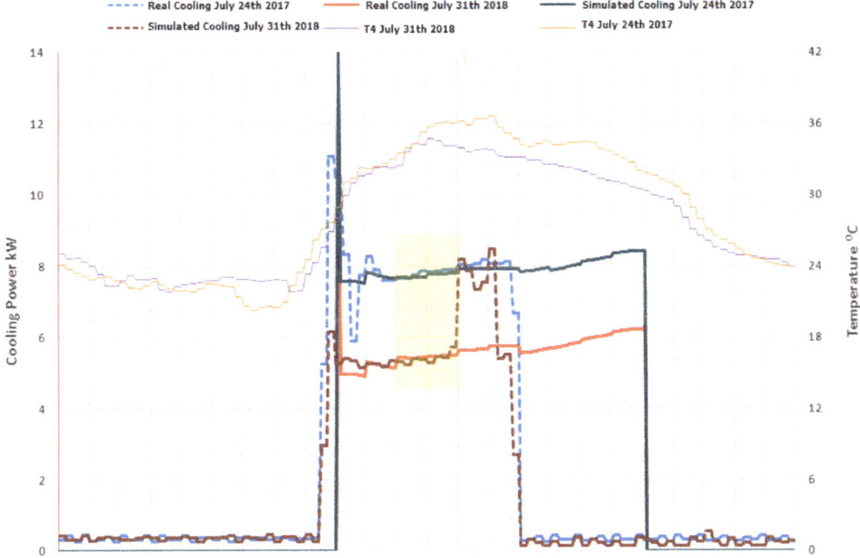

Figure 15. One-day validation of the models for conventional and green roof.

Figure 16 shows temperature evolution in the building outdoors (T4) and indoors (T5), as well as the temperature below the growth medium (T1). As can be observed, top roof temperatures decreased both in summer (40 °C to 35 °C) and winter (10 °C to 5 °C), which requires to compare the benefits for cooling in summer versus the negative effects in heating during winter period.

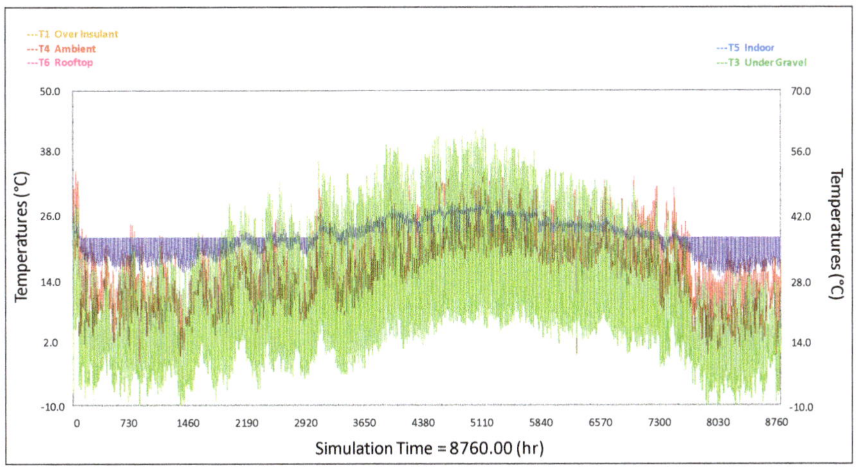

Figure 16. Results for temperatures evolution along the year.

Recorded data for cooling energy demand in both scenarios (conventional and green roof) showed an energy savings of approximately 25% in cooling energy demand, decreasing the maximum peak power demand by 33%. Heating energy demand in both scenarios (conventional and green roof) is presented in Figure 17. In this case, the results show that energy heating demand increased12% in the green roof scenario. Moreover, the maximum energy peak due to heating also rose6% with the green roof in comparison to the conventional rooftop, due to the reduction of solar heat gain reaching the building.

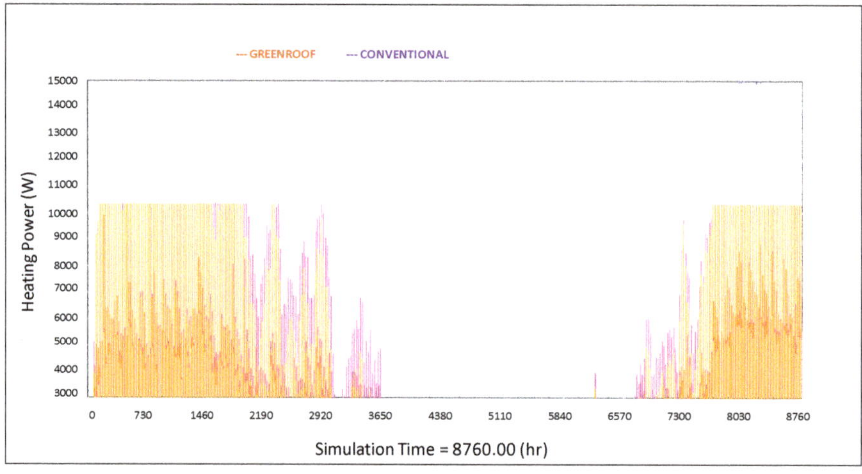

Figure 17. Results of the heating demand in both rooftops.

As a summary, results of the simulation are presented in Tables 4 and 5, which show total energy consumption, saving in cooling and heating mode and the average power saved along the operation time.

Table 4. Annual energy consumption for the conventional and green roof.

Roof Type	Summer (kWh)	Winter (kWh)	Annual (kWh)
Conventional Roof	12,400	6083	18,483
Green Roof	9300	6813	16,113

Table 5. Simulation results in energy savings (green roof vs. conventional rooftop).

Simulation Results	Summer	Winter	Annual
	Cooling Demand Reduction	Heating Demand Reduction	Total Reduction
Total Energy Savings (%)	25%	−12%	19%
Energy Reduction (kWh)	3100	−730	3780
Peak Power Savings (%)	33%	−6%	-

These results are compatible with the experimental values detailed at Table 2, which indicate a net gain in energy saving for the entire year, in the order of 19%. In contrast, an energy demand increase of 6% is noted, due to the requirement for additional heating in the winter period.

5. Conclusions

Consecutive experimental campaigns in a building with a conventional and with a green roof have allowed deducing the impact on the energy efficiency of the building air conditioning system due to green roof installation. In its application to a typical Mediterranean one-story building, insulation effects coming from the presence of the green roof introduced a small deterioration in that energy efficiency for the winter period, but showed clear improvements for the summer one. The global effect along the entire year is a net gain in the order of 19% for the energy consumption, but a 6% increase for the nominal power in the winter period. These results werededuced using a TRNSYS calculation, previously calibrated with the experimental data obtained for summer and winter periods. Therefore, in addition to the beneficial effects on the storm water control by reducing runoff and improving water quality, green roofs are also a significant element to improve energy efficiency in buildings and could help to mitigate urban heat island effect, while increasing urban biodiversity.

Author Contributions: J.C.-C. and E.P.-L. developed the methodology; J.C.-C., E.P.-L. and I.V.-S. prepared the conceptualization and data curation; E.H.-P., D.A.-S., J.C.-C. and E.P.-L. gathered and analyzed the data. J.C.-C. and E.P.-L. wrote the paper. All authors have read and agreed to the published version of the manuscript.

Funding: This work was supported by the European Union's Horizon 2020 research and innovation programme under the project Green Cities for Climate and Water Resilience, Sustainable Economic Growth, Healthy Citizens and Environments with reference 730283.

Conflicts of Interest: The authors declare no conflict of interest.

References

1. Jim, C.Y. Green roof evolution through exemplars: Germinal prototypes to modern variants. *Sustain. Cities Soc.* **2017**, *35*, 69–82. [CrossRef]
2. Dos Santos, S.M.; Silva, J.F.F.; dos Santos, G.C.; de Macedo, P.M.T.; Gavazza, S. Integrating conventional and green roofs for mitigating thermal discomfort and water scarcity in urban areas. *J. Clean. Prod.* **2019**, *219*, 639–648. [CrossRef]
3. Ferrans, P.; Rey, C.V.; Pérez, G.; Rodríguez, J.P.; Díaz-Granados, M. Effect of green roof configuration and hydrological variables on runoff water quantity and quality. *Water* **2018**, *10*, 960. [CrossRef]
4. Gómez, F.; Valcuende, M.; Matzarakis, A.; Cárcel-Carrasco, J. Design of natural elements in open spaces of cities with a Mediterranean climate, conditions for comfort and urban ecology. *Environ. Sci. Pollut. Res.* **2018**, *25*, 26643–26652. [CrossRef] [PubMed]
5. Chen, X.; Shuai, C.; Chen, Z.; Zhang, Y. What are the root causes hindering the implementation of green roofs in urban China? *Sci. Total Environ.* **2019**, *654*, 742–750. [CrossRef] [PubMed]
6. Radhi, H.; Sharples, S.; Taleb, H.; Fahmy, M. Will cool roofs improve the thermal performance of our built environment? A study assessing roof systems in Bahrain. *Energy Build.* **2017**, *135*, 324–337. [CrossRef]
7. Baik, J.J.; Kwak, K.H.; Park, S.B.; Ryu, Y.H. Effects of building roof greening on air quality in street canyons. *Atmospheric Environ.* **2012**, *61*, 48–55. [CrossRef]
8. Shafique, M.; Kim, R.; Rafiq, M. Green roof benefits, opportunities and challenges–A review. *Renew. Sustain. Energy Rev.* **2018**, *90*, 757–773. [CrossRef]
9. Goussous, J.; Siam, H.; Alzoubi, H. Prospects of green roof technology for energy and thermal benefits in buildings: Case of Jordan. *Sustain. Cities Soc.* **2015**, *14*, 425–440. [CrossRef]
10. Yang, W.; Wang, Z.; Cui, J.; Zhu, Z.; Zhao, X. Comparative study of the thermal performance of the novel green (planting) roofs against other existing roofs. *Sustain. Cities Soc.* **2015**, *16*, 1–12. [CrossRef]
11. Foustalieraki, M.; Assimakopoulos, M.N.; Santamouris, M.; Pangalou, H. Energy performance of a medium scale green roof system installed on a commercial building using numerical and experimental data recorded during the cold period of the year. *Energy Build.* **2017**, *135*, 33–38. [CrossRef]
12. Santamouris, M.; Pavlou, C.; Doukas, P.; Mihalakakou, G.; Synnefa, A.; Hatzibiros, A.; Patargias, P. Investigating and analysing the energy and environmental performance of an experimental green roof system installed in a nursery school building in Athens, Greece. *Energy* **2007**, *32*, 1781–1788. [CrossRef]
13. Zhao, X.; Zuo, J.; Wu, G.; Huang, C. A bibliometric review of green building research 2000–2016. *Archit. Sci. Rev.* **2019**, *62*, 74–88. [CrossRef]

14. Lee, E.J. A Study on Correlation between Improvement in Efficiency of PV and Green roof of Public Building. *KIEAE J.* **2013**, *13*, 111–118.
15. Feng, C.; Meng, Q.; Zhang, Y. Theoretical and experimental analysis of the energy balance of extensive green roofs. *Energy Build.* **2010**, *42*, 959–965. [CrossRef]
16. Tang, X.; Qu, M. Phase change and thermal performance analysis for green roofs in cold climates. *Energy Build.* **2016**, *121*, 165–175. [CrossRef]
17. Jim, C.Y.; Tsang, S.W. Ecological energetics of tropical intensive green roof. *Energy Build.* **2011**, *43*, 2696–2704. [CrossRef]
18. Bevilacqua, P.; Mazzeo, D.; Bruno, R.; Arcuri, N. Experimental investigation of the thermal performances of an extensive green roof in the Mediterranean area. *Energy Build.* **2016**, *122*, 63–79. [CrossRef]
19. Susca, T. Green roofs to reduce building energy use? A review on key structural factors of green roofs and their effects on urban climate. *Build. Environ.* **2019**, *162*, 106273. [CrossRef]
20. Ávila-Hernández, A.; Simá, E.; Xamán, J.; Hernández-Pérez, I.; Téllez-Velázquez, E.; Chagolla-Aranda, M.A. Test box experiment and simulations of a green-roof: Thermal and energy performance of a residential building standard for Mexico. *Energy Build.* **2020**, *209*, 109709. [CrossRef]
21. Tian, Z.; Lei, Y.; Gu, X. Building Energy Impacts of Simple Green Roofs in the Hot Summer and Cold Winter Climate Zone: Suzhou as a Study Case. *Procedia Eng.* **2017**, *205*, 2918–2924. [CrossRef]
22. Coma, J.; Pérez, G.; Solé, C.; Castell, A.; Cabeza, L.F. Thermal assessment of extensive green roofs as passive tool for energy savings in buildings. *Renew. Energy* **2016**, *85*, 1106–1115. [CrossRef]

Article

Experimental Study on Energy Efficiency of Multi-Functional BIPV Glazed Façade Structure during Heating Season

Suzana Domjan, Lenart Petek, Ciril Arkar and Sašo Medved *

Laboratory for Sustainable Technologies in Buildings, Faculty of Mechanical Engineering, University of Ljubljana, SI-1000 Ljubljana, Slovenia; suzana.domjan@fs.uni-lj.si (S.D.); lenart.petek508@gmail.com (L.P.); ciril.arkar@fs.uni-lj.si (C.A.)

* Correspondence: saso.medved@fs.uni-lj.si

Received: 29 April 2020; Accepted: 26 May 2020; Published: 1 June 2020

Abstract: Building integrated photovoltaics (BIPV) is technology that can significantly increase the share of renewable energy in final energy supply and are one of essential technologies for the nearly zero-energy buildings (nZEB), new build and refurbished. In the article (a) an experimental semitransparent BIPV glazed façade structure with 60% of PV cell coverage is shown; (b) energy efficiency indicators were developed based on identified impact parameters using experimental data; and (c) multi-parametric models of electricity generation, preheating of air for space ventilation, and dynamic thermal insulation features that enable prediction of solar energy utilization in different climate conditions are shown. The modeled efficiency of electricity production of BIPV was in the range between 8% and 9.5% at daily solar radiation above 1500 Wh/day, while low impact of outdoor air temperature and ventilation air flow rate on PV cell cooling was noticed. Between 35% and 75% of daily solar radiation can be utilized by preheating the air for space ventilation, and 4.5% to 7.5% of daily solar radiation can be utilized in the form of heat gains through opaque envelope walls.

Keywords: nZEB, BIPV; room ventilation; dynamic thermal insulation; multi-parametric model

1. Introduction

Buildings in Europe are responsible for 36% of all greenhouse gas emissions. To fulfil targets presented in the Paris climate agreement, emissions in the building sector must be decreased by 90% [1]. Building integrated photovoltaics (BIPV) in form of façade structures are solutions that can significantly contribute to this goal, as well as increase the share of renewable energy in the final energy supply. As such, BIPV are one of the essential technologies for the nearly zero energy buildings. Regarding to the structure of building stock and for ensuring the cost effectiveness of BIPV in general, solutions for refurbishment of buildings are of great interest [2]. Among several design options, BIPV glazed façade with a natural or forced ventilated air gap has several comparative advantages and are from the architectural perspective upgraded double ventilated façades [3].

One way to fulfil this goal is in the multi-functionality of BIPV solutions. Relative low efficiency of solar energy utilization with PV cells can be improved by solar concentrators or tracking devices, although in case of BIPV applications, it is more convenient to upgrade PV modules to combine power and heat generators to so called photovoltaic thermal building structures (BIPV/T). The liquid heat transfer media can be used to supply the heat to the buildings [4,5], although preheating of ventilation air for building ventilation is a better option because such applications operate as a low exergy system [6]. In [7] opaque PV modules are cooled by air flowing through the forced ventilated air gap and authors report that the overall energy efficiency during the winter months was in the range between 48% to 52% on the monthly basis. Analysis of natural ventilated semitransparent

BIPV designed as a double façade structure is shown in [8] showing energy and environmental advantages of semitransparent BIPV over the opaque PV modules. In [9] authors propose solutions for forced ventilated close loop BIPV/T based on the analytical modeling and point out the need for experimental validation.

The second most common cited advantage of ventilated BIPV structures is the increase of the PV cell efficiency by cooling. Buoyancy driven natural ventilated semitransparent BIPV consisting of see-through a-Si PV cells was studied by [10]. Authors have shown that daily energy output can be increased by 1.9% to 3% due to the lower operating PV cell temperature. Ventilated BIPV façade was studied by [11] and authors claim that the PV modules efficiency can be increased by 2.2% on the annual basis in case of natural ventilation and up to 4.7% to 5.7% in cases of forced ventilation with different air flow rates. Similar results, 2.5% increase in annual electricity production by ventilated façade mounted opaque PV modules, are reported in [12].

Ventilation of buildings significantly contributes to the wellbeing in buildings. Not only bioeffluents, but pollutants such as formaldehyde and odors, can be efficiently removed from indoor air [13]. Proper indoor air quality (IAQ) increases the occupant's productivity as well [14]. Mechanical ventilation systems with heat recovery could efficiently reduce heat demand, but significantly increase the electricity demand, especially in commercial buildings. Nevertheless, in [15] it is shown that in all-glass buildings with BIPV façade structures, electricity demand for ventilation, even in case of the ventilation systems that fulfill present requirements about energy efficiency, is dominant compared to heating, cooling, and lighting systems, measured by primary energy needed. Central mechanical ventilation systems are difficult to adjust to the presence of occupancies and their personal physiological needs. Furthermore, decentralized ventilation can be energy efficient in most European climate regions [16]. Dynamic insulation is a part of the building envelope where outdoor air passes through a porous thermal insulation layer towards the interior and redirects heat loss flux. Research on this technique is shown in [17]. The authors have shown that dynamic thermal transmittance of a ventilated structure having static U 0.3 W/m^{2}K decreases to 0.15 W/m^{2}K at air flow rate 0.75 1/s per m^2 of the building structure area. Similarly, transmission heat losses of the envelope building structure can be decreased if the air gap, designed inside the building structures, is ventilated by outdoor air, and then transferred into the buildings as preheated air. In [18] building ventilated opaque BIPV was studied at a steady state outdoor temperature, solar irradiation, and wind velocity. Indoor heat gains were investigated using computational fluid dynamics (CFD) techniques and compared to static heat losses expressed by thermal transmittance U of the building envelope structure.

In the presented article, a multi-functional modular semitransparent BIPV glazed façade structure was designed, built, in-situ tested, and analyzed during winter-time conditions. M-Si PV cells are built in the BIPV in form of double side glass laminated PV module with 60% PV cell packing factor, which make the BIPV is semitransparent. Modular units can be multiplied according to needs of particular flat/office/building in new, as well as in renovated buildings. Thick temperate glass layers (4 mm each) of BIPV also result in the specific thermal response of the structure. Several aspects of functionality of solar energy utilization are addressed and evaluated, such as (a) electricity production, preheating of air for space ventilation and dynamic thermal insulation performance; (b) overall efficiency of solar energy utilization is determined in form of approximation multi-parametric model, providing a tool for evaluation of such BIPV glazed façade structure in different climate conditions; and (c) all models are developed on the base of diurnal averaging of independent variables, which enables integration in buildings thermal response models.

2. Object of Research and Research Methods

2.1. Semitransparent BIPV Glazed Façade Structure with Forced Ventilated Air Gap

In general term, BIPV are multifunctional devices because they incorporate passive functions of ordinary façade (or roof) structures, for example, precipitation and sound protection with active renewable energy utilization. In this way lower production of electricity compared to self-stand systems, as a consequence of the position of installation defined by the building envelope, can be compensated at least in terms of investment. Nevertheless, in the presented article, the multifunctional nature of the examined BIPV glazed façade structure is evaluated in terms of utilization of solar energy thereby increasing energy efficiency of the building, while improving indoor living comfort. A pilot BIPV glazed façade structure was designed to fit new buildings and could be used for energy refurbishment as well, possibly eliminating the need for additional thermal insulation. It consisted of a transparent glass façade with integrated PV cells and forced ventilated air gap which, beside electricity production, enabled preheating of fresh supply air and it to act as dynamic thermal insulation (Figure 1). In this way, the heat losses of façade envelope decrease as consequence of the lower (static) thermal transmittance U_{st} and because part of transition heat losses preheats the air that flows inside the ventilated air gap. Both effects are evaluated with dynamic thermal transmittance U_{eff}. To increase the efficiency of solar energy utilization, the BIPV was designed as a semitransparent structure with 60% of opaque PV cell area. Another reason for this BIPV design follows from optimization of multilayer glazing according to the natural heating, shading, daylight, and occupancies view towards the outdoor environment [15]. A similar conclusion is presented in [8]. In this way, both the ventilated BIPV on the opaque façade structure and BIPV glazing can be combined with the same architectural appearance of the building. BIPV were produced by [19] and consist of two 4 mm hard glass panes and encapsulated layer. Monocrystalline silicon cells with reference efficiency 18.5% [19] and size 156 × 156 mm are installed in the BIPV glazed structure.

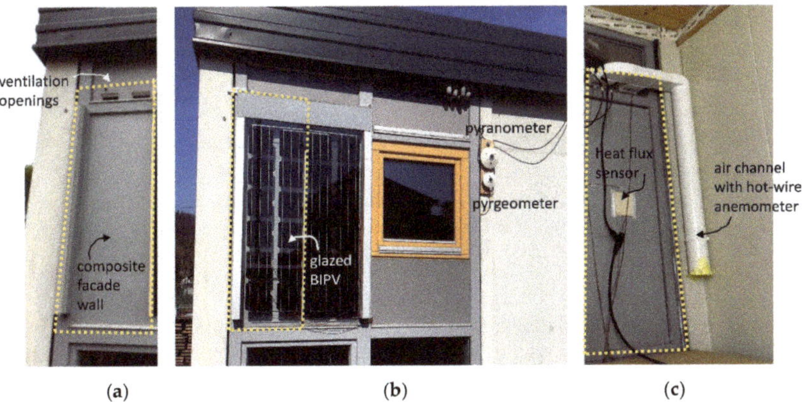

Figure 1. Experimental semitransparent building integrated photovoltaics (BIPV) glazed façade structure (shown by the rectangle section). (**a**) The façade wall with two ventilation openings, each of them was enclosed by a fan; (**b**) BIPV glazed structure installed 80 mm in front of the façade wall, and (**c**) interior of experimental BIPV façade structure—air channels were thermal insulated during the experiment.

The BIPV glazed façade structure was installed in front of the opaque south orientated façade wall of laboratory unit in the way, and an 80 mm thick air gap, which was between BIPV glazed structure and façade wall, was formed. The façade wall (Figure 1a) consisted of a thermal insulation layer (d 0.035 m, λ 0.035 W/mK), lined by an inner and outer layer of solid wood (d 0.005 m, λ 0.14 W/mK) made of two transversely glued soft wood plates. Such building structures are often installed as opaque parapet as part of glazed building façade structures, since its thermal transmittance U_{st} (1.027 W/m^{2}K) does not exceed the common required level for glazed façades (e.g., in Slovenia U_{max} equals 1.3 W/m^{2}K). The surface of the structure has absorptivity of solar irradiation α_s 0.65 and emissivity of IR irradiation ε_{IR} 0.9, which is close to that of concrete façade structures. The section of the experimental BIPV glazed façade structure is 0.435 m wide and 1.167 m high, with area A_{BIPV} 0,508 m^2. All support structures, such as the installing frame and outdoor and indoor air channels were thermal insulated to keep heat transfer close to the 2D problem.

The air gap was forced ventilated by two DC fans with power of 2 W at a supply voltage of 12 V. By changing the supply voltage, ventilation air flow rate $\dot{V}_{a,in}$ was set on a daily basis to the value between 0 m^3/h and 61.5 m^3/h and kept constant all day long. If reverse flow fans were used instead, the overheating protection by forced ventilation of the air gap with cooler indoor air instead of warmer outdoor air, could be applied. The thermal response of the BIPV glazed façade structure in case of the non-ventilated air gap was tested to examine the static thermal transmittance U_{st} of the structure, while other discrete values of ventilation air flow rates were selected based on the indoor air quality (IAQ). It was assumed that the office with a net volume V_n of 35 m^3 and a useful area A_u 14 m^2 will be equipped by the pilot BIPV shown in Figure 1. One person (1.2 met, 1 clo) occupied the office between 8:00 and 17:00 and emitted S_{CO2} 800 mg of CO_2 per minute [20]. No brake or leaving from the office was assumed. When the person started to work, the $CO_{2,8:00}$ concentration was equal to the outdoor concentration 500 ppm. The first order concentration model was used to determine transient $CO_{2,t}$ concentrations assuming constant conservative (decay factor $k = 0$) pollutant source S_{CO2} [21]:

$$C_{CO2,t\to\infty} = \frac{S_{CO2} + C_{CO2,8:00} \cdot \dot{V}_{a,i}}{\dot{V}_{a,i} + \underbrace{\underbrace{k.V_n}_{decay}}_{0}} \left(\frac{\mathrm{mg}}{\mathrm{m}^3}\right) \tag{1}$$

$$C_{CO2,t} = C_{CO2,t\to\infty} + \left(C_{CO2,t=0} - C_{CO2,t\to\infty}\right) \cdot e^{-\frac{\dot{V}_{a,i}}{V_n} \cdot t} \left(\frac{\mathrm{mg}}{\mathrm{m}^3}\right). \tag{2}$$

Taking into account IAQ quality categories as defined in [22], the $CO_{2,17:00}$ concentration that appeared in the office at the end of the workday (at 17:00) did not exceed class III (1350 ppm above outdoor concentration) if office was ventilated with constant air flow rate $\dot{V}_{a,in}$ 19.5 m^3/h, and class I requirements (550 ppm above outdoor concentration) were achieved if office was ventilated with constant air flow rate $\dot{V}_{a,in}$ 62 m^3/h.

2.2. Experiment Setup

The pilot BIPV glazed façade structure was installed on the south oriented façade wall of laboratory building (Figure 1). The indoor environment was heated and cooled with a split air–air heat pump. An appliance built-in control unit was used to control indoor air temperature and because large solar gains were caused by other glazed structures, the laboratory building was often chilled at noon. This resulted in indoor air temperature periodical oscillations, but we want to point out that at least Class II [22] of thermal comfort was achieved during the experiment. BIPV and the laboratory building were equipped with sensors shown in Figure 2. Global solar radiation on the vertical surface was measured with a Kipp & Zonen CMP3 pyranometer (measurement uncertainty ± 5%) [23]. Downward atmospheric long-wave radiation was measured with a Kipp & Zonen CG1 pyrgeometer (measurement uncertainty ± 4%). Other meteorological parameters were measured using a Vantage Pro 2 weather station: ambient air temperature (± 0.5 °C) and wind velocity (± 5%) [24]. The weather station was installed on the roof of the laboratory test building (Figure 2b). Temperatures of air (indoor and in the ventilated air gap) and surface temperatures (BIPV and façade wall) were measured with calibrated K-type thermocouples (± 0.25 °C). Heat flux at the interior surface of the composite façade wall was measured with an AHLBORN FQ A018 C sensor (± 8%) [25]. Air velocity in the middle of the round tube was measured with an Almemo thermoanemometer FV A645 TH3 (± 3%). All measured data was monitored in one-minute intervals, using a data acquisition units Agilent 34970A [26] and AHLBORN Almemo 2290-4, the latter only for the thermoanemometer.

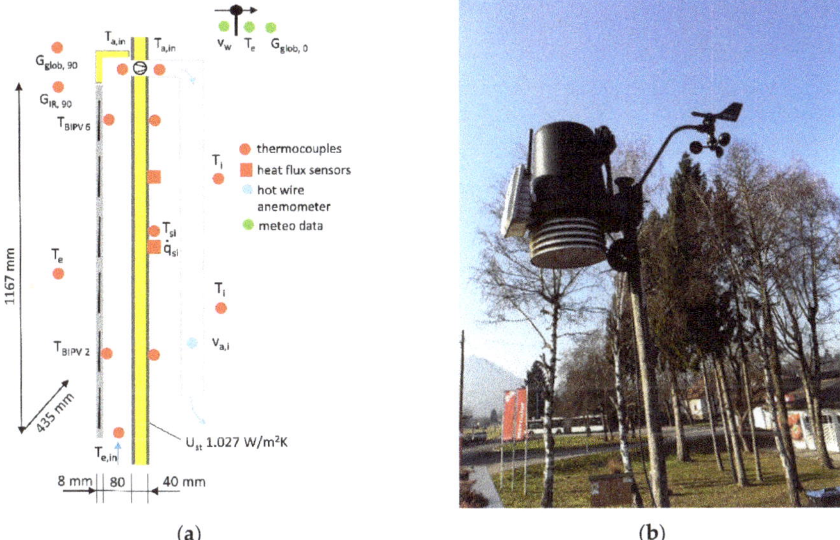

(**a**) (**b**)

Figure 2. Position of sensors (**a**) and meteorological station Vantage Pro 2 (**b**).

2.3. Research Methods

The energy efficiency of functionality mode, as well as total energy efficiency of solar energy utilization, were determined on the basis of an in situ experiment. The efficiency of PV was determined on the basis of measured transmittance of solar irradiation and temperature of the inner surface of BIPV glazing $T_{PV,si}$. The in situ experiment was performed between 25 December 2019 and 15 April 2020. Measured data were gathered in 1-minute intervals (Δt_{meas}). The solar irradiation $G_{glob,90}$ and outdoor air temperature T_e during the experiment are shown in Figure 3. The range of meteorological parameters appearing during the experiments are shown in Table 1.

Energy efficiency indicators were developed as diurnal values by summarizing and averaging the measured data. Although most indicators involve diurnal average values, some of them are developed using shorter time intervals. For example, the efficiency of electricity production indicators involve average meteorological data for the day-time period when $G_{glob,90} > 0$ W/m^2 or the efficiency indicators related to preheating of the ventilation air were developed taking into account the occupancy period in office buildings (8:00 to 17:00).

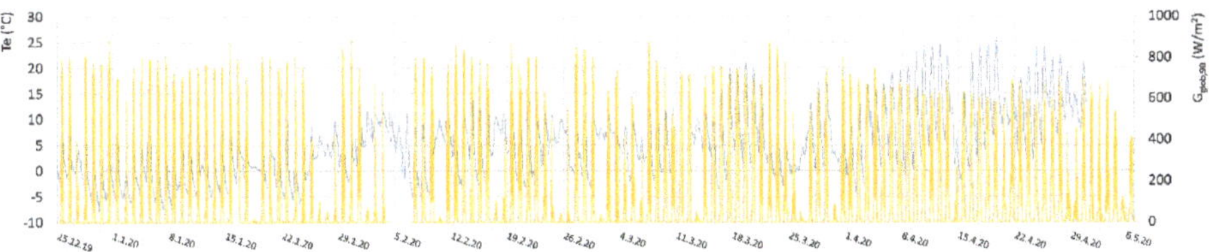

Figure 3. Solar irradiation received by BIPV $G_{glob,90}$ and outdoor air temperature T_e, which occurred during the experiment.

Table 1. Average and extreme values of meteorological variables during the duration of the experiment.

Meteorological Variable	Average Exp. Period	Max Daily av.	Min Daily av.	Max	Min
Outdoor air temperature T_e (°C)	5.75	16.3	−2.8	25.6	−8.7
Solar irradiation $G_{glob,90}$ (W/m^2)	113_{24h}			841	46
Solar radiation $H_{glob,90,day}$ (Wh/m^{2}day)	2706			5410	274
IR sky radiation $H_{IR,90,day}$ (Wh/m^{2}day)	7858			8791	6924
Wind velocity v_w (m/s)	0.49	1.84	0.03	5.6	0.0

In the second evaluation step, statistical methods were used to define the influential parameters for each of the energy efficiency indicators.

In the final evaluation step, influential parameters were involved as independent variables in the statistical evaluation of multi-parametric approximation models developed for each of energy efficiency indicators. These approximation models can be integrated into models for determining the energy performance of buildings on the basis of daily energy balance and can be used for climate conditions at least in the range of values of meteorological parameters as listed in Table 1.

3. Energy Efficiency Indicators

Although experimental data was gathered in one-minute intervals, the energy efficiency of solar energy utilization with the pilot BIPV glazed façade structure, and the indicators of each operation mode of the BIPV façade structure are presented as daily average or integral values. As such, indicators can be implemented in a monthly calculation procedure, which is still commonly used in engineering practice, and can be used for assessment of nZEB as well [27]. The indicators are schematically presented in Figure 4. In addition to those solar energy utilization indicators for which the approximation models were developed, some other are shown to emphasize the advantages of the pilot BIPV glazed façade structure.

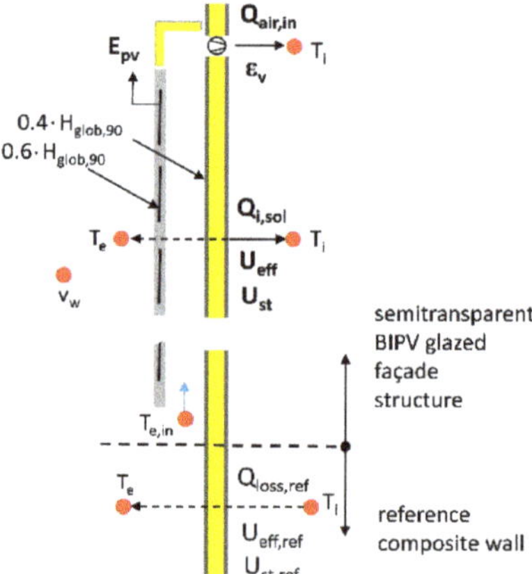

Figure 4. Scheme of indicators of the efficiency of solar energy utilization with the pilot semitransparent BIPV glazed façade structure; diurnal electricity production E_{PV}; diurnal heat supplied with preheated air for space ventilation $Q_{a,in}$; and heat gains through the opaque façade wall $Q_{i,sol}$ were the basis for developing approximation models of energy efficiency, while others, such as static U_{st}, dynamic thermal transmittance U_{eff}, and preheating efficiency ε_v, are used to emphases the advantage of the pilot semitransparent BIPV glazed façade structure.

3.1. Electricity Production

The amount of diurnal produced electricity was determined by analytical model considering the measured values of influenced parameters in each time step of observation Δt_{meas}. Daily amount of produced electricity is determined by equation:

$$E_{PV} = \frac{1}{60}\sum_{tsr\prime}^{tss\prime} A_{PV} \cdot n_{PV} \cdot K_T \cdot K_G \cdot G_{glob,\beta}\Delta t_{meas} \left(\frac{\text{Wh}}{\text{day}}\right), \quad (3)$$

where *tsr′* and *tss′* are sunrise and sunset time relative to the BIPV structure, respectively, indicating the time frame when PV cells produce electricity. K_T is the efficiency factor that corresponds to corrected PV cell efficiency and includes temperature coefficient β, which depends on PV cell technology. Value −0.46%/K was assumed for m-Si cells [28–30]. A_{PV} and n_{PV} are the area of an individual PV cell (0.156 × 0.156 m) and the number of PV cells in the BIPV [15,19]. Because BIPV has relatively thick glass layers (4 + EPA + 4 mm, λ 0.76 W/mK), the PV cell temperature T_{PV} was modelled by combining the heat transfer model and measurements of surface temperature on the outer and inner glass. The surface glass temperatures were measured behind the 2nd row and 6th row of the PV cell ($T_{BIPV,2}$, $T_{BIPV,6}$). It was found that in the case of ventilated air gap, there was not a significant difference between both temperatures, and an average value was used as the representative surface glass temperature. According to the CFD computer simulations, the combined surface heat transfer coefficient $h_{r+c,e}$ 15 W/m^{2}K and $h_{r+c,i}$ 6 W/m^{2}K were assumed and PV cell temperature T_{PV} is approximated in the following way:

$$K_T = \eta_{ref} \cdot \left(1 + \beta \cdot \left(\left(\overbrace{T_{PV,si} + 0.0064 + 0.0013 \cdot G_{glob,\beta}}^{T_{PV}}\right) - 25\right)\right) \quad (-), \quad (4)$$

where $T_{PV,si}$ is temperature of the BIPV surface behind the PV cell towards the air gap. The reference efficiency η_{ref} was taken from producer data [19] and is equal to 0.185. Solar irradiation correction factor K_G considers the decrease of PV cell efficiency at low level of solar irradiation [15]:

$$K_g = 1 \text{ if } G_{glob,90} \geq 200\frac{\text{W}}{\text{m}^2} \text{ and } \frac{0.029 \cdot \ln\left(G_{glob,90}\right) - 0.0037}{\eta_{ref}} \text{ if } G_{glob,90} < 200\frac{\text{W}}{\text{m}^2}. \quad (5)$$

To investigate the impact of ventilation of the air gap on PV cell overheating, diurnal overheating hours *OHH* was introduced. *OHH* is defined as diurnal sum of the difference between modeled PV cell temperature and PV cell reference temperature 25 °C:

$$OHH = \frac{1}{60}\sum_{tsr\prime}^{tss\prime} \delta \cdot (T_{PV} - 25) \cdot \Delta t_{meas}\ \delta = 1 \text{ if } (T_{PV} - 25) > 0 \text{ otherwise } \delta = 0 \left(\frac{\text{Kh}}{\text{day}}\right). \quad (6)$$

As an example, Figure 5 shows measured data for two selected days—the clear sky and overcast cloudy day, and diurnal variables involved in energy efficiency modeling—daily solar radiation received by BIPV $H_{glob,90}$, average daily outdoor air temperature during PV cell operation $T_{e,avg,PV}$, average wind velocity during PV cell operation time $v_{W,avg,PV}$ (m/s), and $\dot{V}_{a,in}$ ventilation air flow rate.

In the case presented in Figure 5a, BIPV was not ventilated, while in the case shown in Figure 5b BIPV was ventilated with air flow rate $\dot{V}a, in$. Concerning the surface glass temperature behind the PV cells ($T_{BIPV,2}$ and $T_{BIPV,6}$), it can be seen that temperatures differ only in the case of non-ventilated (closed) air gap (mark c) as consequence of buoyancy driven convection—by solar irradiation during day-time and by heat flux due heat losses during the night-time. This causes counter flow pattern during the night-time and higher $T_{BIPV,6}$ when compared to $T_{BIPV,1}$.

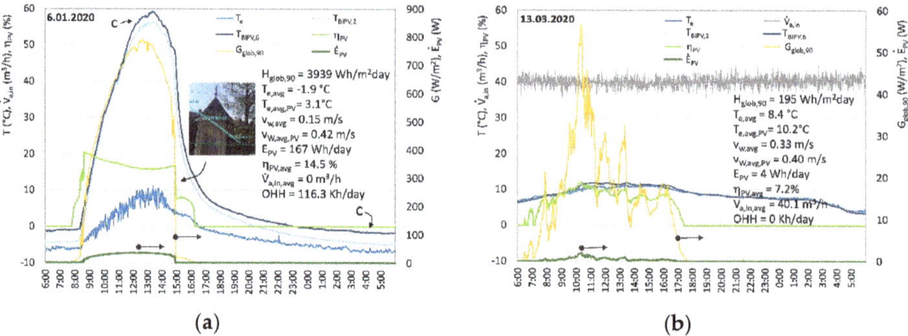

Figure 5. Graphs showing instant η_{PV} and average diurnal PV cell efficiency $\eta_{PV,avg}$, instant electricity power $\dot{E}_{PV}$, and diurnal production of electricity E_{PV} (**a**), PV cell temperatures (glass surface temperatures behind PV cell towards ventilated air gap) in the 2nd ($T_{BIPV,2}$) and the 6th row ($T_{BIPV,6}$) in BIPV, overheating hours (*OHH*) and selected meteorological data—solar irradiation $G_{glob,90}$, diurnal solar radiation $H_{glob,90}$, and instant and daily average outdoor air temperature (T_e, $T_{e,avg}$) (**b**).

3.2. Preheating of Ventilation Air

The air gap formed by BIPV was force ventilated by outdoor air. Two by two temperature sensors were installed at the outlet openings, on both sides of fans. It was found that an increase of the air temperature caused by fans can be neglected. Data on air velocity in the center of the supply pipe, gathered by a hot-wire anemometer, was used to determine volumetric air flow using a continuity equation. Because air flow is turbulent even in case of lowest flow rate set, the volume (and mass) air flow rate was determined by averaging velocity using the Blasius formula and continuity equation. Daily heat transferred into the building by preheated air was determined by the sum of one-minute experimental data over the 24-hour period starting each day at 6:00 in the morning:

$$Q_{a,in} = \frac{1}{60} \sum_{6:00}^{+6:00} \frac{1}{3600} \cdot \rho_a \cdot c_{p,a} \cdot \dot{V}_{a,in} \cdot (T_{a,in} - T_e) \cdot \Delta t_{meas} \left(\frac{\text{Wh}}{\text{day}}\right), \tag{7}$$

where ρ_a is air density, $c_{p,a}$ specific heat capacity of air, $\dot{V}_{a,in}$ is volume air flow rate, $T_{a,in}$ is supply air temperature, and T_e is outdoor temperature. Preheating efficiency of ventilation air, value that can be compared to the heat recovery efficiency in case of mechanical ventilation with recovery unit, is defined by averaging supply air, outdoor air, and indoor air temperatures over the occupied office hours (8:00–17:00), assuming constant ventilation air flow rate during observation period:

$$\varepsilon_v = \left(\frac{T_{a,in,avg} - T_{e,avg}}{T_{i,avg} - T_{e,avg}}\right)_{8:00}^{17:00} \cdot 100(\%). \tag{8}$$

Preheating efficiency ε_v can be above 100% if $T_{a,in,avg} > T_{i,avg}$, and $T_{e,avg} < T_{i,avg}$. Figure 6 shows an example of experimental data for the selected days. The average air inlet temperature during work-hours $T_{a,in,avg}$, the integrated solar radiation $H_{glob,90}$ received by the BIPV, and heat transferred by air into the building $\dot{Q}_{a,in}$ are shown as well. Because average supply air temperature $T_{a,in,avg}$ in Figure 6b is above average indoor air temperature $T_{i,avg}$ while average outdoor temperature $T_{e,avg}$ is below $T_{i,avg}$, the air preheating efficiency is above 100%.

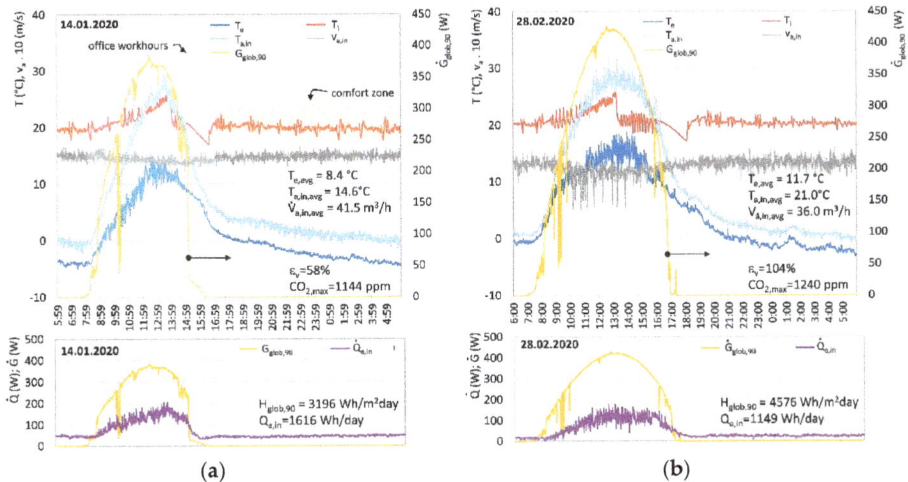

Figure 6. Graphs showing outdoor and indoor air temperatures and the temperature of supplied preheated air (T_e, T_i, and $T_{a,in}$); velocity, ventilation air volume flow rate, and heat flux by preheated air (v_a, $\dot{V}_{a,in}$, and $\dot{Q}_{a,in}$); average ventilation air temperature during office occupant hours $T_{a,in,avg}$; air preheating efficiency (ε_v); as well as solar irradiation $G_{glob,90}$; and diurnal solar radiation $H_{glob,90}$; as well as diurnal heat transferred into the ventilated space by preheated air $Q_{a,in}$; (**a**) for 14 January 2020 and (**b**) for 28 February 2020.

3.3. Dynamic Thermal Insulation

Because building envelope structure is ventilated and air is supplied to the indoor space, part of the heat losses can be recovered and the BIPV structure acts as dynamic thermal insulation. Efficiency of heat loss recovery is determined by comparing daily actual heat losses to theoretical ones at steady state conditions, taking into account reference thermal transmittance of the composite façade wall, which was upgraded with the BIPV structure—U_{st} 1.027 W/m^{2}K (Figure 1). Transmission heat losses of the BIPV structure were evaluated by static thermal transmittance U_{st}, while effective thermal transmittance U_{eff} was evaluated based on the transient thermal response of the BIPV structure. U_{st} was determined by instant indoor T_i and outdoor T_e air temperatures and heat flux was measured on the internal surface of the structure $\dot{q}_{si}$ using measured values for the period between 23:00 and 6:00+ o'clock in each day, to minimize the impact of accumulated solar energy. At that period, the heat transfer was close to the steady state, because the temperature difference ($T_i - T_e$) was almost constant. With the U_{st} value, the impact of the double skin façade, as well as the forced ventilation of the air gap, was evaluated and compared to that of the U_{st} of the reference building envelope structure (Figure 2a). The dynamic thermal transmittance U_{eff} was defined in the similar way, the only difference was in the evaluation period, which was in this case all day long (6:00 to 6:00+1 day). The following equations were used:

$$U_{st} = \frac{1}{7}\frac{1}{60}\sum_{23:00}^{+6:00}\frac{\dot{q}_{si}}{(T_i - T_e)} \cdot \Delta t_{meas}\left(\frac{\mathrm{W}}{\mathrm{m^2K}}\right), \tag{9}$$

$$U_{eff} = \frac{1}{24}\frac{1}{60}\sum_{6:00}^{+6:00}\frac{\dot{q}_{si}}{(T_i - T_e)} \cdot \Delta t_{meas}\left(\frac{\mathrm{W}}{\mathrm{m^2K}}\right). \tag{10}$$

The impact of unsteady parameters ($H_{glob,90}$, T_e, and v_W), as well as impact of ventilation air flow rate $\dot{V}_{a,in}$, are considered by the effective thermal transmittance U_{eff}. Examples of evaluation are shown in Figure 7.

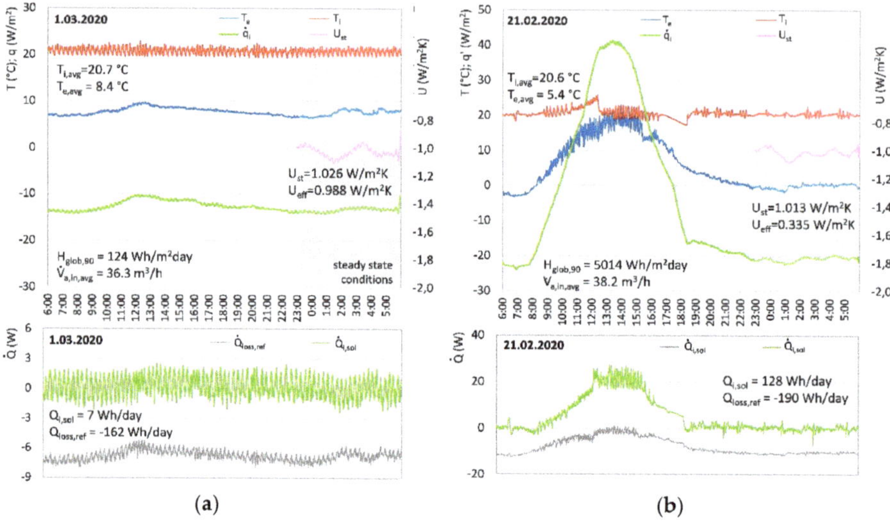

Figure 7. An example of measured values used for the evaluation of the energy efficiency of dynamic thermal insulation of the BIPV glazed façade structure. (**a**) In the case of heavy cloudy day (1 March 2020), because the air gap was ventilated, U_{st} was equal to reference thermal transmittance of composite façade structure (U_{ref} 1.027 W/m^{2}K) and U_{eff} was very close to the U_{st}; (**b**) even in the case of clear sky weather (21 February 2020), the U_{st} was very close to the reference value, while dynamic thermal transmittance was much lower, indicating a significant difference in diurnal transmission heat loss $Q_{i,sol}$ from the pilot BIPV glazed façade structure.

3.4. Overall Efficiency of Solar Energy Utilization

Overall efficiency of solar energy utilization by the pilot BIPV glazed façade structure includes energy gains related to production of electricity, preheating of ventilation air, and decreased transmission heat losses due to the dynamic thermal insulation. In fact, the latter is not achieved solely by utilization of solar energy but also due to lower thermal transmittance of the BIPV structure, nevertheless all indicators are normalized to diurnal received solar radiation $H_{glob,90}$. The overall efficiency is defined by the sum of partial efficiencies in the following way:

$$\eta_{sol.BIPV} = \eta_{PV,BIPV} + \eta_{a,BIPV} + \eta_{i,sol,BIPV} = \left(\frac{E_{PV} + Q_{a,i} + \left| \left(Q_{loss,ref} - Q_{loss,i} \right) \right|}{H_{glob,90} \cdot A_{BIPV}} \right) (-), \tag{11}$$

where:

$$\eta_{PV,BIPV} = \frac{E_{PV}}{H_{glob,90} \cdot A_{BIPV}} = \eta_{PV} \cdot \left(\frac{\overbrace{A_{PV} \cdot n_{PV}}^{0,6 \cdot A_{BIPV}}}{A_{BIPV}} \right) = K_T \cdot K_G \cdot \left(\frac{\overbrace{A_{PV} \cdot n_{PV}}^{0,6 \cdot A_{BIPV}}}{A_{BIPV}} \right) (-), \tag{12}$$

$$\eta_{a,BIPV} = \frac{Q_{a,in}}{H_{glob,90} \cdot A_{BIPV}} = \frac{0.34 \cdot \frac{1}{60} \sum\limits_{8:00}^{17:00} \dot{V}_{a,in} \cdot (T_{a,in} - T_e) \cdot \Delta t_{meas}}{H_{glob,90} \cdot A_{BIPV}} (-), \tag{13}$$

$$\eta_{i,sol,BIPV} = \frac{\left|\left(\overbrace{U_{st,ref}}^{1{,}027\mathrm{W/m^2K}} \cdot \left(T_{i,avg} - T_{e,avg}\right) \cdot 24 - \left(\frac{1}{60} \sum_{6:00}^{+6:00} \dot{q}_{si} \cdot \Delta t_{meas}\right)\right) \cdot A_{BIPV}\right|}{H_{glob,90} \cdot A_{BIPV}} \ (-). \quad (14)$$

The constant 0.34 replaces the product of air density and specific heat capacity. Figure 8 shows an example of overall efficiency of solar energy utilization $\eta_{sol,BIPV}$ determined by measured data in two days during the experiment period.

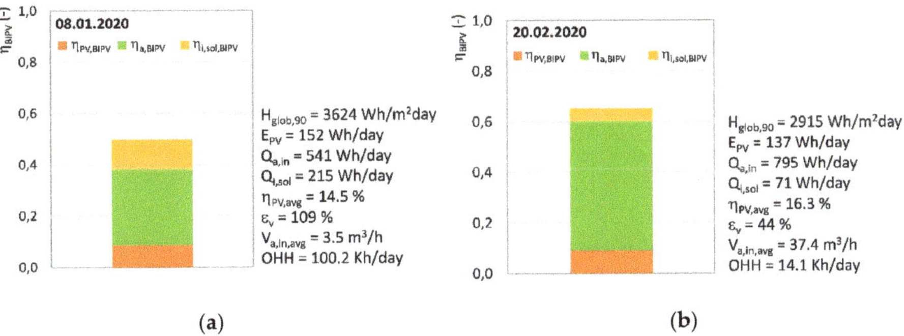

(a) (b)

Figure 8. Overall efficiency of solar energy utilization η_{BIPV} of the pilot BIPV glazed façade structure (**a**) on 8th January 2020 and (**b**) on 20th February 2020; in the case (**a**) the ventilation air flow rate $\dot{V}_{a,in}$ was 3.5 m^3/h, and consequently, the utilization of solar energy for preheating of the ventilation air decreased, while the temperature of the supply air $T_{a,in}$ increased above indoor air temperature and efficiency ε_V was above 100%; the case (**b**) was the opposite because $\dot{V}_{a,in}$ was much higher and utilization of solar energy was higher, while preheating air efficiency ε_v was significantly lower.

4. Results and Discussion

4.1. Parametric Study

The impact of daily solar radiation, average daily outdoor air temperature, and ventilation air flow rate on the static thermal transmittance U_{st} was analyzed and shown in Figure 9a. The U_{st} was determined from measured data, gathered between 23:00 and 6:00 the next morning. It can be concluded that the U_{st} is practically independent of daily solar radiation $H_{glob,90}$ (c2), and values in the middle of the air flow rate range are ~ 0.04 W/m^{2}K lower when compared to that of the reference composite façade wall (c1) due to the increased thermal resistances of the BIPV and air gap. No significant impact of mid-range daily average outdoor air temperatures $T_{e,av}$ can be seen either. This result corresponds to the fact that composite wall is light-weight with limited potential for storing the (solar) heat. This also confirms that the thermal response through the previous day does not affect the heat response of the BIPV the next day, leading to the conclusion that energy efficient indicators can be determined by averaging data over the proposed time period (6:00 to 6:00+). The impact of the air flow rate $V_{a,i}$ can be noticed only when the air gap was not ventilated (U_{st} is lowered to the range between 0.80 to 0.85 W/m^{2}K) and at its highest air flow rate, at which U_{st} increases to 10 W/m^{2}K (c3). Some additional data would increase the credibility of this finding.

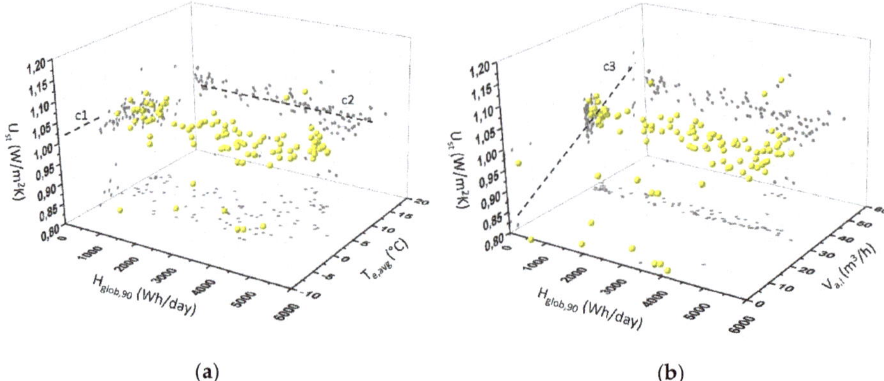

(a) (b)

Figure 9. Static thermal transmittance U_{st} of the glazed BIPV façade structure with a forced ventilated air gap; (**a**) impact of the daily solar radiation $H_{glob,90}$ received by the BIPV façade structure and average daily outdoor air $T_{e,avg}$ (equal to the temperature of the ventilation air at the inlet of the air gap); (**b**) impact of the solar radiation and ventilation air flow rate $V_{a,i}$.

The dynamic thermal transmittance U_{eff} shows a significantly greater dependence on the influencing parameters (Figure 10). In theory, at low daily solar radiation ($H_{glob,90}$ < 300 Wh/day) dynamic thermal transmittance approaches static one (c1) regardless of the outdoor air temperature. At higher daily solar radiation U_{eff} decrease linearly (c4), while it increases with the decreasing of the outdoor air temperature $T_{e,avg}$ (c2). The dynamic thermal transmittance U_{eff} of BIPV is 0.2 W/m^{2}K or lower if daily average outdoor temperature is above ~ 9 °C and the solar radiation is above 4000 Wh/day, and when the outdoor temperature $T_{e,avg}$ is above ~ 5 °C, the daily solar radiation $H_{glob,90}$ will be not less than 2500 Wh/day, if BIPV is not ventilated (c3). The slope of U_{eff} decrease is higher in case of non-ventilated BIPV (c5) when compared to the slope of U_{eff} decrease in the case of ventilated BIPV (c4). The decrease of the U_{eff} with increased ventilation air flow rate $V_{a,i}$ is more evident at higher air flow rates (C6), while it is not seen at daily solar radiation below 2000 Wh/day. Negative U_{eff} were observed at the highest air flow rates. The wind velocity v_w at the experiment location was so low (Table 1) during the whole period of experiment that it cannot be treated as impact parameter.

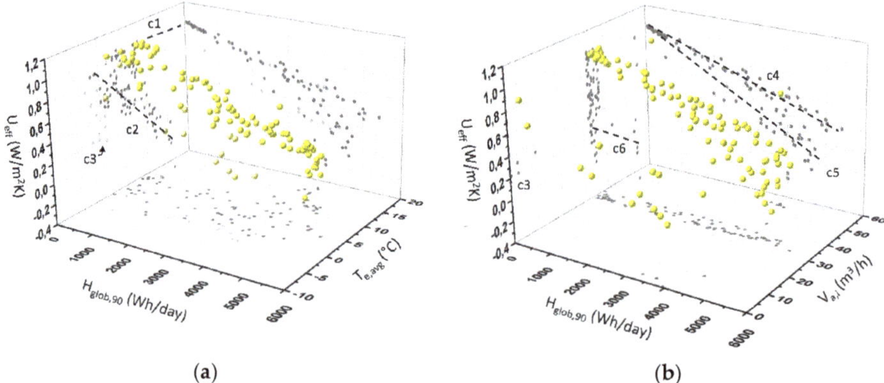

(a) (b)

Figure 10. Dynamic thermal transmittance U_{eff} of the BIPV façade structure with a forced ventilated air gap; (**a**) impact of the daily solar radiation $H_{glob,90}$ received by the BIPV façade structure and average daily outdoor air $T_{e,avg}$; (**b**) impact of the solar radiation and ventilation air flow rate $V_{a,i}$.

The average daily efficiency of electricity production is shown in Figure 11. One must note that values are defined for the BIPV façade structure as whole, while PV cells only cover 60% of BIPV structure. The daily efficiency $\eta_{PV,BIPV}$ increases slightly (c2) with daily solar radiation $H_{glob,90}$ above 2500 Wh/day, while at such conditions efficiency is almost temperature independent (c1). The reason can be in the design of glazed BIPV, which contains two relatively thick glass panes (2 × 4 mm). By measurement it was determined that absorptivity of the transparent area of BIPV is ~ 19.5%. It was also discovered that an increase of the ventilation air flow rate $V_{a,i}$ causes only minor increase of efficiency because of the PV cells cooling (c3). It can be that at higher air flow rate the fully developed flow occurs at a larger distance from the inlet opening. Nevertheless, as mentioned before, the difference between PV cell temperatures (meaning as measured—the temperature of the inner glass of the BIPV behind the PV cell) in the 2nd and 6th rows only differ for < 1–1.5 °C during clear sky conditions at air flow rates above 15 m^3/h, while temperature differences up to 5.5°C were noticed in similar weather conditions in case of buoyancy convection in closed air gap. This finding indicates that additional research will be useful in the future. In this case as well, we found no evidence of impact of the wind velocity v_w on the PV cell efficiency as well.

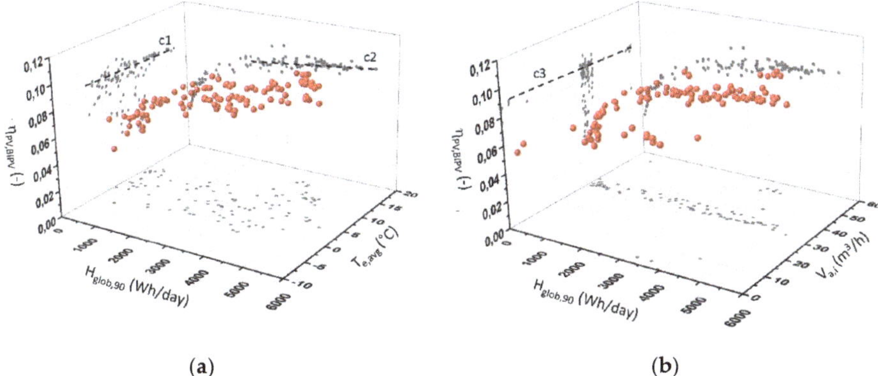

(a) (b)

Figure 11. Average daily efficiency of electricity production by glazed BIPV determined considering that surface area of PV cell in glazed BIPV is 60%; (**a**) impact of the daily solar radiation $H_{glob,90}$ and average daily outdoor air $T_{e,avg}$; (**b**) impact of the solar radiation and ventilation air flow rate $\dot{V}_{a,in}$.

If supply air is predominantly preheated by heat losses through composite façade wall (c1 in Figure 12), the efficiency of solar energy utilization $\eta_{a,BIPV}$ for preheating of the supply air for space ventilation may rise over the value of one. In practice, this is the case in days with low daily solar radiation $H_{glob,90}$ (threshold is at ~500 Wh/day). Such cases appear at daily average outdoor air temperatures $T_{e,avg}$ below 8 °C (c2). Where daily solar radiation $H_{glob,90}$ is above that threshold value, the $\eta_{a,BIPV}$ is in the range 0.50 to 0.75 showing a slightly negative trend due to the increased heat losses of the BIPV glazed façade structure (c3). Consequently, at daily solar radiation above 4000 Wh/day, it will be slightly below the 0.5. This means that the decrease of ventilation heat losses is comparable to the mechanical ventilation with heat recovery. Results also show a slight increase of efficiency with the ventilation air flow rate $\dot{V}_{a,in}$ (c4), while no significant impact of the wind velocity v_w on the $\eta_{a,BIPV}$ can be found from experimental results.

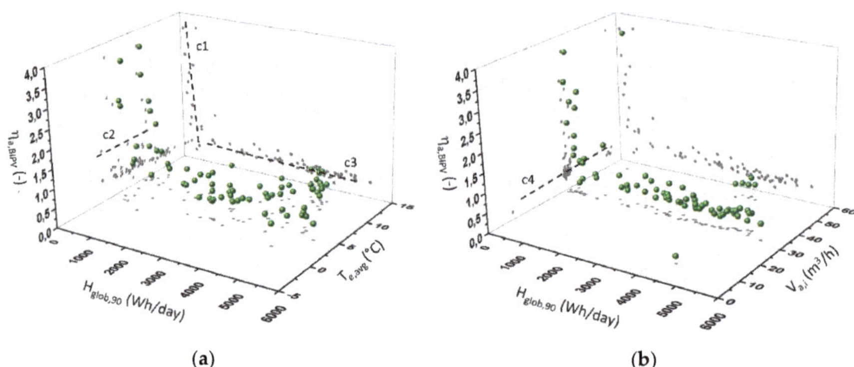

Figure 12. Average daily efficiency of solar radiation utilization for preheating of the supply air for space ventilation; (**a**) impact of the daily solar radiation $H_{glob,90}$ and average daily outdoor air $T_{e,avg}$; (**b**) impact of the solar radiation and ventilation air flow rate $\dot{V}_{a,in}$.

Solar radiation that passed through glazed BIPV façade structure was absorbed on the opaque composite façade wall. Consequently, heat loss decreased or even turned into the heat gain. If solar heat gains exceed steady-state heat loss on the daily basis, the value of $\eta_{i,sol,BIPV}$ will be greater than zero. The heat gains are defined by the product of the $\eta_{i,sol,BIPV}$ and daily solar radiation $H_{glob,90}$. Figure 13 shows how $\eta_{i,sol,BIPV}$ depends on influence parameters: daily solar radiation $H_{glob,90}$, daily average outdoor temperature $T_{e,avg}$, and ventilation air volume flow rate $V_{a,i}$. One must note, that only 40% of the total BIPV structure is transparent. The $\eta_{i,sol,BIPV}$ slightly rises with daily solar radiation if $H_{glob,90}$ is larger than 500 Wh/day, where it will be in the range between 0.06 and 0.08 (c1). In case of non-ventilated BIPV, it is significantly higher (c2)—between 0.10 and 0.12. The efficiency rise varies slightly with the outdoor air temperature $T_{e,avg}$ as well (c3). At lower solar radiation, the $\eta_{i,sol,BIPV}$ rises up to 0.30. Obviously, at such low solar energy potential, increased thermal resistance of the BIPV structure contributes more significantly to decreased heat losses than solar radiation itself. Ventilation air flow rate has a negative and quite small impact on the $\eta_{i,sol,BIPV}$, but it must be considered in a multi-parametric regression model. A small negative impact of wind velocity v_w was also noticed. Furthermore, it was noticed that heat flux that enters the building has a small-time delay because of the low thermal capacity of the composite façade wall (Figure 6b).

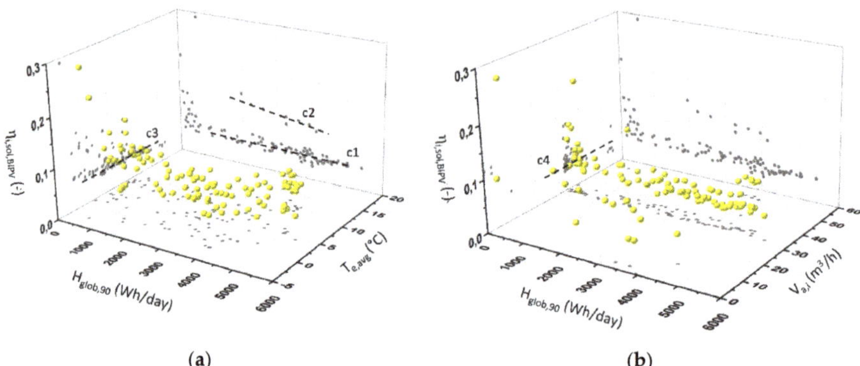

Figure 13. Average daily solar heat gains through composite façade wall behind the glazed BIPV façade structure expressed by efficiency $\eta_{i,sol,BIPV}$; (**a**) impact of the daily solar radiation $H_{glob,90}$ and average daily outdoor air $T_{e,avg}$; (**b**) impact of the solar radiation and ventilation air flow rate $\dot{V}_{a,in}$.

4.2. Multi-Parametric Model of Overall Efficiency of Solar Energy Utilization

In the previous section it was shown which variables have the greatest impact on the overall efficiency of solar energy utilization of the BIPV façade structure. To be able to predict energy efficiency of such structure in different climate conditions based on the diurnal data, multiple linear regression models were developed for each component of overall efficiency of solar energy utilization—$\eta_{PV,BIPV}$, $\eta_{a,BIPV}$, and $\eta_{i,sol,BIPV}$. Statistical regression analysis was made within MS Excel using built-in LINEST function, which fits the data using the least squares method. The level of significance for the regression coefficients of each predictor (independent variable) were tested using Student's t-tests with built-in T.DIST.2T function. Only terms with p-value < 0.05 were used in the final multiple linear regression models. Developed models are of the form:

$$\eta_{PV,BIPV} = 0.01207 \cdot \ln\left(H_{glob,90}\right) - 0.000414 \cdot \left(\overbrace{T_{PV,ref}}^{25^\circ C} - T_{e,avg} \right) - 0.000147 \cdot \dot{V}_{a,i,avg}, \tag{15}$$

$$\eta_{a,BIPV} = \frac{340.775}{H_{glob,90}} - 0.0783 \cdot T_{e,avg} + 0.02807 \cdot \dot{V}_{a,i,avg}, \tag{16}$$

$$\eta_{i,sol,BIPV} = \frac{7.867}{H_{glob,90}} + 0.00152 \cdot \left(T_i - T_{e,avg}\right) + 0.0078 \cdot \ln\left(\dot{V}_{a,i,avg}\right) - 0.0072 \cdot v_{w,avg}, \tag{17}$$

where all independent variables are daily integrals or average values. The accuracy of the developed multiple linear regression models of solar energy utilization efficiencies was tested with widely used indices [31,32]: the adjusted coefficient of determination R^2_{adj}, the normalized mean bias error *(NMBE)* and the coefficient of variation of the root-mean-square error *CV(RMSE)* which are defined by the following equations:

$$R^2_{adj} = 1 - \frac{n-1}{n-(p+1)} \cdot \left(1 - R^2\right), \tag{18}$$

$$NMBE = \frac{1}{\overline{M}} \cdot \frac{\sum_{i=1}^{n} (M_i - P_i)}{n-p} \cdot 100, \tag{19}$$

$$CV(RMSE) = \frac{1}{\overline{M}} \cdot \sqrt[2]{\frac{\sum_{i=1}^{n} (M_i - P_i)^2}{n-p}} \cdot 100. \tag{20}$$

Detailed explanation as well as calibration criteria of ASHRAE and other agencies can be found in [31]. Figure 14 presents the comparison of individual efficiencies of overall efficiency of solar energy utilization obtained from measured values and determined with developed multiple linear regression models (Equations (15)–(17)). The statistical parameters are also shown on the figures.

It can be seen that R^2_{adj} exceeds the recommended value of 0.75 for all three regression models. Also, *NMBE* is within ± 5%, which is the calibration criteria in the case of monthly calculation methods. Only *CV(RMSE)*, which should be below 15%, is slightly higher in the case of $\eta_{a,BIPV}$, which is probably is the consequence of the narrow range of ventilation air flow rates. Nevertheless, we can conclude that the developed multi-parametric regression models are adequate and can be used to evaluate the energy efficiency performance of the analyzed BIPV façade structure within the ranges of meteorological conditions that appeared during the experiments. It should be noted that the conditions during the experiment were quite typical for a moderate and continental climate.

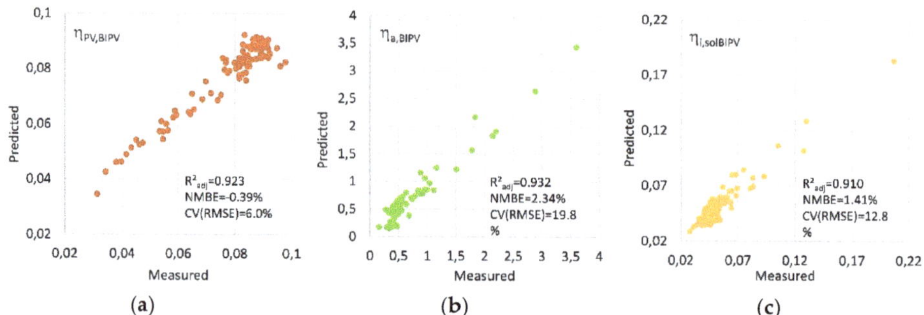

Figure 14. Correlation of components of overall efficiency of solar energy utilization $\eta_{sol,BIPV}$ determined by experimental results and multi-parametric regression models; statistical indicators are also shown for (**a**) $\eta_{PV,BIPV}$, (**b**) $\eta_{a,BIPV}$, and (**c**) $\eta_{i,sol;BIPV}$.

5. Conclusions

In this paper research on energy efficiency of the glazed BIPV façade structure with forced ventilated air gap is presented. Today, BIPV is considered to be one of the most important technologies by which, especially in the urban environment, the standards for the nearly or even zero energy buildings can be achieved. The studied BIPV element is designed as a modular unit that can be used in new, and with even greater advantages, in renovated buildings. The energy efficiency indicators were defined for each of the solar energy utilization modes—electricity production, energy savings due to decreased transmission heat losses and due to decreasing ventilation heat losses by preheating air for space ventilation. Results based on all-winter season experimental results had shown that up to 10% of daily solar radiation could be utilized for electricity and heat supply while preheating of the air utilizes up to 75% of daily received solar radiation. In the case of no clear sky conditions (low daily solar radiation) the efficiency of heat utilization increases further due to the dynamic thermal insulation effect. This are general figures. For day-to-day analyses in different climate conditions, the multi-parametric regression models were developed.

In follow-up research the impact of orientation of the BIPV façade structure with forced ventilated air gap will be studied and multi-parametric models developed, which will consider the thermal transmittance of the opaque envelope wall as an independent variable. The study of the impact of the other techniques for decreasing the PV cell temperature, like phase change materials, will also be interesting.

Author Contributions: Conceptualization, S.M., C.A., and S.D.; methodology, S.M.; validation, S.D., C.A., and L.P.; writing—original draft preparation, S.D. and C.A.; writing—review and editing, S.M. All authors have read and agreed to the published version of the manuscript.

Funding: The authors acknowledge the financial support from the Slovenian Research Agency (research core funding No. P2-0223 (C)).

Conflicts of Interest: The authors declare no conflict of interest.

Nomenclature

A	m^2	area
α	(-)	absorptivity
β	(-)	temperature coefficient
C	mg/m^3, ppm	pollutant concentration
c_p	(J/kgK)	specific heat capacity
d	m	thickness
$\Delta\tau_{meas}$	min	measuring interval

$\dot{E}$	W	electrical power
E	Wh/day	diurnal production of electricity
ε	(-)	emissivity
$\dot{G}$	W	solar irradiation power
G	W/m^2	solar irradiation
h	W/m^{2}K	surface heat transfer coefficient
H	Wh/m^{2}day	diurnal solar radiation
k	1/day	pollutant decay factor
η	(%)	efficiency
K_G	(-)	solar irradiation factor
K_T	(-)	temperature factor
λ	W/mK	thermal conductivity
M	(-)	measured value
n	(-)	number of
OHH	(Kh/day)	diurnal overheating hours
P	(-)	Predicted value
$\dot{q}$	(W/m^2)	density of heat flux
$\dot{Q}$	(W)	heat power
Q	(Wh/day)	diurnal delivered heat
ρ	(kg/m^3)	density
S	mg/min	source of pollutant
T	°C	temperature
U	W/m^{2}K	thermal transmittance
v	m/s	velocity
V	m^3	volume
$\dot{V}$	m^3/h	flow rate
7, 24	h/day	constants
60	min/h	constant
Index		
a		air
avg; avg,day		average, diurnal average
avg, PV		diurnal average during PV electricity production
BIPV		pilot building integrated PV glazed façade structure
e		outdoor, external
eff		effective, dynamic
g		glass
glob,90		solar on vertical surface
glob,90,day		solar on vertical surface diurnal
i		indoor, internal
in		inlet
IR		infrared, long wavelength
IR,90,day		infra-red on vertical surface diurnal
loss		heat loss
max		maximum, maximum at the end of working hours
n		net
PV		photovoltaic
PV,si		on inner glass surface of BIPV structure
r+c		combined radiative and convection
ref		reference, reference structure
s		solar, short wavelength
si		internal surface
sol		generated by solar energy
st		static
v		ventilation
w		wind
2, 6		second row, sixth row of PV

References

1. Sombsthay, A.; Beauvais, A.; Moser, D.; Tecnalia, E.B.; Gaymard, G.; Delmer, G.; Gurdenli, L.; Machado, M.; Po, R.; Krawietz, S.; et al. *Solar Skins: An Opportunity for Green Cities*; SolarPower Europe and ETIP PV: Brussels, Belgium, 2019.
2. Farghaly, Y.; Hassan, F. A Simulated Study of Building Integrated Photovoltaics (BIPV) as an Approach for Energy Retrofit in Buildings. *Energies* **2019**, *12*, 3946. [CrossRef]
3. Ramirez-Balas, C.; Fernandez-Nieto, E.; Narbona-Reina, G.; Sendra, J.J.; Suarez, R. Thermal 3D CFD Simulation with Active Transparent Façade in Buildings. *Energies* **2018**, *11*, 2264. [CrossRef]
4. Yang, X.; Zhou, J.; Yuan, Y. Energy Performance of an Encapsulated Phase Change Material PV/T System. *Energies* **2019**, *12*, 3929. [CrossRef]
5. Good, C.; Andersen, I.; Hestnes, A.G. Solar Energy for net Zero Energy Buildings—A Comparison between Solar Thermal, PV and Photovoltaic–Thermal (PV/T) Systems. *Solar Energy* **2015**, *122*, 986–996. [CrossRef]
6. Agathokleous, R.A.; Kalogirou, S.A. Double Skin Facades (DSF) and Building Integrated Photovoltaics (BIPV): A Review of Configurations and Heat Transfer Characteristics. *Renew. Energy* **2016**, *89*, 743–756. [CrossRef]
7. Agrawal, B.; Tiware, G.N. Optimizing the Energy and Exergy of Building Integrated Photovoltaic Thermal (BIPVT) Systems under Cold Climatic Conditions. *Appl. Energy* **2010**, *87*, 417–426. [CrossRef]
8. Saadon, S.; Gaillard, L.; Menezo, C.; Giroux-Julien, S. Exergy, Exergoeconomic and Enviroeconomic Analysis of a Building Integrated Semi-Transparent Photovoltaic/Thermal (BISTPV/T) by Natural Ventilation. *Renew. Energy* **2020**, *150*, 981–989. [CrossRef]
9. Vats, K.; Tiwari, G.N. Performance Evaluation of a Building Integrated Semitransparent Photovoltaic Thermal System for Roof and Façade. *Energy Build.* **2012**, *45*, 211–218. [CrossRef]
10. Peng, J.; Lu, L.; Yang, H.; Ma, T. Comparative Study of the Thermal and Power Performances of a Semi-Transparent Photovoltaic Façade under Different Ventilation Modes. *Appl. Energy* **2015**, *138*, 572–583. [CrossRef]
11. Shahrestani, M.; Yao, R.; Essah, E.; Shao, L.; Oliveira, A.C.; Hepbasli, A.; Biyik, E.; del Caño, T.; Rico, E.; Lechón, J.L. Experimental and Numerical Studies to Assess the Energy Performance of Naturally Ventilated PV Façade Systems. *Sol. Energy* **2017**, *147*, 37–51. [CrossRef]
12. Martín-Chivelet, N.; Gutiérrez, J.C.; Alonso-Abella, M.; Chenlo, F.; Cuenca, J. Building Retrofit with Photovoltaics: Construction and Performance of a BIPV Ventilated Façade. *Energies* **2018**, *11*, 1719. [CrossRef]
13. Kosonen, R.; Tan, F. The Effect of Perceived Indoor air Quality on Productivity loss. *Energy Build.* **2004**, *36*, 981–986. [CrossRef]
14. Mofidi, F.; Akbari, H. Integrated Optimization of Energy Costs and Occupants' Productivity in Commercial Buildings. *Energy Build.* **2016**, *129*, 247–260. [CrossRef]
15. Domjan, S.; Arkar, C.; Begelj, Ž.; Medved, S. Evolution of All-Glass Nearly Zero Energy Buildings with Respect to the Local Climate and Free-Cooling Techniques. *Build. Environ.* **2019**, *160*, 106183. [CrossRef]
16. Kim, M.K.; Baldini, L. Energy Analysis of a Decentralized Ventilation System Compared with Centralized Ventilation Systems in European Climates: Based on Review of Analyses. *Energy Build.* **2016**, *111*, 424–433. [CrossRef]
17. Imbabi, M.S.-E. A Passive–Active Dynamic Insulation System for All Climates. *Int. J. Sustain. Built Environ.* **2012**, *1*, 247–258. [CrossRef]
18. Lai, C.-M.; Lin, Y.-P. Energy Saving Evaluation of the Ventilated BIPV Walls. *Energies* **2011**, *4*, 948–959. [CrossRef]
19. Union Glass, S.r.i, Union Glass High-Technology Glass. Available online: http://unionglass.it/index.php (accessed on 15 January 2019).
20. Emmerich, S.J.; Persily, A.K. *State-of-the-Art Review of CO2 Demand Controlled Ventilation Technology and Application*; NIST: Gaithersburg, MD, USA, 2001.
21. Masters, G.M.; Ela, W.P. *Introduction to Environmental Engineering and Science*, International ed.; Prentice-Hall: Englewood Cliffs, NJ, USA, 1991.
22. *Energy Performance of Buildings—Ventilation of Buildings—Part 1: Indoor Environmental Input Parameters for Design and Assessment of Energy Performance of Buildings Addressing Indoor Air Quality, Thermal Environment, Light and Acustics*; EN 16798-1:2019; CEN: Brussels, Belgium, 2019.

23. Kipp & Zonen. Available online: www.kippzonen.com (accessed on 16 May 2020).
24. Davis Instruments. Available online: www.davisinstruments.com (accessed on 16 May 2020).
25. Ahlborn Mess—und Regelungstechnik GmbH. Available online: www.ahlborn.com (accessed on 16 May 2020).
26. Agilent Technologies, Inc. Available online: www.agilent.com (accessed on 16 May 2020).
27. The International Organization for Standardization. *Energy Performance of Buildings—Energy Needs for Heating and Cooling, Internal Temperatures and Sensible and Latent Heat Loads—Part 1: Calculation Procedures*; CEN: Brussels, Belgium, 2017.
28. Singh, P.; Ravindra, N.M. Temperature Dependence of Solar Cell Performance—An Analysis. *Sol. Energy Mater. Sol. Cells* **2012**, *101*, 36–45. [CrossRef]
29. Dash, P.K.; Gupta, N.C. Effect of Temperature on Power Output from Different Commercially available Photovoltaic Modules. *Int. J. Eng. Res. Appl.* **2015**, *51*, 148–151.
30. Kamuyu, W.C.L.; Lim, J.R.; Won, C.S.; Ahn, H.K. Prediction Model of Photovoltaic Module Temperature for Power Performance of Floating PVs. *Energies* **2018**, *11*, 447. [CrossRef]
31. Ruiz, G.R.; Bandera, C.F. Validation of Calibrated Energy Models: Common Errors. *Energies* **2017**, *10*, 1587. [CrossRef]
32. Domjan, S.; Medved, S.; Černe, B.; Arkar, C. Fast Modelling of nZEB Metrics of Office Buildings Built with Advanced Glass and BIPV Facade Structures. *Energies* **2019**, *12*, 3194. [CrossRef]

Article

Seasonal Energy Flexibility Through Integration of Liquid Sorption Storage in Buildings

Luca Baldini * and Benjamin Fumey

Empa–Swiss Federal Laboratories for Materials Science and Technology, 8600 Dubendorf, Switzerland; benjamin.fumey@empa.ch

* Correspondence: luca.baldini@empa.ch

Received: 30 April 2020; Accepted: 4 June 2020; Published: 8 June 2020

Abstract: The article estimates energy flexibility provided to the electricity grid by integration of long-term thermal energy storage in buildings. To this end, a liquid sorption storage combined with a compression heat pump is studied for a single-family home. This combination acts as a double-stage heat pump comprised of a thermal and an electrical stage. It lowers the temperature lift to be overcome by the electrical heat pump and thus increases its coefficient of performance. A simplified model is used to quantify seasonal energy flexibility by means of electric load shifting evaluated with a monthly resolution. Results are presented for unlimited and limited storage capacity leading to a total seasonal electric load shift of 631.8 kWh/a and 181.7 kWh/a, respectively. This shift, referred to as virtual battery effect, provided through long-term thermal energy storage is large compared to typical electric battery capacities installed in buildings. This highlights the significance of building-integrated long-term thermal energy storage for provision of energy flexibility to the electricity grid and hence for the integration of renewables in our energy system.

Keywords: long-term thermal energy storage; seasonal thermal energy storage; thermochemical energy storage; liquid sorption storage; power-to-heat; seasonal energy flexibility; seasonal load shifting; virtual battery effect

---

1. Introduction

There is an increasing need for energy storage in buildings to allow for integration of intermittent on-site renewable energy sources or the provision of energy flexibility to the electric grid. Thermal storage thereby plays an important role as it allows cheaply offsetting large amounts of energy. Thermal storage capital cost is low when compared to electric batteries and thus allows for economical energy storage also over longer time periods thus operation with substantially fewer cycles over its lifetime. At its extreme, seasonal storage is possible whereby the low number of cycles put a challenge on acceptable investment cost of the storage technology [1]. Seasonal storage is of great importance for significantly increasing renewable fraction in the operation of buildings. Without seasonal storage, excess energy available in summer cannot be made available for coverage of heat demand in winter such that renewable fraction, especially for space heating remains strongly limited. With seasonal energy storage, excess energy available from on-site renewable production or electricity from the grid can be absorbed in summer to be made available in winter when space heating demand is largest. Assuming a heat pump to provide space heating by using electricity, long-term storage thus contributes to significant load shifting from the winter into the summer. Power to heat conversion with the thermal energy storage together can be considered as a virtual electric battery and the electricity offset as the virtual battery effect. This consequently leads to an increase of integrated on-site renewables or provides a significant amount of energy flexibility to the electric grid. As the flexibility offered represents a seasonal rather than a short-term shift, it is referred to here as seasonal energy flexibility.

Currently, available sensible water storages or storages based on phase change materials suffer from a continuous heat loss such that seasonal storage is possible only at a large scale. To enable compact, long-term thermal energy storage on a building scale, higher volumetric energy storage densities are required, along with high in/out storage efficiencies, meaning low thermal losses during storage period. A class of energy storage known as thermochemical energy storage [2] promises to fulfill these criteria. Within this class there are different sub-classes often categorized into chemical reactions and sorption processes [3–6]. Depending on the type of reaction and the specific materials used energy densities between 200 and 2000 kWh/m^3 can be reached [6,7]. From measurements performed for a liquid sorption storage using aqueous sodium hydroxide (NaOH) as sorbent and water (H_2O) as refrigerant, a concentration difference from 50 to 27 wt.% NaOH was achieved for evaporation and absorber inlet temperatures of 5–20 °C and 25–38 °C, respectively, leading to a maximum theoretical volumetric energy density of 435 kWh/m^3 with reference to the diluted sorbent volume [8]. For comparison, volumetric energy storage density of water is at around 55 kWh/m^3, assuming a temperature difference of 50 K. Most important for thermochemical energy storage is the fact that energy is stored rather by means of a chemical potential than by actual thermal energy such that no continuous heat losses occur during the storage phase but only at conversion (charging/discharging). Generalized indication of storage density for thermochemical energy storage is problematic as it depends on many operational parameters, i.e. mainly temperatures. This fact is very often overlooked even in the research field of sorption energy storage. For a cross comparison of different thermochemical storage types, materials, and processes, thus a common reference framework as suggested, e.g., for the building application, in [9], is needed. A generalized metric provided in [10] can be used for performance comparison across different storage processes reported in literature but does not replace performance metrics such as volumetric energy density, asking for a uniform basis for evaluation.

There are several possibilities of integrating sorption storage into building energy systems, largely depending on the type of process, i.e. open or closed sorption process [5,11–13]. In an open process, absorbate for discharging is provided at atmospheric conditions, typically by ambient air and it is again released to the ambient in charging. In a closed process, phase change is typically taking place at subatmospheric conditions. The absorbate is provided through evaporation using a low temperature source in discharging and is again condensed using a low temperature sink in charging. The coupling between the storage and the heat source/sink happens through a heat exchanger. This article strictly focuses on a closed liquid sorption process using NaOH/H_2O as sorption couple. An overview of examples of open and closed sorption storage processes is presented in [10].

Unlike a heat battery directly storing and releasing heat, sorption storage rather works as a heat pump, thus needing contact to two thermal reservoirs for charging or discharging. In charging a high temperature heat source is required for desorption (evaporation of water from the liquid sorbent, e.g., NaOH) as well as a low temperature heat sink for condensation of the water vapor extracted from the sorbent. In discharging, a low temperature heat source is needed for evaporation of water and a medium temperature sink, typically the building, for absorption of the water vapor. For integration of a sorption storage into the building energy system, it can be coupled with solar thermal collectors for the high temperature source as well as with a ground heat exchanger for low temperature heat source and sink [14]. More interesting from an energy flexibility perspective is the coupling of the storage with the electric grid through a compression heat pump. The heating system can then be looked at as a double-stage heat pump with one stage being a compression heat pump and another stage being a thermal or chemical heat pump, i.e. the sorption storage. Similar hybrid concepts combining compression and sorption cycles are presented in [15–17]. As an electricity source, on-site PV and/or the electric grid can be used. In charging mode, excess electricity can be received, while in discharging mode a smaller temperature lift is expected from the compression heat pump as it only represents one stage of the hybrid concept. Consequently, a higher heat pump efficiency and thus lower electricity consumption is expected. This article strictly focuses on the second option of double-stage heat pumping in order to address the seasonal energy flexibility offered to the electricity grid through building integration of a liquid sorption storage.

2. Materials and Methods

2.1. Sample Building

For the assessment of the seasonal energy flexibility by using a liquid sorption storage, a sample single family home (SFH) with 140 m^2 of floor area is considered as defined by the IEA SHC Task 44 /HPP Annex 38 reference framework [18]. Out of three SFH buildings presented there, the SFH45 was picked, showing an area specific annual space heating demand of 46.255 kWh/(m^2·a). From three different climatic locations proposed, only the one for Strasbourg, France was considered. Results from detailed building simulations are provided as monthly values. This monthly resolution is sufficient to capture seasonal energy flexibility. From simulation results presented, monthly space heating loads, the instantaneous design heat load and average supply and return temperatures of the floor heating system were used (Table 1). Monthly average ambient temperatures for the location of Strasbourg were taken from [19] and are also shown in Table 1. Ambient temperature data was relevant for the evaluation of the air-source heat pump as well as for the sorption storage performance.

Table 1. Monthly average space heating load, space heating supply and return temperatures for SFH45 under climate of Strasbourg from [18] and ambient temperatures from [19].

Input Variables	Jan.	Feb.	Mar.	Apr.	May	Jun.	Jul	Aug.	Sep.	Oct.	Nov.	Dec.
SH load (kWh/m^2)	11.99	8.18	4.53	0.96	0.025	0	0	0	0	1.55	7.62	11.4
Tsup (°C)	30.1	29.1	27.4	26.1	27.2	0	0	0	0	26.4	28.1	29.4
Tret (°C)	22.2	21.4	20.7	20.4	20.5	0	0	0	0	20.5	21	22
Tamb (°C)	0.9	2.4	6.1	9.7	13.8	17.2	19.2	18.6	15.7	10.7	5.3	2.1

2.2. System Description

For the building integration of the closed, liquid sorption storage, a combination with a compression heat pump was chosen. In this way the sorption-based thermal energy storage is coupled to the electricity grid through the compression heat pump. The heat and mass exchanger (HMX) of the sorption storage is shown in Figure 1 with the absorber/desorber representing the left chamber and the evaporator/condenser the right chamber in the figure, respectively.

The heat pump is a water/water heat pump with an additional water/air heat exchanger. The schematic of storage integration is shown in Figure 2a in charging operation and Figure 2b in discharging operation. The H-shaped component represents the HMX. The sorption storage tanks consisting of absorbate (water), concentrated sorbent (NaOH) and a diluted sorbent tank, respectively, are not shown here. For more details of the sorption storage system with all components included, it is referred to [14,21].

In charging operation (Figure 2a), the compression heat pump is providing cold to the condenser of the HMX, while providing heat to the desorber of the HMX. In order to balance the mismatch between the evaporator and condenser, power provided by the heat pump the condenser side is additionally connected to the water/air heat exchanger for excess heat rejection to the ambient. Alternatively, excess heat can be used for domestic hot water production instead. In discharging (Figure 2b), the heat pump is used to extract heat from the ambient air using the water/air heat exchanger while providing heat to the evaporator of the HMX. In the absorption process taking place in the absorber of the HMX, heat is released to the building for space heating purposes.

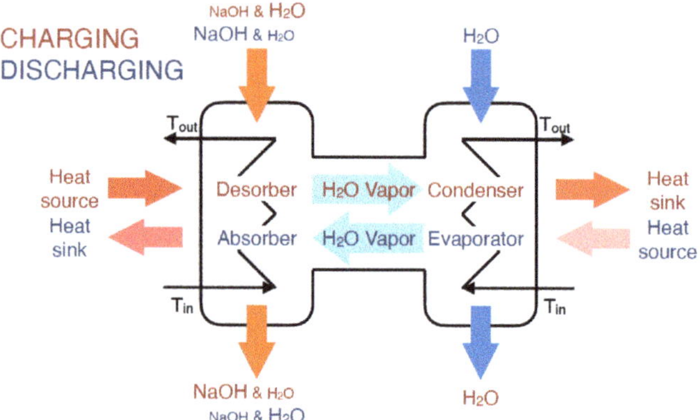

Figure 1. Heat and mass exchanger (HMX) with absorber/desorber on the left side and evaporator/condenser on the right side being connected to each other for water vapor exchange. During charging, diluted sorbent enters the desorber from the top leaving it as concentrated sorbent at the bottom. Thereby, water is evaporated by the external heat source and transported to the condenser, where it changes back to its liquid state, releasing heat to a respective sink. In charging, concentrated sorbent enters the absorber at the top, leaving it as diluted sorbent at the bottom. Thereby, water being evaporated by a low temperature heat source in the evaporator is absorbed, releasing useful heat for space heating or domestic water production. Heat exchange between sorbent and heat transfer fluid is followed in counterflow, allowing for the optimal exploitation of available temperatures. Adapted from [20].

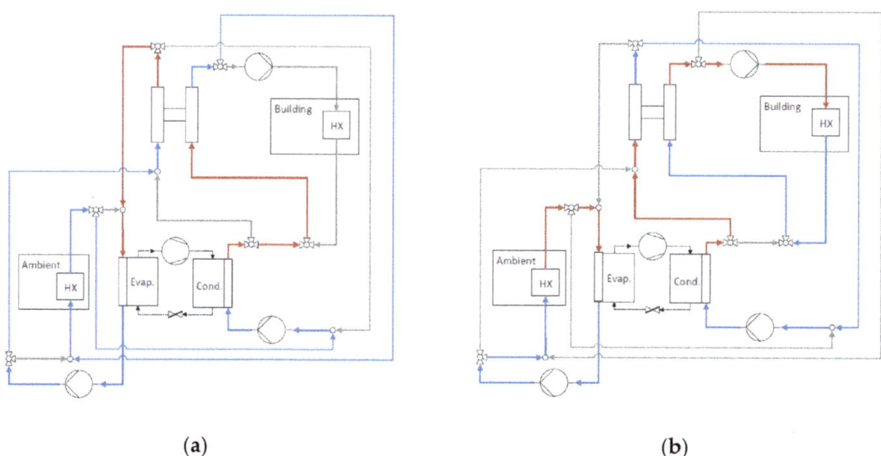

Figure 2. Schematic of the sorption storage integration together with a compression heat pump, where the H-shaped component represents the HMX: (**a**) Charging mode: Heat pump provides high temperature for desorption and low temperature for condensation; (**b**) Discharging mode: Heat pump provides low temperature heat for evaporation.

2.3. Liquid Sorption Storage Modelling

The liquid sorption process is modeled, assuming thermodynamic equilibrium. This means that for a given absorbate vapor pressure and temperature of the liquid sorbent the equilibrium concentration is determined. In charging operations, a fixed outlet sorbent concentration of 50 wt.% NaOH and a maximum desorption temperature of 55 °C are chosen. For this state the water vapor pressure present in the desorber, as well as in the condenser, is determined and with it the resulting condensation temperature. In discharging, the source temperature for the evaporation determines the absorbate pressure in the evaporator and absorber respectively and with it the sorbent concentration depending on its temperature. Minimum sorbent concentration in discharging is thus determined by the absorbate pressure given by the evaporator and the return temperature from the space heating system of the building entering the absorber. NaOH properties depending on temperature or concentration are calculated using available mathematical correlations [22].

For simplicity, during charging, heat provided to the desorber is assumed to be equal to heat rejected by the condenser, meaning that heat losses appearing in the desorber compensate for the neglected heat of solution.

2.4. Seasonal Flexibility Analysis

Seasonal flexibility offered by a building integrated sorption storage is evaluated based on its ability to use grid electricity in summer during charging and to reduce electricity demand in winter during discharging of the storage. Electricity is not stored seasonally but rather the potential to reduce electricity demand in winter/heating season. The electric load shift achieved is referred to as the virtual battery effect as it acts similarly to storing this part of electricity over the season. The seasonal energy flexibility must be separated in two parts, i.e., the additional electricity absorbed during charging (negative flexibility) and the electricity savings in discharging (positive energy flexibility). These energy flexibilities are quantified with reference to the regular heat pump operation without sorption storage integrated. The electricity consumed by the heat pump is calculated as the energy provided by the heat pump divided by its coefficient of performance (COP). The COP is calculated as the ideal Carnot COP based on thermal reservoir temperatures multiplied by a fixed isentropic efficiency (η_{isen}) of 0.5.

Negative energy flexibility provided with storage is equal to the electricity utilized by the heat pump to charge the storage, given a certain time (Δt_{excess}) with excess electricity available from the electricity grid (Equation (1)). During charging, the heat pump must provide the entire temperature lift between the HMX desorber and condenser and sensible heat stored in the charged, concentrated sorbent is lost ($\eta_{charging}$). The temperature of the desorber is fixed at 55 °C and the saturation pressure of water vapor at this temperature and a targeted sorbent concentration of 50 wt.% determines the condensation temperature needed in the HMX. The inlet to the condenser of the HMX provided by the evaporator of the heat pump then needs to be lower by the amount of the temperature difference between the absorbate and the heat transfer fluid ($\Delta T_{hx,HMX}$) and an additional temperature difference of the heat transfer fluid across the HMX condenser ($\Delta T_{HMX,ev}$). The evaporation temperature of the heat pump is then below the inlet temperature on the HMX condenser by $\Delta T_{hx,hp}$. In the condenser of the compression heat pump there is no temperature difference assumed between condensation temperature and temperature of the heat transfer fluid leaving the condenser. This assumption is justified because of the desuperhating taking place in the condenser, allowing even for secondary fluid temperatures above condensation temperature at the outlet of a counterflow-type condenser. This affects Equations (2) and (4) where condensation temperature of the heat pump is equal to the HMX desorber inlet temperature ($T_{desorption} + \Delta T_{hx,HMX}$) and T_{supply} of the space heating, respectively.

$$E_{flex,charging} = P_{el,hp,charging}\Delta t_{excess} = \eta_{charging}\frac{Q_{hp,\ desorption}}{COP_{hp,charging}}\Delta t_{excess} \tag{1}$$

$$COP_{hp,charging} = \eta_{isen}\frac{T_{desorption} + \Delta T_{hx,HMX}}{\left(T_{desorption} + \Delta T_{hx,HMX}\right) - \left(T_{sat,p_{desorb}} - \Delta T_{hx,HMX} - \Delta T_{HMX,ev} - \Delta T_{hx,hp}\right)} \tag{2}$$

In discharging, the energy flexibility is calculated as the difference in electricity consumed during heating season ($\Delta t_{heating}$) by the single-stage and double-stage heat pump including storage respectively (Equation (3)). In case a sorption storage is included the temperature lift is to be provided by the compression heat pump is reduced, thus increasing its COP.

$$E_{flex,discharging} = (P_{el,ref,heating} - P_{el,storage,discharging})\Delta t_{heating} = Q_{heating}\left(\frac{1}{COP_{ref,heating}} - \frac{1}{COP_{storage,discharging}}\right)\Delta t_{heating} \quad (3)$$

In standard operation without storage installed, the heat pump COP ($COP_{ref,heating}$) is calculated based on supply temperatures (T_{supply}) of the space heating system and the evaporation temperature (Equation (4)). The latter is determined by the ambient air temperature reduced by the temperature difference between the air inlet of the water/air heat exchanger and the water outlet of the heat pump evaporator (ΔT_{air}) and the temperature difference between the water and refrigerant ($\Delta T_{hx,hp}$) in the evaporator.

$$COP_{ref,heating} = \eta_{isen}\frac{T_{supply}}{T_{supply} - (T_{amb} - \Delta T_{air} - \Delta T_{hx,hp})} \quad (4)$$

When combined with the sorption storage the evaporation temperature of the heat pump remains the same but the condensation temperature of the heat pump being the inlet temperature of the HMX evaporator ($T_{HMX,ev,in}$) changes, leading to a different COP ($COP_{storage,discharging}$) as expressed in Equation (5). The inlet temperature to the evaporator side of the HMX provided by the compression heat pump as the first stage shown in Equation (6) is determined by the required supply temperature (T_{supply}) of the space heating and the available temperature lift from the sorption storage ($\Delta T_{50,max}$) between the maximum sorbent temperature and the evaporation temperature. Further, there is the temperature difference of the heat exchanger, once between the sorbent and heat transfer fluid ($\Delta T_{hx,HMX}$) and once between the absorbate and the heat transfer fluid ($\Delta T_{hx,HMX}$) and the temperature difference across the evaporator of the HMX ($\Delta T_{HMX,ev}$). The temperature lift provided by the sorption storage ($\Delta T_{50,max}$) depends on the maximum sorbent concentration of 50 wt.% as shown in Figure 3. In this analysis, a departure from ideal equilibrium condition is assumed, leading to a reduction of the effectively provided temperature lift.

$$COP_{storage,discharging} = \eta_{isen}\frac{T_{HMX,ev,in}}{T_{HMX,ev,in} - (T_{amb} - \Delta T_{air} - \Delta T_{hx,hp})} \quad (5)$$

with

$$T_{HMX,ev,in} = T_{supply} + \Delta T_{hx,HMX} - \Delta T_{50,max} + \Delta T_{hx,HMX} + \Delta T_{HMX,ev} \quad (6)$$

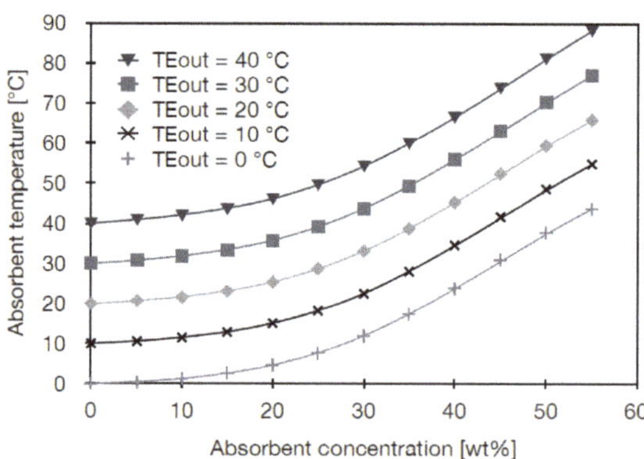

Figure 3. Ideal temperature lift provided between HMX evaporator temperature (TEout) and the maximum HMX absorber temperature by the sorption storage depending of maximum sorbent concentration. Adapted from [20].

The positive energy flexibility provided during discharging is depending on the amount of energy stored. This amount is determined during the charging phase. Maximum energy storage capacity is determined by the installed heat pump capacity multiplied by the time with available excess electricity from the electricity grid (Δt_{excess}). In discharging mode, the fraction of space heating demand that can be covered by the stored energy is determined in order to identify the system operating in double-stage mode. When storage capacity is exhausted it switches to standard single-stage operation mode. This way the positive energy flexibility offered with limited storage capacity is calculated and confronted with the ideal case of unlimited storage capacity.

In order to determine the volumetric energy density of the sorption storage with reference to the diluted sorbent and the volume required to store the calculated energy, terminal sorbent concentrations were evaluated based on system temperatures. Specifically, for the HMX evaporator temperature, the saturation pressure of water vapor is calculated, determining the achieved terminal concentration in the HMX absorber based on absorber inlet temperatures. These temperatures are determined by the return temperatures from the space heating system plus a temperature difference ($\Delta T_{hx,HMX}$) between the heat transfer fluid and the sorbent. An average terminal sorbent concentration is then calculated over the heating period in order to calculate the sorbent mass needed to store the energy provided during charging. For this purpose an energy balance, a species mass (NaOH), and total mass balance over the HMX absorber is applied resulting in an expression for the inlet mass of concentrated sorbent depending on the known ratio of inlet to terminal concentration and the specific enthalpies of concentrated ($h_{sorbent,in}$) and diluted ($h_{sorbent,out}$) sorbent, respectively, and enthalpy of condensation for water (h_{lg}) given in Equation (7).

$$M_{sorbent,in} = \frac{Q_{heating}}{h_{sorbent,in} + \left(\frac{c_{sorbent,in}}{c_{sorbent,out}} - 1\right)h_{lg} - \left(\frac{c_{sorbent,in}}{c_{sorbent,out}}\right)h_{sorbent,out}} \quad (7)$$

The outlet mass of diluted sorbent then becomes

$$M_{sorbent,out} = \left(\frac{c_{sorbent,in}}{c_{sorbent,out}}\right)M_{sorbent,in} \quad (8)$$

The volumetric energy density of the sorption storage is calculated dividing the leaving sorbent mass ($M_{sorbent,out}$) by the sorbent density at the terminal concentration ($c_{sorbent,out}$) assuming a fixed temperature of 25 °C, representing the average absorber inlet temperature.

In Table 2, parameter values used for the evaluation of energy flexibility are summarized.

Table 2. Parameter values used for evaluation of energy flexibility provided with the building integrated sorption storage.

Input Parameters					
η_{isen}	0.5	$c_{sorbent,in}$	0.5	$\Delta T_{hx,HMX}$ (K)	3
$\eta_{charging}$	0.9	$T_{desorption}$ (°C)	55	$\Delta T_{hx,hp}$ (K)	3
Δt_{excess} (h)	720	nominal heat pump/HMX capacity (kW)	4.07	$\Delta T_{HMX,ev}$ (K)	3
ΔT_{air} (K)	10	$\Delta T_{50,max}$ (K)	25		

3. Results

In the following, results are presented for the parameter settings according to Table 2. These include COP improvements, electricity demands, electric load shifting, i.e. energy flexibilities provided when using a building-integrated sorption storage. Results are presented for the different cases of (a) unlimited storage capacity and (b) limited storage capacity with reference to no storage installed. As an extension to the base case using input parameters according to Table 2 with 720 hours of available excess electricity, a simulation case assuming an increased number of 1080 hours of available excess electricity was considered as well.

3.1. Coefficient of Performance (COP) Improvement and Electric Load Shifting, i.e., Seasonal Energy Flexibilities

Sorption storage acts as a thermal heat pump, thus providing a certain temperature lift depending on maximum sorbent concentration. For a NaOH concentration of 50 wt.%, a temperature lift of around 38 K is ideally provided (Figure 3). This maximum theoretical temperature lift is determined by the difference in water vapor saturation temperature for pure water and sorbent solution respectively. Available temperature lift in reality is significantly lower due to imperfections of the process (deviations from equilibrium conditions) leading to an assumed temperature lift available of 25 K based on experience from experiments performed [8]. When using a sorption storage along with a heat pump, the temperature lift to be provided by the heat pump is reduced with comparison to heat pump operation without storage. According to Figure 4, heat pump COP is more than doubled when the total temperature lift required becomes small at moderate ambient air temperatures and is increased by about 50% during the coldest months in January and December.

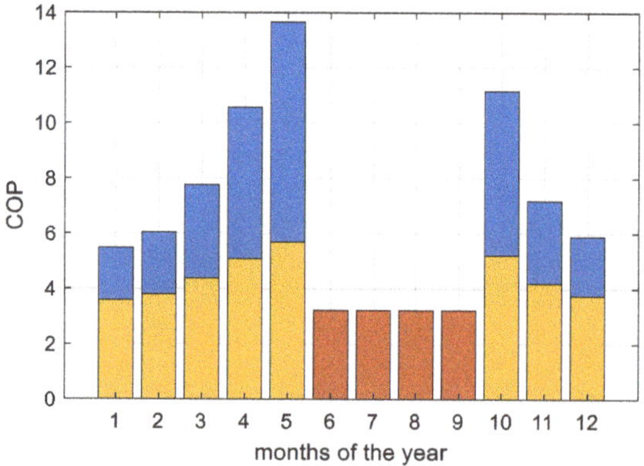

Figure 4. Coefficient of performance (COP) with (blue) and without (orange) sorption storage in heating operation and charging operation (red).

Improved COP with integrated sorption storage directly translates into electricity demand reduction in discharging representing positive energy flexibility to the grid. In Figure 5, electricity demands are shown for (a) no storage capacity limit and (b) limited storage capacity. The limitation in storage capacity is determined by the available heat pump capacity, the charging efficiency, and the total time of available excess electricity. In the proposed integration of the sorption storage with the heat pump, the latter provides the high temperature source and the low temperature sink at the same time when operating in charging mode. As a consequence, heat pump capacity available for charging is dictated by the evaporator power. The difference between condenser and evaporator power of the heat pump cannot be used by the sorption storage and is rejected over the water/air heat exchanger to the ambient air or used for domestic hot water production. With unlimited storage, capacity electricity demand is reduced during the entire heating operation (Figure 5a) while with the limited storage capacity, a demand reduction can be achieved only in January and a little bit in February (Figure 5b). When more excess electricity from the grid is available, charging duration and thus storage capacity can be increased such that more significant reduction in electricity demand can also be achieved in February (Figure 5c). This increase in available excess electricity does of course not have any effect on the theoretical case of unlimited storage capacity, which assumes full coverage of space heat demand by the sorption storage.

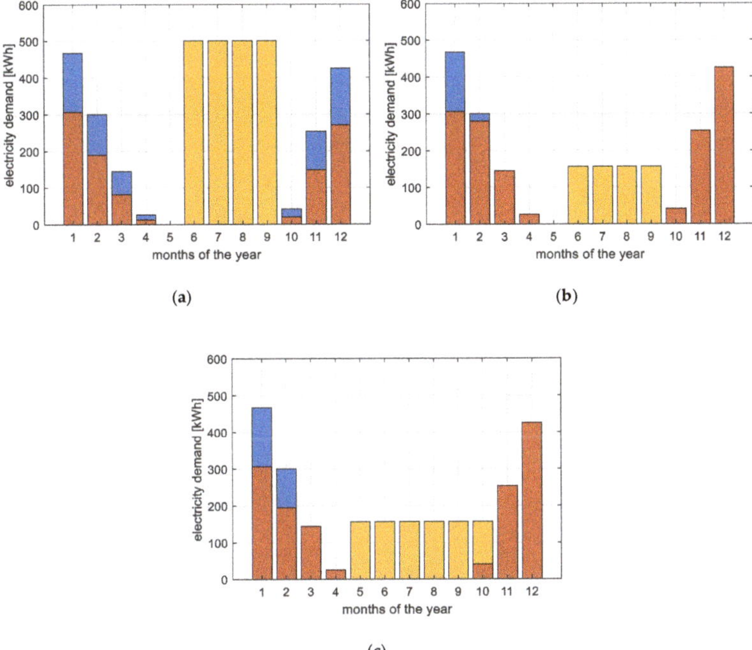

Figure 5. Monthly electricity demand of heat pump for space heating operation (discharging) with (red) and without (blue) sorption storage and charging of sorption storage (orange). For base case with 720 hours of excess electricity: (**a**) with unlimited storage capacity; (**b**) with limited storage capacity and alternative case with 1080 hours of excess electricity available; (**c**) with limited storage capacity.

Electric load shifted between seasons are shown in Figure 6. With unlimited storage capacity total seasonal load shift/flexibility amounts to 631.8 kWh/a, while with a limited storage capacity it reduces to 181.7 kWh/a for the base case with 720 hours of available excess electricity. This leads to 38% and 10.9% of electric load shifted for unlimited and limited storage capacity, respectively. When assuming 1080 hours of available excess electricity, the seasonal load shift increases to 266.7 kWh/a.

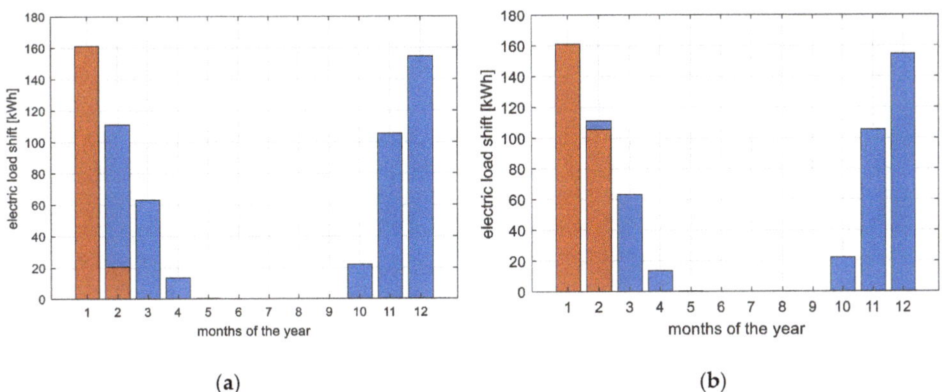

Figure 6. Electric load shift of heat pump operation with storage. With limited (red) and unlimited (blue) storage capacity for: (**a**) base case with 720 h of excess electricity and (**b**) case with 1080 h of excess electricity available.

3.2. Storage Capacity and Size

Based on temperatures assumed and resulting sorbent concentration lift, a volumetric energy storage density of 312.5 kWh/m^3 is achieved with reference to the diluted sorbent. With available charging capacity of the heat pump and assumed operation hours, a total energy storage capacity of 1822.6 kWh results, leading to a storage volume of the diluted sorbent of 5.8 m^3.

4. Discussion

The analysis performed captures seasonal load shifting/flexibility for a single-family home with an integrated sorption storage. It thus represents the electricity saved in heating operation because of seasonal thermal energy storage, called virtual battery effect. It acts the same as using an electric battery to store electricity seasonally. The latter is not done as it is economically not viable to use electric batteries for seasonal storage. This is hence a valuable feature of thermal energy storage. The virtual battery capacity achieved with the limited storage capacity is equal to 181.7 kWh. When comparing this with a standard capacity of an electric battery installed in single family homes of around 8 kWh, the sorption storage represents a 22.7 fold battery capacity. When comparing to the unlimited sorption storage capacity, a 79 fold battery capacity results. The limited sorption storage capacity leads to a storage volume of the diluted sorbent of 5.8 m^3. When no dead volume is assumed this represents the full storage volume needed. When using fixed storage tanks for concentrated and diluted sorbent as well as for the absorbate a total storage volume of 11.6 m^3 would be required. In either case, this size could be fitted in an SFH, taking up a floor area of 2.9 and 5.8 m^2, respectively, assuming cubic tanks with a height of 2 m.

As shown in Figure 3, COP of the electric heat pump can be improved substantially when combined with a sorption storage. In the model a fixed isentropic efficiency was used, leading to some overestimation at very low temperature differences between evaporation and condensation. In reality, isentropic efficiency is not constant but varies across the operation range of a heat pump. When designed accordingly, however, high and relatively constant isentropic efficiencies can be achieved at low temperature lifts for a limited operation window [23].

COP improvement is the main driver for electric load shifting and reflects the effect of the double-stage heat pumping when introducing a sorption storage along with an electric heat pump. The other important part is the available energy storage capacity. It was differentiated between limited and unlimited capacity. Energy flexibility offered with limited storage capacity is smaller than expected because of limitations present in charging. When sizing a heat pump for a low-energy house such as the SFH45, installed capacity is rather small. In a hybrid operation of heat pump and sorption storage, as suggested, all heat provided to the building or the storage needs to go through the heat pump, making it to become the bottleneck. Another restriction comes from the time with available excess electricity from the grid. This is a given and strongly depends on the specific electricity generation system considered.

Different possibilities are seen to increase energy flexibility. Heat pump and HMX capacity could be increased, leading to overcapacity installed. When using a capacity-controlled heat pump, this would not negatively impact space heating operation but only system cost. Alternatively, only HMX size could be increased, dimensioning it rather for the charging than for the discharging operation being determined by the space heating demand of the building. In order to make use of larger HMX capacity, an additional heat source such as solar thermal collectors would be needed. While this would provide additional heat for charging it would not increase the negative energy flexibility. As it leads to larger energy storage capacity and hence seasonal load shift, it would still increase positive energy flexibility offered during heating season. Further, extended time of available excess electricity from the grid could be assumed, leading to larger energy storage capacity. The latter could be justified as times with excess electricity will increase with an increasing share of renewables in the electricity grid. Currently, excess is assumed for the months of June to September with an average of 6 h a day. When considering energy production profiles of solar thermal collectors or PV installations, excess production with reference to space heating demand could already appear in April and May as well as in October. When assuming, for example, 6 instead of 4 months with excess electricity, raising the number of hours by 50%, storage capacity and consequent seasonal electric load shift would rise to 266.7 kWh/a or 16.1% with reference to no storage. This increase of time with excess electricity assumed can be seen as a proxy for any measures discussed above to increase charging and thus energy storage capacity.

Looking at the overall storage potential of building integrated sorption storage for the case of Switzerland, assuming that an average storage capacity of 266.7 kWh is installed in every one of the 1.7 millions of domestic buildings (including multi-family homes), a total figure of 0.46 TWh would result, representing 5.04% of existing electric storage capacity in Switzerland provided by seasonal hydro storage [24].

Building-level sorption storage has substantial potential to provide energy flexibility to the electricity grid and support the integration of renewables. For a better prediction of available energy flexibility a more detailed analysis based on an improved model of the building and the sorption storage with higher temporal resolution is needed. Further, an extended analysis of design parameters and their influence on storage performance together with an optimized operation for minimal electricity demand and CO_2 emissions are desirable. These additions to the current study will be addressed in future research.

Author Contributions: Conceptualization, L.B.; Formal analysis, L.B.; Investigation, L.B. and B.F.; Methodology, L.B.; Writing – original draft, L.B.; Writing – review & editing, L.B. and B.F. All authors have read and agreed to the published version of the manuscript.

Funding: This research work was financially supported by the Swiss Innovation Agency Innosuisse grant Nr. 1155002545 and is part of the Swiss Competence Centre for Energy Research SCCER HaE. Supplementary funding was received from the Swiss Federal Office of Energy SFOE grant Nr. SI/501605-01 in the frame of the IEA SHC Task 58/ECES Annex 33 participation.

Conflicts of Interest: The authors declare no conflict of interest. The funders had no role in the design of the study; in the collection, analyses, or interpretation of data; in the writing of the manuscript, or in the decision to publish the results.

Nomenclature

$E_{flex,charging}$	Negative energy flexibility offered during charging (kWh)
$E_{flex,discharging}$	Positive energy flexibility offered during discharging (kWh)
$P_{el,hp,charging}$	Electric power of heat pump during charging (kW)
$P_{el,storage,discharging}$	Electric power of heat pump during discharging (kW)
$P_{el,ref,heating}$	Electric power of heat pump without storage during heating (kW)
Δt_{excess}	Time of excess electricity available (hours)
$\Delta t_{heating}$	Time with space heating demand (hours)
$\eta_{charging}$	Charging efficiency (%)
η_{isen}	Isentropic efficiency of heat pump (%)
$Q_{hp,desorption}$	Heat provided by heat pump during charging of storage (kWh)
$Q_{heating}$	Space heating demand (kWh)
$COP_{hp,charging}$	Heat pump COP during charging
$COP_{storage,discharging}$	Heat pump COP during discharging
$COP_{ref,heating}$	Heat pump COP without storage during heating
$T_{desorption}$	Sorbent temperature in HMX desorber during charging (°C)
T_{sat,p_desorb}	Saturation temperature of water at desorption pressure (°C)
T_{supply}	Supply temperature of space heating system (°C)
T_{amb}	Ambient air temperature (°C)
$T_{HMX,ev,in}$	Inlet temperature at HMX evaporator (°C)
$\Delta T_{hx,HMX}$	Temp. difference primary/secondary fluid in HMX heat exchanger (K)
$\Delta T_{hx,hp}$	Temperature difference primary/secondary fluid in hp heat exchanger (K)
$\Delta T_{HMX,ev}$	Temperature difference of HTF across HMX evaporator/condenser (K)

ΔT_{air}	Temperature difference ambient air inlet to evaporator water outlet (K)
$\Delta T_{50,max}$	Maximum temperature lift sorption storage (K)
$M_{sorbent,in}$	Concentrated sorbent mass entering HMX absorber in discharge (kg)
$M_{sorbent,out}$	Diluted sorbent mass leaving HMX absorber in discharge (kg)
$h_{sorbent,in}$	Specific enthalpy of sorbent entering HMX absorber (kJ/(kg·K))
$h_{sorbent,out}$	Specific enthalpy of sorbent leaving HMX absorber (kJ/(kg·K))
h_{lg}	Specific enthalpy of condensation for water (kJ/(kg·K))
$c_{sorbent,in}$	NaOH concentration of sorbent entering HMX absorber (kg/kg)
$c_{sorbent,out}$	NaOH concentration of sorbent leaving HMX absorber (kg/kg)

Abbreviations

COP	Coefficient of performence
HMX	Heat and mass exchanger
hp	Heat pump
H_2O	Water
NaOH	Sodium hydroxide
PV	Photovoltaics
SFH	Single family home
SFH45	Single family home with specific space heating demand of 45 kWh/m^2

References

1. Rathgeber, C.; Hiebler, S.; Lävemann, E.; Pablo, D.; Ana, L.; Jaume, G.; Alvaro de, G.C.; Laia, M.; Luisa, F.C.; Andreas, K.-H.; et al. IEA SHC Task 42 / ECES Annex 29—A Simple Tool for the Economic Evaluation of Thermal Energy Storages. *Energy Procedia* **2016**, *91*, 197–206. [CrossRef]
2. Zondag, H. Chapter 6–Sorption Heat Storage. In *Solar Energy Storage*; Sørensen, B., Ed.; Academic Press: Cambridge, MA, USA, 2015; pp. 135–154.
3. N'Tsoukpoe, K.E.; Liu, H.; Le Pierrès, N.; Luo, L. A review on long-term sorption solar energy storage. *Renew. Sustain. Energy Rev.* **2009**, *13*, 2385–2396. [CrossRef]
4. Abedin, A.H. A Critical Review of Thermochemical Energy Storage Systems. *Open Renew. Energy J.* **2011**, *4*, 42–46. [CrossRef]
5. Ding, Y.; Riffat, S. Thermochemical energy storage technologies for building applications: A state-of-the-art review. *Int. J. Low Carbon Technol.* **2012**, *8*, 106–116. [CrossRef]
6. Bales, C. Chemical and sorption heat storage. In Proceedings of the DANVAK Seminar, Lyngby, Denmark, 14 November 2006.
7. Hadorn, J.C. *Thermal Energy Storage for Solar and Low Energy Buildings: State of the Art*; Universidad de Lleida: Lleida, Spain, 2005.
8. Fumey, B.; Weber, R.; Baldini, L. Liquid sorption heat storage—A proof of concept based on lab measurements with a novel spiral fined heat and mass exchanger design. *Appl. Energy* **2017**, *200*, 215–225. [CrossRef]
9. Frazzica, A.; Brancato, V.; Dawoud, B. Unified Methodology to Identify the Potential Application of Seasonal Sorption Storage Technology. *Energies* **2020**, *13*, 1037. [CrossRef]
10. Fumey, B.; Weber, R.; Baldini, L. Sorption based long-term thermal energy storage—Process classification and analysis of performance limitations: A review. *Renew. Sustain. Energy Rev.* **2019**, *111*, 57–74. [CrossRef]
11. Heier, J.; Bales, C.; Martin, V. Combining thermal energy storage with buildings—A review. *Renew. Sustain. Energy Rev.* **2015**, *42*, 1305–1325. [CrossRef]
12. Solé, A.; Martorell, I.; Cabeza, L.F. State of the art on gas—Solid thermochemical energy storage systems and reactors for building applications. *Renew. Sustain. Energy Rev.* **2015**, *47*, 386–398. [CrossRef]
13. Krese, G.; Koželj, R.; Butala, V.; Stritih, U. Thermochemical seasonal solar energy storage for heating and cooling of buildings. *Energy Build.* **2018**, *164*, 239–253. [CrossRef]
14. Fumey, B.; Weber, R.; Gantenbein, P.; Daguenet-Frick, X.; Williamson, T.; Dorer, V. Development of a Closed Sorption Heat Storage Prototype. *Energy Procedia* **2014**, *46*, 134–141. [CrossRef]
15. Aydin, D.; Casey, S.P.; Chen, X.; Riffat, S. Numerical and experimental analysis of a novel heat pump driven sorption storage heater. *Appl. Energy* **2018**, *211*, 954–974. [CrossRef]

16. Wu, W.; You, T.; Wang, J.; Wang, B.; Shi, W.; Li, X. A novel internally hybrid absorption-compression heat pump for performance improvement. *Energy Convers. Manag.* **2018**, *168*, 237–251. [CrossRef]
17. Palomba, V.; Dino, G.E.; Frazzica, A. Coupling sorption and compression chillers in hybrid cascade layout for efficient exploitation of renewables: Sizing, design and optimization. *Renew. Energy* **2020**, *154*, 11–28. [CrossRef]
18. Dott, R.; Haller, M.Y.; Ruschenburg, J.; Ochs, J.; Bony, J. The Reference Framework for System Simulations of the IEA SHC Task 44/HPP Annex 38 Part B: Buildings and Space Heat Load. Available online: https://task44.iea-shc.org/publications (accessed on 27 April 2020).
19. Climate Strasbourg—Monthly Temperatures. Available online: https://de.climate-data.org (accessed on 20 April 2020).
20. Fumey, B.; Weber, R.; Gantenbein, P.; Daguenet-Frick, X.; Williamson, T.; Dorer, V. Closed Sorption Heat Storage based on Aqueous Sodium Hydroxide. *Energy Procedia* **2014**, *48*, 337–346. [CrossRef]
21. Köll, R.; van Helden, W.; Fumey, B. European Union Seventh Framework Program Project COMTES—Combined Development of Compact Thermal Energy Storage Technologies—Deliverable 5.1: Description of Experimental Systems. Johansen, J.B., Furbo, S., Eds.; Available online: https://cordis.europa.eu/project/id/295568 (accessed on 27 April 2020).
22. Olsson, J.; Jernqvist, Å.; Aly, G. Thermophysical properties of aqueous NaOH−H2O solutions at high concentrations. *Int. J. Thermophys.* **1997**, *18*, 779–793. [CrossRef]
23. Gasser, L.; Flück, S.; Kleingries, M.; Meier, C.; Bätschmann, M.; Wellig, B. High efficiency heat pumps for low temperature lift applications. In Proceedings of the 12th IEA Heat Pump Conference, Rotterdam, The Netherlands, 12–14 May 2017.
24. Storage Lakes for a Sucessful Energy Transition (translated). *Factsheet 2019*; Swiss Association of Hydropower Economy. Available online: https://www.swv.ch/fachinformationen/wasserkraft-schweiz/ (accessed on 27 April 2020).

Article

Sensitivity Analysis of Window Frame Components Effect on Thermal Transmittance

Giorgio Baldinelli [1,*], Agnieszka Lechowska [2], Francesco Bianchi [3] and Jacek Schnotale [2]

1 Department of Engineering, University of Perugia, Via Duranti, 67-06125 Perugia, Italy
2 Department of Environmental and Power Engineering, Cracow University of Technology, ul. Warszawska 24, 31-155 Cracow, Poland; agnieszka.lechowska@pk.edu.pl (A.L.); j.schnotale@gmail.com (J.S.)
3 Department of Physics and Geology, University of Perugia, Via Pascoli, 06123 Perugia, Italy; francesco.bianchi@unipg.it
* Correspondence: giorgio.baldinelli@unipg.it; Tel.: +39-075-585-3868; Fax: +39-075-585-3697

Received: 6 May 2020; Accepted: 6 June 2020; Published: 9 June 2020

Abstract: Standard ISO 10077-2 gives the procedure to calculate thermal transmittances of window frames in 2D numerical simulations. It also introduces some examples of frame geometrical models with all necessary input data and the solutions so as to perform validation of the applied numerical tools. In the present paper, the models prepared with a commercial finite volume software of a PVC window frame were first positively validated with the results given in the Standard. An experimental test was then implemented to confirm the simulated data, with satisfactory agreement. The numerical code was used on one of the frames provided by the Standard to perform a sensitivity analysis of all the components and boundary conditions playing a role on the definition of the frame thermal transmittance, such as surface heat transfer coefficients, values of the solid thermal conductivity, emissivity and insulation properties of air gaps. Results demonstrate that the air gap properties represent the most influential parameters for the definition of the PVC window frames thermal transmittance, followed by the surface heat transfer coefficients and the PVC thermal conductivity. The rubber and the steel properties show a negligible effect on the whole frame performance. This procedure could constitute a design tool to guide the efforts of window manufacturers for the achievement of high performance products.

Keywords: window frames; numerical analysis; hot box; sensitivity analysis

---

Highlights

- Numerical simulations of different window frames given in Standard ISO 10077-2 were performed.
- Measurements of window frame U-value in hot box apparatus with satisfactory agreement with simulations were performed.
- A surface response module in Ansys Fluent software was performed, indicating a strong impact of air gap properties on the window frame U-value.
- Numerical sensitivity analysis method proved a reliable tool to optimise window frames performance.

1. Introduction

The massive use of insulating materials on opaque walls makes windows the weakest part of the building envelope in terms of heat loss [1,2]. The transparent components performance have been strongly improved in recent years, both from the glazing and frame points of view [3–6]. The researchers also examined the edge-of-the-glass region to create better joint solutions [7,8] but the efforts from researchers and manufacturers to lower the windows thermal transmittance are still needed [9].

Window frame thermal transmittance (U_f values) can be obtained both in numerical simulations and in measurements performed in the calorimetric chamber. Both methods are described in standards ISO 10077-2 [10] and EN 12412-2 [11], respectively. The computational approach could be implemented by software such as Frame Simulator [12], Therm [13], etc.

The standard ISO 10077-2, gives some examples of geometries of window frames with all the necessary input data and solutions (calculated U_f values). The Standard obliges the users of the method to validate the applied calculation programs and to perform the calculations given in the examples. The users should achieve a level of discrepancy lower than 3% between the results of the calculations and the values given in the Standard. In the first part of the present work, the software Ansys Fluent [14] was validated with the ISO 10077-2 approach to calculate U_f values. Fluent was chosen due to the capability of the code to implement optimization routines that allow a relatively quick calculation of a high number of frame design cases under investigation.

The robustness of the software results was further confirmed by measurements performed in standardized conditions in the calorimetric chamber, assessing the differences between the simulation and measurement results of an actual window frame thermal transmittance. After these steps, a sensitivity analysis [15] of the different parameters playing a role in the frame thermal transmittance was conducted. The investigation covered the boundary conditions (surface heat transfer coefficients on cold and warm sides), as well as material thermal conductivities and surface emissivity. The software allows automatic variation of the above-mentioned parameters within a user-defined range, in order to define the most influential ones on the window global performance. As far as the authors' knowledge, the sensitivity analysis approach has not been reported in previous studies and it could constitute a novel tool in the window frames design phase, covering a wide range of cases through simulations with a reasonable computational time.

Both calculations and experiments carried out in this research are valid only for steady-state conditions. The numerical models and the hot box measurements do not take into account the transient effect of boundary conditions (heat transfer coefficients, boundary temperatures, heat accumulation, overheating due to direct exposure to sunlight, etc.), which could play a significant role when the holistic performance of a window is analyzed.

2. Analyzed Frame Models and Simulation Input Data

The validation of the window frame thermal transmittance calculations took into account four chosen example models given in Standard ISO 10077-2 [10]: aluminum frame (example I1 from the Standard), PVC frame (example I3), a wooden frame (example I4) and a roof window wooden frame (example I5). The views of the models are given in standard [10]. An additional frame model was analyzed, which was a window frame made of PVC hosting a triple glazing [16] (Figure 1), which was also tested experimentally in a hot box apparatus.

All five models contain air gaps in the frames. The gaps were replaced by a solid with an equivalent thermal conductivity calculated according to Standard ISO 10077-2 [10] indications. Both convective and radiative heat transfer in the gaps were taken into account. The Standard [10] introduces the method of calculation including conductivity of air gaps when they are rectangular as well as when they are non-rectangular. In the latter case, the equivalent width, thickness and area of non-rectangular air gaps are calculated, and they are treated as rectangular, following the standard. The knowledge of surface temperatures of the gaps is required which implies an iterative process. In the first step, the gap surface temperatures were assumed and after heat exchange calculations they were corrected from the temperature distribution and the calculations were repeated again. After calculations of the equivalent values of a gap conductivity, the simulations treat all materials as solids. The material thermal conductivities are shown in Table 1. Surface emissivity was assumed as equal to 0.90. When calculating window frame thermal transmittance, glazing has to be replaced with an insulation panel of the same thickness as glazing and a visible height of 190 mm.

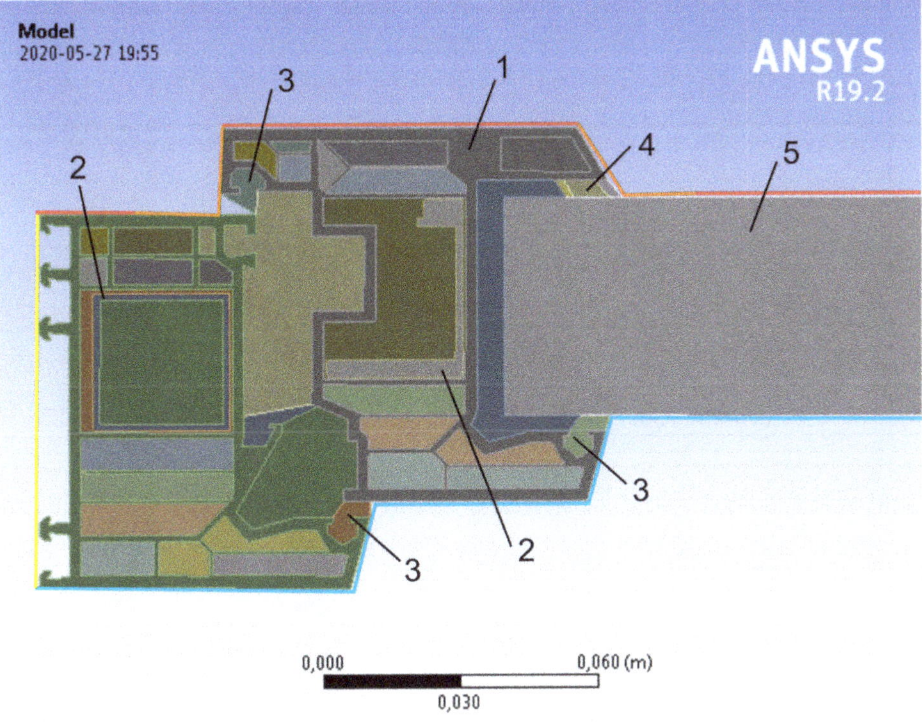

Figure 1. Model of real PVC frame with the following materials: 1, PVC; 2, steel; 3, ethylene-propylene diene monomer (EPDM); 4, butyl rubber; 5, insulation panel. Colored lines define the boundary conditions.

Table 1. Material thermal conductivities.

Material	Thermal Conductivity W/(mK)
Wood	0.130
PVC	0.170
EPDM	0.250
Aluminum	160.0
Polyamide	0.300
Steel	50.0
Butyl rubber	0.24
Insulation panel	0.035

Both the end of the insulation panel and the frame part adjacent to the wall were treated as adiabatic. On the internal (warm) side of the frame, the temperature of air θ_i was fixed at 20 °C, while on the external (cold) frame side, the air temperature θ_e was considered equal to 0 °C, as suggested by the Standard [10]. As the Standard [10] assumes different values of heat transfer coefficients on boundaries, the choices taken in the domain are shown in Figure 1. The yellow boundary line shows adiabatic conditions on the end of the frame (left hand side). Blue lines show external boundary conditions with heat transfer coefficient equal to 25 W/(m^{2}K), while red and orange lines represent internal boundary conditions with the values 7.7 W/(m^{2}K) and 5 W/(m^{2}K), respectively.

Ansys Fluent [14] program uses the finite volume method. The calculations were performed in steady-state conditions, with second order up-wind discretization scheme used for the energy equation.

The temperature distributions for five given geometries are presented in Figures 2–6.

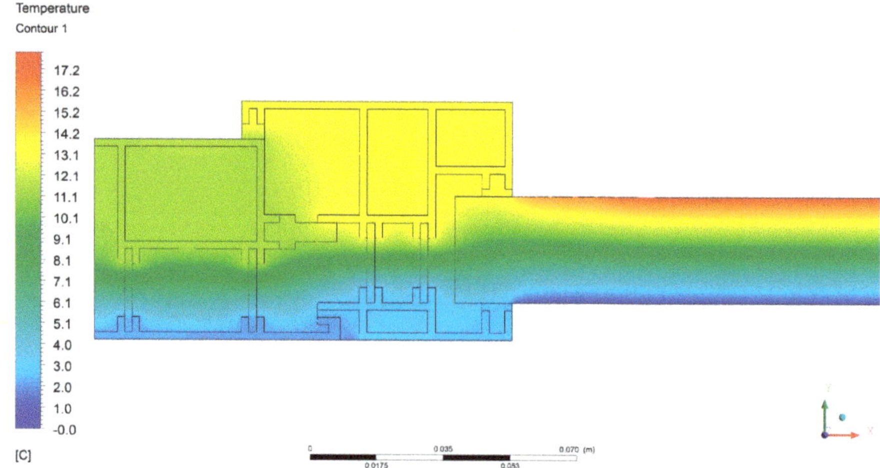

Figure 2. Temperature distribution on aluminum frame, example I1 from ISO 10077-2.

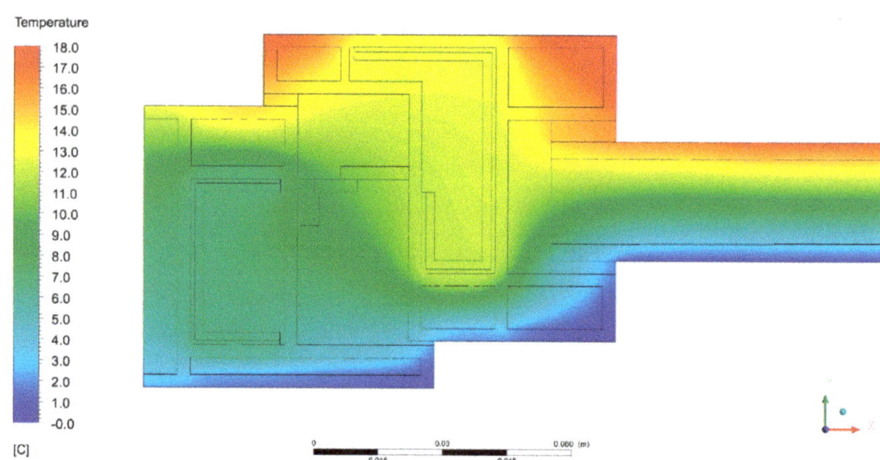

Figure 3. Temperature distribution on PVC frame, example I3 from ISO 10077-2.

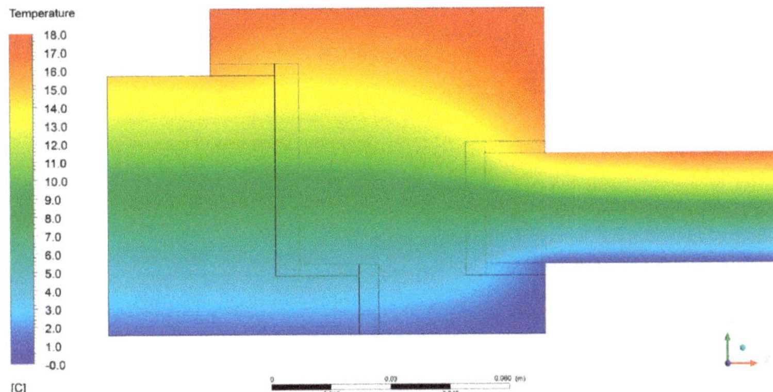

Figure 4. Temperature distribution on wooden frame, example I4 from ISO 10077-2.

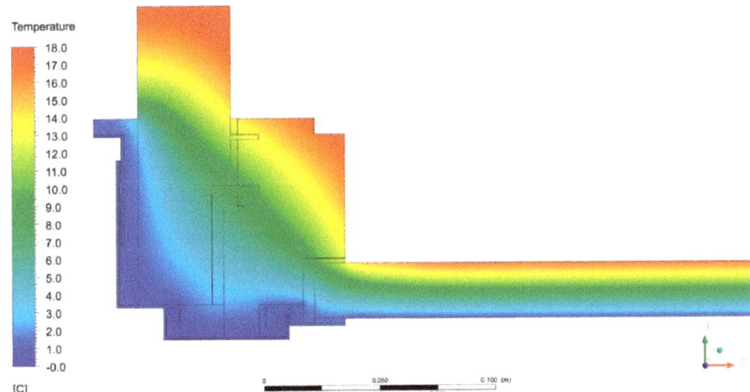

Figure 5. Temperature distribution on roof window frame, example I5 from ISO 10077-2.

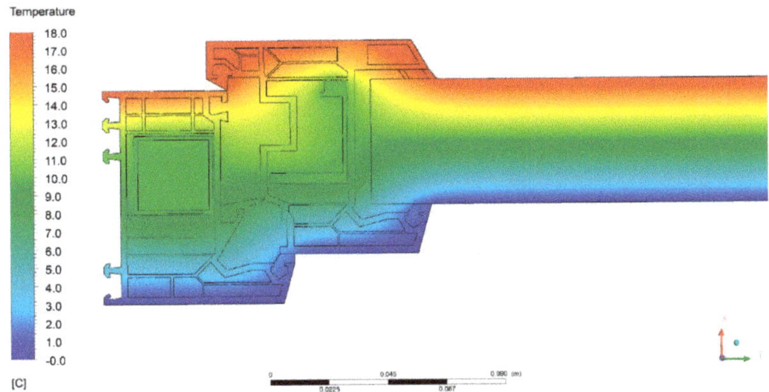

Figure 6. Temperature distribution on real PVC frame.

As can be seen in Figures 2–6, the temperature distribution in the part of glazing is one-dimensional (parallel isotherms), while in the frame the heat flux changes its direction. The warmer the frame on its internal side, the lower the value of frame heat losses. It can be seen that the aluminum one (example I1) is the coldest on its internal side, indicating the worst performance in terms of thermal transmittance (see the results in Table 2).

Table 2. Comparison of the simulation results and the results from the standard ISO 10077-2.

Frame	Source	L_f^{2D}, W/(mK)	U_f, W/(m²K)	$\frac{\lvert U_{f,simul} - U_{f,stand} \rvert}{U_{f,stand}} \cdot 100\%$
I1	Simulations	0.550	3.22	0.2%
	Standard	0.550 ± 0.007	3.22 ± 0.06	
I3	Simulations	0.421	2.05	1.0%
	Standard	0.424 ± 0.006	2.07 ± 0.06	
I4	Simulations	0.345	1.35	0.7%
	Standard	0.346 ± 0.001	1.36 ± 0.01	
I5	Simulations	0.402	2.02	2.9%
	Standard	0.408 ± 0.007	2.08 ± 0.08	
Real PVC frame	Simulations	0.261	1.098	Not applicable

3. Simulation Results

The window frame thermal transmittance was calculated from the following equation [10]:

$$U_f = \frac{L_f^{2D} - U_p b_p}{b_f}, \tag{1}$$

where:

$$L_f^{2D} = \frac{\Phi}{\theta_i - \theta_e}. \tag{2}$$

The results of the numerical calculations of five models are given in Table 2.

Results show that each simulation result differs less than 3% from the values given in the Standard [10]. Thus, it can be stated that the software Ansys Fluent [14] could be applied as a tool for window frame thermal transmittance calculations. The lower row of Table 2 is supplemented with the results of the simulations of the fifth model, whichh is a real PVC window frame hosting a triple glazing.

4. Window Frames Thermal Transmittance Measurements

The second stage of the work was focused on the measurements of window frame thermal transmittance and comparison of the experimental results with the simulations for validation.

The frame measurements should meet the demands of the Standard EN 12412-2 [11] and should be performed in a calorimetric chamber (Figure 7). This device consists of two spaces separated with a surround panel, which consists of a cold chamber and a warm chamber. The specimen to be tested was mounted into the surround panel, which was made of thick insulation material with known thermal conductivity. In the case of window frame thermal transmittance measurements, the glazing was substituted by an insulation panel with (known) low thermal conductivity, and the same thickness as the glazing.

Figure 7. The view of the Laboratory of Thermal Engineering calorimetric chamber situated in Cracow University of Technology, Poland.

In the cold chamber, the air was kept in steady-state thermal conditions (0 °C), through a control and data acquisition system. The same approach was used in the warm chamber, where the air temperature set point was fixed to 20 °C. The test specimen cold side faced a baffle. A set of ventilators was installed on the top of the baffle to push the air upwards in controlled forced convection conditions on the space between the baffle and the specimen. The heart of the measurement stand is the hot box system. It is a smaller guarded chamber embracing the specimen on its warm side. In the lower hot box part, ventilators were installed in order to move the air downwards along the tested specimen, to create steady-state natural convection conditions. The hot box was equipped with internal radiators. The idea of the measurement is that the radiators provide the amount of heat demanded to ensure the temperature difference between the internal and the external hot box walls is 0 K, i.e., no heat is transferred through the hot box walls. That means that all heat power from the radiators is supplied to the specimen, to the surround panel and to the flanking losses between them [16,17].

After more than three hours of steady-state conditions in the whole apparatus, the temperatures of air, surface temperatures of baffle, reveal, insulation filling and surround surfaces were measured on both cold and warm sides, along with the heat power supplied by the radiators. Then, the calculations given by the Standard EN 12412-2 [11] were performed to obtain the value of the window frame thermal transmittance.

The view of the tested window frame filled with the insulation panel and mounted into the surrounding panel in the calorimetric chamber of the Laboratory of Thermal Engineering at Cracow University of Technology is shown in Figure 8.

The uncertainty of the frame section was calculated using the error propagation rule from the equation:

$$\Delta U_f = \sqrt{\left(\frac{\Delta U_{m,t} \partial U_f}{\partial U_{m,t}}\right)^2 + \left(\frac{\Delta A_t \partial U_f}{\partial A_t}\right)^2 + \cdots + \left(\frac{\Delta \theta_n \partial U_f}{\partial \theta_n}\right)^2}, \quad (3)$$

where the uncertainties of all the quantities of Equation (3) had to be calculated according to the error propagation rule. The measured frame thermal transmittance results equaled to 1.063 W/(m^{2}K), with an uncertainty of about 8%.

The discrepancy between the measured and simulated Ansys Fluent values of the frames thermal transmittance was about 3%. The good agreement of measurements with the simulations enforces the reliability of the finite volume analysis for the evaluation of window frames thermal performance.

Figure 8. View of the tested window frame in the surround panel on the warm side.

The measurement results as well as the calculation results based on the Standard EN 12412-2 [11] are given in Table 3.

Table 3. Measurement and calculation results of real frame thermal transmittance.

Quantity	Value	Unit
Air temperature, cold side	0.7	°C
Baffle temperature, cold side	0.6	°C
Reveal temperature, cold side	1.2	°C
Insulation filling temperature, cold side	1.4	°C
Surrounding panel temperature, cold side	1.1	°C
Air velocity, cold side	1.5	m/s
Air temperature, warm side	19.8	°C
Baffle temperature, warm side	19.5	°C
Reveal temperature, warm side	18.0	°C
Insulation filling temperature, warm side	18.1	°C
Surrounding panel temperature, warm side	19.0	°C
Air velocity, warm side	0.1	m/s
Heat power in the hot box	33.2	W
Density of heat flow rate through the specimen	12.5	W/m^2
Surround panel thermal resistance	5.6	m^2K/W
Total surface heat transfer resistance	0.26	m^2K/W
Environmental temperature, warm side	19.5	°C
Environmental temperature, cold side	0.7	°C
Frame thermal transmittance	1.063	$W/(m^2K)$
Frame thermal transmittance uncertainty	0.089	$W/(m^2K)$

5. Sensitivity Analysis of the Influence of Input Data on the Value of Window Frames Thermal Transmittance

In this section, after the simulation code validation, analysis of the influence of input data on the results of window frames thermal transmittance value was performed. The geometry presented in Figure 2 (frame D3 from the Standard [10]) was chosen for the analysis. This is a PVC frame with steel reinforcements, EPDM seals and 11 air gaps within its geometry.

The analysis was performed using Ansys DesignXplorer [14]. This tool enabled us to parameterize the desired input data in Fluent in the assumed range of variations in order to achieve the response surface and to analyze the optimized outputs. The response surface can be used to predict the performance of the data without needing to run the actual simulation. This model was applied for the sensitivity analysis. In the present study, the sensitivity analysis method called "screening" was applied. By means of this approach, it was determined how the change of the input values of a certain parameter influenced the output. The sensitivity is a quotient where the nominator is the difference between the output values Y(X + D) and Y(X), where D is a percentage variation of the input X and it is the term on the denominator. The higher the sensitivity, the higher the impact of the analyzed variable.

In order to prepare the optimization process, it was necessary to create variable input parameters in a given range instead of constant values. The parametric calculations took into account thermal conductivities of the solid materials as well as the surface heat transfer coefficients both on cold and warm frame sides. Moreover, the thermal conductivities of air gaps in the frame were treated as changeable parameters as well. The Standard [10] introduces the method of calculating the equivalent thermal conductivities of air gaps within the frame geometries, considering one-dimensional convective and radiative heat transfer through the gas gaps. As a result, the gap thermal equivalent conductivity treats the gas as a solid material and these values constituted the input data in Ansys Fluent [14] calculations for the frame thermal transmittance. In order to perform the sensitivity calculations, it was necessary to create user defined functions (UDF) in C^{++} to calculate equivalent conductivities of air gaps in Fluent, based on pre-run established gap temperatures and emissivity, which were imported to the UDF by a get-parameter function. These values were then parameterized in Design Explorer. The output parameters were the heat fluxes on the boundary surfaces and a frame U-value that was set as the optimized parameter. The ranges of the variable parameters used for the sensitivity analysis are listed in Table 4.

Table 4. Ranges of parameters variation in the sensitivity analysis.

Parameter	Design Value	Range of Variation
P1: surface heat transfer coefficient, cold side, W/(m^2K)	25.0	10.0–40.0
P2: surface heat transfer coefficient, warm side, W/(m^2K)	7.7	5.0–10.0
P3: PVC thermal conductivity, W/(mK)	0.16	0.10–0.20
P4: surface emissivity of air gaps	0.90	0.10–0.90

The preliminary sensitivity calculations indicated that thermal conductivities of steel and EPDM have no influence on the results, mainly because of the low amount of these materials in the whole geometry. For this reason, among all the solids, only PVC conductivity was varied in the sensitivity calculation process.

The sensitivities of four output parameters described in Table 4, as calculated in Ansys Fluent [14], are given in Figure 9. A higher parameter sensitivity value means that it has a bigger impact on the output results.

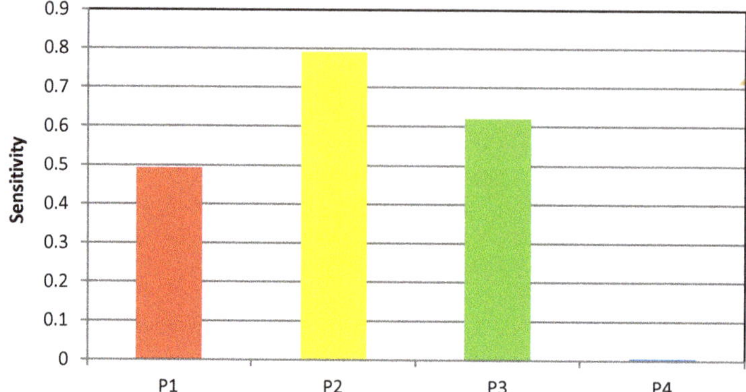

Figure 9. Sensitivities of parameters P1–P4, described in Table 4.

The results indicated that the values of heat transfer coefficients on cold and warm frame sides have dominant impact on the results of frame thermal conductivities together with the PVC thermal conductivity, while the air gaps emissivity alone has a negligible effect on the whole performance.

The surface heat transfer coefficients depend mainly on the air velocity and, at a lower level, on the inner and outer surfaces emissivity. Therefore, their prominent modifications are linked to the convection heat transfer, an effect not tunable by the window frame construction.

The rationing is not to include parameters not directly connected to the building component on the evaluation of its performance. Thus, in the last part of the sensitivity analysis, they were kept constant.

For the same reason, the temperature variations on both cold and warm sides were not investigated, despite the fact that inner and outer temperatures change significantly depending on building location and weather conditions. These temperatures vary quickly in time, especially in the hot season.

On the other hand, the effect of the air gaps equivalent thermal conductivity variation had a substantial impact on the window frame thermal transmittance. It was varied between the worst condition (air in the gap and surface emissivity equal to 0.90) and the most efficient one (which represents the situation of air gaps filled with a super insulation material). Intermediate values may represent filling air gaps with different insulation materials like wool, aerogel or different gases like argon, and giving the surfaces different emissivity [18].

The results of this part of the optimization process indicated that the values of equivalent thermal conductivities of gaps (P5) have the strongest influence on thermal transmittance of window frames with respect to the PVC thermal conductivity (Figure 10).

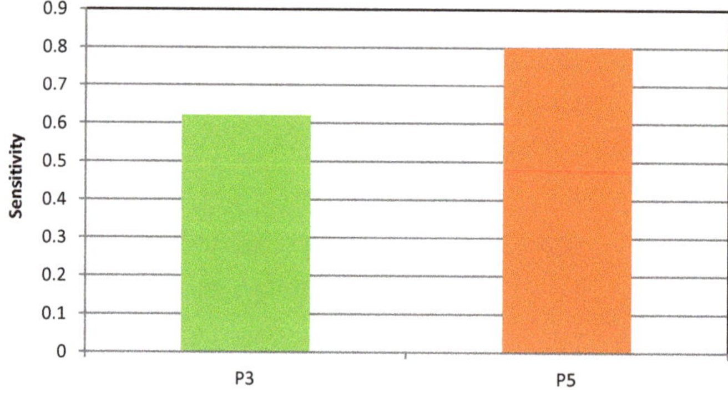

Figure 10. Sensitivities of PVC thermal conductivity (P3) and multiplier of air gap equivalent conductivity (P5).

Summing up the sensitivity analysis results, both window frame construction and external environmental boundaries considerably influence its thermal performance, with the gaps playing the most important role among the parameters where a designer could intervene.

6. Conclusions

The finite volume analysis implemented by a commercial code for the windows frame thermal transmittance numerical assessment proved reliable (discrepancy lower than 3%) both from comparison with Standard indications and from experimental validation obtained from a calorimetric chamber apparatus.

The software has been used to perform a sensitivity analysis of the input data (material conductivities, internal and external surface heat transfer coefficients, surface emissivity) on the results of a PVC triple glazing window frame thermal transmittance. The objective consisted of evaluating the influence of each parameter on the global performance of the frame.

The method proved particularly effective, as it allows one to change the input parameters, spanning between reasonable input values, then looking at the changing results. The procedure is implemented with a relatively low computational effort and it seems original, at least in the field of window thermal properties assessment.

The specific outcomes for the case analyzed indicate that it is worth focusing on air gap properties, which could lead to significant improvements if adequately treated. For instance, filling the gaps with an insulation material, substituting the air with a lower conductivity gas such as argon and covering the gap walls with low-emissivity coatings. Beyond the peculiarity value of these results, the proposed numerical optimization methodology could represent a useful tool for window designers and manufacturers to address their energies towards the most fruitful actions for the enhancement of the building components performance.

Author Contributions: G.B., A.L., F.B. and J.S. have contributed equally to the paper. All authors have read and agreed to the published version of the manuscript.

Funding: This research received no external funding.

Conflicts of Interest: The authors declare no conflict of interest.

Nomenclature

A_f	Frame area, m^2
A_{fi}	Area of the frame filling, m^2
A_t	Surface of project of measurement area, m^2
b_f	Width of the frame section, m
b_p	Visible width of the insulation panel, m
L_f^{2D}	Thermal conductance of the entire model, W/(mK)
U_f	Frame thermal transmittance, W/(m^2K)
$U_{m,t}$	Measured thermal transmittance of frame and insulation filling, W/(m^2K)
U_p	Insulation panel thermal transmittance, W/(m^2K)
$\Delta\theta_n$	Environmental temperature difference, K
$\Delta\theta_{s,fi}$	Surface temperature difference on warm and cold sides of the frame filling, K
Φ	Linear heat flow rate, W/m
Λ_{fi}	Thermal conductance of the frame filling, W/(m^2K)
θ_e	External air temperature, °C
θ_i	Internal air temperature, °C

References

1. Gustavsen, A.; Grynning, S.; Arasteh, D.; Jelle, B.P.; Goudey, H. Key elements of and material performance targets for highly insulating window frames. *Energy Build.* **2011**, *43*, 2583–2594. [CrossRef]

2. Ihm, P.; Park, L.; Krarti, M.; Seo, D. Impact of window selection on the energy performance of residential buildings in South Korea. *Energy Policy* **2012**, *44*, 1–9. [CrossRef]
3. Larsson, U.; Moshfegh, B.; Sandberg, M. Thermal analysis of super insulated windows (numerical and experimental investigations). *Energy Build.* **1999**, *29*, 121–128. [CrossRef]
4. van der Bergh, S.; Hart, R.; Jelle, B.P.; Gustavsen, A. Window spacers and edge seals in insulating glass units: A state-of-the-art review and future perspectives. *Energy Build.* **2013**, *43*, 263–280. [CrossRef]
5. Thalfeldt, M.; Kurnitski, J.; Voll, H. Detailed and simplified window model and opening effects on optimal window size and heating need. *Energy Build.* **2016**, *127*, 242–251. [CrossRef]
6. Yang, Z.; Katsura, T.; Aihara, M.; Nakamura, M.; Nagano, K. Investigation into Window Insulation Retrofitting of Existing Buildings Using Thin and Translucent Frame-Structure Vacuum Insulation Panels. *Energies* **2018**, *11*, 298. [CrossRef]
7. Koo, S.Y.; Park, S.; Song, J.-H.; Song, S.-Y. Effect of Surface Thermal Resistance on the Simulation Accuracy of the Condensation Risk Assessment for a High-Performance Window. *Energies* **2018**, *11*, 382. [CrossRef]
8. Park, S.; Song, S.-Y. Evaluation of Alternatives for Improving the Thermal Resistance of Window Glazing Edges. *Energies* **2019**, *12*, 244. [CrossRef]
9. Baldinelli, G.; Asdrubali, F.; Baldassarri, C.; Bianchi, F.; D'Alessandro, F.; Schiavoni, S.; Basilicata, C. Energy and environmental performance optimisation of a wooden window: A holistic approach. *Energy Build.* **2014**, *79*, 114–131. [CrossRef]
10. ISO 10077-2. *Thermal Performance of Windows, Doors and Shutters—Calculation of Thermal Transmittance—Part 2: Numerical Method for Frames*; International Organization for Standardization: Geneva, Switzerland, 2017.
11. EN 12412-2. *Thermal Performance of Windows, Doors and Shutters—Determination of Thermal Transmittance by Hot Box Method—Frames*; European Standarization Organization: Brussels, Belgium, 2003.
12. Available online: https://www.dartwin.it/sw/frame-simulator/ (accessed on 20 February 2020).
13. Available online: https://windows.lbl.gov/tools/therm/software-download (accessed on 20 February 2020).
14. ANSYS, Inc. *Products 2019 R1*; ANSYS: Canonsburg, PA, USA, 2019.
15. Zajas, J.; Heiselberg, P. Parametric study and multi objective optimisation of window frame geometry. *Build. Simul.* **2014**, *7*, 579–593. [CrossRef]
16. Lechowska, A.; Schnotale, J.; Baldinelli, G. Window frame thermal transmittance improvements without frame geometry variations: An experimentally validated CFD analysis. *Energy Build.* **2017**, *145*, 188–199. [CrossRef]
17. Lechowska, A.; Schnotale, J. Thermal transmittance of multi-layer glazing with ultrathin internal partitions. In Proceedings of the from BS2015: 14th Conference of International Building Performance Simulation Association, Hyderabad, India, 7–9 December 2015.
18. Asdrubali, F.; Baldinelli, G.; Bianchi, F. Influence of cavities geometric and emissivity properties on the overall thermal performance of aluminum frames for window. *Energy Build.* **2013**, *60*, 298–309. [CrossRef]

Article

Coloured BIPV Technologies: Methodological and Experimental Assessment for Architecturally Sensitive Areas

Martina Pelle [1,2], Elena Lucchi [1,*], Laura Maturi [1], Alexander Astigarraga [1] and Francesco Causone [2]

1 Institute for Renewable Energy, EURAC Research, 29100 Bolzano, Italy; martina.pelle@eurac.edu (M.P.); laura.maturi@eurac.edu (L.M.); alexander.astigarraga@eurac.edu (A.A.)
2 Department of Energy, Politecnico di Milano, Via Lambruschini 4, 20156 Milano, Italy; francesco.causone@polimi.it
* Correspondence: elena.lucchi@eurac.edu; Tel.: +39-0471-055-653

Received: 20 June 2020; Accepted: 20 August 2020; Published: 1 September 2020

Abstract: Energy flexibility in buildings is gaining momentum with the introduction of new European directives that enable buildings to manage their own energy demand and production, by storing, consuming or selling electricity according to their need. The transition towards a low-carbon energy system, through the promotion of on-site energy production and enhancement of self-consumption, can be supported by building-integrated photovoltaics (BIPV) technologies. This paper investigates the aesthetic and technological integration of hidden coloured PV modules in architecturally sensitive areas that seem to be the best possibility to favour a balance between conservation and energy issues. First, a multidisciplinary methodology for evaluating the aesthetic and technical integration of PV systems in architecturally sensitive area is proposed, referring to the technologies available on the market. Second, the experimental characterisation of the technical performance specific BIPV modules and their comparison with standard modules under standard weather condition are analysed, with the aim of acquiring useful data for comparing the modules' integration properties and performance. For this purpose, new testbeds have been set up to investigate the aesthetic integration and the energy performances of innovative BIPV products. The paper describes the analyses carried out to define the final configuration of these experimental testbeds. Finally, the experimental characterisation at standard test conditions of two coloured BIPV modules is presented and the experimental design for the outdoor testing is outlined.

Keywords: building-integrated photovoltaics; BIPV; hidden coloured BIPV module; BIPV integration; photovoltaic; PV

1. Introduction

The latest surveys on the current energy use scenario in Europe reported that buildings account for almost 40% of energy consumptions and 36% of carbon dioxide emissions (CO_2 emissions) [1]. The European Union (EU) has adopted a comprehensive regulatory framework to meet the commitments stated in the Paris Agreement [2] and to facilitate the transition from fossil fuels to cleaner energy production. The Energy Performance of Buildings Directive (EPBD) and the European Directive on Energy Efficiency (EED) provide the roadmap for the transformation of the existing building stock into nearly zero-energy buildings (nZEB) [3,4] through the definition of specific measures to improve the energy efficiency of buildings, reduce the energy use and enhance the decarbonisation in the construction sector. Furthermore, the recast version of the Renewable Energy Directive [5] set the target for renewable energy source (RES) penetration in the European energy mix to 32% by 2030. Energy flexibility in buildings is, moreover, gaining momentum through the introduction of self-consumers and collective self-consumers concepts in regulations which, on the one hand, empower users to be prosumers instead of merely consumers and, on the other hand, enable buildings to manage their energy demand and production, by storing, consuming or selling electricity according to their need [5,6]. In this context, building integrated photovoltaics (BIPV) technologies can support the transition towards a low-carbon energy system, promoting on-site energy production and enhancing self-consumption, if integrated into the overall building/district energy system and coupled with electric or thermal storages [7]. BIPV constitute indeed a solution to incorporate RES in the built environment by integrating solar photovoltaics (PV) technologies in the building envelope. More precisely, BIPV systems have a dual purpose: they serve as building envelope and as power generation systems at the same time, harvesting solar energy for on-site energy production [8]. This is a relevant characteristic, especially for the decarbonisation of energy systems in densely built environments, where the on-site energy production is difficult to exploit due to urban constraints, which hinder traditional ground-mounted PV installation. In addition, the solar potential for roof-mounted PV is low if compared to multi-property and multi-story building energy demand [9]. Despite the high potential for BIPV applications, there is the need to overcome technical, social and economic barriers to reach a larger scale of BIPV applications, improving their economic profitability [10–12]. Besides the cost, which has consistently decreased recently, the main limit to the spread of BIPV has so far been their aesthetic, since they are often considered anaesthetic by users and architects [13]: "[...] *When we hear about photovoltaics, however, the image that is invoked in our mind is a blue or black element that usually seems to "overload" the aesthetic image of a building* [14]. Research institutes, universities and industries are working together to design and produce a novel generation of BIPV solutions which will transform the visual appearance of standard PV modules into a more "architecturally pleasing" one [15] to be integrated into sensitive environments. Crystalline silicon modules, thin films, coloured solar cells, homogenised black appearance and integration of high-resolution images are just a few examples of the new possibilities offered by the PV market. However, the turning point in the acceptance of BIPV applications has been the development of hidden coloured BIPV modules. This module typology (which includes several different technologies, as discussed in Section 4) can hide the PV cells behind coloured patterns which hinder the perception of the original material of the cells. In this way, the modules appear very similar to standard construction materials [9]. As a result, a wide range of colours is currently available on the market, and this wide selection enables PV to be integrated also in traditional roofs, façades, and shading systems (Section 4). Nonetheless, the production of these modules is predominantly customisable for a specific installation. Modules customisation ensure a variety of new aesthetic and technical possibilities that facilitate the use of BIPV technologies also in densely built environments. This solution appears appropriate also for architecturally sensitive areas (i.e., historical centres, vernacular and historic buildings, natural and cultural landscapes), thanks to the aesthetical and technological advances related to low-rate reflection, mimetic appearance, compact shape and geometric flexibility [16–19]. Thus, these aesthetic improvements unlock the solar potential of a large set of vertical and horizontal envelope surfaces currently not exploited, leading the building stock to energy flexibility and self-sufficiency [20,21]. Conversely, the extreme requirements of customisation results in a fragmented market scenario and a high variety of colours, shapes, sizes, finishing, mechanical robustness and electricity generation efficiencies. Therefore, there is an urgent need to better frame the hidden coloured BIPV technologies available on the market, while building a common dialectic within the stakeholders along the entire value chain, to boost the BIPV market penetration. Section 2 indicates the aims and methodology pursued in this paper and outlines its structure.

2. Aims and Methodology

This paper investigates the aesthetic and technological integration of hidden coloured PV modules in architecturally sensitive areas. In these areas, there is an increasing debate on the possible integration of PV systems, respecting heritage constraints as well as preserving historic and natural values.

Coloured BIPV modules seem the best possibility to favour a balance between conservation and energy issues (Section 4.1). In parallel, these technologies have shown a relatively recent market growth, and their applications are significantly increasing due to their flexibility. Therefore, there is an urgent need to better understand the overall technical performance of these promising products against the real requirements of the built environment, with the purpose of providing better modelling of their behaviour and fostering energy flexibility in buildings [22,23]. Despite the high technological readiness level (TRL) of BIPV systems, it is essential to further investigate their performance and reliability in operational environment to increase the users' trust and boost its market penetration.

One aspect related to the novelty of this research concerns the interdisciplinarity between aesthetic and technical aspects. The first one refers mainly to the work of conservators, designers, heritage and public authorities that need clear criteria for the methodological assessment of BIPV systems in architecturally sensitive areas. The second one refers to engineers, manufactures, supplies, installers and power utilities that need clear data from BIPV experimental assessment. For this reason, the paper is divided in two parts. First, a multidisciplinary methodology for evaluating the aesthetic and technical integration of PV systems in architecturally sensitive area is proposed, referring to the most promising technology (Sections 3 and 4). More precisely, in Section 3, the key concept of integration is discussed, with particular focus on the international guidelines for the application of BIPV in architecturally sensitive areas, which has been crucial to identifying the criteria for ensuring the aesthetic integration of BIPV technologies (Section 3.1). Then, once recognising that the visual integration is fundamental for broader BIPV deployment, and in particular that the geometrical uniformity and the colour of the cells play a key role in this respect, the analysis focuses on the state of the art of the existing technologies for hidden coloured PV modules. To this purpose, a deep technical review of the BIPV technological solutions available on the market is realised (Section 4). On these bases, an evaluation matrix has been developed with the aim of steering the market analysis required to choose the technologies to be tested within the project (Section 4.1). This evaluation process has been used to identify a set of different coloured BIPV technologies for their experimental characterisation. Second, the experimental characterisation of the technical performance specific BIPV modules and their comparison with standard modules under standard weather condition are analysed, with the aim of acquiring useful data for comparing the modules' integration properties and performance (Section 5). To this purpose, a new facility specifically developed for BIPV system has been set up for investigating the aesthetic integration and the energy performances of innovative PV products, focusing on the three integration concepts: (i) technology (i.e., innovative integration substructures); (ii) aesthetic (i.e., appealing PV modules); and (iii) energy integration (i.e., plug and play concepts). The description of the experimental facilities is provided in Section 5.1.1 (outdoor testbeds) and Section 5.1.2 (indoor laboratory). In Section 5.2, the indoor experimental results are presented, while the experimental design of the outdoor testbeds and the first qualitative results of the experimental camping is provided in Section 5.3. Finally, Section 6 presents the conclusions and lesson learnt. These facilities also allow to tackle specific challenges of BIPV to develop better and more performing systems, as well as to show among all the stakeholders involved in the value chain (i.e., manufactures, designers, supplies, installers, national public authorities, local planning authorities, power utilities, owners, final users and financial bodies) the benefits offered by these innovative BIPV panels.

This work has been made in the framework of the research project *BIPV UPpeal* that aims at accelerating BIPV market penetration by showing to the main stakeholders the benefit of integrating PV systems in architecturally sensitive areas. This permits to concretely show the new aesthetic and technical possibilities of BIPV systems (e.g., with testbeds, case studies and databases) but also to create a network among the professionals. Otherwise, innovative coloured BIPV technologies are selected to be tested in these testbeds following the evaluation criteria defined in the framework of the EU project *BIPV meets history* that aims at creating a value chain for the use of BIPV in heritages, according to the international guidelines and ad hoc working tables [19].

3. PV Integration Concepts

Scientific literature adopted several definitions of BIPV, avoiding a univocal consensus for the PV and the building sectors. PV modules are considered integrated by the standard EN 50583-1:2016 [24] are if "[...] they form a construction product providing a function", as defined in the European Construction Product Regulation CPR 305/2011 [25]. This definition refers mainly to the idea of multi-functionality, according to which a BIPV module must have additional functions for the building envelope, besides the energy production (e.g., structural integrity; thermal insulation; solar shading; daylighting control; noise, fire and weather protection; safety; and security). However, this definition of BIPV does not seem exhaustive [15]. Different working groups of the International Energy Agency (IEA) on BIPV (i.e., PVPS Task 7 [15], IEA Task 41 [26], Task 59 [27] and Task 51 [28]), have underlined the importance of "formal/aesthetic" integration, beyond the multi-functionality concept. Particularly, the IEA-SHC Task 41 [26] defines architectural integration quality as the result of a controlled and coherent integration of the solar collectors from functional, constructive and formal (or aesthetic) points of view simultaneously. Other IEA working groups (Task 59 [27] and Task 51 [28]) also encourage an ad-hoc BIPV design to preserve original shapes, features and values in heritage and existing sites, favouring the aesthetic integration too. The balance of technical and aesthetic aspects is indeed a priority for BIPV system in terms of architectural functionalities and construction requirements (e.g., visual impact, dimensional flexibility, colour selection, easy mounting, safety and reliability, fire security, climate resistance, hygrothermal risks, thermal stability, maintenance and durability) [19,29]. From the analysis of the literature and standard definitions of BIPV, three integration levels can be identified [30]: (i) aesthetic; (ii) technological/functional; and (iii) energy (see Figure 1). Aesthetic integration refers to the capability of the PV solution to be included in the linguistic and morphological rules governing building's architectural language. The technological/functional level is strictly connected to the standard EN 50,583 definition [24], referring to the PV system capability to replace traditional building components. Finally, the energy integration refers to the ability of PV to be efficiently integrated into the overall energy system of the building/district through the "energy-matching" approach [31], thus interacting with the building loads to maximise self-consumption towards the implementation of efficient energy communities.

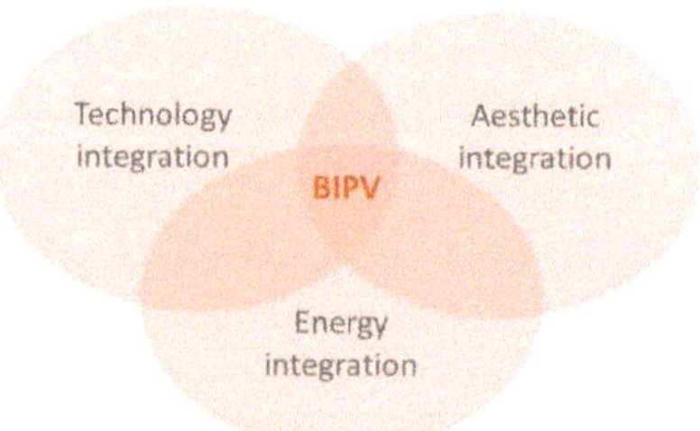

Figure 1. Multilevel BIPV integration aspects.

3.1. *PV Integration in Architecturally Sensitive Areas*

Special care must be given to PV integration in historical and heritage buildings as well as in architecturally and naturally sensitive areas (e.g., historical centres, heritage sites, archaeological areas or heritage landscapes). Consequently, several countries published national guidelines defining the architectural criteria for RES installation [32–38] according to national legislation, local authorisation processes and specific heritage features. These tools are addressed to the specialists in the field of design, architecture, engineering and energy consulting as well as to the public authorities involved in the energy issues, preservation of monuments and release of building permits [34]. The criteria for PV installation are always devoted to the protection of historic and distinctive materials, features, spaces, finishes, construction techniques, traditional craftsmanship and spatial relationships. Therefore,

PV panels must not create permanent losses or transformations in the historic fabric, significant architectural obstructions or disjointed and multi-roof solutions [32–35,37,39]. Both PV and BIPV technologies can be inserted on new constructions, deteriorated historic buildings or elements (i.e., a damaged roof), non-historic buildings, new additions or adjacent constructions, while matching the original designs, colours and texture [32,33,35,39]. Missing features of historical buildings can also be replaced with PV panels or BIPV systems by documentary and physical evidence, using similar colours and textures [35,39]. PV technologies can also be used in industrial buildings and 20th century architecture, where they express the idea of "*material innovation*" [39]. In listed buildings and their settings, the heritage authorities must evaluate the impact of PV systems on the historic and natural values [33,35,36,39].

According to the theory of restoration in architecture, the evaluation criteria in these sensitive areas can be summarised in visibility, technical compatibility and reversibility of the systems [34]. Visibility is the most important aspect in heritage buildings and natural areas. It refers to the minimisation of the visual impact of PV technologies, and thus the preservation of the original features, colours, texture, shapes, geometries, proportions and spatial relationships [33,35,39]. Overall, PV panels are not permitted on roofs and façades on the side of the building which is the most viewed generally from a public thoroughfare, a road or natural site or above the principal elevation [32,35]. On the contrary, PV panels can be located on hidden roof planes, for example in internal valley or street, behind parapets, new additions or outbuilding [35]. In this case, the new roof has to be hidden with existing roof ridge lines and flush [35]. Two main aesthetical parameters are considered for the integration of PV systems in architecturally sensitive areas: geometrical uniformity and colours of the cells. First, the visual impact of PV panels depends on the coverage of the surface by PV modules where 100% coverage is preferable for a more uniform appearance [32]. Therefore, BIPV products may be appropriately suited for historic buildings [35,39]. The acceptability of PV technologies is more difficult in these contexts because it requires also the respect of building lines, the grouping of PV panels, the reduction of the spaces among the panels and the accurate design [32,33,35,38–41]. Second, the chromatic integration with traditional materials is strongly suggested for PV camouflage, using terra-cotta cells for clay roof tiles, anthracite or green-grey cells for slate or stone, white cells for plaster or high-resolution images as marble or wood [39]. The aspects related to low-reflectance, camouflage of the PV cells, texturization and aesthetic pattern of PV modules are not considered in these guidelines but are very important for heritage authorities [19]. Compatibility refers to the protection of the integrity of the property and its environment guaranteeing the technical compatibility between old and new materials, avoiding hygrothermal (e.g., moisture accumulation on the back-side), structural (e.g., falling and excessive deflection) and energy (e.g., reduction of the efficiency and thermal bridges) risks [35]. Reversibility refers to the possibility of removing PV or BIPV system in the future, without affecting the essential form and integrity of the historic property and its environment. Removals and replacements of PV panels should be considered in the design phase to minimise loss or damage of original fabric [32]. The previous guidelines report only general principles for the aesthetic and technical integration of PV systems, but not references to specific technologies. On the contrary to these guidelines, some Italian working tables of the EU project *BIPV meets history* show that the heritage authority prefers BIPV systems instead of PV modules applied to a building element (i.e., roof or façade) or traditional material with PV panels (i.e., PV tile) [19]. Hidden coloured PV modules, semi-transparent PV-active layers and/or textured PV modules seem very promising for the integration in heritage and architecturally sensitive areas [19].

4. Existing Technologies for Hidden Coloured PV Modules

Broader architectural application claims for improvements in the aesthetic rendering of BIPV modules [23,42,43]. The turning point in the aesthetic acceptance of BIPV applications has been the development of modules that can hide the PV cells behind coloured patterns which hinder the perception of the original material of the cells, making the modules appear as standard construction components. This kind of coloured BIPV modules have shown relatively recent market growth, and their application is considerably increasing. Nonetheless, colouring the modules hinders the PV performance, due to the optical and physical behaviour of the coloured layers, which can cause the reflection of portions of solar radiation that would be otherwise converted into electricity [44]. Some theoretical studies, focusing mainly on monochromatic colours, were conducted to define the relations between the modules' colour and their power losses. The research highlighted a rather low level of power loss ranging between 7% and 10% [45]. However, BIPV applications imply a wider colour range and the finishing layer could present textures, uneven surfaces, fouling and time-related performance decay. Hence, there is the need to improve the awareness on coloured BIPV technologies with regards to the electrical behaviour of a large variety of coloured modules, which should guarantee reliable power output during their operation. Different customisation techniques to obtain coloured or textured BIPV modules are currently used for modules available on the market, including: (i) solar cells with anti-reflection coating; (ii) semi-transparent and/or coloured PV-active layers; (iii) layers or interlayers containing solar filters, coloured or patterned coatings; (iv) coloured polymeric encapsulant films; and (v) printed, coated or alternative finished front glass [46,47]. Hereafter, a brief overview of these technologies is provided. The study is not exhaustive, but it aims at defining the peculiarities of each typology in terms of technical potential for each category. The same nomenclature used in this section is reported in Table 1 to categorise some commercial products that have been considered in the frame of the BIPV market analysis (Section 4.1).

4.1. Solar Cells with Anti-Reflection Coating

Coloured solar cells can be produced by means of deposition on the cells' surface of a hydrogenated amorphous silicon nitride SiN_x:H layer, which serves for both passivation and antireflection coating (ARC). SiN_x:H layer is deposited by means of plasma-enhanced chemical vapour deposition [48]. Once this nitride layer is optimised in thickness and refractive index, solar cells assume the classic blue hue of standard PV modules. Other production techniques of the passivation and ARC are possible, using, for example, double anti-reflective coating (DARC) realised by electron beam (e-beam) evaporation techniques to deposit an additional layer of SiO_2 on the SiN_x:H layer. Various colours can be obtained by tuning the SiO_2 layer thickness, without any variation on the coloured solar cells' conversion efficiency [49,50]. Even if a quite large range of colours can be obtained by using this process (blue, yellow, bronze, green and purple), the coloured cells appear to be iridescent and highly variable with viewing angle and incident light polarisation [51].

4.2. Semi-Transparent and/or Coloured Pv-Active Layers

Coloured or semi-transparent PV-active layers can exhibit semi-transparency or colour tunability according to the absorption spectrum of the specific materials used as active layer. In this technology category organic solar cells (OSCs), dye-sensitised solar cells (DSSCs) and perovskite solar cells (PSCs) are included [52]. Distinct colour appearance in OSCs can be obtained, for example, by varying the materials used in the donor-acceptor combinations or adding coloured dye compounds to the active layer. In PSCs, colour tuning is imputable to band gap modification or the inclusion of dyes in the photoactive layer or other layers (e.g., hole transporting layer) [53]. The colours achievable through these techniques are manifold, but the system efficiency is affected by the optical behaviour of the coloured layer.

4.3. Layers or Interlayers Containing Solar Filters, Coloured or Patterned Coatings

Another option to obtain coloured BIPV modules is by using layers or interlayers containing solar filters, coloured or patterned coatings [47,54], that can be laminated into the modules. In addition, encapsulant components and/or back sheet layers can be coloured or printed with semi-transparent ink. The degree of customisation for this BIPV typology is very high, and, consequently, the efficiency is highly affected by the optical properties of the coloured/patterned layers.

Table 1. Commercially available BIPV products, divided by typology: (i) solar cells with anti-reflection coating; (ii) semi-transparent or coloured PV-active layers; (iii) layers or interlayers containing solar filters, coloured or patterned coatings; (iv) coloured polymeric encapsulant films; and (v) printed, coated or alternative finished front glass.

Commercial Information			Functional/Technological Integration				Aesthetic Integration			Energy Integration	
Product	**Producer**	**Typology**	**Technological Unit**	**Technical Element Class**	**Flexibility**	**Shape**	**Dimension [mm]**	**Colour**	**Reflectance**	**Efficiency [%]**	**Nominal Power [Wp/m²]**
Coloured solar cells	LOF Solar	i	Multifunctional	Multifunctional	High	Rectangular	Custom	>10	na	14.5–18	120–165
Colorblast	Kameleon Solar	v	Multifunctional	Multifunctional	High	Custom	Custom	~6000	na	8–15	80–150
Kromatix	SwissINSO	v	Multifunctional	Multifunctional	High	Rectangular	1000 × 1640 664 × 1587	Custom	<20%	11.9–18	120–285
Kaleo	CSEM	iii	Multifunctional	Multifunctional	High	Rectangular	Custom.	Custom	na	Depending on coverage	na
Solaris Heritage (tile)	Freesuns	v	Roofing system	Roof	Medium	Rectangular	Custom.	2	na	up to 12.8	131–156
Solaxess	Solaxess	iii	Multifunctional	Multifunctional	High	Custom	Custom	5	na	11–15	106–147
Solarskin	Sistine solar	v	Roofing system	Roof	Medium	Rectangular	na	Custom	na	Depending on coverage	na
DSD-PV	Dutch Solar Design	iii	Multifunctional	Multifunctional	High		900 × 1200 mm	Custom	na	Depending on coverage	na
Suncol® uniform	Glassfer & Sunage	v	Multifunctional	Multifunctional	High	Custom.	Custom	>15	na	12.7–15	90–250
Suncol® texturing	Glassfer & Sunage	v	Multifunctional	Multifunctional	High	Custom.	Custom	Custom	na	Depending on coverage	90–250
InvisibleCell®	Invent	v	Multifunctional	Multifunctional	High	Rectangular	997 × 1663	Custom	na	Depending on coverage	162
Coloured amorphous silicon PV glass	Onyx solar	iv	Multifunctional	Multifunctional	High	Rectangular	Custom	Custom	na	na	28–58
Soltech Facade	Soltech	ii	Multifunctional	Multifunctional	High	Rectangular	1200 × 600	4	na	1.2–11.2	na
ClearVue PV (LSC)	ClearVue PV	iv	Multifunctional	Multifunctional	High	Rectangular	Custom	various	na	3	na
Solar Roof Tiles	Tesla	v	Roofing system	Roof	Medium	Rectangular	381 × 1143	various	na	10	na

4.4. Coloured Polymeric Encapsulant Films

Polymer materials are usually used in the lamination process as a bonding and protective layer for semiconductors. The most frequent polymers used for this purpose are polyvinyl butyral (PVB) and Ethylene-vinyl acetate (EVA) [55]. Both these polymers can be manufactured in different colours and shades, and, when coupled with amorphous silicon or polycrystalline silicon PV, they lead to coloured PV modules with different degree of transparency and a quite large colour palette (coloured polymeric encapsulant films) [56]. To avoid undesired reflection or absorption of energy in the visible spectrum range, which could lead to the reduction in the efficiency of the modules, the coloured layer is usually provided at the rear side of the PV module [57].

4.5. Printed, Coated or Alternative Finished Front Glass

Modified front glass modules are produced by coupling a glass front sheet with a glass or metal back-sheet by means of lamination with polymeric encapsulants which incorporate c-Si cells. The most common encapsulant materials are ethylene-vinyl acetate (EVA) and polyvinyl butyral (PVB). The front glass sheets can be tinted with different hue or digitally printed to reproduce the appearance of traditional construction materials or any other pattern or image. This procedure ensures the camouflage of the PV cells which are almost completely hidden to the view. The front glass can be also texturized to provide different finishing [58]. Available finishing options on the market are shining, matte or three-dimensional texturized. Although the manifold customisation options offer several aesthetical advantages, the modification of the front glass leads to changes in the optical behaviour of the glass sheet, which could reflect or absorb a portion of the solar spectrum that would otherwise reach the PV cells where it would be used to produce electricity. Hence, the main challenge to be faced in the production of such modules consists in seeking the optimal trade-off between aesthetic and energy efficiency [44,45].

4.6. Evaluation Criteria and Market Analysis

As just reported, BIPV applications claim for multi-functionality properties of the products. The methodology for the evaluation and selection of a BIPV module for a defined project could ground its theoretical roots on this multi-functional integration (as defined in Section 1), identifying the requirements to be satisfied from the three integration levels identified in the literature and standards definition of BIPV (Section 3). The differences in the products of the market and the multifaced technical solutions could make module selection complex for the designer, who needs to be properly provided with the information required to support the design of the three integration aspects. Therefore, in this section, a specific methodology for selecting the BIPV systems in architecturally sensitive areas is proposed. The described methodology guided the selection of two technologies to be installed and tested in the outdoor testbed, in the frame of a broader experimental campaign on coloured BIPV products (see Section 5). The methodology consist in four steps: (i) the identification of three integration levels for BIPV systems (Section 3); (ii) the identification of specific parameters for defining the main characteristics of each integration level, according to the standard UNI 8290-1/2; (iii) the market analysis on commercial hidden coloured PV modules; and (iv) the comparison of different technologies, considering technical elements, flexibility, shapes, dimensions, colours and nominal power.

Once the three integration levels were identified and defined, specific parameters were identified for each of them, with the aim of drawing up an evaluation matrix that served as a base for the selection of the BIPV technologies. Then, a deep market analysis was performed to detect the existing technologies suitable for the experimental campaign. The parameter referred to each integration level are described in detail hereafter, while the results of the market analysis including the analysed technologies are presented in Table 1.

The functional or technological integration refers to the ability of the modules to serve as a building component, thus fulfilling the functional requirements handed over by the original building element.

Typical envelope functions could be rain, snow, wind and solar protection, mechanical strength and reliability, shadowing or daylight admission. In Italy, the UNI 8290-1/2 [59] standard identifies for each technical element (i.e., roof, opaque vertical façade and transparent vertical façade) the corresponding requirements to be satisfied by the envelope components. Since the choice of a BIPV technology to be integrated into the building envelope must consider the functional requirements needed for the selected application, we decided to evaluate BIPV technologies according to the UNI 8290-1/2 [59] classification. Nonetheless, it is common to utilise a single BIPV product (typically coloured c-Si BIPV modules) for different applications by choosing the appropriate mounting system, e.g., as roof tiles or external layer of a ventilated façade. This peculiarity could grant to BIPV products a certain level of flexibility that facilitates the standardisation of the products manufacturing and the procurement design. This aspect has been considered through the identification of flexibility as a specific parameter for the functional integration in the evaluation matrix. As Table 1 highlights, the large variety of modules available on the market reflects the need of flexibility and multi-functionality and enables the use of the same module as elements belonging to a different technical class unit, if coupling with a suited mounting structure. Both multi-functionality and flexibility could be relevant characteristics for enhancing the market penetration of BIPV modules, since they enable the industrial production in series of a larger number of units that would eventually be used as different construction elements. This could represent an important advantage for BIPV industries, since it could potentially reduce the time and the costs of the modules production, encouraging BIPV market penetration.

The aesthetic integration identifies the ability of the product to define morphological and architectural rules which steer the architectural language and composition of buildings [26]. Consequently, the shape, dimension, position, materials, colour and texture of modules are defined in parallel with junction systems and mounting structure, which should be invisible to guarantee a good camouflage of the BIPV technology in the building envelope. The aesthetic evaluation is performed through four main parameters: (i) module's dimension; (ii) shape, defining the morphological integration of the BIPV technology in the building envelope; (iii) colour, which ensures the mimicking of the traditional building envelope material, camouflaging the BIPV module into the building envelope; and (iv) the module's reflectance, which is an essential aspect to be considered to ensure a high-quality aesthetic result and to avoid glare and overheating in the surroundings. As shown in Table 1, the first three parameters are quite common in the producers' specifications, while it is rare to find information about the module's reflectance. Furthermore, the market analysis highlights a common practice within the BIPV producers to provide customised solutions, mainly in terms of dimensions and colours. This is due to the peculiarity if the BIPV applications that, being tailored on the building envelopes, quite often require a specific design for the modules. This high level of customisation on the one hand constitutes the strength and the uniqueness of BIPV products in the PV scene, while on the other hand represents a limit since it hinders the series production of the modules limiting their cost reduction. In addition, the improvements of the BIPV industry in terms of available modules colours, which can range among an impressive hue palette and printing, are remarkable.

The energy integration refers to the ability of BIPV plants to interact with the energy systems at the building level or at the district level, with the aim of maximising the self-consumption. In fact, BIPV products could be used extensively on façades, enlarging the envelope surfaces available for PV installation. As a result, BIPV could lead to a shift in the energy paradigm for buildings: buildings would no longer be a mere energy consumer in the local electric grid, but it could indeed provide load flexibility, by producing, storing and selling electricity to the grid according to mutual needs. At present, within this paper, two preliminary parameters are evaluated in respect to the energy integration, i.e., the module's efficiency and its nominal power per square meter. This information is not easy to find on the producers' technical sheets since they are highly dependent on the chosen colour or texture. Table 1 shows some efficiency and nominal power range which have been retrieved both from technical documentation (if available) or directly from the producers, by means of interviews. In future studies, the "energy integration" concept will be more deeply investigated, since the PV

Integration Lab is conceived to allow experiments on several "energy integration" configurations (i.e., stand alone, grid connected and plug and play). This way, the PV production can be associated to specific building loads (electrical consumption).

As stated before, multifunctionality and flexibility are relevant characteristics for enabling the use of the same BIPV technology in different technical elements, when coupled with the appropriate mounting structure. For this reason, modules' multifunctionality and high flexibility are prioritised in the selection of the technologies involved in the experimental campaign. Consequently, the technologies suited only for roofing systems have been discarded (Table 1). The architectural applications require larger aesthetical possibilities to ensure flexibility in the design. Thus, the customisation of modules' shapes and dimensions is important in architecturally sensitive areas, even if the standardisation of these characteristics would imply advantages for industrial series production. Therefore, we select the technologies that allow the customisation of shape and dimension. the same considerations are applied to colours. Producers that guarantee colour customisation or larger colour palette have been preferred. Then, among those, the producers that provided the higher modules' nominal power (W/m^2) have been selected. As a result of this procedure, two BIPV technologies from Glassfer & Sunage producer have been selected. Among the available colours and customised printed patterns, two module typologies suitable for the installation in sensitive architectural areas have been chosen. Both technologies present a modified front glass. The first one has a tinted front glass in a uniform "terracotta" colour, which is representative of the typical chromatic palette of Italian historical roofs (similar RAL 8015). The second one has a textured front glass, which reproduces the pattern of the terracotta Portuguese tiles through ceramic ink printing. Portuguese tiles are the most used typology of clay roof tile in vernacular and traditional Italian architecture as well as in historical towns.

5. Experimental Characterisation of BIPV Technologies

As emerged in H2020 Project *PV IMPACT* [60], BIPV stakeholders workshops, there is an urgent need by architects and designers to acquire more knowledge on coloured modules, to better understand the current possibilities on the market (as provided in Table 1) and to gather information on their performance and reliability. Although a broad literature exists on the theoretical relationship between colour and efficiency/power generation [44,45], there is a lack of information on the real final performance of coloured modules due to the fragmented techniques used by different PV modules producers. In fact, during the module assembling and the lamination process, the colour could change significantly compared to the initial components colour, obtaining different aesthetical solutions in the final product.

When standard modules became mainstream many scientific publications covered and shared test results, which helped to gain a greater understanding on the topic [61–63]. In this section, a similar approach is chosen to fill the knowledge gap about the technical characterisation at module level of coloured BIPV products in the scientific community. In fact, when it comes to the characterisation of coloured PV modules, several studies analysed coloured glass at material and optical level, while few analysed the electric behaviour of PV coloured modules ready for the market. Among the latter, those available assessed the modules' performance through outdoor tests [58,64]. Therefore, the experimental characterisation of BIPV modules is needed both at standard test conditions (STC) and under real operating conditions. For this reason, the project *BIPV UPpeal* aims at testing several BIPV products in the EURAC Research facilities in the next years. The overarching aim of the research is to test different BIPV technologies to collect useful information for comparing the technical properties of different modules. To do so, at first, the modules will be tested in the indoor laboratories at STC to characterise the performance according to the existing standard procedures. Then, the modules will be tested in the outdoor facilities, under real exposure conditions, to gain information on the dynamic behaviour of the modules under operating conditions in terms of energy performance, functional adequacy and aesthetical appealing. Hereafter, we provide a brief description of the experimental facilities that will be used within the research (Section 5.1) and we present the results of the first experimental indoor campaign (Section 5.2). Then, the description of the experimental design of outdoor testing is provided (Section 5.3).

5.1. Experimental Facilities Description

5.1.1. Outdoor Facilities

The selected modules will be mounted on the testbeds of the PV Integration Lab of EURAC Research (Figure 2). It consists of an outdoor infrastructure which allows testing and evaluating in real conditions the three aspects of BIPV integration: aesthetics, technology and energy. The roof testbed is a rotating roof measuring 20 square metres. It permits tilting up to 60° from the horizontal plane and orienting in any direction to reproduce any exposure condition or any type of cover for roofs. The façade testbed consists of a vertical substructure of about 30 square meters that can host BIPV modules with any mounting systems. The testbeds are connected to a monitoring system for recording the electrical and environmental parameters (such as yield, solar radiation, air temperature and relative humidity) and the energy performance of any type of PV module. The BIPV systems can be connected to storage systems and/or the grid to verify the impact on electricity networks.

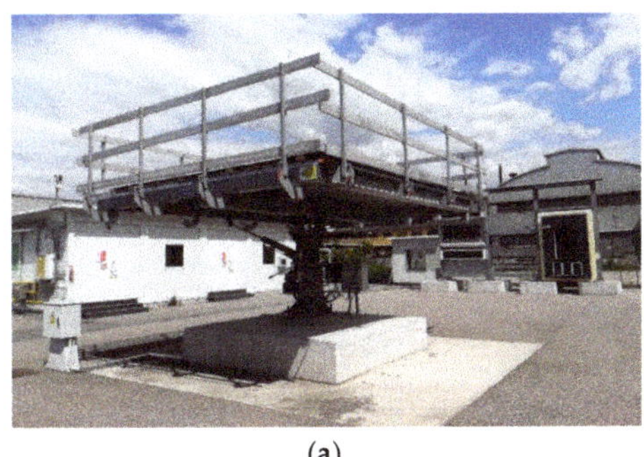
(a)

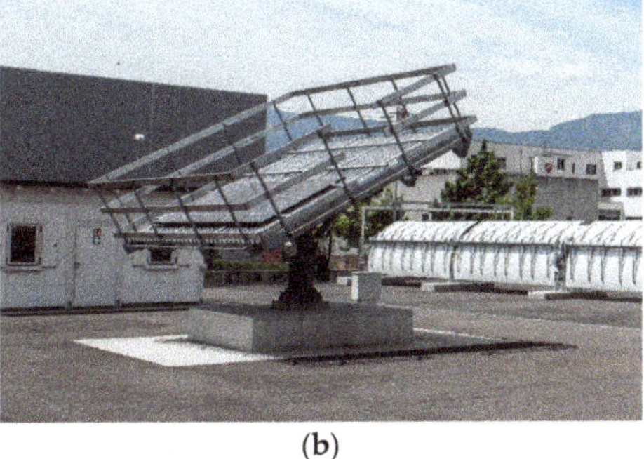
(b)

Figure 2. Outdoor PV integration lab: (**a**) view of roof in horizontal configuration and façade testbeds; and (**b**) detail of rotating configuration of the of the roof testbed.

5.1.2. Indoor Laboratory

The indoor characterisation of BIPV modules is performed by means of the solar simulator (Figure 3a) and electroluminescence (EL) camera (Figure 3b) provided by indoor "Solare PV Lab" of EURAC Research. The pulsed light solar simulator is in class "AAA", according to the international standard IEC 60904-9 [65]. It measures the electrical performance of PV modules, allowing the performance analysis of a PV cell or the comparison among different technologies in controlled conditions. It measures the PV module's IV curve under standard conditions [65]. The measurements detect the energy performance of the module in different combinations of irradiance (0–1000 W/m^2) and temperature (5–75 °C) and its temperature coefficients, in accordance with UNI CEI EN ISO/IEC 17025:2005 [66]. The test accredited is the Performance at STC (MQT 06.1) for PV modules according to the standard IEC 61215:2016 [67]. The electroluminescence camera is a VIS-SWIR InGaAs camera with a quantum efficiency over 60% at 1–1.2 μm and sensor of 640 × 512 pixels. This camera enables the implementation of the test following IEC TS 600904-13:2018 [68] indications.

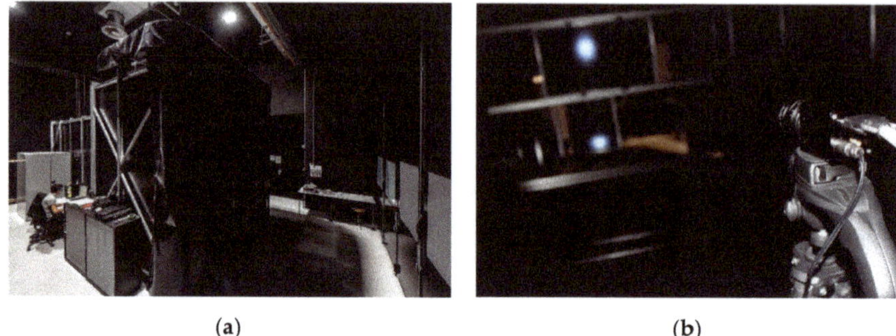

Figure 3. Indoor "Solare PV Lab": (a) panoramic view of solar simulator room; and (b) electroluminescence test execution.

5.2. Indoor Testing of BIPV Technologies: First Results

From the market analysis described in Section 4.1, two modules typologies have been selected to be tested through indoor and outdoor laboratories. The tests described hereafter constitute the first experimental campaign on coloured BIPV products. The first set of chosen modules belong to the Suncol® Tile technology, provided by Glassfer & Sunage, and consist in a sandwich PV panel of two tempered solar glass sheets within which a layer of monocrystalline cells is laminated by means of polymeric encapsulant films. The two sets of modules differ in dimensions, number of cells and customisation of the coloured front glass. The detail of the two modules typologies is given in Table 2, while Figure 4 shows the modules' samples. These modules have been selected for their high flexibility and aesthetical integration for architecturally sensitive areas, as explained in Section 4.1. Two different customisation techniques have been selected to compare their aesthetic impact and their energy performance: one has a full colour tinted front glass and one has a printed tile pattern. In both cases, an anti-reflection coating has been applied on solar cells. Furthermore, the modules are provided with an invisible mounting system and module frame that guarantees the reduction of the visual disturbance as well as their reversibility.

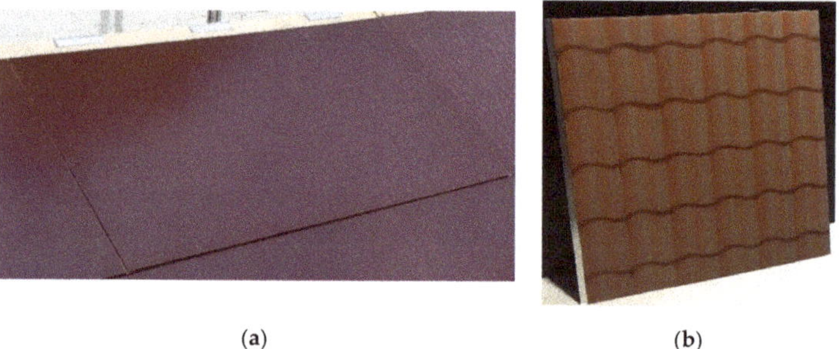

Figure 4. Tested modules samples: Suncol® Tile-Terracotta Simil RAL 8015 (a); and Suncol® Tile-Texturing Simil roof tile (b).

Table 2. Detail of the two tested modules typologies.

	Suncol® Tile-Terracotta	Suncol® Tile-Texturing
Solar tempered front glass	Simil RAL 8015	Simil roof tile
Active layer	18 monocrystalline cells	36 monocrystalline cells
Solar tempered back glass	Black printed	Black printed
Dimensions [m × m]	1 × 0.575	1 × 1.05

The integration of solar cells into coloured modules could results in losses in the irradiance reaching the solar cell within the module, since the colours and materials used in the glass modification have an impact on glass transmittance and light spectrum reaching the underlying cells. Some studies investigated the influence that coloured layers could have in the energy performance of c-Si PV modules, with both theoretical and experimental analyses [44,45]. One main result of these studies is that usually the theoretical colour-related power losses are lower than the actual ones, due to undesired reflections in the near infrared spectrum (NIR, wavelengths > 780 nm) that could appear. Therefore, as a first analysis for the characterisation of the two selected technologies, we decided to investigate the energy parameters of the two modules' set, with the aim of evaluating the power losses due to the front glass modification. The expected power loss due to the variations in the optical properties is analysed through the cell-to-module (CTM_x) factor (Equation (1)), where X is the considered electrical parameter:

$$CTM_x = \frac{X_{module}}{\sum_{i=1}^{n} X_{cell,i}} \quad (1)$$

This ratio is calculated using the electrical parameters of the bare cells before the assembly of the module. To calculate the CTM factor, firstly the electric performance of the two modules typologies in standard test conditions have been investigated in the indoor laboratory, with the solar simulator. The results present a normal shape of the current-voltage curves (Figure 5). The electrical parameters obtained in the standard test conditions (STC) performance test are presented in Table 3, where the power at maximum point (P_{mpp}), short-circuited current (I_{sc}) and open voltage (V_{oc}) are highlighted. Regarding the maximum power linearity respect to irradiance levels at 100–1000 W/m^2, the test shows acceptable performances comparable to those in commercial transparent glass photovoltaic modules (Figure 6).

Table 3. Electrical parameters at standard test conditions of the tested module at maximum power point (P_{mpp}); current at maximum power point (I_{mpp}); voltage at maximum power point (V_{mpp}); open circuit voltage (V_{oc}); and short-circuit current (I_{sc}).

	P_{mpp} (W)	I_{mpp} (A)	V_{mpp} (V)	V_{oc} (V)	I_{sc} (A)
Suncol® Tile Terracotta	77	8.49	9.8	12.06	8.3
Suncol® Tile Texturing Roof tile	133.6	6.71	19.9	23.96	7.19

Table 4 provides CTM factors for both the types of PV modules and a reference monocrystalline module with clear front glass. The results show power losses in accordance with those calculated by Peharz and Ulm [45] using a numerical model for RAL colours between 8000 and 8050, which is about −21% in comparison with zero reflective devices. We obtained −20.7% for Suncol® Tile Terracotta and 31.2% for Suncol® Tile Texturing Roof tile. Thus, our test shows agreement with the numerical model for the PV modules of uniform terracotta colour. These performance losses are expected: they depend on the layers superposed to the c-Si cells (EVA polymeric encapsulant and glass pane) and solar cells and strings interconnection. This behaviour is enhanced by the colours and ceramic ink used to customise the modules' front glass pane which hinders the PV performance, due to the optical and physical behaviour of the coloured layers, that depends on the modules' hue and coverage percentage [65,66].

Table 4. CTM loss in Suncol® Tile Terracotta and Texturing Roof tile modules.

CTM Loss	P_{mpp} (W)	I_{sc} (A)	V_{oc} (V)
Suncol® Tile Terracotta	−20.7%	−17.4%	−0.60%
Suncol® Tile Texturing Roof tile	−31.2%	−28.4%	−1.30%
Reference monocrystalline module [69]	−3.8	−4.1	-

As a further analysis, we provide hereafter the results of electroluminescence control technique, which is increasingly relevant in the analysis of PV modules quality. It basically shows the path taken by the electrons along the circuitry in the module. Several issues can be detected such as diverse mechanical breaks in cells or disconnected areas. The test was realised with a VIS-SWIR InGaAs camera and focused on the effect of the non-transparent glass in the EL signal emission from the module; therefore, from normal to near 0° incidence shooting was carried out in the indoor lab (Figure 7). The first results in normal incidence demonstrate a good reception of the signal compared to transparent glass technology (Figure 8). For small angles of incidence, no blind spot has been detected in both types of modules and the transmission of the EL signal is still acceptable. Thus, the results of the EL test show a good electrical response of the modules regardless the incident angle of the radiation. This is a critical aspect when performing outdoor operation and maintenance (O&M) activities in real installations where the position of the panel can be diverse depending on the building and limitations for shooting can be multiple (Figure 9).

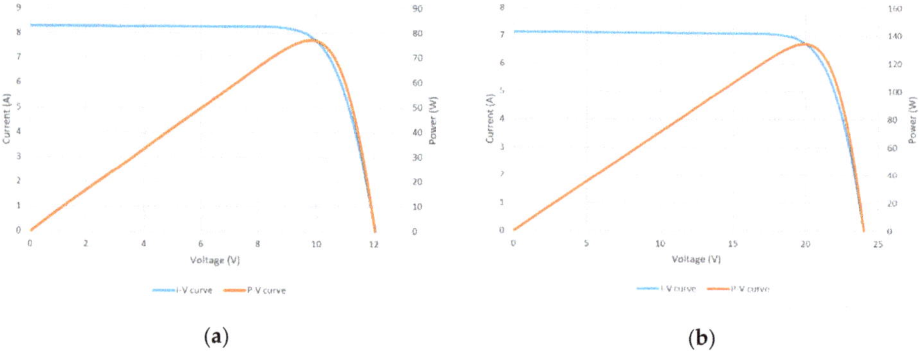

Figure 5. STC performance of PV modules under analysis: (**a**) Suncol® Tile Terracotta; and (**b**) Suncol® Tile Texturing Roof tile.

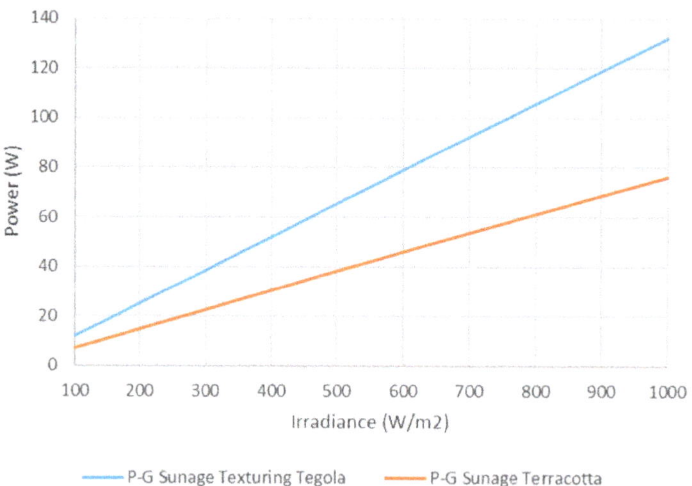

Figure 6. Power at maximum point according to different irradiance levels.

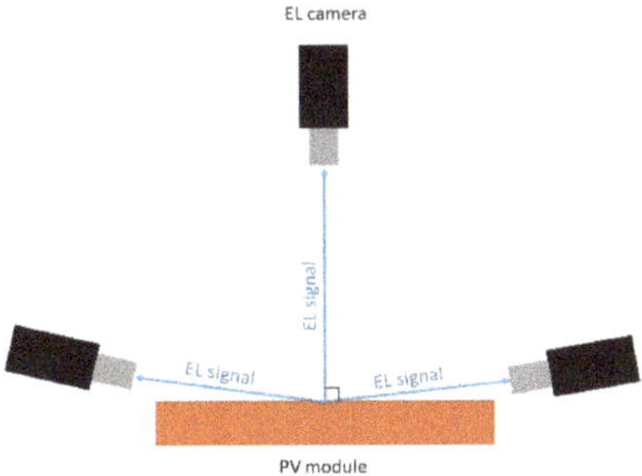

Figure 7. Electroluminescence test at different shooting angles.

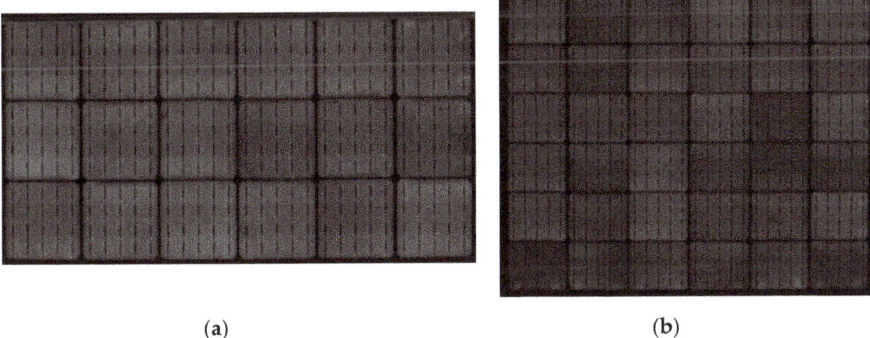

(**a**) (**b**)

Figure 8. Electroluminescence images obtained at normal incidence shooting for: (**a**) Terracotta; and (**b**) Texturing Roof Tile.

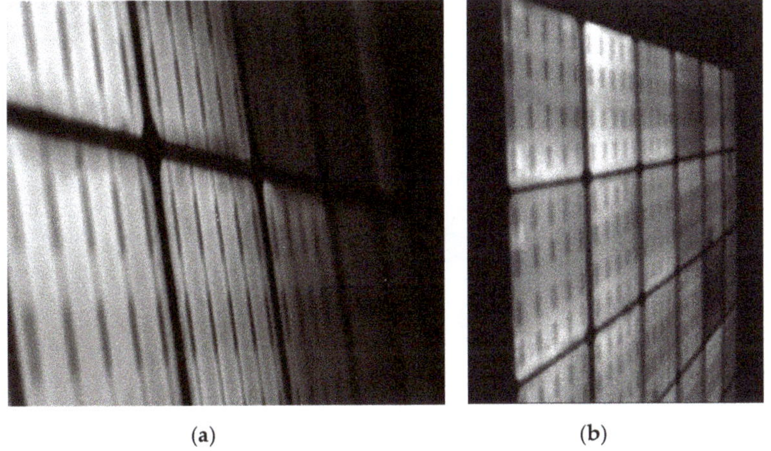

(**a**) (**b**)

Figure 9. Electroluminescence images obtained at small incidence shooting for: (**a**) Terracotta; and (**b**) Texturing roof Tile.

5.3. Experimental Design of Outdoor Testing

The roof testbed described in Section 5.1.1 is conceived with an interdisciplinarity approach focusing on the three integration aspects mentioned in Section 3 (technology, aesthetic and energy). This interdisciplinarity approach makes it unique with respect to other existing BIPV outdoor setups described in [70] that provides an overview of the existing international BIPV R&D testing facilities. Indeed, the proposed experimental set up is not conceived for testing stand-alone modules performance but rather a large portion of envelope BIPV systems, offering the opportunity to investigate the aesthetical and technological integration of real scale installations, involving several modules. The added value of this kind of testbed is that it will offer the opportunity to different stakeholders in the BIPV community (students, designers and the public and heritage authorities involved in the energy issues) to witness first-hand innovative BIPV technologies, better understanding the benefits of coloured PV systems.

A first qualitative study has been performed on the aesthetic integration and technical compatibility. The ensemble of PV modules are inserted to minimise their visual impact, guaranteeing: (i) 100% coverage of the roof surface; (ii) aesthetic and chromatic integration with traditional clay roof tiles (the selection terra-cotta colour, both with and without the texturing pattern); (iii) geometrical and chromatic uniformity inserting two different kinds of PV parallels in parallel rows; (iv) coplanarity with the roof line; (v) colour rendering under different exposure and tilting conditions; and (vi) pattern continuity for the tile texturing PV cells. Furthermore, the PV modules are integrated into the roof testbed through a back attached tile-type mounting system that guarantees successful integration from the aesthetical point of view since no mounting system or module frame is disturbing visual appearance (Figure 10). Considering the technology integration, technical compatibility and system reversibility are evaluated. Technical compatibility refers mainly to the reduction of moisture accumulation on the backside of the panels, while reversibility refers to the use of mounting systems to remove the panels without affecting the integrity of the roof. Figure 11 shows the roof layout (Figure 11a) and its electrical configuration (Figure 11b). From a qualitative point of view, the testbed design has been conducted with particular focus to the functional requirements for the roof outer layers, such as water resistance and moisture accumulation prevention. The first qualitative findings show no accumulation of water or moisture problems to appear. Quantitative measurements will be provided in future works to rigorously evaluate the technology performance in this respect. Moreover, the back attached tile-type mounting system guarantees the reversibility of the system without affecting the original roof, as requested for the preservation of heritage systems.

(a) (b)

Figure 10. Testbed realisation on the outdoor roof testbed. To perform quantitative tests, the two module types Suncol® Terracotta and Suncol® Texturing Roof Tile are connected in two strings to a multi-string (two maximum power point trackers) grid-connected inverter. PT100 temperature sensors are applied on the backside of eight modules, as displayed in Figure 10b.

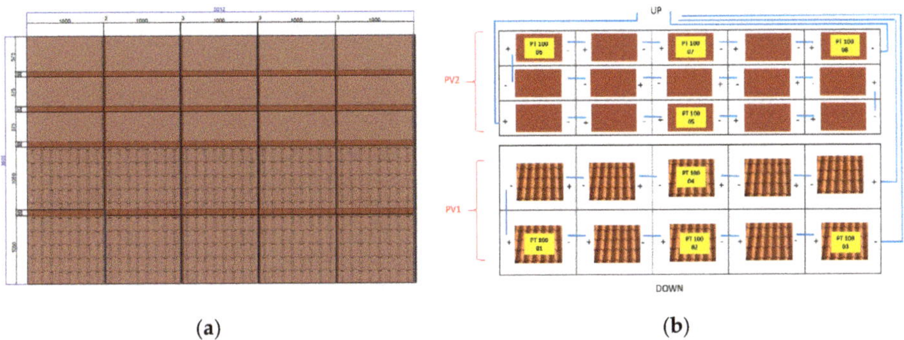

(a) **(b)**

Figure 11. Testbed design: (**a**) roof layout; and (**b**) electrical configuration.

6. Conclusions

This paper outlines the research activities carried out to define the configuration of new testbeds for the experimental characterisation of coloured BIPV technologies. The investigation of BIPV technology is an interdisciplinary activity that calls for heterogeneous expertise, having the BIPV modules multifunctional features related to the aesthetic and technological integration in building envelope and to energy integration both in the building systems and in the electric grid. Being BIPV applications tailored on the building envelopes, the modules and the mounting systems require a specific design. This high level of customisation constitutes the strength and the uniqueness of BIPV products but also a limit, since it hinders the series production of the modules and, thus, it limits their cost reduction. Furthermore, the customisation of coloured BIPV modules that completely hide the PV cells from the view and their expansion on the market would contribute in enhancing the social acceptance of PV in sensitive areas, where the PV technologies have been often considered anaesthetic, and thus unacceptable, by users and architects. Therefore, there is the need to overcome the technical, social and economic barriers to reach a larger scale of BIPV applications and thus improve their economic profitability. The first issue to deal with is the definition of the concept of *integration*, which needs to find a unique expression within BIPV literature and standards. The definition of the criteria that set the application of BIPV in architecturally sensitive areas need especially to be revised. In fact, the existing guidelines refer mainly to PV systems applied on the building envelope (BAPV), without considering the actual potential of the modern coloured BIPV technologies that can "disappear" from the view. Secondly, it is important to better investigate the wide range of available technologies for coloured BIPV, with regards to the electrical behaviour of a large variety of coloured modules. This variety results in a fragmented market scenario where several customisation options are now available. Nonetheless, the customisation process could lead to modifications in the electrical behaviour of the module, due to the presence of one or more coloured layers which cause the reflection or absorption of a portion of the solar spectrum (in the visible range) that would otherwise be converted into electricity, causing a reduction in the modules' yield.

Therefore, to deeply investigate this behaviour, experimental assessment is needed both at STC and under real operating conditions on final assembled BIPV modules. For this reason, we selected two BIPV modules to be tested in EURAC Research's indoor and outdoor facilities. Indoor tests are already completed with satisfactory results and will support outdoor experiment analysis. Besides the technical assessment, real scale BIPV testbeds are useful to test different solution for the mounting systems and to understand the risks related to the installation of BIPV modules. Testbeds are also important because they could show different stakeholders the new and concrete aesthetic and technical possibilities of BIPV systems, improving the users' trust on aesthetic integration, the market penetration and the economic profitability. In fact, BIPV products could sensibly expand the envelope surfaces available for PV installation. As a result, BIPV could lead to a shift in the energy paradigm for buildings that would no longer be a mere energy consumer in the local electric grid, but could indeed provide load flexibility by producing, storing and selling electricity to the grid according to mutual needs. To achieve the objectives of deeper market penetration and higher economic profitability of BIPV applications, further test activities will be carried out in these facilities. In fact, several BIPV systems (including PV modules, substructures and energy systems) will be integrated in both the outdoor roof and façade testbed, with the aim of testing and comparing them with respect to the three integration aspects of aesthetic, technology and energy, since all these aspects are of fundamental importance to enhance the application of BIPV technologies in the built environment.

Author Contributions: Conceptualisation, M.P., E.L. and L.M.; methodology, M.P.; investigation, M.P., E.L., L.M. and A.A.; resources, M.P. and M.L.; writing—original draft preparation, M.P. and E.L.; writing—review and editing, M.P., E.L., L.M., A.A. and F.C.; visualisation, M.P.; and project administration, L.M. All authors have read and agreed to the published version of the manuscript.

Funding: The research leading to these results received funding from the Institute for Renewable Energy of Eurac Research, within the project *"BIPV UPpeal"*. In addition, this research was co-financed by the European Union, European Regional Development Fund, the Italian Government, the Swiss Confederation and Cantons, as part of the Interreg V-A Italy-Switzerland Cooperation Program, within the context of the *"BIPV meets History*! project (grant No. 603882), for the definition of criteria and products for BIPV integration in architecturally sensitive areas. Furthermore, the research received funding from the Program EFRE/FESR Provincia autonoma di Bolzano-Alto Adige 2014–2020, under Project number FESR1042, *"Studio dell'integrazione di reti elettriche e termiche con la flessibilità energetica degli edifici–INTEGRIDS"*, for the purchasing of the inverter, connection and distribution boxes. Finally, it has received funding from the Program EFRE/FESR Provincia autonoma di Bolzano-Alto Adige 2014–2020, under Project number FESR1128, "Use of Industry 4.0 and Internet of Things logics in the photovoltaic sector—PV 4.0", for the purchasing of PT100 temperature sensors.

Acknowledgments: We acknowledge David Moser and Alexandra Troi for the revision and the useful suggestions on the paper structure; Gazmend Luzi for the data provided on PV cells; and Ilaria Alberti for the revision of English text.

Conflicts of Interest: The authors declare no conflict of interest.

References

1. EC Clean Energy for All Europeans. *Clean Energy for All Europeans*; European Commission: Brussels, Belgium, 2008; Available online: https://ec.europa.eu/energy/topics/energy-strategy/clean-energy-all-europeans_en (accessed on 5 April 2020).
2. *United Nation Framework Convention on Climate Change in 21st Conference of the Parties Paris Agreement*; UNFCC: Paris, France, 2015.
3. European Parlament. *DIRECTIVE (EU) 2018/844. Energy Performance of Buildings*; European Parlament: Brussels, Belgium, 2018.
4. European Parlament. *DIRECTIVE (EU) 2018/2002 on Energy Efficiency*; European Parlament: Brussels, Belgium, 2018.
5. European Parlament. *DIRECTIVE (EU) 2018/2001 on the Promotion of the Use of Energy from Renewable Sources*; European Parlament: Brussels, Belgium, 2018.
6. Jensen, S.Ø.; Marszal-Pomianowska, A.; Lollini, R.; Pasut, W.; Knotzer, A.; Engelmann, P.; Stafford, A.; Reynders, G. IEA EBC annex 67 energy flexible buildings. *Energy Build.* **2017**, *155*, 25–34. [CrossRef]
7. Michas, S.; Stavrakas, V.; Spyridaki, N.-A.; Flamos, A. Identifying research priorities for the further development and deployment of solar photovoltaics. *Int. J. Sustain. Energy* **2019**, *38*, 276–296. [CrossRef]
8. Zhang, T.; Wang, M.; Yang, H. A Review of the energy performance and life-cycle assessment of building-integrated photovoltaic (BIPV) systems. *Energies* **2018**, *11*, 3157. [CrossRef]
9. Eder, G.; Peharz, G.; Trattnig, R.; Bonomo, P.; Saretta, E.; Frontini, F.; Polo López, C.S.; Rose Wilson, H.; Eisenlohr, J.; Martin Chivelet, N.; et al. *Coloured BIPV-Market, Research and Development*; International Energy Agency: Paris, France, 2019; Volume 60.
10. Chang, R.; Cao, Y.; Lu, Y.; Shabunko, V. Should BIPV technologies be empowered by innovation policy mix to facilitate energy transitions?—Revealing stakeholders' different perspectives using Q methodology. *Energy Policy* **2019**, *129*, 307–318. [CrossRef]

11. Heinstein, P.; Ballif, C.; Perret-Aebi, L.-E. Building integrated photovoltaics (BIPV): Review, potentials, barriers and myths. *Green* **2013**, *3*, 125–156. [CrossRef]
12. Freitas, S.; Brito, M.C. Solar façades for future cities. *Renew. Energy Focus* **2019**, *31*, 73–79. [CrossRef]
13. Escarre, J.; Li, H.Y.; Sansonnens, L.; Galliano, F.; Cattaneo, G.; Heinstein, P.; Nicolay, S.; Bailat, J.; Eberhard, S.; Ballif, C.; et al. When PV modules are becoming real building elements: White solar module, a revolution for BIPV. In Proceedings of the 2015 IEEE 42nd Photovoltaic Specialist Conference (PVSC), New Orleans, LA, USA, 14–19 June 2015; pp. 1–2.
14. Saretta, E.; Bonomo, P.; Frontini, F. Active BIPV glass facades: Current trends of innovation. In Proceedings of the GPD Glass Performance Days 2017, Tampere, Finland, 28–30 June 2017.
15. IEA-PVPS Task 7 Photovoltaic Power Systems in the Built Environment. Available online: https://iea-pvps.org/research-tasks/photovoltaic-power-systems-in-the-built-environment/ (accessed on 14 June 2020).
16. PV Accept. Available online: http://www.pvaccept.de (accessed on 5 June 2020).
17. 3ENCULT: Efficient Energy for EU Cultural Heritage. Available online: http://www.3encult.eu/en/project/welcome/default.html (accessed on 5 June 2020).
18. EFFESUS: Energy Efficiency for EU Historic Districts' Sustainability. Available online: http://www.effesus.eu (accessed on 5 June 2020).
19. BIPV Meets History: Value-Chain Creation for the Building Integrated Photovoltaics in the Energy Retrofit of Transnational Historic Buildings. Available online: http://www.bipvmeetshistory.eu (accessed on 6 June 2020).
20. Lovati, M.; Salvalai, G.; Fratus, G.; Maturi, L.; Albatici, R.; Moser, D. New method for the early design of BIPV with electric storage: A case study in northern Italy. *Sustain. Cities Soc.* **2019**, *48*, 101400. [CrossRef]
21. Huang, P.; Lovati, M.; Zhang, X.; Bales, C.; Hallbeck, S.; Becker, A.; Bergqvist, H.; Hedberg, J.; Maturi, L. Transforming a residential building cluster into electricity prosumers in Sweden: Optimal design of a coupled PV-heat pump-thermal storage-electric vehicle system. *Appl. Energy* **2019**, *255*, 113864. [CrossRef]
22. Maturi, L.; Belluardo, G.; Moser, D.; Del Buono, M. BiPV system performance and efficiency drops: Overview on PV module temperature conditions of different module types. *Energy Procedia* **2014**, *48*, 1311–1319. [CrossRef]
23. Davis, M.W.; Fanney, A.H.; Dougherty, B.P. Measured Versus predicted performance of building integrated photovoltaics. *J. Sol. Energy Eng.* **2003**, *125*, 21–27. [CrossRef]
24. BSI. *EN 50583-1:2016. Photovoltaics in Buildings. BIPV Modules*; BSI: Brussels, Belgium, 2016.
25. European Parliament and European Council. *Regulation (EU) No 305/2011*; European Parliament and European Council: Brussels, Belgium, 2011.
26. Farkas, K.; Frontini, F.; Maturi, L.; Munari Probst, M.C.; Roecker, C.; Scognamiglio, A.; Zanetti, I. *T.41.A.2: Solar Energy Systems in Architecture—Integration Criteria and Guidelines Subtask A: Criteria for Architectural Integration*; International Energy Agency: Paris, France, 2012.
27. IEA-SHC T59 Deep Renovation of Historic Buildings towards Lowest Possible Energy Demand and CO_2 Emission (nZEB). Available online: http://task59.iea-shc.org (accessed on 10 May 2020).
28. IEA-SHC T51 Solar Energy in Urban Planning. Available online: http://task51.iea-shc.org/ (accessed on 10 May 2020).
29. Franco, G.; Magrini, A. *Historical Buildings and Energy*; Springer International Publishing: Cham, Switzerland, 2017; ISBN 978-3-319-52613-3.
30. Maturi, L.; Adami, J. *Building Integrated Photovoltaic (BIPV) in Trentino Alto Adige*; Green Energy and Technology; Springer International Publishing: Cham, Switzerland, 2018; ISBN 978-3-319-74115-4.
31. European Union's Horizon 2020 Project EnergyMatching. Available online: https://www.energymatching.eu/ (accessed on 15 June 2020).
32. Scotland, H. *Micro-Renewables in the Historic Environment*; Historic Scotland: Edinburgh, UK, 2014.
33. Wohlleben, M.; Moeri, S.; Müller, N.; Schletti, B. *Energieeffizienz am Baudenkmal: Solarenergie*; Kantonale Denkmalpflege Bern und Kantonale Denkmalpflege: Bern-Zürich, Switzerland, 2014.
34. Polo López, C.S.; Lucchi, E.; Franco, G. *Acceptance of Building Integrated Photovoltaic (BIPV) in Heritage Buildings and Landscapes: Potentials, Barrier and Assessment Criteria*; REHABEND: Granada, Spain, 2020; pp. 1–8.
35. National Renewable Energy Laboratory—National Trust for Historic Preservation. *Implementing Solar Photovoltaic Projects on Historic Buildings and in Historic Districts*; National Renewable Energy Laboratory: Golden, CO, USA, 2011; Volume 1.

36. Repubblica e Cantone Ticino. *Interventi Nei Nuclei Storici. Criteri di Valutazione Paesaggistica Nell'ambito Della Procedura Edilizia*; Repubblica e Cantone Ticino: Bellinzona, Switzerland, 2016.
37. Dipartimento Federale Dell'interno (DFI). *Cultura Solare. Conciliare Energia Solare e Cultura Della Costruzione*; DFI: Bern, Switzerland, 2019.
38. Dipartimento Federale Dell'interno (DFI). *Energia e Monumento*; DFI: Bern, Switzerland, 2018.
39. Ministero per i Beni e le Attività Culturali (MiBACT). *Linee di Indirizzo per il Miglioramento Dell'efficienza Energetica del Patrimonio Culturale: Architettura, Centri e Nuclei Storici ed Urbani*; MiBACT: Roma, Italy, 2015.
40. Changeworks. *Renewable Heritage. A Guide to Microgeneration in Traditional and Historic Homes*; Changeworks: Edinburgh, UK, 2009.
41. Achenza, M.; Desogus, G. *Guidelines on Building Integration of Photovoltaic in Mediterranean Area*; UNICA-DICAAR: Cagliari, Italy, 2013.
42. Scognamiglio, A.; Berni, F.; Frontini, C.S.P.L.; Maturi, L. The complex dialogue between photovoltaics and pre-existing: Starting point for a discussion. In Proceedings of the 27th European Photovoltaic Solar Energy Conference and Exhibition, Frunkfurt, Germany, 24–28 September 2012; pp. 4161–4168.
43. Bellazzi, A.; Belussi, L.; Meroni, I. Estimation of the performance of a BIPV façade in working conditions through real monitoring and simulation. *Energy Procedia* **2018**, *148*, 479–486. [CrossRef]
44. Røyset, A.; Kolås, T.; Jelle, B.P. Coloured building integrated photovoltaics: Influence on energy efficiency. *Energy Build.* **2020**, *208*, 109623. [CrossRef]
45. Peharz, G.; Ulm, A. Quantifying the influence of colors on the performance of c-Si photovoltaic devices. *Renew. Energy* **2018**, *129*, 299–308. [CrossRef]
46. Jolissaint, N.; Hanbali, R.; Hadorn, J.C.; Schüler, A. Colored solar façades for buildings. *Energy Procedia* **2017**, *122*, 175–180. [CrossRef]
47. IEA PVPS Task 15 Subtask E. *Report T15-07—Coloured BIPV Market, Research and Development*; IEA: Paris, France, 2019.
48. El amrani, A.; Menous, I.; Mahiou, L.; Tadjine, R.; Touati, A.; Lefgoum, A. Silicon nitride film for solar cells. *Renew. Energy* **2008**, *33*, 2289–2293. [CrossRef]
49. Li, M.; Zeng, L.; Chen, Y.; Zhuang, L.; Wang, X.; Shen, H. Realization of Colored Multicrystalline Silicon solar cells with SiO_2/SiN_x:H double layer antireflection coatings. *Int. J. Photoenergy* **2013**, *2013*, 1–8. [CrossRef]
50. Chen, Y.; Yang, Y.; Feng, Z.; Altermatt, P.P.; Shen, H. Color modulation of c-Si solar cells without significant current-loss by means of a double-layer anti-reflective coating. In Proceedings of the 27th European Photovoltaic Solar Energy Conference and Exhibition, Frunkfurt, Germany, 24–28 September 2012; pp. 2014–2016.
51. Ji, C.; Zhang, Z.; Masuda, T.; Kudo, Y.; Guo, L.J. Vivid-colored silicon solar panels with high efficiency and non-iridescent appearance. *Nanoscale Horiz.* **2019**, *4*, 874–880. [CrossRef]
52. Pascual-San José, E.; Sánchez-Díaz, A.; Stella, M.; Martínez-Ferrero, E.; Alonso, M.I.; Campoy-Quiles, M. Comparing the potential of different strategies for colour tuning in thin film photovoltaic technologies. *Sci. Technol. Adv. Mater.* **2018**, *19*, 823–835. [CrossRef]
53. Calvo, M.E. Materials chemistry approaches to the control of the optical features of perovskite solar cells. *J. Mater. Chem. A* **2017**, *5*, 20561–20578. [CrossRef]
54. Kim, Y.; Son, J.; Shafian, S.; Kim, K.; Hyun, J.K. Semitransparent blue, green, and red organic solar cells using color filtering electrodes. *Adv. Opt. Mater.* **2018**, *6*, 1–10. [CrossRef]
55. Sehati, P.; Malmros, I.; Karlsson, S.; Kovacs, P.; Kovacs, P. *Aesthetically Pleasing PV Modules for the Built Environment*; RISE Research Institutes of Sweden AB: Växjö, Sweden, 2019.
56. Eastman Vanceva Rethink Color—Colored PVB Interlayer Brochure. Available online: https://www.vanceva.com/sites/default/files/pictures/ai-arch-3146_vanceva_brochure.pdf (accessed on 12 June 2020).
57. Onyx Solar Onyx Technical Guide. Available online: http://onyxsolardownloads.com/docs/ALL-YOU-NEED/Technical_Guide.pdf (accessed on 12 June 2020).
58. Wittkopf, S. Building integrated photovoltaic at NEST—Preliminary test bedding results. *J. Phys. Conf. Ser.* **2019**, *1343*, 012090. [CrossRef]
59. UNI 8290—Edilizia Residenziale. *Sistema Tecnologico. Classificazione e Terminologia*; Ente Nazionale Italiano di Unificazion n(UNI): Milano, Italy, 1981.
60. H2020 PV IMPACT. Available online: https://pvimpact.eu/ (accessed on 29 July 2020).

61. Green, M.A.; Dunlop, E.D.; Hohl-Ebinger, J.; Yoshita, M.; Kopidakis, N.; Hao, X. Solar cell efficiency tables (version 56). *Prog. Photovoltaics Res. Appl.* **2020**, *28*, 629–638. [CrossRef]
62. Dunlop, E.D.; Halton, D.; Ossenbrink, H.A. 20 years of life and more: Where is the end of life of a PV module? In Proceedings of the Conference Record of the Thirty-first IEEE Photovoltaic Specialists Conference, Lake Buena Vista, FL, USA, 3–7 January 2005; pp. 1593–1596.
63. Virtuani, A.; Caccivio, M.; Annigoni, E.; Friesen, G.; Chianese, D.; Ballif, C.; Sample, T.; Dunlop, E.D.; Halton, D.; Ossenbrink, H.A. 35 years of photovoltaics: Analysis of the TISO-10-kW solar plant, lessons learnt in safety and performance—Part 1. In Proceedings of the Conference Record of the 46th IEEE Photovoltaic Specialists Conference, Chicago, IL, USA, 16–21 June 2019; Volume 27, pp. 328–339.
64. Tzikas, C.; Valckenborg, R.M.E.; Dörenkämper, M.; van den Donker, M.; Lozano, D.D.; Bognár, Á.; Loonen, R.; Hensen, J.; Folkerts, W. Outdoor characterization of colored and textured prototype PV facade elements. In Proceedings of the 35th European Photovoltaic Solar Energy Conference and Exhibition, Brussels, Belgium, 24–27 September 2018.
65. IEC. *60904-9:2007 Photovoltaic Devices—Part 9: Solar Simulator Performance Requirements*; IEC: Geneva, Switzerland, 2007; p. 30.
66. UNI CEI EN ISO/IEC. *17025:2005—Requisiti Generali per la Competenza dei Laboratori di Prova e di Taratura*; Ente Nazionale Italiano di Unificazion n(UNI): Milano, Italy, 2005.
67. IEC. *61215-1:2016—Terrestrial Photovoltaic (PV) Modules—Design Qualification and Type Approval—Part 1: Test Requirements*; IEC: Geneva, Switzerland, 2016.
68. IEC TS. *60904-13:2018 Photovoltaic Devices—Part 13: Electroluminescence of Photovoltaic Modules*; IEC: Geneva, Switzerland, 2018.
69. Haedrich, I.; Surve, S.; Thomson, A. Cell to module (CTM) ratios for varying industrial cell types. In Proceedings of the 2015 Asia Pacific Solar Research Conference, Melbourne, Australia, 30 November–2 December 2015; pp. 1–6.
70. EA-PVPS T15-02. *BIPV Research Teams BIPV R&D Facilities. An International Mapping*; IEA-PVPS: Paris, France, 2017.

Article

Novel Simulation Algorithm for Modeling the Hysteresis of Phase Change Materials

Anna Zastawna-Rumin [1], Tomasz Kisilewicz [1] and Umberto Berardi [2,*]

[1] Faculty of Civil Engineering, Cracow University of Technology, 31-155 Krakow, Poland; azastawna@pk.edu.pl (A.Z.-R.); tkisilew@pk.edu.pl (T.K.)

[2] Department of Architectural Science, Ryerson University, 350 Victoria st., Toronto, ON M5B 2K3, Canada

* Correspondence: uberardi@ryerson.ca; Tel.: +1-416-979-5000 (ext. 553263)

Received: 20 January 2020; Accepted: 3 March 2020; Published: 5 March 2020

Abstract: Latent heat thermal energy storage (LHTES) using phase change materials (PCM) is one of the most promising ways for thermal energy storage (TES), especially in lightweight buildings. However, accurate control of the phase transition of PCM is not easy to predict. For example, neglecting the hysteresis or the effect of the speed of phase change processes reduces the accuracy of simulations of TES. In this paper, the authors propose a new software module for EnergyPlus™ that aims to simulate the hysteresis of PCMs during the phase change. The new module is tested by comparing simulation results with experimental tests done in a climatic chamber. A strong consistency between experimental and simulation results was obtained, while a discrepancy error of less than 1% was obtained. Moreover, in real conditions, as a result of quick temperature changes, only a partial phase transformation of the material is often observed. The new model also allows the consideration of the case with partial phase changes of the PCM. Finally, the simulation algorithm presented in this article aims to represent a better way to model LHTES with PCM.

Keywords: phase change material; hysteresis; simulations; EnergyPlus; thermal energy storage

---

1. Introduction

The term 'passive buildings' has become common nowadays. However, when this term was coined, it originally related to a building that included passive solar systems [1] and was designed to maximize solar gains [2]. Although the passive solar energy gains in buildings are predictable and controllable, protecting buildings against overheating is often neglected during the design process, leading to the construction of buildings with high cooling loads even in heating-dominated climates [3] and in mild ones [4]. This occurs because the measures to protect buildings against the risk of overheating are commonly poorly understood given the dynamicity of solar heat gains.

The lack of thermal inertia in many modern buildings exacerbates the risk of reaching too high indoor temperatures. In the case of lightweight buildings, there is no opportunity to accumulate the available exterior (solar) energy gains for later use efficiently [5]. This is why phase change materials (PCM) that store large amounts of energy thanks to the latent heat needed for a phase change process have been proposed over the last two decades [6].

Specific PCM algorithms have been added to various kinds of simulation tools, as the dynamic control of PCM and of their phase change process requires to be controlled dynamically. One of the largely adopted tools for PCM simulations is EnergyPlus™, which will be considered in this study, and it is briefly presented in the next section.

2. Simulation Method of Phase Changing Materials in EnergyPlus™

The first EnergyPlus™ version, including a PCM simulation algorithm, was released over a decade ago. Pedersen et al. developed an implicit finite difference thermal model of building surfaces that simulates the performance of PCM using their enthalpy law [7]. This model has been incorporated into EnergyPlus™ and combined with the general transformation based algorithm.

In recent years, many studies have been performed on the validation and verification of the results of the EnergyPlus™ calculations using different procedures: correspondence between the measured and simulated wall surface temperatures [8], heat flux density [9], and internal air temperatures [10], just to cite a few. These works did not lead to modification of the source code of the program, but only to the comparison and assessment of the reliability of the results.

The subjects of the simulation were mainly wall components with PCM. The impact of these components on thermal conditions in the simulated rooms and on the thermal properties of the entire building has been examined in previous research using fiber insulation by Kosny et al. [11] and concrete tiles by Naraid et al. [12]. In most cases, the authors reported a relatively strong agreement between simulation results and test data [13–15]. In many of these studies, the authors benchmarked EnergyPlus against controlled field data [14] and confirmed the value of this software.

Numerous simplifications are often used during numerical modeling. For example, the complex geometry of the products containing PCM is replaced with a homogeneous layer to which the enthalpy data is assigned. It is not surprising that the results of some experimental investigations performed did not find a strong agreement between the test results and the simulation ones [16]. For example, Castell et al. [16] evaluated software simulations in EnergyPlus™ with measurements done inside $3 \times 3 \times 3$ m test cells in Spain. The authors suggested that the reason for the discrepancy of the results could be the inconsistency of the weather data used for calculations and the actual variable thermal conditions. Recent studies indicating strong agreement between simulation and research results in a moderate United States of America climate [17] and in the sub-tropical one of Hong Kong [18] confirmed that the right weather data and material data are key factors for reducing the discrepancy between simulation and test results. Lee et al. showed that the differences between experimental and predicted total heat transfer values were under 5% [19].

The results of calculations in the EnergyPlus™ program were also compared with the results from the HEATING program [20]. Errors in the routing of EnergyPlus™ were, hence, corrected over the years [21]. For example, in version 8, code modifications allowed an acceleration of the calculation process and the inclusion of variable values for the thermal conductivity coefficient.

In EnergyPlus™, the calculation of the energy balance of building surface constructions is based on the conduction transfer function (CTF), but in the case of more advanced constructions, such as PCMs or variable thermal conductivity, a more flexible approach in the form of a conduction finite difference algorithm is used [22]. Because of the implicit solution of the equation set, it is more efficient to set a time step shorter than those used for the CTF solution algorithm [20]. Tabares-Velasco recommended a simulation time step of fewer than three minutes for a more accurate prediction of the behavior of PCM [21]. This approach is critical to allow to fully control the dynamic changes that occur in a PCM during a phase transition. However, one of the critical elements that still remain poorly modeled is the hysteresis of PCM, an aspect that is described in the next section.

2.1. PCM Hysteresis

The hysteresis may be explained as the dependence of the state of a system on its history or the lagging of an effect behind the cause of this state. This effect applies to the phase-change phenomenon. Many researchers have already dealt with the hysteresis effect during simulation modeling, revealing the effects of this phenomenon on simulation results with real conditions [9]. Biswas et al. [23] conducted PCM measurements using heat-flow meter apparatus for freezing and melting cycles and used them to model different simulation scenarios both with and without hysteresis. The results showed that including the hysteresis effect significantly impacts the calculated thermal performance of the PCM layer. Barz and Sommer [24] also confirmed that the phase transition of PCM is significantly affected by hysteresis phenomena. According to their results, the static hysteresis model and the macro kinetic models showed qualitatively consistent results.

Delcroix et al. [25], Mehling et al. [26], Berardi and Gallardo [27] proved that PCM thermal features are sensitive to the testing method and stated that DSC (differential scanning calorimetry) test results are not directly applicable to building simulations as the process of charging and discharging in PCM mixed in other porous materials is hardly represented by testing samples of a few grams. In particular, Delcroix et al. [25] and Mehling et al. [26] analyzed hysteresis as a function of the cooling and heating rates and noticed that the hysteresis tested in DSC is low (around 1 °C) at low cooling and heating rates, while it can reach up to around 7–8 °C at higher rates that can be found in real conditions. The authors did not comment in this context on the whole volume complexity of phase change. Mehling et al. [26] also advised adjustments to measure methods and conditions to PCM application to include the subcooling effect in enthalpy curves.

Mandilaras et al. presented a new hybrid methodology for the precise determination of the effective heat capacity and enthalpy curves of PCMs [28]. They combined DSC testing with dynamic operated heat flow apparatus followed by the numerical optimization of the obtained experimental enthalpy curves. Fateha et al. included the hysteresis effect in their numerical and experimental investigation of an insulation layer with PCM [29]. However, no switching between the melting and freezing curves was explored during the cycle. When using the hysteresis model for enthalpy, they were able to identify a strong agreement between measurements and simulations.

Kumarasamy et al. elaborated on the numerical schemes for the testing and simulation of encapsulated PCM with temperature hysteresis [30]. Their results showed that encapsulation greatly altered the thermal response of PCM in terms of the phase change temperatures and hysteresis. The authors suggested that in such a case, a Computer Fluid Dynamics-based conduction-dominant scheme should be incorporated into the simulation. Moreles et al. investigated the application of PCM, considering the hysteresis of phase change and only complete phase transitions [31]; based on a developed numerical model, a graphical method of optimized PCM selection was, hence, proposed.

All the previously discussed papers were based on former versions of the EnergyPlus™ program in which the PCM parameters did not take into account the hysteresis effect. In the simplified enthalpy–temperature function, only one curve representing both the melting and solidifying process was given. However, the latest releases of EnergyPlus™ 9.2 (released on September 27, 2019) take into account two separate freezing and melting curves with user-specified temperature data [20]. In this algorithm, an actual value of the specific heat that is used in the simulation process is not only dependent on the current state of the PCM but also on the former state as in symbolic Equation:

$$c_p = f(T_{i;new}; T_{i;prev}; PhaseState_{new}; PhaseState_{prev}) \quad (1)$$

where c_p is the specific heat, and T_i is the previous and new i-node temperature. The values of PCM thermal conductivity and density should be entered for the liquid and solid states. It should be noted that the current description of the new algorithm related to hysteresis is unclear in the program documentation, as it is not entirely clear what 'temperature difference' refers to in this context and which PCM data should be introduced [20].

One of the first papers on the new hysteresis algorithm embedded in EnergyPlus™ was written by Goia et al. [32]. Unfortunately, the hysteresis algorithm was only briefly highlighted in this paper, and the authors' input data was not presented. The simulation results were compared against experimental data obtained from the small-scale dynamic tests performed with the heat-flow apparatus. It was expected that a new algorithm that enables modelling of hysteresis would closely follow the real mode of phase change and thus the precision of the PCM simulations. Goia et al. confirmed these expectations stating that the numerical results were significantly better than those obtained with more conventional models. However, in the case of an incomplete phase change of PCM, the obtained results were regarded as 'questionable'.

2.2. PCM Use Efficiency

One of the basic items of technical information of PCM is the transition temperature, usually given as one specified value. It is common knowledge that the real physical course of the PCM occurs within a temperature range of a few degrees. Additionally, the transition temperature range is usually different for the freezing and melting curves because of hysteresis. The total transition from solid to liquid and vice versa does not occur when the real temperature fluctuations around the PCM and do not cover the whole range of both phase change courses. In these conditions, the large amounts of heat are neither stored nor released and the operating results of the system do not meet expectations. Berardi and Soudian discussed this aspect extensively, suggesting the use of night cooling to obtain full solidification of the melted material [33] and reported the frequency of PCM activation and the percentage of time with full solidification.

The relatively wide phase-change temperature ranges and the thermal conditions in a room not covering this range can be the cause of only a partial solidification or melting. Such situations occur in practice when the entire transformation range goes beyond the minimum or maximum air temperature in the interior or when the time of maintaining a sufficiently low temperature is too short, and the material only partially solidifies.

In the situation of partial phase change, it is necessary for the simulation algorithm to break the initial process and switch to the second enthalpy curve without reaching the end of phase change. Moreles et al. stated that even all the numerical models tested were not fully able to replicate the behavior of PCM layers if the PCM did not melt or re-solidify completely [31].

The transition process between the heating and cooling enthalpy–temperature curves in the case of incomplete phase change was investigated by Delcroix et al. [34]. In their experiment, the PCM-equipped wall sample was quickly transferred from a cold to a hot environment and conversely. For the interrupted cooling process, the PCM followed an enthalpy curve that was very close to the heating curve. In the case of the interrupted heating process, PCM followed a new curve that was located between the heating and cooling curve. It was revealed that the transition process was sensitive to the actual values of the boundary conditions. In another paper, Delcroix et al. [25] showed that varying heat transfer rates have a significant impact on the phase change temperature range and the hysteresis between heating and cooling curves. Higher rates increase the hysteresis and shift the phase change temperature towards colder temperatures. The authors suggested adjusting a PCM testing method to the perspective conditions of its application. All these results of testing and observations mean that a precise description of the material properties is difficult.

The problems highlighted above are known to people involved in PCM research, but they are poorly understood by designers or potential investors. The effective design of a building in which PCM is applied requires extensive and precise information about the material used, and second, a simulation tool that would allow effective modelling of such a phenomenon.

The introduction of the hysteresis modelling and temperature-dependent enthalpy in the EnergyPlus™ program improved the simulation capabilities of this tool. However, the above-highlighted aspect of incomplete PCM transition should be taken into account when modelling phase change phenomena.

3. New Simulation Algorithm

The hysteresis effect is usually shown in the form of two curves describing the course of the same phenomenon. The difficulty associated with modeling the hysteresis is due to the need to introduce two different functions describing the material enthalpy within the phase change range. The selection of the proper curve (Figure 1) must be based on the history of changes and the identification of the direction of the current changes. Such an algorithm would be appropriate if it could be assumed that the phase transitions occurred in the entire volume of the material.

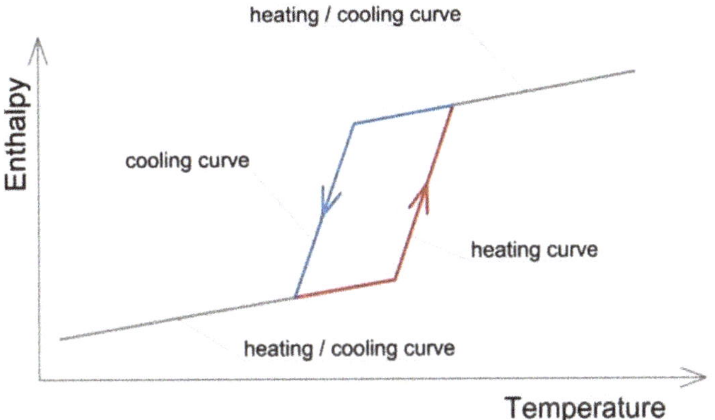

Figure 1. Phase change hysteresis process represented as a function of temperature.

As outlined in the previous section, the real course of the phase changes is more complicated, and, in many cases, the process of melting/solidification is incomplete. A hysteresis algorithm should enable the direct transition from the melting curve to the freezing curve and vice versa. The concept of a 'state' was used by Zastawna [35] to create a new algorithm. Figure 2 shows the transition from the 'cooling curve state' to the 'heating curve state' defined as the occurrence of the third kind of state called the 'transition curve state'. The states are represented by the colors: blue—cooling curve; red—heating curve; purple-transition curve.

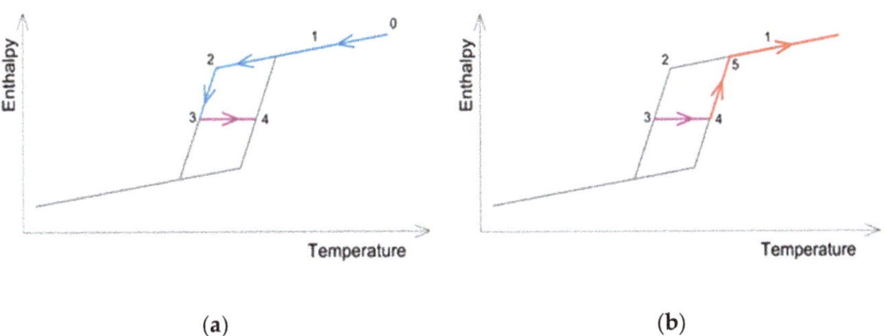

Figure 2. Intermediate transition curve between solidification (**a**) and melting (**b**) curves.

When the cooling phase of the completely melted material has begun (curve 0–1–2, Figure 2a) and before the phase change starts (point 2), the heat is released in a sensible way. After the phase change process has started, the energy is released in both latent and sensible ways (curve 2–3). At this point (point 3), the heating of the material begins again. Thus far, the enthalpy value follows the cooling curve state. The re-start of heating results in the termination of the cooling curve state and the start of the transition curve state (curve 3–4). For the transition curve, the enthalpy value is determined from either the heating curve or the cooling curve.

Rose et al. [36] suggested an instantaneous transition between the curves and a horizontal movement of the state until it reaches either the melting or solidifying curve. Bony et al. [37] assumed a sloped transition curve between the heating and cooling curves but a horizontal transition for the sub-cooling effect. Any change in the system temperature during the transition phase should obviously be associated with a change in enthalpy. The method of transition in the form of a horizontal line adopted in the article is a simplification.

While the hysteresis curves can be measured, the shape and slope of the transition curve are completely unknown, and experimental data are not available as they depend on many parameters, including the current entropy value. Curves describing the full transformation hysteresis are also a kind of idealization of this phenomenon and only describe it in an approximate way. Among others, heating and cooling curves are not usually parallel to each other, so any decision regarding the transition curve slope would always be approximate. Because the impact of simplification in the form of a horizontal transition line is small, and because of this, the modeling algorithm becomes significantly simpler, it was decided to use a horizontal transition curve in the model.

At point 4, the process of melting of the partially solidified material restarts. At this moment, the transition curve state ends, and the heating curve state will restart (curve 4–5 in Figure 2b). After completion of the phase change, the material again accumulates heat only in a sensible way (curve 5–1) while remaining in the heating curve state.

Equivalent specific heat of a PCM in case of the heating curve is calculated in the following way:

$$c_p(T) = \frac{H_{i,h}^{j} - H_{i,h}^{j-1}}{T_i^j - T_i^{j-1}} \tag{2}$$

where $H_{i,h}^{j}$ is the heating enthalpy of a discrete node 'i' at a time-step 'j' (and respectively '$j-1$') [J/kg]. In the same way for the cooling curve:

$$c_p(T) = \frac{H_{i,c}^{j} - H_{i,c}^{j-1}}{T_i^j - T_i^{j-1}} \tag{3}$$

where $H_{i,c}^{j}$ is the cooling enthalpy of a discrete node 'i' at a time-step 'j' (and respectively '$j-1$') [J/kg].

For the transition from cooling to heating curve 3–4, Equation or, if $H_{i,tr}^{j} > H_{i,h}^{j}$ then:

$$c_p(T) = \frac{H_{i,tr}^{j} - H_{i,tr}^{j-1}}{T_i^j - T_i^{j-1}} \tag{4}$$

where $H_{i,tr}^{j}$ is the transition enthalpy, taken from transition curve and $H_{i,tr}^{j} = H_{i,tr}^{j-1}$

If $H_{i,tr}^{j} \leq H_{i,h}^{j}$ then

$$c_p(T) = \frac{H_{i,h}^{j} - H_{i,tr}^{j-1}}{T_i^j - T_i^{j-1}} \tag{5}$$

For the transition from heating to cooling curve 4–3, Equation or, if $H_{i,tr}^{j} < H_{i,c}^{j}$ then:

$$c_p(T) = \frac{H_{i,tr}^{j} - H_{i,tr}^{j-1}}{T_i^j - T_i^{j-1}} \tag{6}$$

and $H_{i,tr}^{j} = H_{i,tr}^{j-1}$

If $H_{i,tr}^{j} \geq H_{i,c}^{j}$ then:

$$c_p(T) = \frac{H_{i,c}^{j} - H_{i,tr}^{j-1}}{T_i^j - T_i^{j-1}} \tag{7}$$

Appendix A shows the data that should be entered into the newly created PCM hysteresis algorithm. Each temperature corresponds to two enthalpy values separately for the heating and for the cooling curves.

4. Results

To assess the correctness of the new EnergyPlus™ algorithm, the simulation results have been compared to the results of measurements conducted in the climate chambers shown in Figure 3. A wall, located between the warm and cold chamber, was made of 15 cm expanded polystyrene covered with the standard gypsum panel on the warm side. Two material samples were glued to the gypsum panel. One of them was a modified gypsum panel (50 × 60 cm) containing 30% by mass of PCM Micronal. The second one was a reference standard gypsum panel of the same size and thickness (12.5 mm) as the PCM panel. The principle of the presented research was based on a comparison of surface temperature fluctuations and energy accumulation in the PCM and the reference standard samples. In this way, the authors wanted to examine the efficiency of the added PCM wall cladding as a form of overheating protection in a hot summer environment.

Figure 3. Full-scale climate chambers (cold chamber on the left and warm chamber on the right).

The melting temperature of the PCM panel, as declared by the manufacturer, was 23 °C.

Before the experiment, a sample of PCM was tested in the calorimeter DSC214 (Polyma–Netzsch), and measured latent heat was equal to 127.7 J/g. The PCM 23 effective phase change temperature range measured was very wide: 17.8 °C ÷ 31.5 °C. Both gypsum boards were initially tested for thermal conductivity using the Laser Comp Fox 314 plate apparatus. The tests were performed for different temperature ranges referring to the melting temperature of PCM. In the first case, the average temperature below the phase change temperature range was adopted and the value of thermal conductivity λ was 0.159 W/mK. In the second case, only a part of the material was in the liquid state (the temperature of the heating plate was 40 °C and the cooling plate was 20 °C). In these conditions, the λ-value was equal to 0.162 W/mK. In the third case, the temperature of both plates exceeded the phase change value, so all the PCM material contained in the tested plate was in the liquid state during the test, and the λ-value was equal to 0.164 W/mK. The observed small differences in the thermal conductivity coefficients (max. 2.8%) can be treated as negligible and were not included in further analysis.

Three temperature sensors were placed on the surface of both gypsum panels (K-type thermocouples, measurement class 2, sensors attached to the surface with adhesive paper tape). The heat flow on both surfaces of the sample plates was measured by means of Ahlborn 118SI silicone heat-flow transducers (sensor size 120 × 120 mm), measurement class A. Air temperature was measured by means of the Pt 100 and Pt 1000 sensors, measurement class B, protected with an aluminum foil against thermal radiation. All the measurements and data recording were performed by the Ahlborn Almemo 5690 data acquisition system and the Data-Control 4.2 program.

Temperature fluctuations in the warm chamber corresponded to the conditions that occur in a room during a hot summer day. However, the technical capabilities of the control system only allowed stepped (non-continuous) changes. Four testing cycles were conducted as reported in Table 1.

Table 1. Testing cycles of the boundary conditions.

Temperature Cycle	Air Temperature in the Warm Chamber [°C]	Air Temperature in the Cold Chamber [°C]	Peak Temperature Duration [h]
0129	18 ÷ 36	18	4
0130	18 ÷ 37	18	3.5
0131	18 ÷ 30	18 ÷ 46	4
0203	18 ÷ 30	16 ÷ 32	3

The temperature in the warm chamber was gradually changing during the daily cycle within the range 18 ÷ 37 °C. The daily maximum temperature was maintained for 4 hours at the same level, which was long enough to enable the phase change in the whole volume of PCM. During the whole night, a constant temperature of 18 °C was maintained in the chamber; therefore, after each cycle, it was possible to discharge the total amount of energy stored in the tested materials during the day to the surrounding environment. The air temperature in the cold chamber was maintained at a constant 18 °C in two tests, and in the other two tests, it was variable in a similar way as was the case in the warm chamber. The last two cycles were close to the conditions that occur during hot summer days. A simulation model of the climate chambers with the sample wall between the chambers was created in the EnergyPlus™ software. As in the real conditions, the modeled wall consisted of a 15 cm layer of expanded polystyrene and gypsum board, to which a layer of phase-changing material was attached. It was assumed that the conditions of heat exchange between the outer shell of the chamber and the surrounding environment were adiabatic. A dynamically changing cycle of air temperature inside the warm chamber was assumed. This was consistent with the cycle of the experimental measurements. The same model was used as the reference variant of the standard gypsum board.

Measured and calculated temperature fluctuations during one selected day of cycle 0129 are presented in Figure 4. The red curve shows the daily internal air temperature cycle that was a driving force of the fluctuations and the boundary condition both in the experiment and the simulation.

The scheduled maximum temperature of 36 °C was maintained for more than three hours. The curves obtained from numerical calculations and the measurements were, in general, very close to each other, with the exception of the surface temperature of the reference gypsum plate. The relatively low thermal capacity of the standard material in which heat is stored only in a sensible way resulted in a significantly higher temperature of the gypsum surface when the internal air temperature was rising and much lower temperature during the cooling stage. The gypsum plate cooled quicker than the PCM. The results of the simulation by means of both algorithms (with and without the hysteresis effect) fit perfectly with each other and very closely followed the measurements during the heating stage. This means that the enthalpy characteristic introduced to the simplified algorithm without hysteresis followed the heating curve of the PCM.

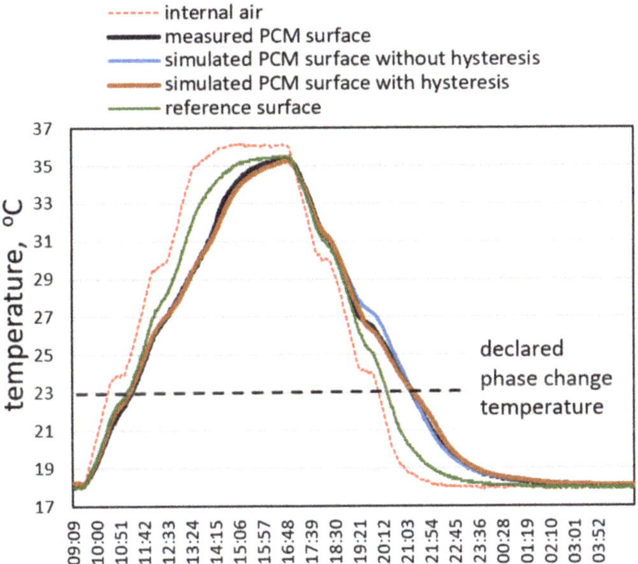

Figure 4. Measured versus simulated temperature fluctuations of the phase change materials' (PCM) surface.

As could be expected, significant discrepancies may be observed during the cooling stage, when the solidification process starts. Within the wide range of phase change, but still, before the phase change temperature declared by the producer (23 °C) was reached, the model without hysteresis (orange curve) showed a surface temperature higher than the other models due to the single characteristic curve. The new model that takes into account the effect of hysteresis closely followed the measured values. Below the phase change temperature, this relationship was reversed.

Figure 5 shows the heat flux values on the internal surface of the PCM board obtained from measurements and computer simulations. One simulation was conducted on the model without taking into account phase change hysteresis and the second one with this effect. The cycle begins with the discharge phase of the wall's thermal capacity (negative flux value). Then, as a result of the internal temperature increase in the tested space, heat flux released from the wall decreased, and then the process of recharging (plus flux value) began. The significant differences between both simulation models may be noticed in the second phase of the research cycle, especially in cooling mode. The maximum instantaneous heat flux value was high and equal to 62 W/m^2. During the whole measuring period, the PCM panel accumulated around 180 Wh/m^2.

To assess the accumulation efficiency of the PCM board, Figure 6 shows the results of measurement and simulation of heat flux on the internal surface of the reference gypsum board. The heat exchange on the surface of the standard panel took place according to a similar scenario as before, but the instantaneous heat flux values were definitely lower. In the case of material without phase change, the graph obtained from the simulation was very close to the measurement results. The maximum momentary heat flux value was, in this case, equal to 32 W/m^2 and accumulated during the whole cycle amount of energy was only 36 Wh/m^2.

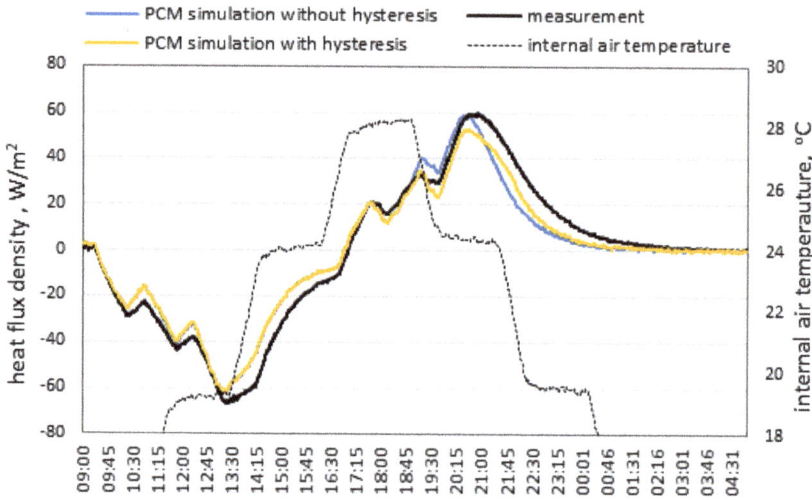

Figure 5. Measured versus simulated heat flux density on the internal surface of the PCM panel.

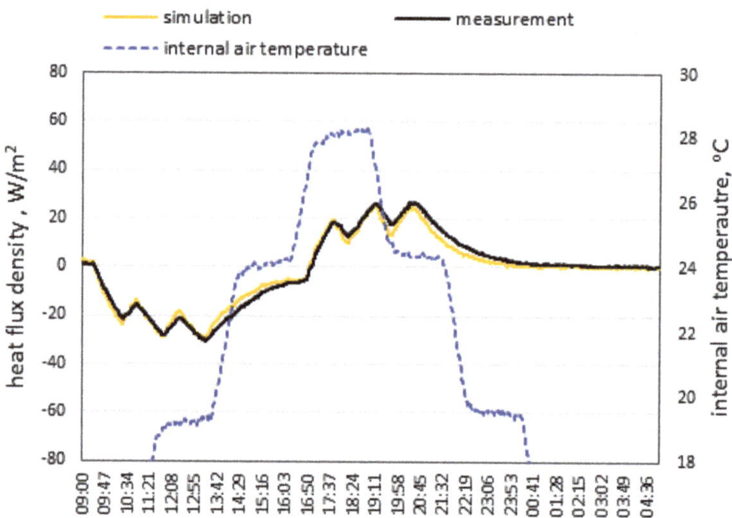

Figure 6. Measured versus simulated heat flux density on the internal surface of the standard gypsum panel.

During the full phase transition cycle, the entire volume of PCM material could accumulate 5 times more energy than a standard material of the same volume. The application of PCM wall panels substantially enlarged the storage capacity of the wall and also enabled a decrease in cooling power demand. All these measures support the improvement in space thermal comfort.

5. Discussion of the Simulation Precision

Figure 7 presents the temperature difference between the results of measurements and calculations for the two variants of the simulation algorithm. The maximum momentary error of simulation occurred in the case of the algorithm without hysteresis and was equal to 0.96 K.

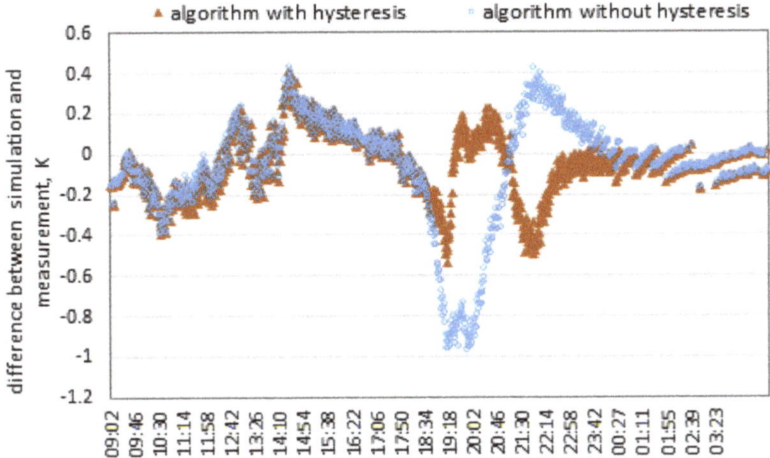

Figure 7. Absolute momentary errors of simulation.

The average results of relative error calculation for all the temperature testing cycles are shown in Figure 8. In the case of cycle 0130, the mean relative error of the variant without hysteresis was 5.38 times larger than that of the variant with hysteresis. The lowest disproportion may be observed in the case of cycle 0203, the mean relative error was 1.71 times larger than in the case of the variant with hysteresis. The average ratio between the errors of both compared algorithms in the four tested cycles of measurements was 3.35. It should be noted that the errors related to the algorithm with hysteresis were close to the errors of the simulation of the standard materials (reference plate), i.e., precision of the applied simulation software.

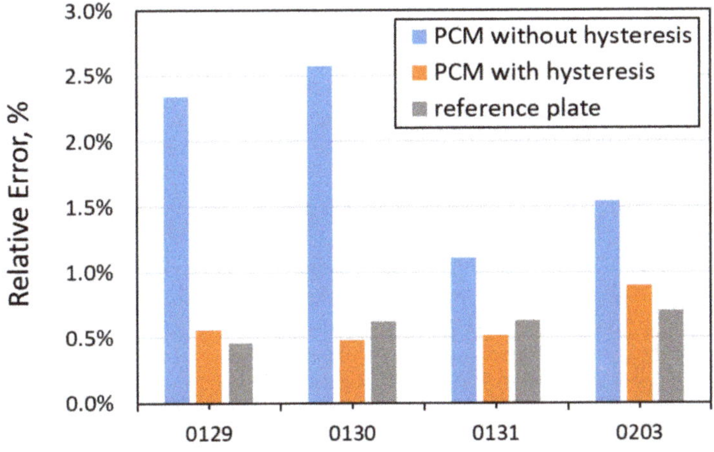

Figure 8. Mean error of surface temperature estimation in four cycles of measurements.

Figure 9 shows mean absolute errors of simulation of the heat flux density at the internal surface of tested samples. The inclusion of the hysteresis effect in the simulation algorithm reduced simulation error in three cycles of simulation only to a small extent. Better accuracy of heat flux evaluation could be expected only in the case of standard building material, in which only sensible heat accumulation takes place. Relatively high fluctuations can be caused on the one hand by the large variation in the value of heat fluxes and on the other hand, by much lower, than in case of temperature, precision of the heat flux measurement (5%). However, the absolute error values allowed a rough estimation of the room's cooling load and may be used as a simple designing tool.

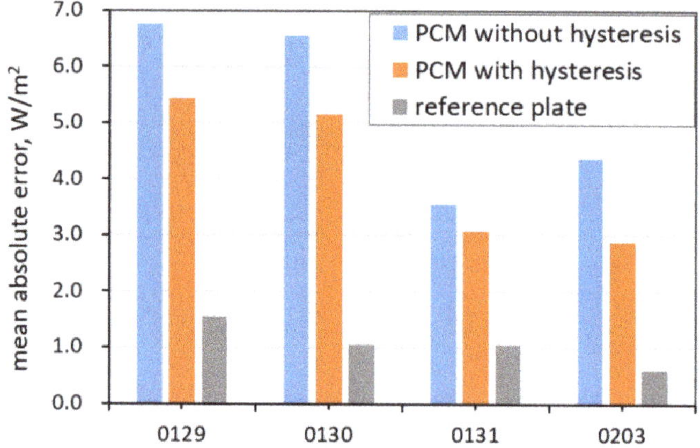

Figure 9. Mean error of surface temperature estimation in four cycles of measurements.

6. Conclusions

Numerous experimental tests confirmed that the hysteresis of the phase change process has a noticeable effect on heat accumulation in PCM. It is, therefore, necessary to include this phenomenon in commonly used simulation programs. This article presented the principles of a novel EnergyPlus™ simulation module, which enables the hysteresis effect of phase change in building materials to be taken into account.

In the developed algorithm, apart from the basic difference associated with a different course of the heating and the cooling phenomenon of the material, it is possible to model an incomplete phase change too. This feature of the algorithm seems to be a unique and important aspect of the developed algorithm because incomplete transition quite often appears in the practical applications of PCM and adversely affects the achieved energy effects. The verification of the new calculation module confirmed its usefulness in simulation analyses. The relative calculation error of surface temperature was less than 1%, while in the case of the simplified model without hysteresis, the simulation error would be even a few times higher. We conducted experimental research and created a simulation model that confirmed the practical effects of PCM application as a passive measure of overheating protection. The addition of 30% phase changing material to the gypsum board increased its effective heat capacity several times and doubled the maximum cooling power of PCM wall cladding.

However, the practical effects of PCM application are often much lower than the expectations associated with them. This may be related to several factors that are poorly understood by building designers. One of these is the phase change temperature, declared by the material manufacturer as a specified numerical value, while in reality, the transformation process covers a temperature range of a few degrees. The incomplete phase transition is largely related to the fact that the phase change does not take place at one temperature but in a fairly wide temperature range. Effective phase change temperature ranges of the PCM 23 applied in this research, measured in the DSC apparatus, was 17.8 °C–31.5 °C. As a result, the heat capacity of PCM can be only partially used. Another reason for poor performance is an insufficient temperature fluctuation or too short a heating or cooling stage, during which the material is not fully charged or discharged.

Both experimental tests in the climate chamber, as well as simulation analysis, confirmed the effectiveness of PCM enriched gypsum board as an energy accumulator.

Author Contributions: Conceptualization, A.Z.-R., T.K., U.B.; methodology, A.Z.-R., T.K.; software, A.Z.-R., validation, T.K.; writing—review and editing, A.Z.-R., T.K., U.B. All authors have agreed to the published version of the manuscript.

Funding: This research was funded by NSERC.

Conflicts of Interest: The authors declare no conflict of interest.

Appendix A

Example of the separate data input for melting and solidifying curves.
Material Property:Phase Change,
PCM plate, !- Name
-40, !- Temperature 1 {C}-temperature beyond phase change range
0, !- Enthalpy-m 1 {J/kg} – melting enthalpy
0, !- Enthalpy-s 1 {J/kg} – solidification enthalpy
23, !- Temperature 2 {C}
2005, !- Enthalpy-m 2 {J/kg}
3005, !- Enthalpy-s 2 {J/kg}
25, !- Temperature 3 {C}
6570, !- Enthalpy-m 3 {J/kg}
12714, !- Enthalpy-s 3 {J/kg}
27, !- Temperature 4 {C}-temperature within phase change range
12714, !- Enthalpy-m 4 {J/kg} – melting enthalpy
29200, !- Enthalpy-s 4 {J/kg} – solidification enthalpy
28, !- Temperature 5 {C}
25300, !- Enthalpy-m 5 {J/kg}
29300, !- Enthalpy-s 5 {J/kg}
29, !- Temperature 6 {C}
30000, !- Enthalpy-m 6 {J/kg}
30300, !- Enthalpy-s 6 {J/kg}
30, !- Temperature 7 {C}
33000, !- Enthalpy-m 7 {J/kg}
31300; !- Enthalpy-s 7 {J/kg}

References

1. Mazria, E. *The Passive Solar Energy Book*; Rodale Press: Emmaus, PA, USA, 1978.
2. Kisilewicz, T. Overheating—An unexpected side-effect of heating decreased demand. In Proceedings of the 2nd Central European Symposium on Building Physics, Vienna, Austria, 9–11 September 2013; pp. 125–130.
3. Kisilewicz, T. *Effect of Insulation, Dynamic and Spectral Properties of the Building Walls on the Thermal Balance of the Low Energy Buildings*; CUT Publishing Office, 364: Krakow, Poland, 2008.
4. D'Ambrosio Alfano, F.R.; Olesen, B.W.; Palella, B.I.; Riccio, G. Thermal comfort: Design and assessment for energy saving. *Energy Build.* **2014**, *81*, 326–336. [CrossRef]
5. Pedersen, C.O. Advanced zone simulation in EnergyPlus: Incorporation of variable properties and phase change material (PCM) capability. In Proceedings of the Building Simulation 2007, Beijing, China, 3–6 September 2007.
6. Ozdenefe, M.; Dewsbury, J. Dynamic Thermal Simulation of a PCM Lined Building with EnergyPlus. In Proceedings of the WSEAS International Conference on Energy and Environment, Kos, Greece, 14–17 July 2012; pp. 359–364.
7. Zhuang, C.; Deng, A.; Chen, Y.; Li, S.; Zhang, H.; Fan, G. Validation of Veracity on Simulating the Indoor Temperature in PCM Lightweight Building by EnergyPlus. In *Life System Modelling and Intelligent Computing*; Springer: Berlin/Heidelberg, Germany, 2010; pp. 486–496.
8. Tardieu, A.; Behzadi, S.; Chen, J.; Farid, M. Computer simulation and experimental measurements for an experimental PCM-impregnated office building. In Proceedings of the 12th Conference of International Building Performance Simulation Association, Sydney, Australia, 14–16 November 2011.

9. Kuznik, F.; Virgone, J. Experimental investigation of wallboard containing phase change material: Data for validation of numerical modelling. *Energy Build.* **2009**, *41*, 561–570. [CrossRef]
10. Kosny, J.; Stovall, T.; Shrestha, S.; Yarbrough, D. Theoretical and Experimental Thermal Performance Analysis of Complex Thermal Storage Membrane Containing Bio-Based Phase-Change Material (PCM). In Proceedings of the Thermal Performance Exterior Envelopes Whole Buildings XI International Conference, Clearwater Beach, FL, USA, 5–9 December 2010.
11. Kosny, J.; Kossecka, E.; Brzezinski, A.; Tleoubaev, A.; Yarbrough, D. Dynamic thermal performance analysis of fiber insulations containing bio-based phase change materials (PCMs). *Energy Build.* **2012**, *52*, 122–131. [CrossRef]
12. Naraid, J.; Jin, W.; Ghandehari, M.; Wilke, E.; Shukla, N.; Berardi, U.; El-Korchi, T.; Dessel, V. Design and application of concrete tiles enhanced with microencapsulated phase-change material. *J. Archit. Eng.* **2016**, *22*. [CrossRef]
13. Berardi, U.; Soudian, S. Benefits of latent thermal energy storage in the retrofit of Canadian high-rise residential buildings. *Build. Simul.* **2018**, *11*, 709–723. [CrossRef]
14. Shrestha, S.; Miller, W.; Stovall, T.; Desjarlais, A.; Childs, K.; Porter, W.; Bhandari, M.; Coley, S. Modelling PCM-enhanced insulation system and benchmarking EnergyPlus against controlled field data. In Proceedings of the 12th International Conference Building Performance Simulation Association, Sydney, Australia, 14–16 November 2011; pp. 800–807.
15. Zalba, B.; Marın, J.M.; Cabeza, L.F.; Mehling, H. Review on thermal energy storage with phase change: Materials, heat transfer analysis and applications. *Appl. Therm. Eng.* **2003**, *23*, 251–283. [CrossRef]
16. Castell, A.; Medrano, M.; Cabeza, L.F. Analysis of the simulation models for the use of PCM in buildings. In Proceedings of the Effstock 2009—11th International Conference on Thermal Energy Storage, Stockholm, Sweden, 14–17 June 2009; pp. 1–8.
17. Campbell, K.R. *Phase Change Material as a Thermal Storage Device for Passive Houses*; Portland State University: Portland, OR, USA, 2011.
18. Chan, A.L. Energy and environmental performance of building façades integrated with phase change material in subtropical Hong Kong. *Energy Build.* **2011**, *43*, 2947–2955. [CrossRef]
19. Lee, K.O.; Medina, M.A.; Sun, X. Development and verification of an algorithm to predict heat transfer through building walls integrated with phase change materials. *Build. Phys.* **2016**, *19*, 1–19. [CrossRef]
20. U.S. Department of Energy. *Documentation: Engineering Reference and Input Output Reference*; EnergyPlus Version 9.2.0; U.S. Department of Energy: Washington, DC, USA, 2019.
21. Tabares-Velasco, P.C.; Christensen, C.; Bianchi, M. Verification and validation of EnergyPlus phase change material model for opaque wall assemblies. *Build. Environ.* **2012**, *54*, 186–196. [CrossRef]
22. Loonen, R.; Favoino, F.; Hensen, J.; Overend, M. Review of current status, requirements and opportunities for building performance simulation of adaptive facades. *J. Build. Perform. Simul.* **2017**, *10*, 205–223. [CrossRef]
23. Biswas, K.; Shukla, Y.; Desjarlais, A.; Rawal, R. Thermal characterization of full-scale PCM products and numerical simulations, including hysteresis, to evaluate energy impacts in an envelope application. *Appl. Therm. Eng.* **2018**, *138*, 501–512. [CrossRef]
24. Tilman Barz, A. Sommer, Modelling hysteresis in the phase transition of industrial-grade solid/liquid PCM for thermal energy storages. *Int. J. Heat Mass Transf.* **2018**, *127*, 701–713. [CrossRef]
25. Delcroix, B.; Kummert, M.; Daoud, A.; Bouchard, J. Influence of experimental conditions on measured thermal properties used to model phase change materials. *Build. Simul.* **2015**, *8*, 637–650. [CrossRef]
26. Mehling, H.; Barreneche, C.; Solé, A.; Cabeza, L.F. The connection between the heat storage capability of PCM as a material property and their performance in real scale applications. *J. Energy Storage* **2017**, *13*, 35–39. [CrossRef]
27. Berardi, U.; Gallardo, A. Properties of concretes enhanced with phase change materials for building applications. *Energy Build.* **2019**, *199*, 402–414. [CrossRef]
28. Mandilaras, I.D.; Kontogeorgos, D.A.; Founti, M.A. A hybrid methodology for the determination of the effective heat capacity of PCM enhanced building components. *Renew. Energy* **2015**, *76*, 790–804. [CrossRef]
29. Fateha, A.; Klinker, F.; Brütting, M.; Weinläder, H.; Devia, F. Numerical and experimental investigation of an insulation layer with phase change materials (PCMs). *Energy Build.* **2017**, *153*, 231–240. [CrossRef]

30. Kumarasamy, K.; An, J.; Yang, J.; Yang, E. Novel CFD-based numerical schemes for conduction dominant encapsulated phase change materials (EPCM) with temperature hysteresis for thermal energy storage applications. *Energy* **2017**, *132*, 31–40. [CrossRef]
31. Moreles, E.; Huelsz, G.; Barrios, G. Hysteresis effects on the thermal performance of building envelope PCM-walls. *Build. Simul.* **2018**, *11*, 519–531. [CrossRef]
32. Goia, F.; Chaudhary, G.; Fantucci, S. Modelling and experimental validation of an algorithm for simulation of hysteresis effects in phase change materials for building components. *Energy Build.* **2018**, *174*, 54–67. [CrossRef]
33. Berardi, U.; Soudian, S. Experimental investigation of latent heat thermal energy storage using PCMs with different melting temperatures for building retrofit. *Energy Build.* **2019**, *185*, 180–195. [CrossRef]
34. Delcroix, B.; Kummert, M.; Daoud, A. Thermal behavior mapping of a phase change material between the heating and cooling enthalpy-temperature curves. *Energy Procedia* **2015**, *78*, 225–230. [CrossRef]
35. Zastawna-Rumin, A. The Analysis of the Application Efficiency of Phase Change Materials in Partitions of Polish Low-Energy Buildings. Ph.D. Thesis, Cracow University of Technology, Krakow, Poland, 2018.
36. Rose, J.; Lahme, A.; Christensen, N.U.; Heiselberg, P.; Hansen, M.; Grau, K. Numerical Method for Calculating Latent Heat Storage in Constructions containing Phase Change Material. In Proceedings of the Building Simulation, Glasgow, Scotland, UK, 27–30 July 2009; pp. 400–407.
37. Bony, J.; Citherlet, S. Numerical Model and Experimental Validation of Heat Storage with Phase Change Materials. *Energy Build.* **2007**, *39*, 1065–1072. [CrossRef]

Article

Application of Rough Set Theory (RST) to Forecast Energy Consumption in Buildings Undergoing Thermal Modernization

Tomasz Szul * and Stanisław Kokoszka

Faculty of Production and Power Engineering, University of Agriculture in Krakow, 30-149 Kraków, Poland; s.kokoszka@urk.edu.pl

* Correspondence: t.szul@urk.edu.pl; Tel.: +48-12-662-46-47

Received: 8 February 2020; Accepted: 7 March 2020; Published: 11 March 2020

Abstract: In many regions, the heat used for space heating is a basic item in the energy balance of a building and significantly affects its operating costs. The accuracy of the assessment of heat consumption in an existing building and the determination of the main components of heat loss depends to a large extent on whether the energy efficiency improvement targets set in the thermal upgrading project are achieved. A frequent problem in the case of energy calculations is the lack of complete architectural and construction documentation of the analyzed objects. Therefore, there is a need to search for methods that will be suitable for a quick technical analysis of measures taken to improve energy efficiency in existing buildings. These methods should have satisfactory results in predicting energy consumption where the input is limited, inaccurate, or uncertain. Therefore, the aim of this work was to test the usefulness of a model based on Rough Set Theory (RST) for estimating the thermal energy consumption of buildings undergoing an energy renovation. The research was carried out on a group of 109 thermally improved residential buildings, for which energy performance was based on actual energy consumption before and after thermal modernization. Specific sets of important variables characterizing the examined buildings were distinguished. The groups of variables were used to estimate energy consumption in such a way as to obtain a compromise between the effort of obtaining them and the quality of the forecast. This has allowed the construction of a prediction model that allows the use of a fast, relatively simple procedure to estimate the final energy demand rate for heating buildings.

Keywords: building energy consumption; building load forecasting; energy efficiency; rough set theory; thermal improved of buildings

---

1. Introduction

In the climatic conditions of central and eastern Europe, the heat used to heat rooms is a basic item in the building's energy balance and significantly affects its operating costs. The accuracy of the assessment of heat consumption in an existing building and identification of the main components of heat loss depends largely on whether the energy and economic effects assumed in the thermal modernization project are achieved.

Activities aimed at improving the thermal efficiency of buildings, and thus affecting the thermal comfort of users, usually consist in increasing their energy standard. In the case of buildings that existed before taking action to save energy, there is a need to prepare an energy audit, the purpose of which is to obtain adequate knowledge of the existing energy consumption profile of a given building or buildings complex, determine the manner and amount of energy that can be obtained, and notify about the results.

The objective of the audit is to determine the amount and structure of consumed energy and to identify and then recommend specific solutions that are energy profitable. The audit may identify modernization operations that are profitable in the investigated building and which products and technical solutions are the most favourable. Since 1999, there has been a program in Poland that supports actions aiming at reduction of energy consumption in existing buildings. It is set forth in the act on supporting thermal efficiency improvement and renovations [1], which is in line with the provisions of the EU Directive on the energy performance of buildings [2]. As a part of this program, a majority of apartment buildings are modernized—according to BGK [3], approximately 41,000 buildings were modernized. According to the statutory provisions, the auditing of buildings is based on the following stages: analysis of the present condition of buildings, verification of the assumed parameters based on the data on the real energy consumption, identification of possible facilitation along with determination of costs of their realization, calculation of savings resulting from facilitations (at the assumption that modernized partitions should have heat transmission ratios U lower or equal to the border values set forth in binding industry resolutions) and economic analysis of profitability, where it is assumed that investment expenditures incurred for a given facilitation should be returned within 10 years. Energy calculations in audits are made with a balance method (monthly or annual), where the auditor used the so-called diagnostic method that consists of estimation of parameter values. Current energy consumption is determined on this basis. When estimating energy consumption after thermal efficiency improvement, additionally economic aspects of the suggested energy saving solutions are included. This method is faulty due to the possibility of its use, which enables forecasting thermal energy consumption of single buildings. For the estimation of energy consumption for a bigger group of buildings, this method is too time consuming and requires a considerable work input.

Thus, it is necessary to look for other methods that will be suitable for analysis of the adopted thermal modernization measures and determine their impact on future heat consumption in existing buildings.

There are many methods to forecast current and future energy needs of buildings, which can be divided into three main groups [4–6]:

- Engineering methods—use a transparent process based on solving physics equations to describe the energy behaviour of buildings [5–7]
- Based on data—called "black box" models—are mainly based on statistical analyses of time series and machine learning algorithms to develop an energy model of a building [5,8–14]
- Hybrid methods—a combination of methods based on physics and data [15,16].

Engineering methods determine the thermal balance of the building, taking into account the use and efficiency of the heating/cooling and hot water preparation system. These models are used to determine the energy performance as well as to create forecasts of thermal comfort in buildings (user comfort). They are also used to determine indoor environment quality indicators. These models can be divided into two groups: dynamic and static. The dynamic models are based on the guidelines contained in EN 15251 [17], which defines the indoor environment conditions for individual rooms (thermal conditions for winter, thermal conditions for summer, air quality, criteria for ventilation, lighting and acoustics). They also take into account the influence of changing external conditions such as temperature, solar radiation intensity, wind speed, and others. They are mainly used in newly constructed (highly efficient) energy-saving and passive buildings [18–20]. Dynamic methods, determining the heat balance in short periods (usually 1 h), allow for a more accurate consideration of the effects related to the storage of heat or cold energy in the building's structural elements. Static models are based on the European standard EN 13790 [21], which is also supplemented by the EN 12831 standard [22]. Based on this method, heat balance is determined for a long calculation period (usually one month or heating season). Such models can be found in the literature [23–25]. This method is also used when preparing energy audits. Statistical methods are mainly regression models that correlate energy consumption or energy index with independent variables, which can be either quantitative or qualitative. Empirical models are constructed on the basis of historical results, which means that before training the model, we have to collect enough historical data to achieve sufficient precision. Statistical regressions have been widely used both for research at the design stage and after—during the use of the building. Regression models are used to forecast energy consumption based on detailed data, such as the age of the building and its main geometrical and thermo-humidity parameters, such as for example: shape factor, area of the opaque and glazed partitions, heat transfer coefficient, and indoor and outdoor temperatures, etc. Model calculations were carried out in analyses of heat consumption in systems supplying entire housing estates or towns, as well as at the level of a single building [9,26]. In some simplified models, regression is used to correlate energy consumption with climatic variables (e.g., a degree-days) in order to obtain the energy performance index [4,5,27–38]. Although the simplicity of regression is generally an advantage of these models, it is also a disadvantage because most regression techniques cannot deal with non-linear phenomena that occur during the use of a building. In models based on artificial intelligence, artificial neural networks and their derivatives are most often used. This type of models is based on solving non-linear problems and is an effective approach to this complex issue. The detailed division of methods and input variables for modelling is discussed in the paper [4]. These are: weather data grouping all data related to outdoor conditions, indoor environment in the building (temperature, humidity, etc.); occupancy and behaviour of occupants; time indicators that provide information about the functioning of the building and its energy behavior; past time steps, which take into account the potential impact of past events on current and projected energy demand states of the building; and building characteristics with information about active and passive systems. In recent years, artificial neural networks have been used to analyse the energy consumption of different types of buildings for different processes, such as heating/cooling, electricity consumption, heat loss through partitions and optimisation of energy consumption and estimation of performance parameters. The use of artificial neural networks and machine learning methods for modelling energy consumption can be found in the works of many authors [9,39–43]. Most of the presented calculation methods can be successfully used to estimate energy consumption for heating/cooling and hot water preparation and to determine the energy performance for different thermal parameters of partitions, the way buildings are used, and weather variables [4,5]. The authors of these works mainly focus on energy modelling in buildings such as offices, hotels, schools, universities, etc. However, few works concern single and multi-family residential buildings. In particular, there is a lack of studies on real buildings [4], data for which are difficult to obtain. Individual energy demand in residential buildings is more difficult to estimate due to the lack of data on occupancy of buildings and the complexity of the inhabitants' behaviour. Forecasting models focus mainly on estimating energy consumption, thermal comfort in existing (or simulated), newly built, energy efficient and passive facilities, for which it is possible to obtain reliable data on the insulation of building partitions, ventilation air streams, and number of inhabitants [18–20,44]. Most of the calculation methods presented are successfully suited for the estimation of energy consumption when determining the energy efficiency of buildings, for different parameters of thermal barriers, use, and variable weather conditions. Nevertheless, it is necessary to look for other solutions that could be used in the case of real buildings characterized by different availability of data describing the object from the thermal and operational point of view [14]. Another extremely important aspect of assessing energy consumption is the fact that the use of residential buildings often differs from the intended project. This is often due to the fact that the auditor's calculations are in many cases based on unreliable and inaccurate data, which significantly affects the accuracy of the assessment of current and future energy consumption of a building. At each stage of the audit calculations, some characteristics are likely to be inaccurately estimated, most often the physical parameters of buildings and the way the building is used, due to the difficulty of collecting all the figures that characterize the building and its surroundings with sufficient precision. This applies in particular to the value of heat transfer coefficient U through the building envelope. In such cases, the auditor carries out the examination of partitions and then assesses the equivalent thermal resistance of the partition. Even methodologically correctly calculated resistances can be wrong, as it is necessary to determine the thermal conductivity and thickness of individual layers, which is not always possible. A frequent problem in thermal calculations is the lack of complete architectural and construction documentation of the analyzed objects. In addition, there are other factors that affect the accuracy of the calculations, which are due to e.g., moisture, ageing of the material, etc. Uncertainty as to the correct assessment of the heat transfer coefficient, as well as other important parameters, such as the volume of ventilation air flow, influences the result of the final assessment of heat demand for both the existing buildings and the buildings for which thermal modernization measures are planned. It is therefore advisable to test new methods that could be used for a rapid technical analysis of measures taken to improve energy efficiency and to determine their impact on future heat consumption in existing buildings. These tools should allow decision-makers to assess the potential for real energy savings resulting from planned actions to improve the thermal performance of buildings. One of these methods may be the Rough Set Theory (RST) method [45], which is developed to analyze inaccurate, generic and undefined data. The more so because this method—according to the literature review—has not been used so far in forecasting energy consumption in buildings [4,8,15,43]. Therefore, the aim of the research was to determine the usefulness of a model based on RST for estimating thermal energy consumption in buildings undergoing thermal improvement. Due to the different availability and accuracy of data describing the building, the used input variable configurations will be tested during model construction in such a way as to achieve a compromise between the effort of the auditor to obtain them and the quality of the forecast.

2. Materials and Methods

Before the main objective of the study was achieved, analyses were carried out to establish a potential list of explanatory variables (conditional attributes). During the research, a database was created covering 109 buildings from the end of the last century that were thermal improved in the years 2010–2015. These buildings had energy audits prepared, on the basis of which the optimum variants of thermal modernization were selected, the partitions that should be modernized were indicated, and the appropriate thicknesses of layers of thermal insulation materials were selected. The analyzed buildings were described with many parameters.

For experimental reasons, most relevant characteristics have been selected. Some of them are measured and others calculated, as pointed out in Table 1.

Table 1. Characteristics of the selected values that influence the reduction of annual energy demand for buildings subjected to thermal efficiency improvement.

No.	Parameter	Abbreviation	Average	Median
1	calculated surface of heated floors from interior measurements, [m^2]	A_f	1567.4	1524.2
2	calculated area of temperature-controlled rooms (heated surface), [m^2]	A_H	1764.0	1565.4
3	calculated from exterior measurements surface of roof projection area (net), [m^2]	A_r	467.0	383.1
4	calculated from exterior measurements total walls' surface (net) area, [m^2]	A_w	1096.6	979.8
5	calculated surface of floor from interior measurements (floor over basement or floor on the ground), [m^2]	A_{fl}	395.4	360
6	calculated from exterior measurements total windows area, [m^2]	A_{tw}	290.5	251.1
7	calculated from exterior measurements the heated volume of building, [m^3]	V_e	6391.6	5408.8
8	shape coefficient of buildings (the ratio surface to volume), [m^2·m^{-3}], [m^{-1}]	S/V_e	0.46	0.42
9	number of stores, [pc.]	N_{Os}	4.3	4
10	number of residential flats, premises [pc.]	N_{Op}	32.4	29
11	number of living persons per building [N_b]	N_{Opb}	73.9	64
12	calculated thermal transmittance of walls components, [W·m^{-2}·K^{-1}]	U_w	1.12	1.16
13	calculated thermal transmittance of peak walls components, [W·m^{-2}·K^{-1}]	U_{pw}	1.0	0.94
14	calculated thermal transmittance of roof projections components, [W·m^{-2}·K^{-1}]	U_r	1.24	0.72
15	calculated thermal transmittance of floor components on the ground, [W·m^{-2}·K^{-1}]	U_g	1.62	1.41
16	calculated thermal transmittance of floors components (floor over basement), [W·m^{-2}·K^{-1}]	U_f	1.13	1.1
17	calculated thermal transmittance of windows (commercial data), [W·m^{-2}·K^{-1}]	U_{win}	1.82	1.6
18	calculated heating consumed power, [kW]	Φ_h	189.2	161.2
19	measured, the annual heat consumption for building heating converted (according to formula 3) to the conditions of the standard heating season, [MWh·year^{-1}]	$Q_{K,H0}$	506.6	475.3

The average value of parameters describing the examined buildings is comparable to the data contained in the building typology "TABULA" for Poland. These parameters are typical for an "average building" of a multi-family residential building built in 1967–1985 [46,47]. Therefore, the surveyed group of buildings can be considered as representative.

The data, selected after preliminary selection, were used to build sets of input variables based on which the usefulness of a method based on rough set theory (RST) for estimating the energy consumption of a building after the performance of thermal modernization measures in it was checked.

These variables were used to develop four sets of input data with different degrees of influence on energy consumption and difficulty in obtaining them, which are summarized in Table 2. A very limited set of indicators has been selected for the first set of variables explaining changes in energy consumption for heating a building after its thermal modernization. It included the amount of demand for thermal power to heat the building before the thermal modernization, the actual unitary indicator of final energy consumption for heating, and information about improvement actions taken—that is, which of the partitions will be insulated. From the next set of variables, the information concerning the demand for thermal power of its heating was eliminated and replaced with information concerning characteristic dimensions of individual components of the building, that is, area of individual building partitions, area and cubic capacity of the building, shape coefficient of buildings, and indicators characterizing a given building (number of people using the building, number of dwellings). The previous set of variables contained information which can be relatively easily obtained for any residential building, but it did not contain a very important parameter that would characterize the thermal insulation of individual partitions in the current state. Next set of variables has been supplemented with heat transfer coefficients for individual partitions. Collecting such an extensive range of information allows for precise characteristics of the object, but requires a lot of work on its reliable preparation. Because the above set of variables was very extensive and gathering such an extensive range of data is time-consuming, in the last set of input data, only the variables having direct influence on heat losses in the building, such as heat transfer coefficients through partitions and the fields of partition surfaces through which heat losses occur, are left. Sets of variables 2, 3 and 4 also included information on improvement measures taken—which of the partitions are to be isolated.

The groups of variables presented in Table 2, constituting conditional attributes, were used to build a model of prediction of final energy demand for building heating based on the Rough Set Theory. It was introduced in the 1980s by Professor Zdzisław Pawlak [45]. It is used as a tool to synthesize advanced and effective methods of analysis and to reduce datasets [48]. The rough sets serve as a methodology in the process of discovering knowledge in databases. It is a tool used to describe inaccurate, uncertain knowledge; to model decision-making systems; and for approximation reasoning [49]. The deduction methodology using the Approximate Collection Theory refers only to the qualitative nature of object characteristics. This causes limitations and difficulties when we deal with the occurrence of features in a quantitative form, not a qualitative one. The specificity of the attributes of the surveyed buildings shows a great variety of ways of encoding the given characteristics, which mostly occur in the quantitative form. In this case, the integration of the valued tolerance relation proves helpful [50]. It allows to implement more flexibility in data mining into the approximate set theory and to analyze observations expressed in quantitative form. The classic assumption of RSTs is based on the concept of the indistinguishability relationship as an exact relationship of equivalence, i.e., objects will only be indistinguishable if they have similar attributes (system 0–1). The introduction of a valued tolerance relation to RST allows to determine the upper and lower approximation of the crop with different degrees of indifference ratio [51]. This allows for the comparison of two sets of data and gives a result between 0 and 1, which is the level of indistinguishability. This range is a membership function derived from the assumptions of fuzzy harvest theory. The closer the result is to one, the more similar (indistinguishable) the objects are in terms of the analyzed attribute, and the closer to 0 the more distinguishable they are [50,51]. Detailed description of the prediction model based on quantitative variables has been presented in [51,52]. The general course of the construction of the model using the approximate sets theory is presented in Figure 1.

Table 2. Sets of input variables for analyzed predictive models. Sets of variables (before thermal modernization) (Recorded in the form of 0–1 information—whether the peak wall, external wall, floors, ground floors, windows, and flat roof are to be thermal modernized).

	Parameter-Condition Attributes
Set I	Φ_h – calculated heating consumed power, [kW] FE_0 – index of final energy demand for heating before modernization, [kWh·m^{-2}·year^{-1}]
Set II	V_e – calculated from exterior measurements the heated volume of building, [m^3] S/V_e – shape coefficient of buildings (the ratio surface to volume), [m^2·m^{-3}], [m^{-1}] A_f – calculated surface of heated floors from interior measurements, [m^2] A_w – calculated from exterior measurements total walls' surface (net) area, [m^2] A_r – calculated from exterior measurements surface of roof projection area (net), [m^2] A_{tw} – calculated from exterior measurements total windows area, [m^2] A_{in} – calculated from interior measurements total (net internal area), [m^2] No_{pb} – number of living persons per building, [N_b] No_p – number of residential flats, premises, [pcs.] FE_0 – index of final energy demand for heating before modernization, [kWh·m^{-2}·year^{-1}]
Set III	U_w – calculated thermal transmittance of walls components, [W·m^{-2}·K^{-1}] U_{pw} – calculated thermal transmittance of peak walls components, [W·m^{-2}·K^{-1}] U_r – calculated thermal transmittance of roof projections components, [W·m^{-2}·K^{-1}] U_f – calculated thermal transmittance of floors components (floor over basement), [W·m^{-2}·K^{-1}] U_{win} – thermal transmittance of windows (commercial data), [W·m^{-2}·K^{-1}] U_g – calculated thermal transmittance of floor components on the ground, [W·m^{-2}·K^{-1}] V_e – calculated from exterior measurements the heated volume of building, [m^3] S/V_e – shape coefficient of buildings (the ratio surface to volume), [m^2·m^{-3}], [m^{-1}] A_f – calculated surface of heated floors from interior measurements, [m^2] A_w – calculated from exterior measurements total walls' surface (net) area, [m^2] A_r – calculated from exterior measurements surface of roof projection area (net), [m^2] A_{tw} – calculated from exterior measurements total windows area, [m^2] A_{in} – calculated from interior measurements total (net internal area), [m^2] No_{pb} – number of living persons per building, [N_b] No_p – number of residential flats, premises, [pc.] FE_0 – index of final energy demand for heating before modernization, [kWh·m^{-2}·year^{-1}]
Set IV	U_w – calculated thermal transmittance of walls components, [W·m^{-2}·K^{-1}] U_{pw} – calculated thermal transmittance of peak walls components, [W·m^{-2}·K^{-1}] U_r – calculated thermal transmittance of roof projections components, [W·m^{-2}·K^{-1}] U_f – calculated thermal transmittance of floors components (floor over basement), [W·m^{-2}·K^{-1}] U_{win} – thermal transmittance of windows (commercial data), [W·m^{-2}·K^{-1}] U_g – calculated thermal transmittance of floor components on the ground, [W·m^{-2}·K^{-1}] A_f – calculated surface of heated floors from interior measurements, [m^2] A_w – calculated from exterior measurements total walls' surface (net) area, [m^2] A_r – calculated from exterior measurements surface of roof projection area (net), [m^2] A_{tw} – calculated from exterior measurements total windows area, [m^2] FE_0 – index of final energy demand for heating before modernization, [kWh·m^{-2}·year^{-1}]

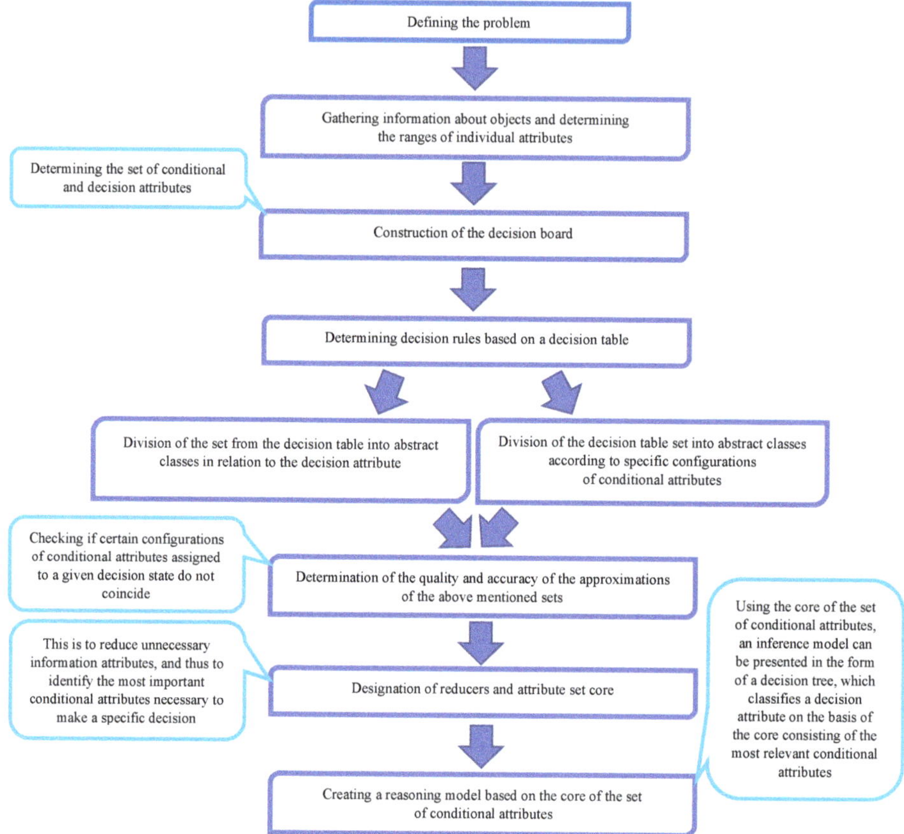

Figure 1. Diagram for building of a model of inference based on the core of a set of conditional attributes using the theory of rough sets.

After selecting the possible list of independent variables (conditional attributes), the developed database was divided into a didactic set, to which 80% of the tested buildings were randomly selected, and a test set created from other objects. A schematic view of the work areas with individual blocks, illustrating the methodology for determining the energy demand indicator for heating after improvements based on the inference model built, is presented in Figure 2.

The evaluation of the quality of the developed model was assessed for individual groups of variables. For assessment of past due forecasts, the mean absolute error (MAE), mean absolute percentage error (MAPE)—also known as mean absolute percentage deviation (MAPD) [53,54]—as well as mean bias error (MBE), and coefficient of variance of the root mean square error (CV RMSE), was used; these are accepted as statistical calibration standards by ASHRAE Guideline 14-2002 [14,55,56]:

$$MAE = \frac{1}{n_g}\sum_{m=1}^{n_g}\left|O_r - O_{pr}\right| \qquad m = 1,2,3\ldots,n_g \tag{1}$$

$$MAPE = \frac{1}{n_g}\sum_{m=1}^{n_g}\left|\frac{O_r - O_{pr}}{O_r}\right|\cdot 100\%\ m = 1,2,3\ldots,n_g \tag{2}$$

$$MBE = \frac{\sum_{m=1}^{n_g} (O_r - O_{pr})}{\sum_{m=1}^{n_g} O_r} \cdot 100\% \; m = 1,2,3\ldots,n_g \quad (3)$$

$$CV\ RMSE = \frac{\sqrt{\sum_{m=1}^{n_g} \frac{(O_r - O_{pr})^2}{n_g}}}{\frac{1}{n_g} \sum_{m=1}^{n_g} O_r} \cdot 100\% \; m = 1,2,3\ldots,n_g \quad (4)$$

where: O_r—real value of the index of final energy demand for heating after modernization (FE1); O_{pr}—the forecast value of the index of final energy demand for heating after modernization; and n_g—is the number of buildings covered by the study.

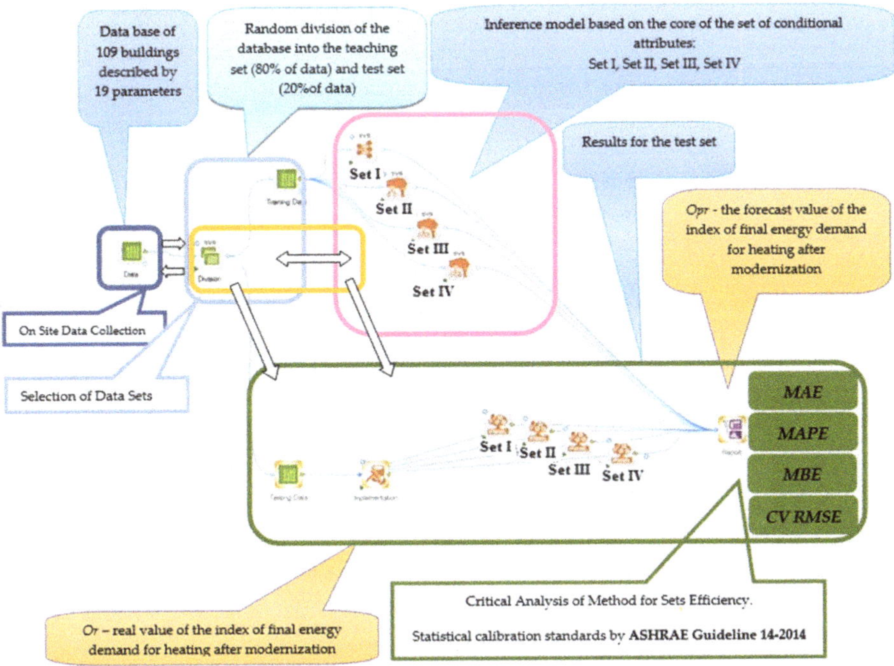

Figure 2. View of the working space.

According to the ASHRAE Guideline, models are considered to be calibrated when MBE values are within ± 10% and CV RMSE values are within 30% [14,57].

The novelty of this research is an attempt to develop a universal model based on the rough set theory describing the final energy demand indicator for heating buildings and to use, for this purpose, groups of variables characterized by different degrees of difficulty in obtaining them. The algorithm of building a model for forecasting energy demand presented in the article allows decision-makers to assess the potential for real energy savings resulting from planned actions to improve thermal performance of existing buildings. It is important to underline the fact that, as the literature review shows, it has not yet been used in the energy assessment of buildings. The presented method based on Rough Set Theory (RST) should not be considered as competition for statistical analysis, but as an optional choice of method for data analysis. Bearing in mind that a common disadvantage of using classical statistical and data-based analyses is their time-consuming, costly nature (equipment and the collection of a sufficient, generally large number of representative observations), and the great complexity of the procedures used, which consists of so-called preliminary analyses (i.e., checking the assumptions of randomness of variables, examining the probability distribution and correct interpretation of statistical analysis results), as well as the ability to conduct and interpret statistical tests. In many cases, this results in the data being matched to the model, not the model to the data, as it should be in reality. When using rough set methods, the observations "speak for themselves" and are not corrected in any way, either before the application of the method or during the analysis. Moreover, the method based on RST is not limited, unlike regression models, by the number of sets of representative observations (both small and large sample of observations), nor is the construction of a statistical model required; decisions are made on the basis of dependencies: if certain conditions are met, a specific decision is taken (according to Boolean application). The presented method does not impose complicated rules of control of the features taken into account and the results of analyses. Only two main coefficients are used to control the significance of conditional attributes in relation to the decision attribute and the created decision rules: quality and approximation accuracy—easy to apply and interpret.

3. Results and Discussion

3.1. Analysis of Final Energy Consumption for Heating Buildings

The analyzed buildings are heated from the municipal heating network. Therefore, information on actual heat consumption for heating in the three heating seasons before and after the thermal modernization was obtained. On this basis, calculations of final energy demand for heating were made, and then the energy characteristics of objects in the state before and after thermal modernization were determined. To exclude seasonal fluctuations, the actual energy consumption values obtained were converted (corrected) to standard season conditions (multi-year average). The data concerning the heating season degree days (from the years 2010–2018 and the multiannual average) based on which the calculations were carried out were taken from the climate database Eurostat [58] for the Małopolska region.

The amount of final energy consumption was calculated using the formula:

$$Q_{K,H} = \sum_{i=1}^{3} \frac{HDD(t_b)_i}{HDD(t_b)_0} \cdot Q_{K,Hi} \cdot \frac{1}{3} \quad (5)$$

where: $Q_{K,H}$ —the final energy demand for the heating season, [kWh]; $HDD(t_b)_0$—the number of degree days in a standard heating season, [°Cd]; $HDD(t_b)_i$—the number of degree days for the "i" of this year, [°Cd]; $Q_{K,H\,i}$—final energy consumption for heating in a measurement period for the "i" of this year, [kWh].

The index of final energy demand for heating before and after the implementation of the improvement was calculated according to the formula:

$$FE = \frac{Q_{K,H}}{A_H} \quad (6)$$

where: FE—index of final energy demand for heating, [kWh·m^{-2}·year^{-1}]; $Q_{K,H}$—the final energy demand for the heating season, [kWh]; A_H—calculated area of temperature-controlled rooms (heated surface), [m^2].

The energy characteristics of the analyzed group of buildings in the state before FE0 and after FE1 thermal modernization are shown in Figure 3. Figure 4 shows the structure of buildings in terms of energy consumption.

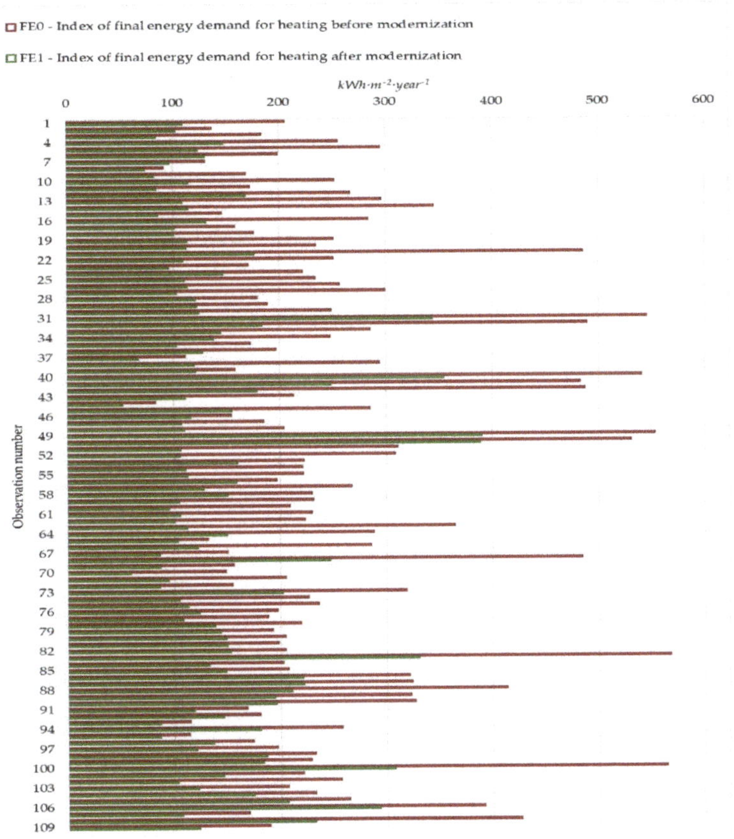

FE0 - Index of final energy demand for heating before modernization

FE1 - Index of final energy demand for heating after modernization

$kWh \cdot m^{-2} \cdot year^{-1}$

Observation number

Figure 3. Index of final energy demand for heating of buildings before and after implementing improvements.

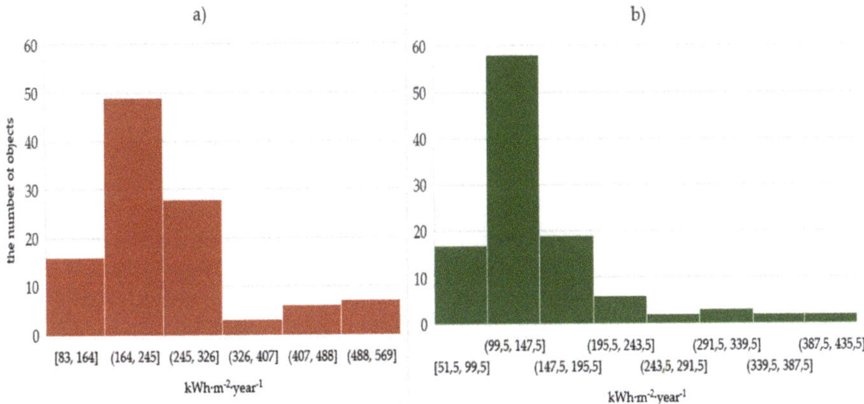

Figure 4. Structure of buildings due to the size of the energy demand indicator for heating (**a**) before (**b**) after the improvement.

Index of final energy demand for heating of buildings is very varied. In buildings before modernization, it was between 83–569 [kWh·m^{-2}·year−^{1}], while after the improvement, it was 51–389 [kWh·m^{-2}·year^{-1}].

The average value of the index before modernization was 287 [kWh·m^{-2}·year^{-1}]; after the improvement, it decreased to 144 [kWh·m^{-2}·year^{-1}]. Average final energy consumption for heating an "average building" in Poland determined on the basis of a standard reference calculation based on EN ISO 13790/seasonal method [21] is: for buildings before thermal modernization 265.6 [kWh·m^{-2}·year^{-1}], while energy consumption after the improvement is 77.9 [kWh·m^{-2}·year^{-1}].

Nearly half of the buildings before the thermal modernization were characterized by energy consumption in the range 164–245 [kWh·m^{-2}·year^{-1}]. In the group of improved buildings, the largest group is made up of buildings where the energy demand for heating is between 99–147 [kWh·m^{-2}·year^{-1}].

Comparing the average values of the FE0 actual energy consumption index in the studied group of buildings with theoretical values, it can be seen that for existing buildings they are similar, whereas for improved buildings they differ. The projected energy consumption in a thermally improved "average building" based on standard reference calculations [46] is twice lower than the actual values.

3.2. Modeling the Consumption of Thermal Energy in Buildings Undergoing Energy Modernization

In the first part of the prediction model construction, the database was randomly divided into a teaching set and a test set. The input data characterizing the buildings were then divided into groups of variables Set I to Set IV. For individual groups from the learning set, the reducts and the core of the set of attributes were determined, which were used to create an inference model. The built model was subjected to a critical analysis (on the test set) in terms of accuracy and quality of the built forecast. The working space, on the basis of which the analyses were performed, is shown in Figure 2.

The results of quality and accuracy calculations of the constructed models depending on the selected set of input variables are shown in Table 3.

Table 3. Assessment of model of energy demand indicator for heating based on studied set of input variables (Set I to Set IV).

Assessment Parameters	Sets of variables			
	Set I	*Set II*	*Set III*	*Set IV*
MAE [kWh·m^{-2}·year^{-1}]	22.8	23.1	25.3	18.1
MAPE [%]	15.4	16.4	17.8	14
MBE [%]	−0.73	1.68	−16	-9.6
CV RMSE [%]	21.7	18.2	32.2	18.8

Analysis of the mean absolute values of the actual deviation of the final energy demand indicator for heating of residential buildings from the predicted value indicates that the model based on set IV has the lowest deviation value (18.1 kWh·m^{-2}·year^{-1}).

The highest error value (about 25.3 kWh·m^{-2}·year^{-1}) was shown by the model based on set III. The error value for the model based on the set I and II is similar and amounts to about 23 kWh·m^{-2}·year^{-1}.

The analysis of the MAPE error on the test set confirmed the usefulness of the model for determining the size of the energy demand indicator for heating for the building after thermal modernization. The obtained average error values for the selected Sets ranged from 14% to 17.8%.

The values of the two other meters used to assess the correctness of the models are as follows:

The MBE error values indicate that the energy consumption forecasts obtained from the model are overestimated for sets I, III and IV. The lowest value of the indicator (−0.73%) is obtained for set I, the highest (−16%) for set III. The model based on set II undervalued the actual values on average by

1.68%. The assessment of the energy demand indicator model expressed in the RSME CV ranges from 18.2% (Set II) to 32.2% (Set III).

Bearing in mind ASHARE's recommendations for model calibration, it can be concluded that three models based on data sets meet the requirements. These are the models based on sets I, II and IV. The best values for the assessment indicators can be observed for the set II model and then for set I. An analysis of the assessment indicators (MBE and CV RSME, which are considered together) showed that the best value of the assessment indicators can be obtained for set II and then for set I. Set IV has a CV RMSE value close to set II, but the MBE value is clearly different from the others.

Taking all four assessment indicators into account, it can be seen that the best fit values are found for the model using the first set of variables.

4. Conclusions

The research was carried out on a group of 109 thermally improved residential buildings. The energy performance of these buildings was determined. The energy performance is based on the actual energy consumption for heating. The calculations were made for the state before and after thermal modernization. The specific sets of important variables characterizing the examined buildings have been identified. The variables were grouped into sets depending on the difficulty of obtaining them. They were used to build a prognostic model based on the rough set theory (RST). The use of this method made it possible to quickly determine the energy saving potential for heating after the completion of the thermal renovation.

The following conclusions can be drawn from the analysis of the indicators' evaluation of the model for forecasting the final energy demand for heating after thermal improvement:

- The model developed based on the Rough Sets Theory (RST) is a universal solution that can be used for estimating thermal energy consumption in buildings undergoing thermal improvement. This is evidenced by the results of the assessment, where, according to the ASHRAE Guide, the calibration targets are set at ±10% (MBE) and less than 30% (CV RMSE). The achievement of these thresholds has been demonstrated for three models. These are models based on sets I, II and IV. The best results can be obtained for the model using sets II and I.
- Taking into account all four evaluation indicators, it was found that the best match between the predicted and real values can be obtained if a limited set of input variables (set I) is used in the model, the value of the deviation of the real value from the predicted value (MAE) is amounts to about 22.8 kWh·m^{-2}·year^{-1}, whereas the accuracy of estimation (MAPE) of the model built on the basis of these data is 15.4%. Similar forecasting results can be obtained by using the data set II, but in this case, a greater number of conditional attributes characterizing the building must be available.
- Analyzing the values of MAE and MAPE indicators, it was found that the best results for forecasting energy consumption after thermal improvement can be obtained using the IV set of input variables. The use of this set of variables to build the model allows obtaining the results with the error (MAE) 18.1 kWh·m^{-2}·year^{-1}. This gives an estimated accuracy (MAPE) of 14%. Despite this, this model is recommended as the third in order because of the high value of the MBE indicator, which clearly differs from the others.
- Forecasting the energy consumption of buildings using a model based on Rough Set Theory (RST) using variables that characterize buildings, allows for estimation accuracy of 14.4–15.9%. However, in further research, it is advisable to test this method on a larger, several hundred elementary set of objects (buildings) from different regions, characterized by different climatic conditions from those in which the research was performed, in order to verify the results obtained.
- The examined group of objects should be used to test other forecasting methods so that the results of the estimation can be compared with each other.

Author Contributions: Conceptualization, T.S.; software, T.S.; data curation, T.S.; investigation, T.S. and S.K.; methodology, T.S.; project administration, T.S. and S.K.; supervision, S.K.; writing—original draft, T.S. and S.K.; writing—reviewing and editing, S.K. and T.S. All authors have read and agreed to the published version of the manuscript.

Funding: This research was financed by the Ministry of Science and Higher Education of the Republic of Poland.

Acknowledgments: We are grateful to Sławomir Kurpaska from the University of Agriculture in Krakow, Poland, Faculty of Production and Power Engineering and Thomas G. Mathia from Laboratoire de Tribologie et Dynamique des Systèmes, École Centrale de Lyon, France for their valuable support during this research.

Conflicts of Interest: The authors declare no conflict of interest.

Nomenclature

RST	Rough Set Theory
MAE	Mean Absolute Error
$MAPE$	Mean Absolute Percentage Error
$MAPD$	Mean Absolute Percentage Deviation
MBE	Mean Bias Error
$CVRMSE$	Coefficient of Variation of the Root Mean Square Error
O_r	real value of the index of final energy demand for heating after modernization (FE1)
O_{pr}	the forecast value of the index of final energy demand for heating after modernization
n_g	number of buildings covered by the study
$Q_{K,H}$	final energy demand for the heating season
$Q_{K,H\,i}$	final energy consumption for heating in a measurement period for the "i" of this year
$HDD(t_b)_0$	the number of degree days in a standard heating season
$HDD(t_b)_i$	the number of degree days for the "i" of this year
FE	index of final energy demand for heating
$FE0$	index of final energy demand for heating before modernization
$FE1$	index of final energy demand for heating after modernization
A_f	calculated surface of heated floors from interior measurements
A_H	calculated area of temperature-controlled rooms (heated surface)
A_r	calculated from exterior measurements surface of roof projection area (net)
A_w	calculated from exterior measurements total walls' surface (net) area
A_{fl}	calculated surface of floor from interior measurements (floor over basement or floor on the ground)
A_{tw}	calculated from exterior measurements total windows area
V_e	calculated from exterior measurements the heated volume of building
S/V_e	shape coefficient of buildings (the ratio surface to volume)
N_{Os}	number of stores
N_{Op}	number of residential flats, premises
N_{Opb}	number of living persons per building
U_w	calculated thermal transmittance of walls components
U_{pw}	calculated thermal transmittance of peak walls components
U_r	calculated thermal transmittance of roof projections components
U_g	calculated thermal transmittance of floor components on the ground
U_f	calculated thermal transmittance of floors components (floor over basement)
U_{win}	calculated thermal transmittance of windows (commercial data)
Φ_h	calculated heating consumed power

References

1. Act of 21 November 2008 on Support for Thermal Modernization and Renovation. Available online: http://prawo.sejm.gov.pl/isap.nsf/download.xsp/WDU20082231459/U/D20081459Lj.pdf (accessed on 3 November 2019).
2. Directive 2010/31/EU of The European Parliament and of the Council of 19 May 2010 on the Energy Performance of Buildings. Available online: https://eur-lex.europa.eu/legal-content/PL/ALL/?uri=CELEX%3A32010L0031 (accessed on 3 November 2019).

3. BGK. Figures of the Thermal Improvement and Repair Fund. 2018. Available online: https://www.bgk.pl/files/public/Pliki/Przedsiebiorstwa/fundusz_kredytu_technologicznego/Dane_liczbowe_FTiR.pdf (accessed on 29 November 2018).
4. Bourdeau, M.; Zhai, X.-Q.; Nefzaoui, E.; Guo, X.; Chatellier, P. Modelling and forecasting building energy consumption: A review of data-driven techniques. Sustain. *Sustain. Cities Soc.* **2019**, *48*, 101533. [CrossRef]
5. Fumo, N. A review on the basics of building energy estimation. *Renew. Sustain. Energy Rev.* **2014**, *31*, 53–60. [CrossRef]
6. Foucquier, A.; Robert, S.; Suard, F.; Stéphan, L.; Jay, A. State of the art in building modelling and energy performances prediction: A review. *Renew. Sustain. Energy Rev.* **2013**, *23*, 272–288. [CrossRef]
7. ASHRAE. *ASHRAE Handbook–Fundamentals—Energy Estimation and Modeling Methods*, SI ed.; American Society of Heating, Refrigerating and Air-Conditioning Engineers, Inc. (ASHRAE): Atlanta, GA, USA, 2009; ISBN 978-1-61583-170-8. Available online: https://app.knovel.com/web/toc.v/cid:kpASHRAE37/viewerType:toc/ (accessed on 5 November 2019).
8. Ahmad, T.; Chen, H.; Guo, Y.; Wang, J. A comprehensive overview on the data driven and large scale based approaches for forecasting of building energy demand: A review. *Energy Build.* **2018**, *165*, 301–320. [CrossRef]
9. Tardioli, G.; Kerrigan, R.; Oates, M.; O'Donnell, J.; Finn, D. Data Driven Approaches for Prediction of Building Energy Consumption at Urban Level. *Energy Procedia* **2015**, *78*, 3378–3383. [CrossRef]
10. Yildiz, B.; Bilbao, J.I.; Sproul, A.B. A review and analysis of regression and machine learning models on commercial building electricity load forecasting. *Renew. Sustain. Energy Rev.* **2017**, *73*, 1104–1122. [CrossRef]
11. Wang, J.; Srinivasan, R.S. A review of artificial intelligence based building energy use prediction: Contrasting the capabilities of single and ensemble prediction models. *Renew. Sustain. Energy Rev.* **2017**, *75*, 796–808. [CrossRef]
12. Deb, C.; Lee, S.E. Determining key variables influencing energy consumption in office buildings through cluster analysis of pre-and post-retrofit building data. *Energy Build.* **2018**, *159*, 228–245. [CrossRef]
13. Zhao, H.; Magoulès, F. A review on the prediction of building energy consumption. *Renew. Sustain. Energy Rev.* **2012**, *16*, 3586–3592. [CrossRef]
14. Chang, C.; Zhu, N.; Yang, K.; Yang, F. Data and analytics for heating energy consumption of residential buildings: The case of a severe cold climate region of China. *Energy Build.* **2018**, *172*, 104–115. [CrossRef]
15. Chalal, M.L.; Benachir, M.; White, M.; Shrahily, R. Energy planning and forecasting approaches for supporting physical improvement strategies in the building sector: A review. *Renew. Sustain. Energy Rev.* **2016**, *64*, 761–776. [CrossRef]
16. Mat Daut, M.A.; Hassan, M.Y.; Hasimah, H.A.; Rahman, A.; Abdullah, M.P.; Hussin, F. Building electrical energy consumption forecasting analysis using conventional and artificial intelligence methods: A review. *Renew. Sustain. Energy Rev.* **2017**, *70*, 761–776. [CrossRef]
17. CEN. Indoor Environmental Input Parameters for Design and Assessment of Energy Performance of Buildings Addressing Indoor Air Quality, Thermal Environment, Lighting and Acoustics; ISO 15251. 2008. Available online: https://standards.globalspec.com/std/1110417/EN%2015251 (accessed on 5 November 2019).
18. Costanzoa, V.; Fabbrib, K.; Piraccini, S. Stressing the passive behavior of a Passivhaus: An evidence-based scenario analysis for a Mediterranean case study. *Build. Environ.* **2018**, *142*, 265–277. [CrossRef]
19. Djamila, H. Indoor thermal comfort predictions: Selected issues and trends. *Renew. Sustain. Energy Rev.* **2017**, *74*, 569–580. [CrossRef]
20. Wang, Y.; Kuckelkorn, J.; Zhao, F.-Y.; Spliethoff, H.; Lang, W. A state of art of review on interactions between energy performance and indoor environment quality in Passive House buildings. *Renew. Sustain. Energy Rev.* **2017**, *72*, 1303–1319. [CrossRef]
21. CEN. European Standard: Energy Performance of Buildings-Calculation of Energy Use for Space Heating and Cooling. ISO 13790:2008. Available online: https://www.iso.org/standard/41974.html (accessed on 5 November 2019).
22. CEN. *European Standard: Heating Systems in Buildings*; ISO 12831-1:2017-08; CEN: Brussels, Belgium, 2017.
23. Ballarini, I.; Corrado, V. Application of energy rating methods to the existing building stock. Analysis of some residential buildings in Turin. *Energy Build.* **2009**, *4*, 790–800. [CrossRef]

24. Crawley, D.B.; Lawrie, L.K.; Winkelmann, F.C.; Buhl, W.F.; Huang, Y.J.; Pedersen, C.O.; Strand, R.K.; Liesen, R.J.; Fisher, D.E.; Witte, M.J.; et al. Energyplus: Creating a new-generation building energy simulation program. *Energy Build.* **2001**, *33*, 319–331. [CrossRef]
25. Rivers, N.; Jaccard, M. Combining top-down and bottom-up approaches to energy–economy modeling using discrete choice methods. *Energy J.* **2005**, *26*, 83–106. [CrossRef]
26. Geysena, D.; De Somera, O.; Johanssonc, C.; Bragec, J.; Vanhoudta, D. Operational thermal load forecasting in district heating networks using machine learning and expert advice. *Energy Build.* **2017**, *162*, 144–153. [CrossRef]
27. Allard, I.; Olofsson, T.; Hassan, O.A.B. Methods for energy analysis of residential buildings in Nordic countries. *Renew. Sustain. Energy Rev.* **2013**, *22*, 306–318. [CrossRef]
28. Asadi, S.; Amiri, S.S.; Mottahedi, M. On the development of multi-linear regression analysis to assess energy consumption in the early stages of building design. *Energy Build.* **2014**, *85*, 246–255. [CrossRef]
29. Asadi, S.; Marwa, H.; Beheshti, A. Development and validation of a simple estimating tool to predict heating and cooling energy demand for attics of residential buildings. *Energy Build.* **2012**, *54*, 12–21. [CrossRef]
30. Caldera, M.; Corgnati, S.P.; Filippi, M. Energy demand for space heating through a statistical approach: Application to residential buildings. *Energy Build.* **2008**, *40*, 1972–1983. [CrossRef]
31. Chou, J.S.; Bui, D.K. Modeling heating and cooling loads by artificial intelligence for energy-efficient building design. *Energy Build.* **2014**, *82*, 437–446. [CrossRef]
32. Fumo, N.; Biswas, R. Regression analysis for prediction of residential energy consumption. *Renew. Sustain. Energy Rev.* **2015**, *47*, 332–343. [CrossRef]
33. Lü, X.; Lu, T.; Kibert, C.J.; Viljanen, M. Modeling and forecasting energy consumption for heterogeneous buildings using a physical–statistical approach. *Appl. Energy* **2015**, *144*, 261–275. [CrossRef]
34. Ma, Z.; Li, H.; Sun, Q.; Wang, C.; Yan, A.; Starfelt, F. Statistical analysis of energy consumption patterns on the heat demand of buildings in district heating systems. *Energy Build.* **2014**, *85*, 464–472. [CrossRef]
35. Praznik, M.; Butala, V.; Zbašnik-Senegačnik, M. Simplified evaluation method for energy efficiency in single-family houses using key quality parameters. *Energy Build.* **2013**, *67*, 489–499. [CrossRef]
36. Sekhar-Roy, S.; Roy, R.; Balas, V.E. Estimating heating load in buildings using multivariate adaptive regression splines, extreme learning machine, a hybrid model of mars and elm. *Renew. Sustain. Energy Rev.* **2018**, *82*, 4256–4268. [CrossRef]
37. Tiberiu, C.; Virgone, J.; Blanco, E. Development and validation of regression models to predict monthly heating demand for residential buildings. *Energy Build.* **2008**, *40*, 1825–1832. [CrossRef]
38. Tsanas, A.; Xifara, A. Accurate quantitative estimation of energy performance of residential buildings using statistical machine learning to OLS. *Energy Build.* **2012**, *49*, 560–567. [CrossRef]
39. Biswas, M.; Robinson, M.D.; Fumo, N. Prediction of residential building energy consumption: A neural network approach. *Energy* **2016**, *117*, 84–92. [CrossRef]
40. Ekici, B.B.; Aksoy, U.T. Prediction of building energy consumption by using artificial neural networks. *Adv. Eng. Softw.* **2009**, *40*, 356–362. [CrossRef]
41. Li, K.; Sua, H.; Chua, J. Forecasting building energy consumption using neural networks and hybrid neuro-fuzzy system: A comparative study. *Energy Build.* **2011**, *43*, 2893–2899. [CrossRef]
42. Kumar, R.; Aggarwal, R.K.; Sharma, J.D. Energy analysis of a building using artificial neural network: A review. *Energy Build.* **2013**, *65*, 352–358. [CrossRef]
43. Seyedzadeh, S.; Rahimian, F.; Glesk, I.; Roper, M. Machine learning for estimation of building energy consumption and performance: A review. *Vis. Eng.* **2018**, *6*, 5. [CrossRef]
44. Mihai, M.; Tanasiev, V.; Dinca, C.; Badea, A.; Vidu, R. Passive house analysis in terms of energy performance. *Energy Build.* **2017**, *144*, 74–86. [CrossRef]
45. Pawlak, Z. *Rough Sets. Theoretical Aspects of Reasoning about Data*; Kluwer Academic Press: Dordrecht, The Netherlands, 2012; Available online: http://bcpw.bg.pw.edu.pl/Content/2026/RoughSetsRep29.pdf (accessed on 5 November 2019).
46. TABULA. Polish building typology. In *Scientific Report*; Narodowa Agencja Poszanowania Energii: Warszawa, Poland, 2012. Available online: https://episcope.eu/fileadmin/tabula/public/docs/scientific/PL_TABULA_ScientificReport_NAPE.pdf (accessed on 15 January 2020).

47. Comparison of Typical Buildings from Evaluation of the TABULA Database and Heat Supply Systems 20 European Countries TABULA Evaluation of the Database. Available online: https://episcope.eu/fileadmin/tabula/public/docs/report/TABULA_WorkReport_EvaluationDatabase.pdf (accessed on 15 January 2020).
48. Nutech Solution-Science for Business. 2005. Available online: http://www.nutechsolutions.com.pl/ (accessed on 10 October 2019).
49. Nguyen, H.S. Tolerance Rough Set Model and Its Applications in Web Intelligence. In Proceedings of the IEEE/WIC/ACM International Joint Conferences on Web Intelligence (WI) and Intelligent Agent Technologies (IAT), Atlanta, GA, USA, 17–20 November 2013; IEEE CS: Washington, DC, USA, 2013; pp. 237–244. [CrossRef]
50. Nguyen, D.V.; Yamada, K.; Unehara, M. Extended Tolerance Relation to Define a New Rough Set Model in Incomplete Information Systems. *AFS* **2013**, *2013*, 372091. [CrossRef]
51. Renigier-Biłozor, M. Zastosowanie teorii zbiorów przybliżonych do masowej wyceny nieruchomości na małych rynkach (Application of rough set theory for mass valuation of real estate in small markets). *Acta Sci. Pol. Adm. Locorum* **2008**, *7*, 35–51.
52. Szul, T.; Nęcka, K.; Knaga, J. Application of Rough Set Theory to Establish the Amount of Waste in Households in Rural Areas. *Ecol. Chem. Eng. S* **2017**, *24*, 311–325. [CrossRef]
53. Dittmann, P. *Prognozowanie w Przedsiębiorstwie*; Wolters Kluwer Polska Sp. z o.o.: Kraków, Poland, 2008.
54. Cieślak, M. *Prognozowanie Gospodarcze*; Wydawnictwo Naukowe PWN: Warszawa, Poland, 1999.
55. Ruiz, G.R.; Bandera, C.R. Validation of Calibrated Energy Models: Common Errors. *Energies* **2017**, *10*, 1587. [CrossRef]
56. ASHRAE. *ASHRAE Guideline 14-2002 for Measurement of Energy and Demand Savings*; American Society of Heating, Refrigeration and Air Conditioning Engineers: Atlanta, GA, USA, 2002.
57. ASHRAE. *American Society of Heating, Ventilating, and Air Conditioning Engineers (ASHRAE). Guideline 14-2014, Measurement of Energy and Demand Savings*; Technical Report; American Society of Heating, Ventilating, and Air Conditioning Engineers: Atlanta, GA, USA, 2014.
58. European Union Statistics, Eurostat. Cooling and Heating Degree Days by NUTS 2 Regions—Annual Data. Available online: http://data.europa.eu/88u/dataset/7kv8vguICyNRJYqLRzzFw (accessed on 30 November 2019).

Article

Development and Implementation of Fault-Correction Algorithms in Fault Detection and Diagnostics Tools

Guanjing Lin, Marco Pritoni, Yimin Chen and Jessica Granderson *

Lawrence Berkeley National Laboratory, Berkeley, CA 94706, USA; gjlin@lbl.gov (G.L.); mpritoni@lbl.gov (M.P.); yiminchen@lbl.gov (Y.C.)

* Correspondence: JGranderson@lbl.gov; Tel.: +1-510-486-6792

Received: 29 April 2020; Accepted: 19 May 2020; Published: 20 May 2020

Abstract: A fault detection and diagnostics (FDD) tool is a type of energy management and information system that continuously identifies the presence of faults and efficiency improvement opportunities through a one-way interface to the building automation system and the application of automated analytics. Building operators on the leading edge of technology adoption use FDD tools to enable median whole-building portfolio savings of 8%. Although FDD tools can inform operators of operational faults, currently an action is always required to correct the faults to generate energy savings. A subset of faults, however, such as biased sensors, can be addressed automatically, eliminating the need for staff intervention. Automating this fault "correction" can significantly increase the savings generated by FDD tools and reduce the reliance on human intervention. Doing so is expected to advance the usability and technical and economic performance of FDD technologies. This paper presents the development of nine innovative fault auto-correction algorithms for Heating, Ventilation, and Air Conditioning pi(HVAC) systems. When the auto-correction routine is triggered, it overwrites control setpoints or other variables to implement the intended changes. It also discusses the implementation of the auto-correction algorithms in commercial FDD software products, the integration of these strategies with building automation systems and their preliminary testing.

Keywords: fault correction; fault detection and diagnostics; building operation; energy efficiency; field testing

---

1. Introduction

Commercial buildings constitute 18% of the U.S. primary energy consumption [1] and account for $149 billion in annual energy expenditures [2]. Much of this consumption is due to operational waste, representing a tremendous potential for savings. The literature indicates that median whole-building savings of 16% are achieved by commissioning existing buildings [3] and that 5–30% of commercial building energy use is wasted due to problems associated with controls [4–9].

Commercially available fault detection and diagnostics (FDD) tools provide a means of monitoring-based commissioning, through which instances of operational inefficiency can be continuously identified, isolated, and surfaced for resolution by operations and maintenance staff. Today's FDD technology has been documented to enable whole building savings of 8% on average, across users [10]. These technologies integrate with building automation systems (BASs) or can be implemented as retrofit add-ons to existing equipment, and continuously analyze operational data streams across many system types and configurations. This is in contrast to the historically typical variants of FDD that are delivered as original equipment manufacturer-embedded equipment features or handheld FDD devices that rely upon temporary field measurements.

Figure 1 represents an idealized architecture of a BAS, adapted from American Society of Heating, Refrigerating and Air-Conditioning Engineers (ASHRAE) Guideline 13 [11]. Field devices (and controllers) connect to the sensors and actuators in the field. Network controllers typically provide supervisory control capabilities, scheduling, alarms, trending, local data storage, and user interfaces, in addition to some security features. Modern versions of these controllers have the ability to communicate via a BACnet (a data communication protocol for building automation and control network) IP over an IP network. When such functionality is not available, a common integration strategy employs "integration gateways" (e.g., Niagara JACE) that translate from proprietary protocols to standard protocols, such as BACnet IP. For larger installations and campuses, the controllers or gateways are also connected to a BAS server that provides configuration and management, long-term data storage (i.e., databases) and visualization tools. FDD tools can be installed in the local IP network, run from the cloud, or have a combination of cloud and local components. Integration with the BAS is typically implemented through a one-way interface using one of these three FDD–BAS integration pathways:

1. The FDD tool collects data from the central server database (common for large campus-wide installations) via a database application programming interface (API) (e.g., structured query language).
2. The FDD tool collects data from a central server, controller, or gateway using vendor-specific API (e.g., Automated Logic web services).
3. The FDD tool collects data directly via the BACnet IP network shared with other controllers and gateways.

Through these interfaces, system-level operational data are made available to the FDD software. Meter data are often also included. Data are continuously analyzed and detected faults are presented to operational staff through a graphical user interface. Since the BAS is the primary source of data, the FDD is most commonly focused on Heating, Ventilation, and Air Conditioning (HVAC) equipment. However, today's technologies offer extensive libraries of FDD logic and algorithms, and therefore can be applied to lighting and other building end-use systems for which data are available [12].

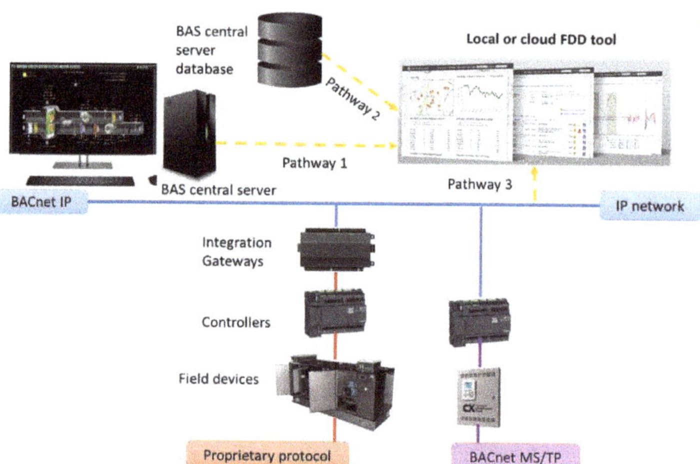

Figure 1. Schematic illustration of the integration of building automation system (BAS) data into fault detection and diagnostics (FDD) products. (BACnet MS/TP - BACnet Master Slave Token Passing protocol).

Although FDD tools are being used to enable cost-effective energy savings, there remains an opportunity to advance the state of the technology. In practice, the need for human intervention to fix faults once they are identified often results in delay or inaction, resulting in additional operations and maintenance (O&M) costs or lost opportunities. Traditionally, FDD generates recommendations and follow-up actions which are implemented by service technicians or other staff. An emerging capability comprises the integration of FDD outputs with facility management "work order" or CMMSs (computerized maintenance management systems). While this makes it possible to automatically generate work orders from the FDD system, human intervention is still required to implement the corrective action. Therefore, this work seeks to develop automated fault-correction approaches and integrate them with commercial FDD technology offerings, thereby closing the loop between the passive diagnostics and active control. This is possible by converting the one-way BAS interface into a two-way interface, as is done with supervisory predictive control technologies that are emerging in the market.

It is not possible to automate the correction of mechanical faults such as failed actuators; however, there is nonetheless a compelling set of operational problems that are detectable in today's FDD offerings and are correctable through the software-based manipulation of the BAS parameters that can be exposed to external applications via BACnet. For example, Fernandez et al. [6] assessed control problems in commercial buildings, as well as their prevalence and whole-building energy impact for key commercial building sectors. Among the most common faults that relate to biased sensors, improper control parameter settings and inefficient schedules have significant impact and high prevalence rates. Automating the correction of these types of faults can increase the savings realized through the use of FDD tools and reduce the extent to which savings are dependent upon human intervention.

The academic and technical literature has extensively covered the development of automated FDD applied to HVAC and lighting systems [13,14]; however, very little has been published on the automatic correction of the identified faults via the actual control system. One set of relevant papers stem from the vast literature on rule-based FDD algorithms. Fernandez et al. [15,16] developed passive and proactive fault auto-correction algorithms for various HVAC components and systems. The methods proposed to correct some faults which include biased air-handler unit (AHU) mixed air (MA), outside air (OA), and return air (RA) temperature and humidity sensors; damper control hunting; minimum outdoor air damper too open/closed; and manual overrides in large HVAC systems. Using the same approach, Brambley et al. [17] extended Fernandez et al. [15,16] by adding correction routines for the biased AHU supply air (SA) temperature and flow rate sensors and the biased variable-air-volume (VAV) box discharge air temperature and flow rate sensors. This project implemented and tested a subset of these algorithms (sensor bias and minimum outdoor air damper position) in a laboratory experiment. This research stopped short of validating the developed solutions in physical buildings or integrating them with existing BAS and commercial FDD products.

Related to the concept of fault correction is a body of work in the building control literature that focuses on fault tolerant control. The purpose of a fault tolerant controller is to the maintain proper operation of a system despite the presence of faults [18,19]. These approaches have been widely adopted in other industries for safety-critical systems such as nuclear power plants, spacecraft and aircraft. In the context of buildings, Padilla et al. [20] developed a model-based strategy which aims to replace defective sensors in AHUs [20] with "virtual sensors." The signal generated from these "virtual sensors" can be used in the AHU control system when the actual physical sensors behave abnormally. Supply air temperature and pressure sensor faults are effectively corrected by using the proposed algorithms. Wang et al. [21] developed a supervisory control scheme that adapts to the presence of a measurement error in an outdoor air flow rate. The method uses neural network models to estimate the correct behavior of the faulty sensor and to maintain indoor air quality while minimizing energy use. Hao et al. [22] employed principal component analysis to develop fault-tolerant control and data recovery in the HVAC monitoring system. Bengea et al. [23] developed a fault-tolerant optimal control strategy for an HVAC system integrating FDD and model predictive control. The output of the FDD algorithm is used to continuously update the model's predictive control algorithm parameters. The approaches described in these papers offer innovations to the state of the art, yet they are not readily implemented in today's buildings control systems. This is because they comprise strategies that are not supported by traditional BAS capabilities. Similarly, while the literature focuses on the development of these advanced controllers, it does not explore their integration with existing FDD technologies. An additional practical challenge is that a large volume of non-faulty data under various operational conditions is typically needed to train the models employed in these solutions.

This paper complemented and extended previous work in three ways. (1) It developed a comprehensive set of fault auto-correction algorithms designed to be integrated with commercial FDD tools. These algorithms target incorrectly programmed schedules, manual control "lock out," sensor bias, control hunting, rogue zone, and suboptimal setpoints/setpoints setback. Typically, commercial FDD tools are developed as a software layer on top of the existing BAS. There exists a natural separation of roles in this arrangement, in which the BAS actively controls the building and the FDD tool observes its operation and provides insights and recommendations to the building manager. The new auto-correction algorithms afford the FDD technology a certain degree of control capability. (2) It conducted preliminary testing and performance validation during which two auto-correction routines were deployed in a commercial FDD tool and tested on two AHUs in a real building. The enhanced FDD tool was able to correct faults successfully. (3) It presented the challenges of the integration of developed auto-correction algorithms into commercial FDD tools along with the solutions through work with three industry partners. New insights were gained by implementing the pseudo-code developed by the research team in real systems and real buildings. Sections 2–4 present the auto-correction algorithms, preliminary testing and the implementation changes and solutions, respectively. Section 5 concludes the paper and describes future work.

2. Fault Auto-Correction Algorithms

To identify the faults that are auto-correctable, we reviewed the existing literature and discussed the topic with 10 subject matter experts who had years of experience in FDD research and application. These experts included a set of FDD technology and service providers from the industry who were participating in this R&D effort as implementation partners. These providers maintained a large footprint in the FDD market and have explicitly included fault correction in their product development roadmaps. It is not possible to automate the correction of mechanical faults such as failed actuators, valve leakage and damper stuck, as they require physical repair or replacement. However, there is nonetheless a compelling set of operational problems that are detectable in today's FDD offerings, and correctable through a software-based manipulation of the BAS parameters that can be exposed to external applications via BACnet. Considering both the possibility of the automated correction and the feasibility of implementation in an FDD platform, a set of nine HVAC system faults were isolated, as shown in Table 1.

Table 1. Summary of the auto-correctable faults of focus in this study.

Fault	Fault Description
1. Schedules are incorrectly programmed	Heating, Ventilation, and Air Conditioning (HVAC) equipment does not turn on/off according to its intended schedule due to an error in the control programming.
2. Override manual control	Operator unintentionally neglects to release what was intended to be a short-term override of setpoints or other control commands (e.g., fan variable frequency drive (VFD) speed, cooling coil valve control command).
3. AHU OA or SA temperature sensor bias	Air-handler unit's (AHU's) outside air (OA) or supply air (SA) temperature sensor measurements have constant bias over time, representing an offset from a correct or true value.
4. Control hunting	The damper, valve, pump, or fan hunting fault due to an improper proportional gain.
5. Rogue zone	A rogue zone continuously sends cooling/heating requests whenever its schedule is on, due to the zone-level equipment problems like a leaky reheat valve, a dysfunctional supply air damper, or an insufficient capacity variable-air-volume (VAV) terminal.

Table 1. *Cont.*

Fault	Fault Description
6. Improve economizer high-lockout temperature setpoint	In the AHU with fixed dry-bulb economizer control, the economizer shall be disabled whenever the outdoor air conditions exceed the economizer high-lockout temperature setpoint. If the setpoint is set too low in the control logic, it will result in missed opportunity to use outdoor air to reduce the mechanical cooling load in mild conditions.
7. Improve zone temperature setpoint setback	Each zone has separate occupied and unoccupied cooling and heating setpoints. If the zoom temperature cooling setpoint is too low or the heating setpoint is too high, the space will be overcooled or overheated, causing unintended energy consumption.
8. Improve AHU static pressure setpoint reset	Non-optimized AHU static air pressure setpoint.
9. Improve AHU SA temperature setpoint reset	Non-optimized AHU supply air temperature setpoint.

For each of the faults in Table 1, correction algorithms were developed (for faults 1, 3 and 5–9) or adapted from the existing literature (for faults 2 and 4). The auto-correctable faults in Table 1 were divided into two categories: faults 1–5 were in the "Fault" category, which indicated problems that violated the intended operation of the equipment (e.g., sensor bias); faults 6–9 were in the "Opportunity" category, which indicated problems that represented potential to improve the current operation of the equipment (e.g., improve a setpoint reset). This distinction was made to differentiate between the intent of the restoring operation to what it was originally intended to be and that of optimal control.

As the objective of this study was to develop automated fault-correction algorithms that could be integrated with commercial FDD and BAS products, the auto-correction algorithms described in this section were decoupled from the fault detection and diagnostics algorithms embedded in the FDD tools. This permitted the applicability of the developed correction algorithms across a variety of FDD technologies that employed different FDD rules and algorithms. Furthermore, it was assumed that the FDD tools were able to detect the faults of focus, as they represented some of the more commonly encountered faults in commercial buildings.

Figure 2 shows the flow chart of the general auto-correction process. In this process, after the FDD algorithm generates a fault flag of a specific fault, the fault auto-correction algorithm is initiated to correct this fault with the approval from the building operator. Control_variable_being_overwritten is the key element in the auto-correction process. The algorithm overwrites this variable (Control_variable_being_ overwritten_current) with a new value (Control_variable_being_overwritten _new). The control_variable_being_overwritten_current is the one identified in the FDD algorithm to be associated with the problematic value (fault) or potential to improve (opportunity). The control_variable_ being_overwritten_new is the same variable that has the correct value (fault) or optimized value (opportunity). All of the auto-correction algorithms developed in this work followed this structure, with different control variables overwritten, and different ways to determine the correct or improved value of the variable.

Each auto-correction algorithm is presented and discussed in the following.

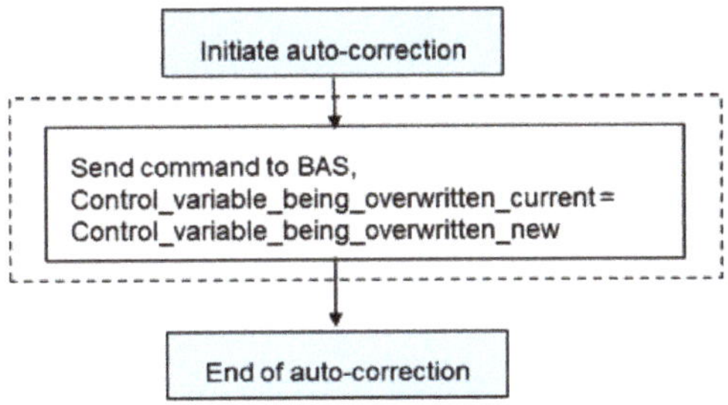

Figure 2. Flow chart of the general auto-correction process.

2.1. Schedules Are Incorrectly Programmed

In this auto-correction algorithm, the control_variable_being_overridden is Equip_Schedule, which is the HVAC equipment on/off times that are programmed in the BAS for weekends, weekdays and holidays. The control_variables_being_overwritten_current are the current schedules read from BAS, which are concluded to be incorrectly programmed (e.g., the AHU starts at 4 a.m. and stops at 8 p.m. on weekdays). The control_variable_being_overwritten_new is Intended_Schedule, which is HVAC equipment on/off times as they are supposed to be (e.g., the AHU starts at 6 a.m. and stops at 7 p.m. on weekdays). The auto-correction algorithm overwrites the Equip_Schedule and replaces it with the Intended_Schedule to enable the system to start and stop as it is supposed to. The Intended_Schedule is prior knowledge that is specified by the building operator or another resource.

2.2. Override Manual Control

As documented in [15], the control_variable_being_overwritten in the auto-correction algorithm is Manual_Override. This variable indicates the equipment (e.g., fan speed, valve control command, damper control command) manual control status or equivalent flag: 1—equipment is in manual control, 0—equipment is in automatic control. When there is the override fault, Manual_Override = 1 in the faulty case. The correction algorithm changes the manual control variable back to automatic (Manual_Override = 0).

2.3. AHU Supply Air or Outside Air Temperature Sensor Bias

Two approaches can be used to correct the AHU supply air or outside air temperature sensor bias fault. In the first approach, the control_variable_being_overwritten is a data point (TempSensorOffset) in the control loop that controls the offset of the controller's temperature input. In the second approach, the control_variable_being_overwritten is the control setpoint associated with the temperature. These two approaches are presented here in detail for the supply air temperature (SAT) sensor bias fault. Figure 3a illustrates the auto-correction workflow for the first method. It is assumed that when the FDD algorithm flags the sensor bias fault it also determines the bias value, which is the difference between the sensor reading value and the actual value. This bias value (SAT_bias) is fed into the auto-correction algorithm. After judging the bias direction, the bias value is directly written to the TempSensorOffset. The adjusted TempSensorOffset is added from the SAT reading value and the conversion results (SAT+TempSensorOffset) enter the controller as input (e.g., adjust the incorrect SAT value with a new offset to provide the correct reading which is fed into the cooling/heating coil valve controller). Figure 3b illustrates the auto-correction workflow for the second method. In this method, the new SAT setpoint (SAT_spt) value can be calculated by adding or subtracting the bias value accordingly, and then be written to the BAS. The auto-correction of the outside air temperature uses the similar two approaches as above. The control_variables_being_overwritten in the second approach are the AHU economizer high (low) lockout temperature setpoints, which are the outside air temperatures above (below) which the outside air damper will return to its minimum position.

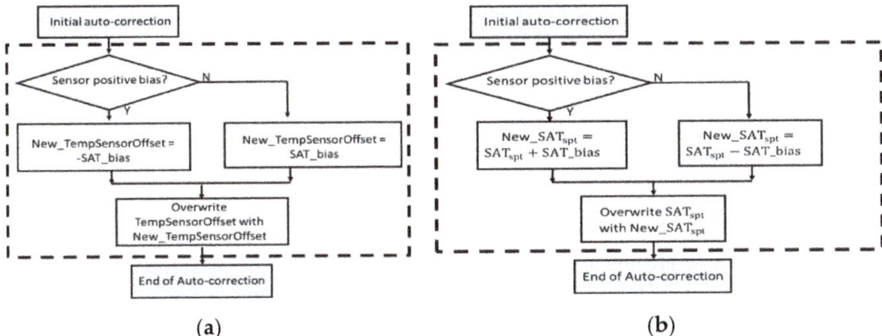

Figure 3. Flowchart of the supply ai temperature (SAT) sensor bias fault auto-correction algorithm. (**a**) Approach 1: overwrite the SAT temperature value, and (**b**) Approach 2: overwrite the SAT setpoint.

2.4. Damper/Valve/Fan/Pump Control Hunting Due to Improper Proportion Gain

In contrast to the other algorithms, the auto-correction of control hunting due to improper proportion gain employs a trial and error procedure [15]. The control_variable_being_overwritten is the proportional–integral–derivative (PID) controller parameter proportion gain (Kp). In the auto-correction process, the Kp is continually adjusted to find out the appropriate value that eliminates the hunting behavior. When the FDD algorithm flags an improper Kp causing the hunting fault, the auto-correction algorithm is initiated. First, a maximum auto-correction duration threshold (T_AC_thresh) is set to avoid an endless auto-correction process. Note that a setting time of an actuator during the control response may be varied due to different actuator control characteristics. For example, for a VAV terminal unit damper, the settling time is typically in the order of one or two minutes, but the settling time for a cooling coil valve may be several minutes. Then, the current value of the Kp is compared to a Kp_threshold, in this case 0.2. This test is meant to avoid an unacceptably long settling time under pure integral control. If the Kp value is above the Kp_threshold, the Kp is decreased by 10% [15]. Then, the algorithm starts a proactive test scenario to see if the hunting issue still persists, by changing the setpoint (T_set) of the damper/valve/fan/pump to trigger the component's movement. If the component is still hunting, the procedure is repeated; otherwise, the procedure is terminated. If the Kp reaches the Kp_threshold and there is still a hunting fault, then it is flagged as an error and the Kp is reset to the original value (Figure 4).

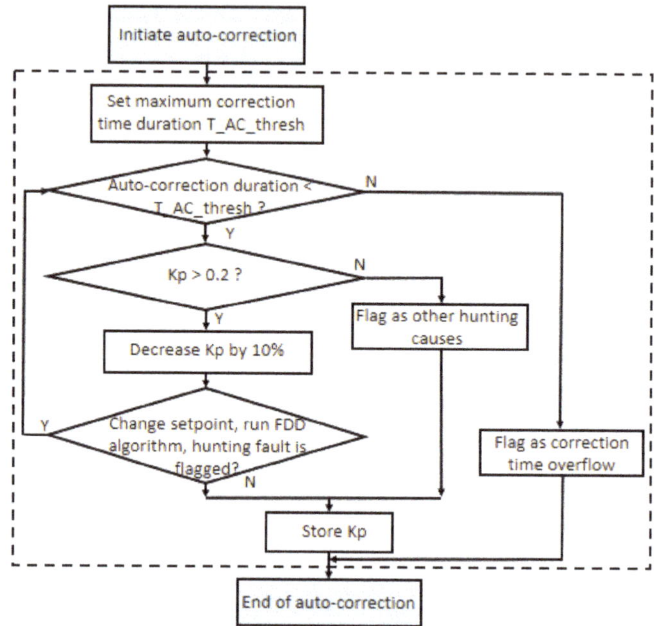

Figure 4. Flowchart of Algorithm 2.4: control hunting due to improper proportion gain (T_AC_thresh - a maximum auto-correction duration threshold, Kp - controller parameter proportion gain).

2.5. Rogue Zone

ASHRAE Guideline 36 [24] defines high-performance control sequences for AHU–VAV systems. The "Trim and Respond" logic (see Sections 2.8 and 2.9) is adopted to reset the supply air temperature and static pressure setpoints at an AHU. The adjustment of these setpoints depends on the number of cooling "requests" generated by downstream zones that are served by the same AHU. For each time step, the change value of setpoint (SPchange) is determined by Equations (1) and (2) below:

$$SP_{change} = SP_{res} \times (R - I) \tag{1}$$

$$R = \sum IM_i Request_i \tag{2}$$

where SPres is a unit respond amount (e.g., 0.06 inches for static pressure setpoint), R is the total number of cooling requests from the downstream zones, I is the defined number of ignored requests, i is the indicator of the downstream zone, IM is the importance multiplier that is used in the control sequence to decide if the cooling requests from the zone level should be used to control the upstream AHU, and *Request* is the cooling request from the zone. Therefore, if there is a rogue zone that continuously sends cooling requests whenever its schedule is on, due to the zone-level equipment problems, the parameter R will always include this request, and it keeps the setpoints in the control loop to its high end. Excluding rogue zones from the corresponding reset control strategies improves operation and increases energy savings. After the zone-level equipment problems that lead to the rogue zone are fixed, the rogue zone is no longer rogue, and all the control variables that are overwritten during the auto-correction process change back to their original value.

Two correction strategies were developed to eliminate the rogue zone impacts (i.e., to ignore the cooling request from the rogue zone). The first is to overwrite I in Equation (1). The auto-correction algorithm increases I by n for each currently identified rogue zone. The value of n is the same as the number of cooling requests determined in the control sequence of that rogue zone. The second is to overwrite the *IM* of the rogue zone in Equation (2). When the FDD tool flags the rogue zone fault, the *IM* of the rogue zone is overwritten to be zero. Therefore, the cooling requests from the rogue zone can be removed.

2.6. Improve Economizer High-Lockout Temperature Setpoint

The previous five algorithms focused on correcting faults to restore the intended operation. This algorithm and the next three serve to provide more optimal control. They implement improved control setpoints or sequences when the FDD tool identifies the opportunity to do so.

After the opportunity to improve the economizer lockout temperature setpoint is identified, the setpoint is overwritten to the recommended value, following the flowchart in Figure 2. The recommended value can be determined based on the high-lockout limit recommended in the energy code [25]. For example, the recommended lockout setpoint is 23.9 °C, 21.1 °C and 18.3 °C, respectively, in the dry climate zone, the cold–humid climate zone, and the hot–humid climate zone.

2.7. Improve Zone Temperature Setpoint Setback

Similar to the algorithm in Section 2.6, this auto-correction algorithm overwrites the zone temperature cooling or heating setpoint during the occupied or unoccupied hours to the recommended values wherever there is an opportunity.

2.8. Improve AHU Static Pressure Setpoint Reset

The auto-correction algorithms for this and the next opportunities are most closely related to optimal controls. Both algorithms correct the fault "continuously" as it continuously adjusts the control variables to optimize the equipment operation (e.g., resets). They are relevant for AHUs without sophisticated reset strategies, such as no reset or simple resets based on return air temperature or outside air temperature.

The auto-correction algorithm uses the ASHRAE Guideline 36 [24] "Trim and Respond" logic for the static pressure setpoint. To optimize the operation of the AHU and minimize discomfort, the static pressure setpoint (SSP_spt) is continually reset using the Trim and Respond logic between a minimum and maximum setpoint (SP_{min} and SP_{max}). When the supply air fan is off, the setpoint is the initial setpoint (SP_0). The reset logic is active while the supply air fan is proven on, starting a delay timer (T_d) after the initial device start command. When active, for every time step T, when the cooling request from the downstream zones (*R*) is less than or equal to a defined number of ignored requests (*I*), the setpoint is trimmed by a trim amount (SP_{trim}), but no less than SP_{min}. If *R* is more than *I*, the setpoint changes by a respond amount, (i.e., $SP_{res} * (R - I)$), but no more than the maximum response per time interval ($SP_{res\text{-}max}$).

2.9. Improve AHU SAT Setpoint Reset

Similar to the algorithm to improve the static pressure setpoint reset, this auto-correction algorithm uses the ASHRAE Guideline 36 [24] "Trim and Respond" logic to reset the SAT setpoint continuously between a minimum and maximum setpoint. The control_variable_being_overwritten is the SAT setpoint.

3. Results: Preliminary Testing

Three commercial FDD providers participating in this research selected a subset of the algorithms that were created by the authors and integrated them into their development product environments for field testing. The partners chose the relevant algorithms for a variety of reasons, including: the expected ease of implementation, the reduction of operational cost, savings potential, and the ability to solve problems common to their customers. The implementation process varied depending on the platform, but generally consisted of the following phases: (1) confirm/add two-way communication functionality between the FDD and the BAS, (2) build an auto-correction interface to communicate with the building operator, (3) translate the algorithms into the FDD programming environment, (4) modify the BAS programming of the specific building to integrate the new control actions sent by the FDD tool, and (5) commission and test the new system. Further details are presented in Lin et al. [26]. This section illustrates the test results of two auto-correction algorithms: "Rogue zone" and "Improve AHU supply air temperature setpoint reset" for one implementation partner. Section 4 summarizes the challenges that were faced by three partners during the implementation process, as well as the solutions that were used by one or more project partners to mitigate them.

In the preliminary testing, the two routines were deployed in a commercial FDD product (SkySpark® by SkyFoundry) and tested on two AHUs in a building in Berkeley, California, US. between March 3 and April 5, 2020. The goal of this preliminary test was to determine whether the enhanced FDD solutions were able to correct faults without adverse operational effects.

3.1. Description of the Testing Site and Equipment

Table 2 summarizes the test site and equipment information. AHU01 and AHU02 are structurally identical. Figure 5 shows the BAS graphics (i.e., native dashboard) for one of the two AHUs.

Table 2. Test site information.

Building Type	Size (m^2)	Building Schedule	HVAC Configuration	BAS Brand and Model
Mixed laboratory and office space	8919	Labs: 24/7 operation, Offices: 4 a.m.–9 p.m., Monday–Sunday	3 chillers and 2 AHUs (AHU01 and AHU02), covering about 90% of the floor area, and the connected zones (n = 83 and n = 80, respectively)	Johnson Controls (JCI) Metasys

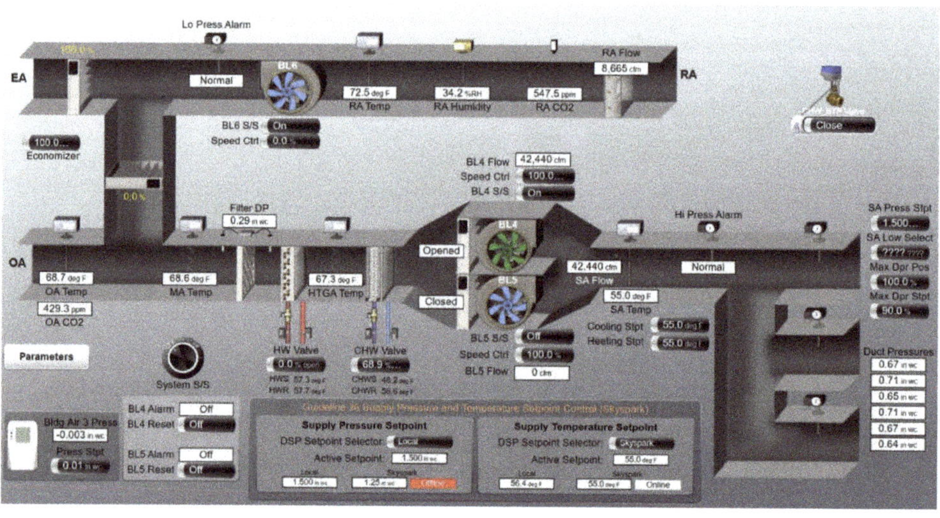

Figure 5. BAS graphics for the AHU02 at the test site. AHU01 has a similar structure.

Both AHU01 and AHU02 were controlled by a control sequence implemented in the native BAS control language and hosted on local controllers. Each AHU was controlled independently. The supply air temperature cooling and heating setpoint was reset based on the algorithm highlighted below in plain English:

- If the AHU is enabled (based on schedules, normally 24/7):
    - Calculate the average cooling demand output for all the zones served by the AHU (cooling demand output is the output calculated by the PI[D] loop based on the proportional, integral [and derivative] component of the difference between zone cooling setpoint and zone temperature).;
    - Constrain the results between min = 3% and max = 12%;
    - Linearly map the average output to a calculated cooling setpoint between 18.3 °C and 12.8 °C. The value of 3% average cooling output is mapped to 18.3 °C, 12% is mapped to 12.8 °C, and all the values in between are calculated linearly;
    - The heating setpoint is fixed to 12.8 °C, except for when the cooling setpoint reaches 12.8 °C. In that case, the heating setpoint becomes 12.2 °C.
- The economizer damper and the chilled water valve are controlled to maintain the cooling supply air temperature setpoint. The heating hot water valve is controlled to maintain the heating supply air temperature setpoint. As a result, when the outside air temperature is between the heating and the cooling setpoints, the air handling unit typically does not cool or heat the air.

The existing SAT control strategy is relatively efficient, compared to common practice in the industry (fixed setpoint or resets based on outdoor temperature or return temperature alone and no deadband). However, the current strategy presents two limitations: (1) it responds to outlier zones or rogue zones, although minimally, as the reset is based on an average cooling demand outputs from all the zones; and (2) its calibration parameters (e.g., min and max average zone feedback of 3% and 12%, respectively) were established via trial and error and personal judgement. Given the limited capabilities of the BAS zone controllers (i.e., field devices in Figure 1), the reset strategy was entirely calculated within the AHU controllers.

The FDD tool connected to the BAS is a commercial product managed by a consultant and the facility manager of the site. The tool allows for custom programming and bi-directional communication to the BAS via the BACnet network. In contrast to the BAS, the FDD tool coding language is a modern scripting language with the ability to use high-level functions that allow the portability of the code between the buildings and equipment. The two auto-correction algorithms were coded using this platform and tested on the two AHUs. In the FDD tool, a zone was identified as a rogue zone when one or more disqualifying conditions were detected for that zone and the zone was sending a request to the AHU. The zone requests are calculated based on zone PID loop output >95%. Disqualifying conditions for cooling requests include:

- Leaky reheat valve (VAV box discharge air temperature (DAT) > AHU SAT + 2.8 °C);
- Supply airflow setpoint not met (<90% or >110% of setpoint and delta > 1.4 cubic meter per minute);
- Zone cooling setpoint too low (lower than 22.2 °C unless exempt).

3.2. *Auto-Correction Code in the FDD Tool*

3.2.1. Code for "Rogue Zones"

The code adopts the first correction strategy in Section 2.5 and overwrites the number of ignored cooling requests from the identified rogue zones. The number of requests and ignored requests are calculated as in Equations (3)–(5):

$$R' = \max(R - I_{total}, 0) \tag{3}$$

$$I_{total} = I_{default} + I_{rogue\_zones} \tag{4}$$

$$I_{rouge\_zones} = \sum_i I_{rouge\_zone\_i} \tag{5}$$

where R' is the number of net cooling requests from the downstream zones of an AHU; R is the number of total cooling requests from the downstream zones; $I_{default}$ is the default number of ignored cooling requests (set by the user); I_{rouge_zones} is the number of ignored cooling requests from the all rogue zones; I_{total} is the sum of the previous two variables; and $I_{rouge\_zone\_i}$ is the number of cooling requests from the ith identified rogue zone. R' is calculated by subtracting the sum of all the rogue zones ignored based on the conditions described above (I_{rogue_zones}) and a default minimum of the ignored zones ($I_{default}$) from all the requests (R). If the equation leads to a negative result, R' becomes zero. R' is used in the SAT reset calculation below.

3.2.2. Code for "Improve AHU Supply Air Temperature Setpoint Reset"

The supply air temperature cooling setpoint (SAT_spt) is continually reset using "Trim and Respond" logic between a minimum and maximum setpoint (SATmin = 12.8 °C and SATmax = 18.3 °C). When the supply air fan is turned on, the initial setpoint is set to SAT_0 = 18.3 °C and the reset logic is active immediately. When active, for every time step t = 5 min, the net cooling request from the downstream zones (R') is calculated using Equations (3)–(5) above. If the R' is above zero, SAT_spt is decreased by a defined respond amount (SATres = 0.06 °C for each request) until the SAT_spt reaches SATmin; if R' is equal to zero, the SAT_spt is increased by a fixed amount (SATtrim= 0.12 °C) until the SAT_spt reaches SATmax.

3.3. *Test Results*

3.3.1. Test Results of "Rogue Zone" Algorithm

The "rogue zone" auto-correction algorithm worked as expected on AHU01 and AHU02 during the testing period. Figure 6 shows the results of the values for the requests and ignored requests calculated on 4 March. The two heat maps in Figure 6 depict each zone (in the vertical axis) plotted against the time of the day, for a single day. For each zone, the darker areas show when the requests (red) and ignored requests (blue) happened during the day. R, I_{rouge_zones}, and R' were reevaluated every five minutes. Taking 11:00 a.m. as an example, the vertical line marks the values of R_i, $I_{rogue\_zone\_i}$, R, I_{total}, and R' at 11:00 a.m.: eight zones (i.e., Rm 4107, Rm 411, Rm 4113, Rm 4203B, Rm 5113, Rm 5115, Rm 5116, and Rm 5204A) sent requests ($R = 8$), three zones (i.e., Rm 4107, Rm 5113, and Rm 5116) were identified as rogue zones ($I_{rogue\_zones} = 3$), and two additional zones were ignored by default ($I_{default} = 2$), summing up to a total of five zones ignored ($I_{total} = 5$). R' is therefore equal to 3, based on Equation (3). This test confirmed that the system correctly calculated and implemented the modified requests, ignoring the rogue zones.

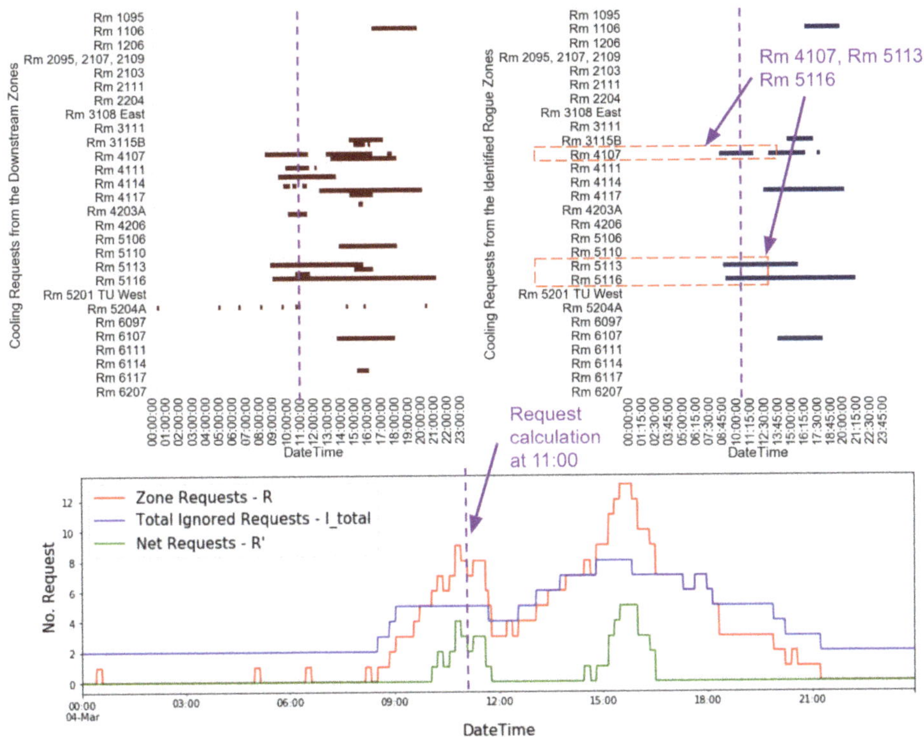

Figure 6. Zone requests per zone (upper left), the ignored requests per zone (upper right), the sum of the requests, the total of ignored requests and the net requests for all the zones of AHU01 on 4 March 2020. The vertical line marks 11 a.m. on all the plots.

3.3.2. Test Results of "Improve AHU SAT Setpoint Reset" Algorithm

The auto-correction algorithm "Improve AHU SAT setpoint reset" successfully changed the SAT setpoint of AHU01 and AHU02 in the BAS. As shown in Figure 7, the SAT setpoint changes followed the routine described in Section 3.2. The supply fan was on for the whole time, since this AHU serves laboratory areas. When R' was larger than zero starting at 10:05 a.m., the algorithm slowly reduced the SAT setpoint by 0.06 °C for each request every five minutes. Starting at 11:50 a.m., the R' remained at zero and the routine slowly increased the SAT setpoint by 0.12 °C every five minutes until it reached SAT_{max} (18.3 °C). The SAT setpoint remained at SAT_{max} until R' was larger than zero at 14:50 p.m. Then, the SAT setpoint again slowly decreased when R' was larger than zero and slowly increased when R' was zero.

Because both the original and corrected logic used feedback loops, a direct comparison of the two was not possible without modeling the dynamic behavior of the system or collecting enough data to perform a system-level evaluation. However, since the original controller was still active (for backup purposes) we could qualitatively compare the time at which each algorithm would start reducing the SAT setpoint in the morning. Figure 8 shows this comparison for two consecutive days in AHU01. The red line represents the corrected setpoint that was calculated by the algorithm and the blue line was the actual temperature that tracked the setpoint. The green line depicts the original logic. As highlighted by the text, the original logic would try to reduce the temperature much earlier than the corrective algorithm. This behavior was consistent across the testing period for both AHUs. For AHU01, during 14 days out of the 34 days, the old logic started earlier, while during the other 20 days the system did not require cooling. Conversely, for AHU02, during 13 days the old logic started earlier, during one day it started later, and the other days it did not require cooling (the test was conducted during a mild spring). Overall, the preliminary test was successful. It showed the uninterrupted operation of the two algorithms in two AHUs for more than a month. The SAT tracked the new setpoint throughout the whole testing period. The new control sequence did not cause any occupant complaints, and it worked more efficiently than the previous one, although a precise savings estimate was beyond the scope of the test.

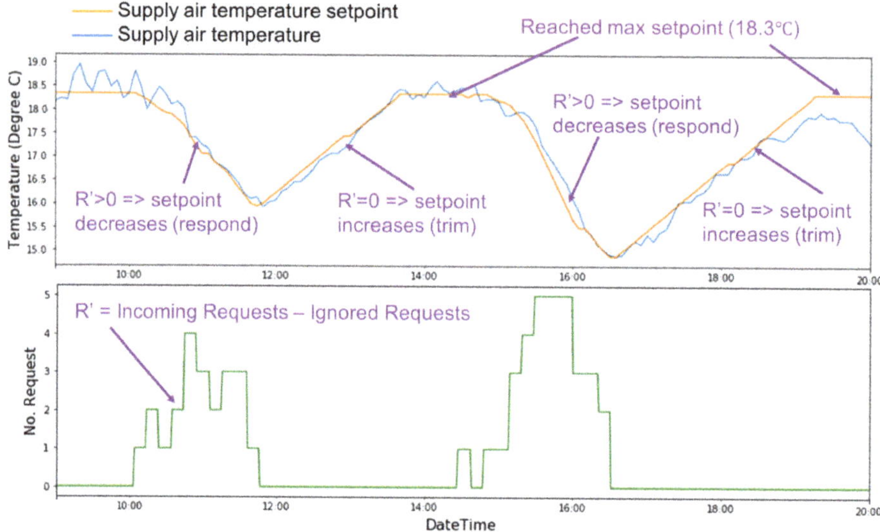

Figure 7. The SAT setpoint of AHU01 after the execution of the auto-correction algorithm (4 March 2020).

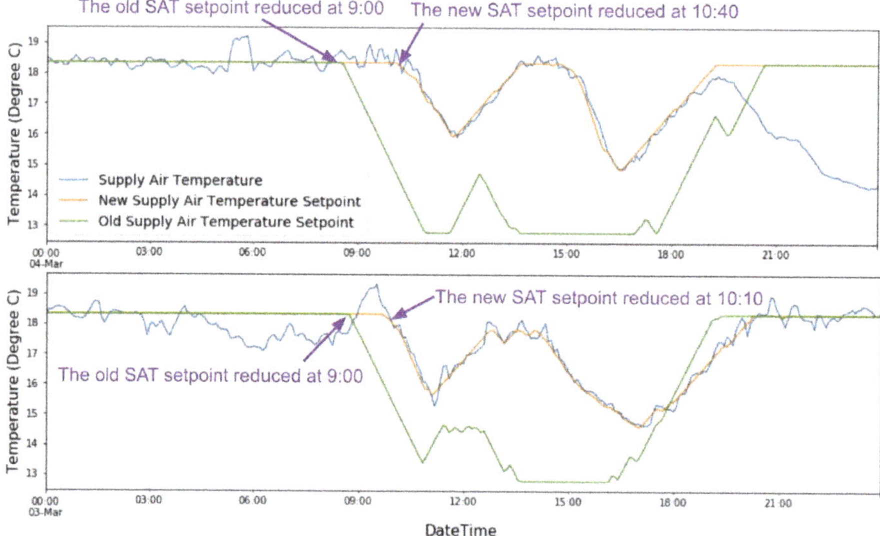

Figure 8. Comparison of the new and the old setpoint control strategies in AHU01 during 3–4 March 2020.

4. Discussion: Implementation Challenges and Solutions

The commercial partners faced a number of challenges when implementing the auto-correction algorithms into the FDD tools. We organized these issues under four areas: (1) developing a secure two-way communication between the FDD tool and the BAS; (2) incorporating operator approval; (3) managing the customizations necessary to the specific BAS/site installation; and (4) managing the potential conflict between the auto-correction and the BAS control actions. This section describes these challenges, as well as the solutions that the partners came up with to mitigate them.

The differences in the solutions described below also stemmed from different FDD software architectures, as well as different BAS network setups across the implementation partners. The FDD products developed by the two partners were based on software platforms that ran from the cloud with centralized fault libraries and analytics engines. These companies used additional hardware and software installed on site to collect data and send it to the cloud. A third implementation partner was the distributor of a third-party software and developed custom algorithms for various customers. This software ran on the local BAS network and had direct access to it, allowing access to external users via a virtual private network connection.

4.1. Develop a Secure Two-Way Communication Between the FDD and the BAS

Opening a two-way communication between the FDD system and the BAS was a challenge seen across all project partners implementing fault auto-correction into their FDD product environment. FDD tools typically read operational data from the BAS, run analytics and flag faults on the software interface. Often, they do not have capabilities to write commands directly onto the BAS. As indicated in Figure 1, the FDD tool commonly collects operational data using three pathways: (1) from the BAS server database, (2) from a central BAS server via API and (3) directly via the BACnet IP network. The first pathway prevents the FDD tool from writing back to the control system, therefore it cannot be used to implement auto-correction procedures. The second one requires BAS-specific interfaces; thus, implementers tend to avoid it. For this reason, when expanding the one-way interface to two-way communication, all the partners selected the third pathway to write back directly via the BACnet IP network.

The project partners mitigated the two-way communication challenge by upgrading their FDD system infrastructure. Figure 9 illustrates the solutions for the two cloud-based FDD systems. The solid line shows the original infrastructure, and the red dashed line shows the upgrade. In the first cloud-based FDD case (Figure 9a), the BACnet stack (a software library allows users to add a native BACnet interface to talk to the devices or applications in the BACnet network) of the FDD data acquisition device was updated to include a "write" function. The local data acquisition device was also updated to make API requests to the cloud FDD platform to retrieve the auto-correction command information. This enabled the FDD system to send the auto-corrective command to the local device and then to the writable properties used to control the BAS. In a cloud-based FDD system of another partner (Figure 9b), the current BACnet library already had writing capabilities. To enable secure communication with the cloud, the system architecture was changed. The standard data acquisition device was paired with a new field device (an auto-correction execution device) specifically designed to execute the new routines and log the interaction with the BAS. The cloud FDD engine initiated the auto-corrective command onto the auto-correction execution device. The device then executed the commands onto the BAS BACnet network and reported back the results to the cloud FDD engine. The BAS data were still acquired by the existing FDD data acquisition field device and delivered to the cloud FDD engine. The third on-premise FDD system was already capable of writing commands via BACnet. It only needed to change a setting in the BAS to authorize the changes.

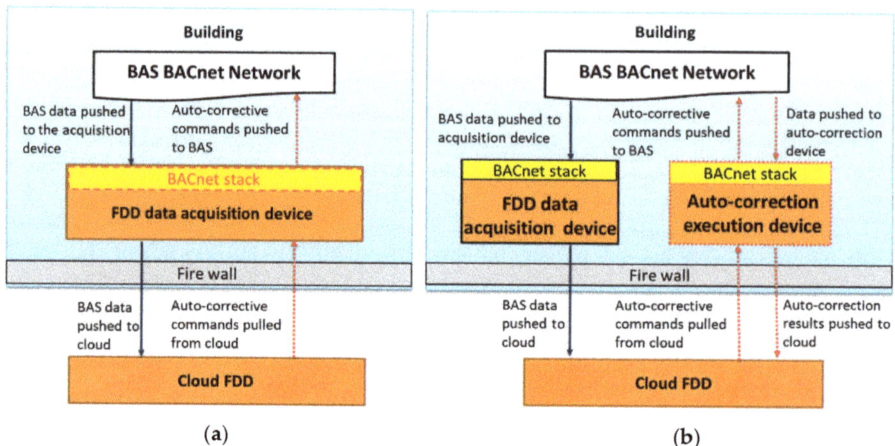

Figure 9. Two-way communication infrastructure for (**a**) the cloud-based FDD system 1, and (**b**) the cloud-based FDD system 2.

4.2. Incorporate Operator Approval

The second challenge faced by the partners was incorporating operator approval. The new auto-correction feature affords the FDD technology a certain degree of control capability. The building operators may be hesitant to trust this new capability and feel a lack of control. To mitigate this challenge, one project partner updated the existing interface to make sure the users were allowed to actively start, interrupt and track the auto-correction activities. Auto-correction enable and disable functionality was added to the user interface (UI), and the name of the control variables, their current value, and the new proposed values were provided to increase the operators' awareness. All the user and system activities of auto-correction are stored in a history log that is available to the user. Figure 10 shows a simplified mockup of the new UI displaying the auto-correction enable/disable functionality, action history and other details. Another partner also developed new interfaces for auto-correction authentication and acknowledgement.

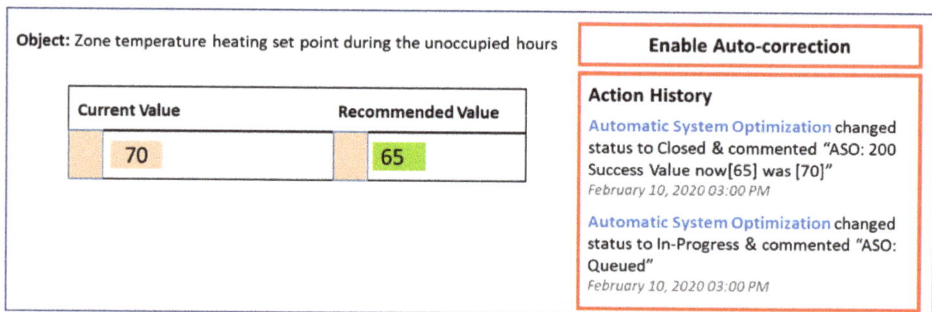

Figure 10. Mockup of the new user interface (UI) developed by one partner, displaying the auto-correction enable/disable functionality, the action history and other details (ASO – Automatic System Optimization).

4.3. Manage BAS and Site-Specific Customizations

The traditional separation of the roles between the FDD and the BAS allowed the FDD tools to develop general algorithms that were independent from some of the details about BAS and the implementation of specific control programs. For instance, an algorithm that detects opportunities to save energy by shortening the AHU schedules did not need to know how these schedules were implemented in the BAS, it just needs to analyze the data produced by them. However, auto-correcting the same schedule meant overriding the operation of the BAS, therefore the developers must know many more details about the specific implementation of the control logic to avoid unintended consequences. The third challenge confronted was the lack of standardization in the BAS control logic, variables and interfaces. The implementation partners reported issues in (1) deciphering the BAS control sequences and identifying the exact control variables to override, (2) gaining access to these variables and (3) gathering data with frequency and timeliness appropriate to the application. An example of the first issue is the implementation of the "override manual control" algorithm described in Section 2.2. Depending upon the BAS, the override can be accomplished via an "override" variable Manual_override (whose value is 1—equipment is in manual control, 0—equipment is in automatic control) or by the setting of the priority level of the BACnet points (e.g., 8—manual operator override, 16—default automatically operation) [27].

Accessing the proper control variable was another part of the challenge. The auto-correction algorithms may require the FDD tools to be able to access the control variables that are not commonly exposed to the outside by the BAS. An example is the PID-tuning parameters required by the "control hunting" algorithm. One implementation partner reported being unable to retrieve these points via BACnet for a site, since the BAS vendor used a proprietary solution. The third issue emerged when a partner implemented the algorithms described in Section 3. These routines need real-time data updated every few minutes, since the algorithms are reevaluated continuously, while the existing BAS was storing it at 15-minute intervals and transmitting it to the FDD tool once a day (to save memory and bandwidth).

To address the challenges, all the partners had to spend significant time to understand and modify the BAS programming and setup, in addition to its interface with the FDD tool. The parameters of the BAS controller, gateway or server were changed to expose the necessary variables, making sure they could be modified when needed. Sampling frequency and data transfer rate were increased to implement some of the algorithms. A partner created a virtual point in the actual codes to accommodate the settings of override in various BASs. The virtual point is a string semicolon delimited list of point IDs that are mapped to any points that need to be changed from override in the BAS. A partner reported that particular care had to be put into matching appropriate data types (e.g., binary, analog with different precisions, arrays) used by the BACnet protocol, to avoid communication errors. All these customizations varied by BAS vendor, hardware vintage and site configuration.

4.4. Manage Control Conflicts between the BAS and the FDD Tool

The last challenge reported pertained to the conflicts between the BAS and the FDD control actions. Algorithms that make one-time changes to the BAS operation (e.g., the incorrectly programmed schedule in Section 2.1) may be overridden by operators or the BAS logic at a later date. It is unclear whether or not the auto-correction procedure should periodically update these variables. Moreover, algorithms that continuously change variables may also conflict with the existing BAS sequence of operation. An example is improving the AHU static pressure setpoint reset (Section 2.8) on a BAS that already has a reset strategy. There is a need to understand which one takes precedence and if the existing control sequences should be turned off.

To address the first issue, one implementation partner used an existing feature of the FDD platform to separately track the active schedule and the most efficient schedule and let the operator decide which one to activate. In addition, it logged all the changes to that schedule to offer more information to the user. For the second issue, another partner set up a fallback mechanism in the BAS, for use with the new FDD auto-correction algorithm that continuously modified the control setpoint. A watchdog was added in the BAS programming to make sure the FDD tool was online. If the FDD tool went offline, the BAS reverted back to the setpoint generated by the original control logic in the case of loss of communication with the FDD tool.

4.5. Other Considerations

The development and deployment of these algorithms stimulated an interesting discussion among the partners and advisors of the project about the role of the FDD and the BAS. Typically, the commercial FDD tools are developed as a software layer on top of the existing BAS. There exists a natural separation of roles in this arrangement, in which the BAS actively controls the building and the FDD tool observes its operation and provides insights and recommendations to the building manager. However, some consider FDD tools as a new generation of BAS that can take over some of its functionalities when it is necessary. At the beginning of the project, one facility manager expressed the desire to implement Guideline 36 sequences [24] on a building being controlled by an obsolete control system. To implement the sequences on that system, significant hardware upgrades and BAS programming labor would be required. The code cannot be easily reused between controllers due to the limitations of the control language, therefore this operation was not scalable and its implementation on multiple systems was hampered. As an alternative, he suggested to host the sequences in the significantly more modern FDD tool (algorithms 2.5, 2.8 and 2.9) and use the existing BAS as a simple tool to collect data and provide direct control over the lower-level hardware. This strategy was eventually implemented and was described in Section 3. Other partners disagreed, objecting that extensive real-time control of the building is outside the scope of FDD tools. Business models of the companies developing the tools may play a role in the way these new functions will be eventually incorporated in the FDD products at the end of this project.

5. Conclusions and Future Work

This paper presented nine algorithms for HVAC systems that were designed to automatically correct faults or improve operations relative to incorrectly programmed schedules, overriding manual control, sensor bias, control hunting, rogue zones and less aggressive setpoints or setpoints setback. It also showed preliminary tests confirming the efficacy of a subset of these algorithms, as tested in a large commercial building. Finally, it discussed challenges faced during the integration of these auto-correction algorithms into three commercial FDD tools and the solutions to these challenges that were adopted by the project partners. The main challenges included: (1) developing a secure two-way communication between the FDD tool and the BAS; (2) incorporating operator approval; (3) managing the customizations necessary to the specific BAS/site installation; and (4) managing the potential conflict between the auto-correction and the BAS control actions. The suggested solutions will help future auto-correction developers address similar challenges.

With respect to automated fault auto-correction, future work will focus on more field testing of the FDD integrated correction algorithms in a cohort of existing buildings. This will include the evaluation of the technical efficacy and the performance of each correction routine, the evaluation of the operations and maintenance benefits for each site in cohort and the characterization of challenges and best practices. A second area of future work will entail the design and execution of a techno-economic analysis to quantify the broader market opportunity to inform ongoing commercialization efforts.

The state of today's FDD technology can be advanced through research focused on enhanced diagnostic (as opposed to detection) approaches and methods for fault prioritization. Complementary work to characterize fault prevalence based on empirical data from the field could also prove valuable in guiding future FDD technology development and implementation efforts. There is also an overarching need to navigate issues related to data management, integration, cybersecurity, and interoperability.

Author Contributions: Conceptualization, J.G., Formal analysis, G.L., M.P. and Y.C.; Methodology, G.L., M.P., Y.C. and J.G.; Writing—original draft, G.L., M.P., Y.C. and J.G.; Writing Review & Editing, G.L. All authors have read and agreed to the published version of the manuscript.

Funding: This research was funded by the Assistant Secretary for Energy Efficiency and Renewable Energy, Building Technologies Office, of the U.S. Department of Energy under Contract No. DE-AC02-05CH11231.

Acknowledgments: This work was supported by the Assistant Secretary for Energy Efficiency and Renewable Energy, Building Technologies Office, of the U.S. Department of Energy under Contract No. DE-AC02-05CH11231.

The authors wish to acknowledge Harry Bergmann for his guidance and support of the research. We also thank the fault detection and diagnostics technology and service providers who participated in this study.

Conflicts of Interest: The authors declare no conflict of interest.

References

1. Energy Information Administration. *Monthly Energy Review April 2019*; EIA: Washington, DC, USA, 2019.
2. Energy Information Administration. *Commercial Buildings Energy Consumption Survey (CBECS)*; EIA: Washington, DC, USA, 2016.
3. Mills, E. Building commissioning: A golden opportunity for reducing energy costs and greenhouse gas emissions in the United States. *Energy Effic.* **2011**, *4*, 145–173. [CrossRef]
4. Katipamula, S.; Brambley, M.R. Methods for fault detection, diagnostics, and prognostics for building systems—A review, part I. *Hvac R Res.* **2005**, *11*, 3–25. [CrossRef]
5. Roth, K.W.; Westphalen, D.; Feng, M.Y.; Llana, P.; Quartararo, L. *Energy Impact of Commercial Building Controls and Performance Diagnostics: Market Characterization, Energy Impact of Building Faults and Energy Savings Potential*; US Department of Energy: Washington, DC, USA, 2005.
6. Fernandez, N.E.; Katipamula, S.; Wang, W.; Xie, Y.; Zhao, M.; Corbin, C.D. *Impacts of Commercial Building Controls on Energy Savings and Peak Load Reduction*; Pacific Northwest National Lab: Richland, WA, USA, 2017.
7. Deshmukh, S.; Glicksman, L.; Norford, L. Case study results: Fault detection in air-handling units in buildings. *Adv. Build. Energy Res.* **2018**, 1–17. [CrossRef]
8. Fernandes, S.; Granderson, J.; Singla, R.; Touzani, S. Corporate delivery of a global smart buildings program. *Energy Eng.* **2018**, *115*, 7–25. [CrossRef]
9. Wall, J.; Ying, G. *Evaluation of Next-Generation Automated Fault Detection & Diagnostics (FDD) Tools for Commercial Building Energy Efficiency—Final Report Part. I: FDD Case Studies in Australia, RP1026*; Low Carbon Living CRC: Boca Raton, FL, USA, 2018.
10. Lin, G.; Kramer, H.; Granderson, J. Building fault detection and diagnostics: Achieved savings, and methods to evaluate algorithm performance. *Build. Environ.* **2020**, *168*, 106505. [CrossRef]
11. ASHRAE. *Guideline 13–2015—Specifying Building Automation Systems*; ASHRAE: Akron, OH, USA, 2015.
12. Granderson, J.; Singla, R.; Mayhorn, E.; Ehrlich, P.; Vrabie, D.; Frank, S. *Characterization and Survey of Automated Fault Detection and Diagnostics Tools*; Report Number LBNL-2001075; Lawrence Berkeley National Laboratory: Washington, DC, USA, 2017.
13. Kim, K.; Rao, P.; Burnworth, J.A. Self-tuning of the PID controller for a digital excitation control system. *IEEE Trans. Ind. Appl.* **2010**, *46*, 1518–1524.
14. Shi, Z.; O'Brien, W. Development and implementation of automated fault detection and diagnostics for building systems: A review. *Autom. Constr.* **2019**, *104*, 215–229. [CrossRef]
15. Fernandez, N.; Brambley, M.; Katipamula, S. *Self-Correcting HVAC Controls: Algorithms for Sensors and Dampers in Air-Handling Units, PNNL-19104*; Pacific Northwest National Laboratory: Richland, WA, USA, 2009.
16. Fernandez, N.; Brambley, M.; Katipamula, S.; Cho, H.; Goddard, J.; Dinh, L. *Self Correcting HVAC Controls Project Final Report PNNL-19074*; Pacific Northwest National Laboratory: Richland, WA, USA, 2009.
17. Brambley, M.; Fernandez, N.; Wang, W.; Cort, K.A.; Cho, H.; Ngo, H.; Goddard, J.K. *Fial Project Report: Self-Correcting Controls for VAV System Faults Filter/Fan/Coil and VAV Box Sections. No. PNNL-20452*; Pacific Northwest National Laboratory (PNNL): Richland, WA, USA, 2011; Volume 20.
18. Isermann, R. *Fault-Diagnosis Systems: An Introduction from Fault Detection to Fault Tolerance*; Springer Science & Business Media: New York, NY, USA, 2006.
19. Zhang, Y.; Jiang, J. Bibliographical review on reconfigurable fault-tolerant control systems. *Annu. Rev. Control* **2008**, *32*, 229–252. [CrossRef]
20. Padilla, M.; Choinière, D.; Candanedo, J.A. A model-based strategy for self-correction of sensor faults in variable air volume air handling units. *Sci. Technol. Built Environ.* **2015**, *21*, 1018–1032. [CrossRef]
21. Wang, S.; Chen, Y. Fault-tolerant control for outdoor ventilation air flow rate in buildings based on neural network. *Build. Environ.* **2002**, *37*, 691–704. [CrossRef]
22. Hao, X.; Zhang, G.; Chen, Y. Fault-tolerant control and data recovery in HVAC monitoring system. *Energy Build.* **2005**, *37*, 175–180. [CrossRef]

23. Bengea, S.C.; Li, P.; Sarkar, S.; Vichik, S.; Adetola, V.; Kang, K.; Lovett, T.; Leonardi, F.; Kelman, A.D. Fault-tolerant optimal control of a building HVAC system. *Sci. Technol. Built Environ.* **2015**, *21*, 734–751. [CrossRef]
24. ASHRAE. *Guideline 36–2018. High. Performance Sequences of Operation for HVAC Systems*; ASHRAE: Akron, OH, USA, 2018.
25. ASHRAE. *ASHRAE/IES Standard 90.1–2016. Energy Standard for Buildings Except Low-Rise Residential Buildings*; ASHRAE: Akron, OH, USA, 2016.
26. Lin, G.; Pritoni, M.; Chen, Y.; Granderson, J. Can We Fix It Automatically? Development of Fault Auto-Correction Algorithms for HVAC and Lighting Systems. *ACEEE* **2020**, in press.
27. BACnet®Primer. What is BACnet? Phoenix Controls: Acton, MA, USA. 2009. Available online: https://www.phoenixcontrols.com/CatalogDocuments/Products/Network%20Integration/BACnet%20Primer%20(MKT-0233).pdf (accessed on 19 May 2020).

Article

The Effect of Deep Energy Retrofit on The Hourly Power Demand of Finnish Detached Houses

Janne Hirvonen [1,*], Juha Jokisalo [1] and Risto Kosonen [1,2]

1 Department of Mechanical Engineering, Aalto University, 00076 Espoo, Finland; juha.jokisalo@aalto.fi (J.J.); risto.kosonen@aalto.fi (R.K.)
2 College of Urban Construction, Nanjing Tech University, Nanjing 211816, China
* Correspondence: janne.p.hirvonen@aalto.fi; Tel.: +358-50-431-5780

Received: 18 February 2020; Accepted: 6 April 2020; Published: 7 April 2020

Abstract: This study examines how the energy renovation of old detached houses affects the hourly power consumption of heating and electricity in Finland. As electrification of heating through heat pumps becomes more common, the effects on the grid need to be quantified. Increased fluctuation and peak power demand could increase the need for fossil-based peaking power plants or call for new investments to the distribution infrastructure. The novelty in this study is the focus on hourly power demand instead of just annual energy consumption. Identifying the influence of building energy retrofits on the instantaneous power demand can help guide policy and investments into building retrofits and related technology. The work was done through dynamic building simulation and utilized building configurations obtained through multi-objective optimization. Deep energy retrofits decreased both the total and peak heating power consumption. However, the use of air-source heat pumps increased the peak power demand of electricity in district heated and wood heated buildings by as much as 100%. On the other hand, peak power demand in buildings with direct electric heating was reduced by 30 to 40%. On the building stock level, the demand reduction in buildings with direct electric heating could compensate for the increase in the share of buildings with ground-source heat pumps, so that the national peak electricity demand would not increase. This prevents the increase of demand for high emission peaking power plants as heat pump penetration rises. However, a use is needed for the excess solar electricity generated by the optimally retrofitted buildings, because much of the solar electricity cannot be utilized in the single-family houses during summer.

Keywords: single-family house; detached house; energy renovation; deep retrofit; power demand; electric heating; ground-source heat pump

---

1. Introduction

The European Union (EU) aims to reduce CO_2 emissions by 80% compared to 1990 levels by the year 2050 [1]. The EU recognizes electrification of heating through heat pumps as one key step towards decarbonization [2]. Since buildings are responsible for 40% of EU's energy demand and emissions, the Energy Performance of Buildings Directive (EPBD) was created to reduce emissions caused by the operation of future buildings [3]. However, despite tightening regulation of new buildings, the existing energy-inefficient building stock that has mostly been built before modern regulations remains the biggest source of building-based emissions. To tackle this issue, the EU has called for national retrofit strategies to effect a positive change in the existing buildings as well [4]. Finnish building code also requires the consideration of energy efficiency whenever renovation tasks are performed on buildings [5]. This is important, as 79% of Finnish buildings have been built before the year 2000 [6] and certain mandatory renovation work provides a chance for lower cost energy efficiency improvements as well.

The influence of building retrofits on energy demand has been examined in many studies. The case of various Italian building types was examined in [7]. The results show the cost-effectiveness of different levels of energy efficiency, showing that energy consumption in single-family houses could go down by up to 77%. Similarly, the energy saving potential in Swedish detached houses was found to be 65–75% in [8]. Here the focus was on standardized buildings built between the years 1961 and 1980, increasing the usability of the suggested actions. Four reference buildings from different regions were dynamically simulated in IDA-ICE, which is a multi-zone simulation tool for evaluating indoor thermal conditions and building energy consumption. Retrofit measures were added step-by-step based on their prevalence in real life, until the energy efficiency matched passive houses. In Ireland, deep energy retrofits in semi-detached houses were only feasible with government grants [9]. This study used six different environmental indicators and included the environmental impact of the materials needed for retrofitting in addition to the operational impact. Thirty-five pre-determined retrofit packages were calculated using quasi-steady state equations in the DEAP software, a web-based tool for producing Building Energy Ratings. In the Finnish context, energy retrofits have been examined for apartment buildings [10,11], where both primary energy demand and CO_2 emissions could be significantly reduced cost-effectively, especially using heat pumps, but also including improvements to the building envelope. In Finnish office buildings CO_2 emissions could be cost-effectively reduced by 50% while preserving thermal comfort [12]. In Finnish detached houses [13] life cycle costs and CO_2 emissions could be most effectively lowered by deep energy retrofits in buildings with direct electric or oil-based heating. In these studies, dynamic simulation with IDA-ICE was combined with multi-objective optimization by a genetic algorithm, to go through hundreds of retrofit packages. Heat pumps in particular stand out as a good heating solution for the future [14]. As heating demand is increasingly met with electricity, short term fluctuations in electric power use can be expected to increase. However, all these studies have reported their results only on an annual level, leaving the seasonal changes unknown and providing no information about the peak power levels on short timescales.

The power levels in shorter time scales are important for the development of the energy generation infrastructure, especially as the share of undispatchable renewable energy increases. Investments into new power plants are based on capacity-based costs and the variable energy prices [15], which are again influenced by the instantaneous power demand and the available energy generation capacity. Increase of peak power demand may call for increased investments to power transmission lines [16] or to demand response services which are used to shift peak demand to lower load hours [17]. Strategies for reducing peak demand under uncertain loads are being developed such as in [18], which highlights the importance of energy flexibility in buildings. Important strategies include the price mechanism effect on occupant's behavior, centralized energy management with demand response and HVAC peak load controls. A review on power system planning studies concluded that when modelling the power grid, smaller details such as heating systems should also be taken into account [19]. On the building side, a Norwegian study did report some monthly results on the deep retrofit of apartment buildings [20], though the main goal was to show the feasibility of evaluating retrofits using only the hottest and coldest months. A Spanish study showed the monthly renewable energy use and changes in peak demand after building retrofits [21]. Peak power demand reduction was also the focus of a study made for Dubai's cooling dominated climate [22]. Hourly power demand after building retrofits has been reported for Finnish apartment buildings [23], but a similar study on detached houses was not found. In many studies on building energy retrofits, heat pumps have been presented as the lowest emission heating solution. This is to be expected as there is a lot of low emission electricity generation in the Nordic countries, such wind power (Denmark and Sweden), nuclear power (Finland and Sweden), and hydro power (Finland, Norway, and Sweden) [24]. However, in a Sweden-based study on building retrofits, new heat pump systems were assumed to increase the total electricity use, thus forcing the use of high emission fossil fuel sources, under the assumption that all existing low emission generation is already in use [25]. This raises the question of whether the demand in other buildings could be reduced to make the existing supply of low emission electricity to stretch further or if the benefits of heat pumps have been exaggerated.

Many articles have been published on emission reduction and energy efficiency improvements in different building types. However, typically only annual changes to emission levels or energy consumption are reported, while few retrofit studies focus on the potential effects on the grid. In this paper, the changes in power levels of Finnish detached houses with different heating systems are examined individually and on the building stock level. The studied scenarios are based on previously optimized emission reducing configurations. The key questions are: What is the impact of deep energy retrofit on the seasonal and peak district heating and electric power demand of Finnish detached houses? How does the excess solar electricity generation of optimized building configurations compare to the hourly demand? What is the potential impact of large-scale building retrofits and electrification of heating on the electric power requirements of the whole detached house building stock?

2. Methods and Materials

2.1. Simulation and Optimization

To estimate the effect of energy retrofits on power demand in single-family houses, Finnish houses of various ages were modelled and simulated in the IDA-ICE dynamic energy simulation software [26], which has been validated, for example, in [27] and [28]. Hourly space heating and ventilation demand were obtained through simulation, while the domestic hot water [29] and electric equipment loads [30] were based on measured data. MATLAB was used for pre-processing tasks before the IDA-ICE simulation and for energy balance and cost calculations after the heat loads were obtained. Retrofitted configurations of the reference buildings were generated through multi-objective optimization, using the MOBO tool [31] with the genetic algorithm NSGA-II [32], as described in Figure 1. Finally, a few optimally retrofitted building configurations were selected for further study. The calculations are based on previous simulation work of the authors [13], where Finnish single-family houses of various ages were retrofitted to reduce emissions.

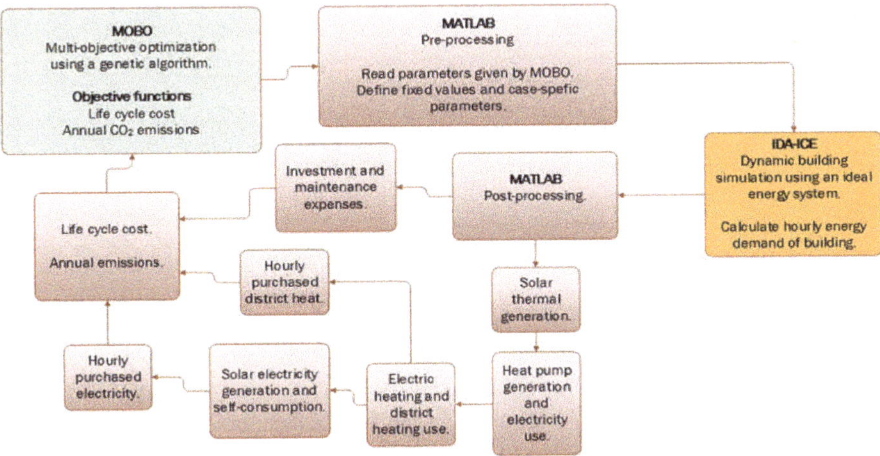

Figure 1. The process chart for the utilized simulation and optimization method.

2.2. Building Descriptions

The base building was a single-family house (SH), which has previously been used as a type building [33], with two storeys and a heated net area of 180 m^2. The studied buildings were split into four age categories, based on their construction year and the building code in effect at the time: SH1

(<1976), SH2 (1976–2002), SH3 (2003–2009), and SH4 (>2010). Each age category has tighter thermal insulation requirements than the previous. The ventilation system for SH1 was natural ventilation and for SH2 it was mechanical exhaust ventilation without heat recovery. SH3 utilized mechanical supply and exhaust ventilation with heat recovery as did SH4, but with a higher heat recovery efficiency. The details of the reference building properties are shown in Table 1.

Table 1. Properties of the reference single-family houses [34].

	Building Age Class	**SH1**	**SH2**	**SH3**	**SH4**
	Construction years	–1975	1976–2002	2003–2009	2010–
U-values of envelope (W/(m²·K))	**External wall**	0.58	0.28	0.25	0.17
	Floor	0.48	0.36	0.25	0.16
	Ceiling	0.34	0.22	0.16	0.09
	Doors	1.4	1.4	1.4	1.0
	Windows	1.8	1.6	1.4	1.0
Total solar heat transmittance (g)		0.71	0.59	0.46	0.46
Direct solar transmittance (ST)		0.64	0.52	0.39	0.39
Air tightness	n_{50}, **(1/h)**	6	4	3.5	2
	q_{50}**m³/(h m²)**	15.6	10.4	9.1	5.2
Ventilation	**Type**	Natural ventilation	Mech. E. vent.	Mech. S.&E. vent.	Mech. S.&E vent.
	Heat recovery temp. eff.	0	0	0.55	0.65
	Ventilation rate (L/s/m²)	0.30	0.33	0.36	0.36
	Total air exchange rate (1/h)	0.41	0.46	0.5	0.5
	SFP (kW/m³/s)	0	1.5	2.5	2
	Heating setpoint (°C)	22	22	21.5	21

In addition, the buildings were divided according to the main heating system in use in the building: district heating (DH), wood/oil boiler, direct electric heating, or ground-source heat pump (GSHP). The share of different heating systems in the building stock within each building age category is shown in Figure 2.

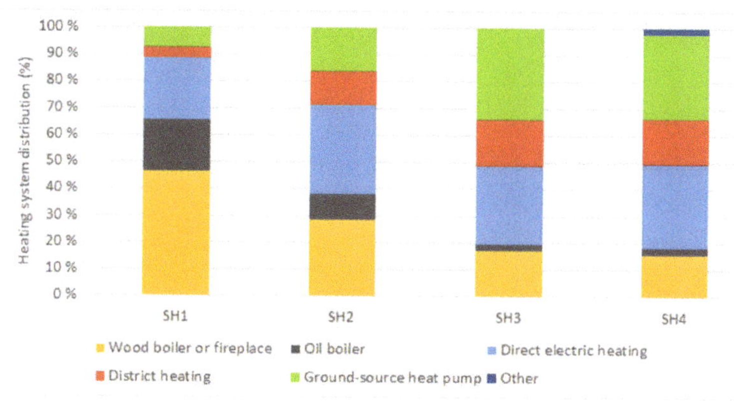

Figure 2. Distribution of main heating systems in the building stock within each building age category.

buildings could be reduced to make the existing supply of low emission electricity to stretch further or if the benefits of heat pumps have been exaggerated.

Many articles have been published on emission reduction and energy efficiency improvements in different building types. However, typically only annual changes to emission levels or energy consumption are reported, while few retrofit studies focus on the potential effects on the grid. In this paper, the changes in power levels of Finnish detached houses with different heating systems are examined individually and on the building stock level. The studied scenarios are based on previously optimized emission reducing configurations. The key questions are: What is the impact of deep energy retrofit on the seasonal and peak district heating and electric power demand of Finnish detached houses? How does the excess solar electricity generation of optimized building configurations compare to the hourly demand? What is the potential impact of large-scale building retrofits and electrification of heating on the electric power requirements of the whole detached house building stock?

2. Methods and Materials

2.1. Simulation and Optimization

To estimate the effect of energy retrofits on power demand in single-family houses, Finnish houses of various ages were modelled and simulated in the IDA-ICE dynamic energy simulation software [26], which has been validated, for example, in [27] and [28]. Hourly space heating and ventilation demand were obtained through simulation, while the domestic hot water [29] and electric equipment loads [30] were based on measured data. MATLAB was used for pre-processing tasks before the IDA-ICE simulation and for energy balance and cost calculations after the heat loads were obtained. Retrofitted configurations of the reference buildings were generated through multi-objective optimization, using the MOBO tool [31] with the genetic algorithm NSGA-II [32], as described in Figure 1. Finally, a few optimally retrofitted building configurations were selected for further study. The calculations are based on previous simulation work of the authors [13], where Finnish single-family houses of various ages were retrofitted to reduce emissions.

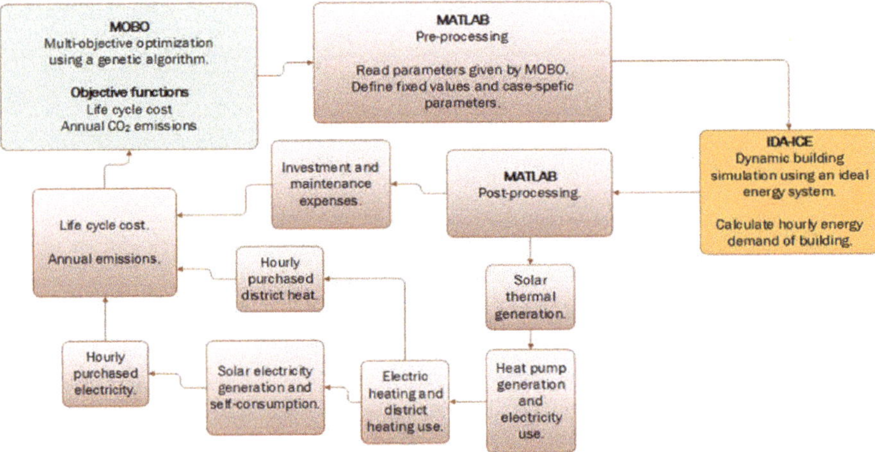

Figure 1. The process chart for the utilized simulation and optimization method.

2.2. Building Descriptions

The base building was a single-family house (SH), which has previously been used as a type building [33], with two storeys and a heated net area of 180 m^2. The studied buildings were split into four age categories, based on their construction year and the building code in effect at the time: SH1

The domestic hot water (DHW) use profile was based on measured data from Finnish buildings [29] and was normalized to 35 kWh/m^2 per year [35], though distribution losses of 0.5 W/m^2 were assumed. While the buildings of different ages had different space heating demands, the DHW demand was the same for all buildings. Lighting and electric appliance profiles were based on measured profiles from 1630 Finnish households [30], and the consumption and internal gains were normalized to 5.3 kWh/m^2 and 15.9 kWh/m^2 per year, for lighting and equipment, respectively, to match the Finnish building code [35].

In total, 75% of the Finnish building stock is located in the Southern and Western Finland (Zones I and II), which is why the building simulation used the TRY2012 Helsinki-Vantaa weather data that describes these regions [36]. The average annual air temperature is 5.6 °C with an annual solar insolation of 970 kWh/m^2 on a horizontal surface. The heating degree day value is 3952 Kd (at indoor temperature of 17 °C) in this heating dominated climate [37].

2.3. Retrofitted Building Configurations

Figure 3 shows the possible retrofit paths for each building. It was assumed that buildings with district heating or direct electric heating would keep their main heating system the same, while other retrofits were performed. Buildings with oil or wood boilers would switch to (or keep using) wood boilers or ground-source heat pumps. No retrofits were made for buildings originally equipped with a GSHP. The retrofit measures used in the optimization were the increased thermal insulation of the building envelope, replacing old windows with more efficient ones, installing a new ventilation system with heat recovery or variable air volume ventilation (VAV), replacing old water radiators with low temperature radiators, installing solar thermal or photovoltaic (PV) solar electric systems, and installing a ground-source heat pump (GSHP) or an air-to-air heat pump (AAHP).

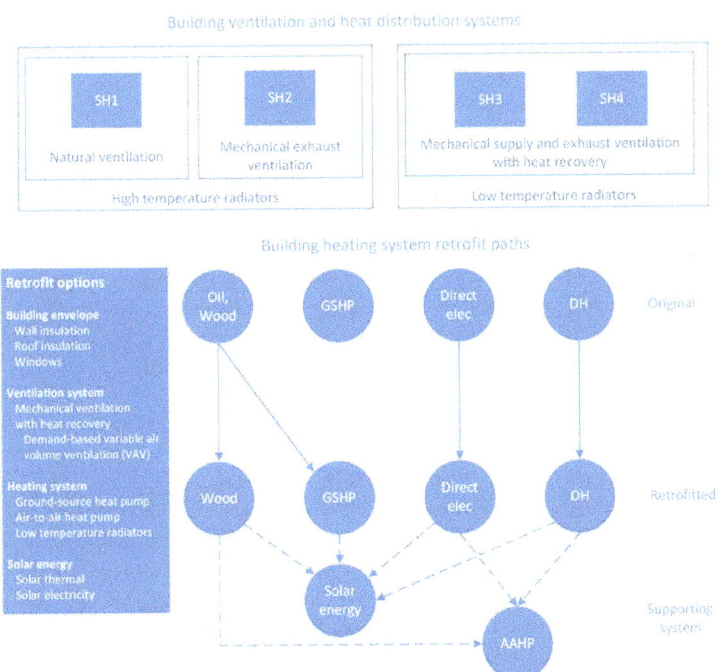

Figure 3. Top: The ventilation and heat distribution systems for the reference buildings. Bottom: The retrofit paths considered.

In this study, three versions of the buildings are presented for each configuration: the original unrenovated case (Ref), a building retrofitted to minimum cost levels (D), and a building retrofitted to significant emission reductions (B). The original optimization study included dozens of Pareto optimal solutions for each building type, but these two optimized levels were selected to limit the amount of cases to be presented while still providing a range of feasible solutions.

Table 2 shows the building configurations of the oldest building type SH1, obtained in the earlier optimization study. It shows the emissions and energy demand as well as the thermal insulation levels and other system properties. The same information for the rest of the building age classes can be found from Tables 3–5.

3. Results

3.1. Buildings With District Heating

This section shows the specific power demand of heating and electricity use for all building configurations that use district heating. Both the original and the retrofitted cases are presented. Figure 4 (left side) shows the hourly duration curves of district heating use in all applicable buildings. The blue lines show the district heating demand in the reference cases, highlighting the improvements in energy efficiency along with the tightening building code (new buildings consume less energy). The green lines show the district heating demand for buildings that have been retrofitted to the minimum cost level, D. This includes the use of an air-to-air heat pump, a slightly improved building envelope and some solar thermal capacity. For the older building types SH1 and SH2, a major drop in both peak DH demand (−33% and −23%, respectively) and average DH demand is seen. The differences are the smallest where the demand is the lowest. Finally, the red lines show the DH demand for buildings that have been significantly retrofitted to level B. This includes a further improved envelope and a large solar thermal capacity. For half of the year, no district heating is needed at all because of abundant solar energy. The peak DH demand goes down by 30 to 50% in all cases, as generally does the DH demand.

The duration curves give no indication on the temporal distribution of power demand. Having several hourly chronological lines would make the figure too cluttered to read, which is why Figure 4 (right side) shows the weekly maximum and minimum DH power consumption instead. However, instead of the single highest demand hour, the sustained peak power demand, that is, the average of the top (or bottom) 5% of weekly demand is shown. Each week was sorted according to hourly power demand and then the 8 hours (5% of 168 weekly hours) with the highest (or lowest) demand were averaged to obtain the plotted values. Together the values show how much the power demand varies within shorter periods during the whole year and give an indication on the required energy storage capacities. In the B cases, the summertime heating demand goes to zero, thanks to solar energy.

Similar to the heating demand, Figure 5 presents the electricity consumption of all district heated buildings. It can be seen on the left side that the minimum electricity consumption for the Reference and D cases remains slightly positive. However, in case B, the solar PV capacity grows so high that the exported excess solar power exceeds the maximum purchased power as much as six times. This increases the load on the grid, but could support the energy needs of other users in the grid. However, such an influx of power might require strengthening of the local grid, which was not taken into account in the optimization. The use of air-source heat pumps in the retrofitted buildings increased the peak electric power demand by as much as 60% compared to the reference case, but the absolute value of the increase is much lower than the power exported from the PV system. The reported values include only the heating use of AAHP, as cooling energy was not considered in this study. The right side of the figure shows the sustained peak (and bottom) power demand. This is the average power of the 5% of weekly hours (8 h) with the highest (or lowest) demand.

Table 2. The building configurations and other properties of the oldest single-family house Table 1976. SH1. Ref is an unmodified case while D and B are retrofitted cases [13]. For ground-source heat pump (GSHP) capacity, the number in parentheses is the ratio of HP thermal capacity to the maximum space heating demand. For ventilation, the number in parenthesis is the mechanical ventilation heat recovery efficiency.

SH1	Emissions	LCC	Electricity Demand	DH / Boiler Demand	U-Values (W/m^2K)				ST Cap	PV Cap	Ventilation Type	Distribution Temperature	GSHP Capacity	AAHP Capacity
Solution	$kg\text{-}CO_2/m^2/a$	€/m^2/25a	kWh/m^2	kWh/m^2	Walls	Roof	Doors	Windows	m^2	kW_p	-	°C	kW_{th}	kW_{th}
DH Ref	41.3	497.7	20.6	234.3	0.58	0.34	1.4	1.8	0	0	Natural	70	0	0
DH D	24.2	449.6	34.4	117.5	0.2	0.1	1.4	1.8	2	0	Natural	70	0	2
DH B	13.8	544.9	23.7	62.7	0.1	0.1	1.4	0.6	18	7	Natural	70	0	3
Oil Ref	61.6	729.8	20.6	234.3	0.58	0.34	1.4	1.8	0	0	Natural	70	0	0
Wood Ref	94.4	491.9	20.6	234.3	0.58	0.34	1.4	1.8	0	0	Natural	70	0	0
Wood D	52.3	452.6	34.4	117.5	0.2	0.1	1.4	1.8	2	0	Natural	70	0	2
Wood B	28.7	552.4	25.0	62.2	0.1	0.09	1.4	0.6	16	5	Natural	70	0	5
Elec Ref	32.6	768.5	224.8	0	0.58	0.34	1.4	1.8	0	0	Natural	-	0	0
Elec D	12.1	559.5	81.6	0	0.15	0.09	1.4	1.8	6	9	Mech (75%) + VAV	-	0	2
Elec B	8.4	617.7	56.4	0	0.1	0.09	1	0.6	14	8	Mech (75%) + VAV	-	0	3
GSHP Ref	13.0	381.8	89.5	0	0.58	0.34	1.4	1.8	0	0	Natural	70	10 (68%)	0
GSHP D	8.1	437.3	55.1	0	0.2	0.12	1.4	1.8	0	10	Natural	70	7 (68%)	0
GSHP B	5.1	550.2	34.8	0	0.1	0.1	1.4	0.6	8	9	Natural	40	8 (110%)	0

Table 3. The building configurations and other properties of the 1976–2002 single family house type, SH2. Ref is an unmodified case while D and B are retrofitted cases [13]. For GSHP capacity, the number in parentheses is the ratio of HP thermal capacity to the maximum space heating demand. Mech E stands for mechanical exhaust ventilation without heat recovery.

SH2	Emissions	LCC	Electricity Demand	DH / Boiler Demand	U-Values (W/m^2K)				ST Cap	PV Cap	Ventilation Type	Distribution Temperature	GSHP Capacity	AAHP Capacity
Solution	kg-$CO_2/m^2/a$	€/m^2/25a	kWh/m^2	kWh/m^2	Walls	Roof	Doors	Windows	m^2	kW_p	-	°C	kW_{th}	kW_{th}
DH Ref	32.2	435.0	23.2	177.0	0.28	0.22	1.4	1.6	0	0	Mech E	70	0	0
DH D	24.0	421.4	38.6	112.6	0.19	0.08	1.4	1.6	0	0	Mech E	70	0	3
DH B	13.3	528.4	25.5	58.1	0.1	0.08	1	0.6	18	7	Mech E	70	0	5
Oil Ref	46.6	605.1	23.2	177.0	0.28	0.22	1.4	1.6	0	0	Mech E	70	0	0
Wood Ref	71.3	428.4	23.2	177.0	0.28	0.22	1.4	1.6	0	0	Mech E	70	0	0
Wood D	57.4	424.8	39.9	121.1	0.28	0.08	1.4	1.6	0	0	Mech E	70	0	3
Wood B	29.1	536.8	26.5	59.1	0.12	0.08	1.4	0.6	20	5	Mech E	70	0	5
Elec Ref	25.8	639.9	179.2	0	0.28	0.22	1.4	1.6	0	0	Mech E	-	0	0
Elec D	15.5	530.2	104.6	0	0.19	0.08	1.4	1.6	6	7	Mech E	-	0	5
Elec B	10.5	616.6	70.4	0	0.08	0.08	0.8	0.6	20	7	Mech E	-	0	4
GSHP Ref	10.7	339.0	74.6	0	0.28	0.22	1.4	1.6	0	0	Mech E	70	8 (76%)	0
GSHP D	8.3	404.3	56.7	0	0.28	0.08	1.4	1.6	0	10	Mech E	70	7 (71%)	0
GSHP B	5.3	551.7	35.7	0	0.08	0.08	1	0.8	12	7	Mech E	40	6 (93%)	0

Table 4. The building configurations and other properties of the 2003–2009 single family house type, SH3. Ref is an unmodified case while D and B are retrofitted cases [13]. For GSHP capacity, the number in parentheses is the ratio of HP thermal capacity to the maximum combined space and ventilation heating demand. For ventilation, the number in parenthesis is the mechanical ventilation heat recovery efficiency.

SH3	Emissions	LCC	Electricity Demand	DH / Boiler Demand	U-Values (W/m^2K)				ST Cap	PV Cap	Ventilation Type	Distribution Temperature	GSHP Capacity	AAHP Capacity
Solution	kg-$CO_2/m^2/a$	€/m^2/25a	kWh/m^2	kWh/m^2	Walls	Roof	Doors	Windows	m^2	kW_p	-	°C	kW_{th}	kW_{th}
DH Ref	28.2	401.8	28.4	148.3	0.25	0.16	1.4	1.4	0	0	Mech (55%)	40	0	0
DH D	22.2	391.1	34.0	106.6	0.25	0.08	1.4	1.4	2	0	Mech (55%) + VAV	40	0	1
DH B	12.2	492.4	24.9	52.3	0.1	0.09	1.4	0.8	14	7	Mech (75%) + VAV	40	0	5
Oil Ref	39.0	555.1	28.4	148.3	0.25	0.16	1.4	1.4	0	0	Mech (55%)	40	0	0
Wood Ref	59.7	409.1	28.4	148.3	0.25	0.16	1.4	1.4	0	0	Mech (55%)	40	0	0
Wood D	47.7	407.4	34.0	106.6	0.25	0.08	1.4	1.4	2	0	Mech (55%) + VAV	40	0	1
Wood B	24.8	516.6	26.7	52.1	0.1	0.07	1	0.6	10	2	Mech (75%) + VAV	40	0	5
Elec Ref	23.8	605.9	166.6	0	0.25	0.16	1.4	1.4	0	0	Mech (55%)	-	0	0
Elec D	14.0	512.4	94.2	0	0.17	0.07	1.4	1.4	6	9	Mech (55%) + VAV	-	0	3
Elec B	10.2	578.5	68.2	0	0.1	0.07	1.4	0.6	14	8	Mech (75%) + VAV	-	0	4
GSHP Ref	9.3	316.2	66.0	0	0.25	0.16	1.4	1.4	0	0	Mech (55%)	40	7 (77%)	0
GSHP D	6.9	378.3	47.0	0	0.25	0.1	1.4	1.4	0	10	Mech (55%) + VAV	40	6 (70%)	0
GSHP B	4.9	527.3	33.2	0	0.08	0.07	1.4	0.6	10	9	Mech (75%) + VAV	40	7 (117%)	0

Table 5. The building configurations and other properties of the post 2010 single family house type, SH4. Ref is an unmodified case while D and B are retrofitted cases [13]. For GSHP capacity, the number in parentheses is the ratio of HP thermal capacity to the maximum combined space and ventilation heating demand. For ventilation, the number in parenthesis is the mechanical ventilation heat recovery efficiency.

SH4	Emissions	LCC	Electricity Demand	DH / Boiler Demand	U-Values (W/m^2K)				ST Cap	PV Cap	Ventilation Type	Distribution Temperature	GSHP Capacity	AAHP Capacity
Solution	kg-$CO_2/m^2/a$	€/m^2/25a	kWh/m^2	kWh/m^2	Walls	Roof	Doors	Windows	m^2	kW_p	-	°C	kW_{th}	kW_{th}
DH Ref	22.7	348.0	26.8	115.8	0.17	0.09	1	1	0	0	Mech (65%)	40	0	0
DH D	17.1	347.1	31.6	77.6	0.17	0.09	1	1	4	0	Mech (65%) + VAV	40	0	1
DH B	10.8	469.4	23.6	45.2	0.07	0.07	0.8	1	18	7	Mech (75%) + VAV	40	0	4
Oil Ref	30.4	475.8	26.8	115.8	0.17	0.09	1	1	0	0	Mech (65%)	40	0	0
Wood Ref	46.7	364.5	26.8	115.8	0.17	0.09	1	1	0	0	Mech (65%)	40	0	0
Wood D	35.7	368.5	31.6	77.6	0.17	0.09	1	1	4	0	Mech (65%) + VAV	40	0	1
Wood B	21.8	495.5	23.5	45.8	0.07	0.06	0.8	1	20	7	Mech (65%) + VAV	40	0	5
Elec Ref	20.4	538.8	142.6	0	0.17	0.09	1	1	0	0	Mech (65%)	-	0	0
Elec D	12.2	457.1	82.0	0	0.17	0.07	1	1	6	9	Mech (65%) + VAV	-	0	3
Elec B	9.4	561.3	62.8	0	0.08	0.06	1	0.6	14	7	Mech (75%) + VAV	-	0	5
GSHP Ref	8.0	291.7	57.2	0	0.17	0.09	1	1	0	0	Mech (65%)	40	6 (83%)	0
GSHP D	6.3	344.0	43.4	0	0.17	0.09	1	1	0	10	Mech (65%)	40	5 (69%)	0
GSHP B	4.7	513.6	32.1	0	0.08	0.07	1	1	20	7	Mech (75%) + VAV	40	14 (226%)	0

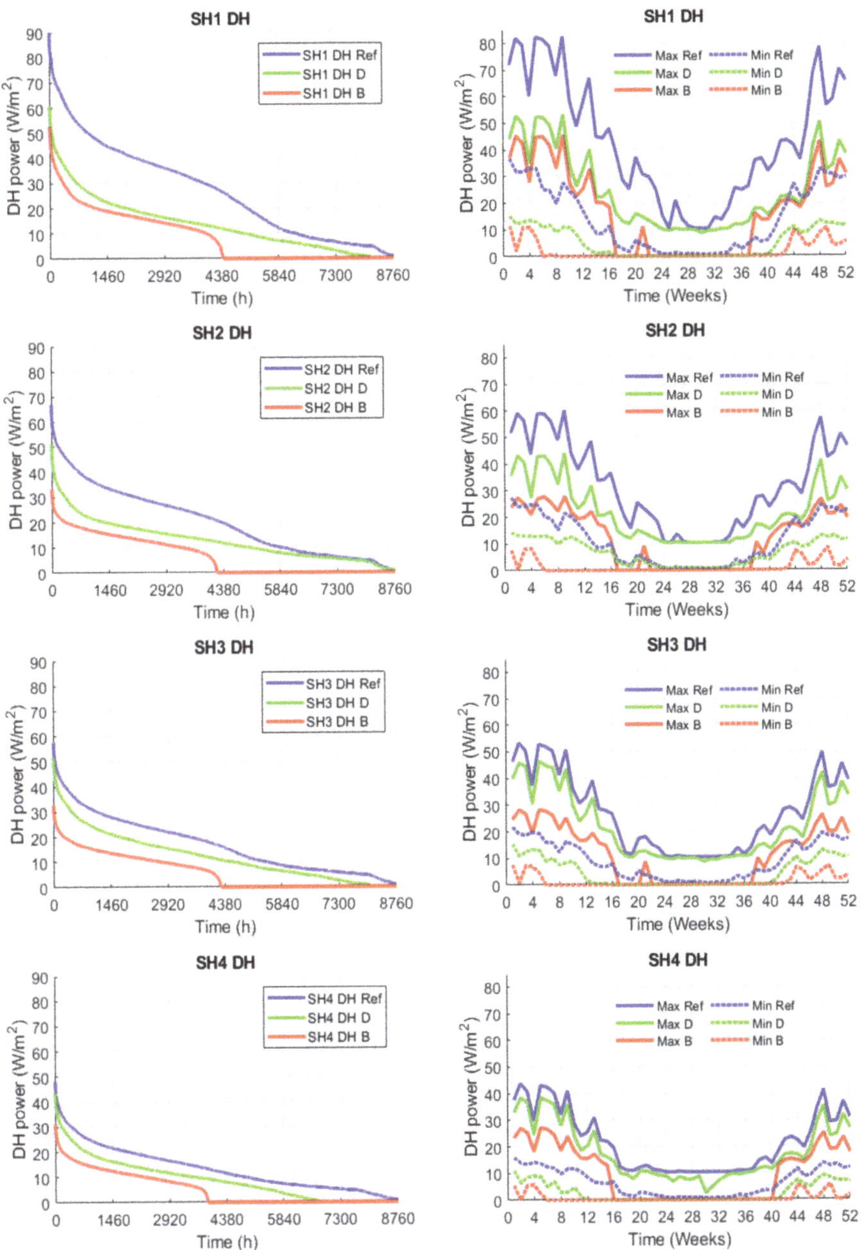

Figure 4. District heating use of all district heated houses. On the left: The hourly duration curve of district heating (DH) power demand for the whole year. On the right: The sustained peak/bottom DH power demand. Solid lines depict the weekly top 5% of demand and dotted lines the weekly bottom 5% of demand.

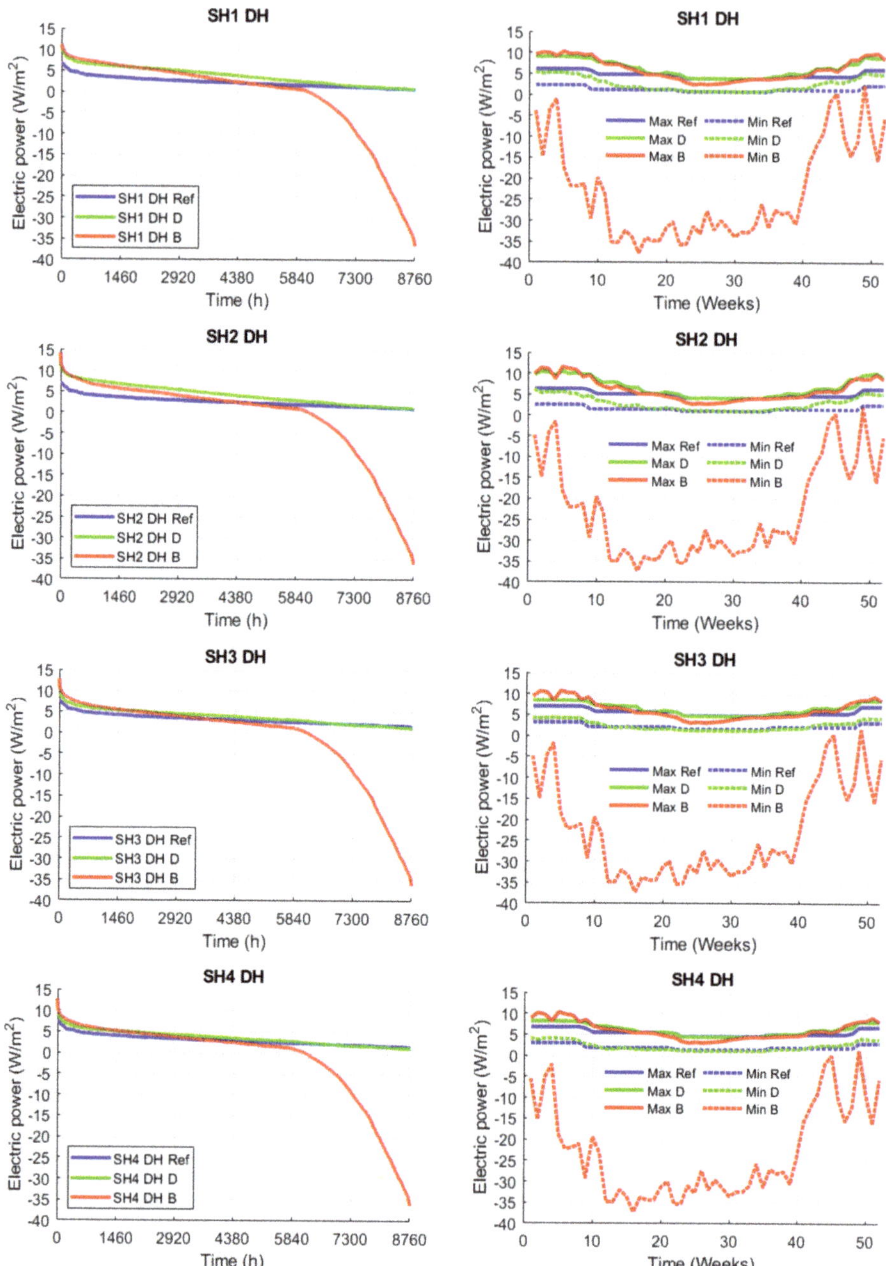

Figure 5. Electric power demand of the district heated buildings. Negative values represent exports of excess electricity back to the grid. On the left: The hourly duration curve of power demand for the whole year. On the right: Sustained peak/bottom power demand. Solid lines depict the weekly top 5% of demand and dotted lines the weekly bottom 5% of demand.

3.2. Buildings with Wood Boilers

The wood boiler use in the wood heated buildings closely matches that of district heating in the previous section. The electricity use for buildings with wood-based heating is also similar, as shown in the duration curves in Figure 6. The retrofitted cases have higher maximum power demand, because part of the wood-based heating was shifted to the air-source heat pumps in both the D and B cases. In the minimum demand side, we see that the B cases have very large exports of electricity due to the large solar panel arrays installed in the high investment cases.

The sustained weekly peak power demands are also shown in Figure 6 (right side). The sustained peak electricity demand in summer is the lowest for the B cases, due to self-consumption of solar electricity. In winter, the B cases have the highest peak demand, because of higher capacity air-source heat pumps. Solar panels produce very little power in the Finnish winter and do not influence peak demand.

3.3. Buildings with Direct Electric Heating

Figure 7 shows the duration curves of electricity use for buildings with direct electric heating. Electric heating in the oldest building (SH1) increases the peak demand over ten times compared to the non-electrically heated buildings. However, with electric heating the retrofits significantly reduce electricity demand, unlike in the DH and wood boiler cases. Here the absolute values of the maximum demand and the maximum solar energy exports are of the same scale in the retrofitted cases. Qualitatively, all the different age classes have similar sets of duration curves.

Figure 7 also shows the sustained peak and minimum power levels (the average of 5% of weekly max/min hours) for all cases with direct electric heating. During summer, the peak demand in the retrofit B case is similar to the minimum demand in the reference case for all age classes. In winter, the sustained peak demand in SH1 was 70, 50 or 40 W/m^2 for the Ref, D and B cases, respectively, showing a great reduction due to the retrofits. The difference was smaller for the newer building, such as SH4, where the sustained peak demands were 45, 36, or 30 W/m^2, for the Ref, D and B cases, respectively.

3.4. Buildings with Ground-Source Heat Pumps

Use of the GSHP produced a major decrease in peak electric power demand compared to direct electric heating. Duration curves of the GSHP cases are shown in Figure 8. Comparing the original GSHP systems to the retrofit scenarios, in the case of SH1, the D level retrofit reduced peak demand from 55 to 43 W/m^2 while the capacity ratio (HP power vs. space heating demand) was 68% for both cases. In SH2, SH3, and SH4 there was no difference in peak power when comparing the reference buildings with original GSHPs and the D level retrofits. The power demand increased significantly during the peaks compared to the base level due to capacity constraints of the GSHP. The systems did not cover 100% of heating demand and electric backup heating was needed, thus significantly increasing demand during peak hours. With the level B retrofits, which included significantly improved thermal insulation of the building envelope, the heat pump size was sufficient (93% of space heating demand in SH2 and over 100% for the rest) and peak power was in check. As in the other cases, the PV arrays were oversized and the power exported to the grid was comparable to and even higher than the demand from the grid.

The seasonal variance of the electric power demand is shown on the right side of Figure 8, which shows the weekly top and bottom 5% of power flow. In SH1, retrofit D reduced sustained winter peak demand from 50 to 36 W/m^2, while in SH2 and SH3 there was no difference. In SH4, retrofit D actually had higher peak demand, because the GSHP was sized down from 83% capacity ratio to 69% and thus the electric backup heater saw more use. Retrofit B shows major decreases in power demand during the entire heating season for all building age classes. It also significantly lowers the absolute variance between the peak demand in high and low demand time periods. For example, during weeks 1 to 5 in the case of SH2, the peak demand changes by 15 W/m^2 in the reference and retrofit D, but in retrofit B the change is only 5 W/m^2. This is due to the sizing difference of the GSHP systems. In the retrofit B cases the GSHP capacity is close to peak demand, which means that electric backup heating with high power demand is not needed. In both the retrofit levels D and B, the exports of surplus solar electricity happen at high power. Especially in level B, the peak power in exports in summer is 1.5 to 3 times as much as the winter demand peak. Depending on the strength of the distribution grid, the optimal solution may not be feasible after all.

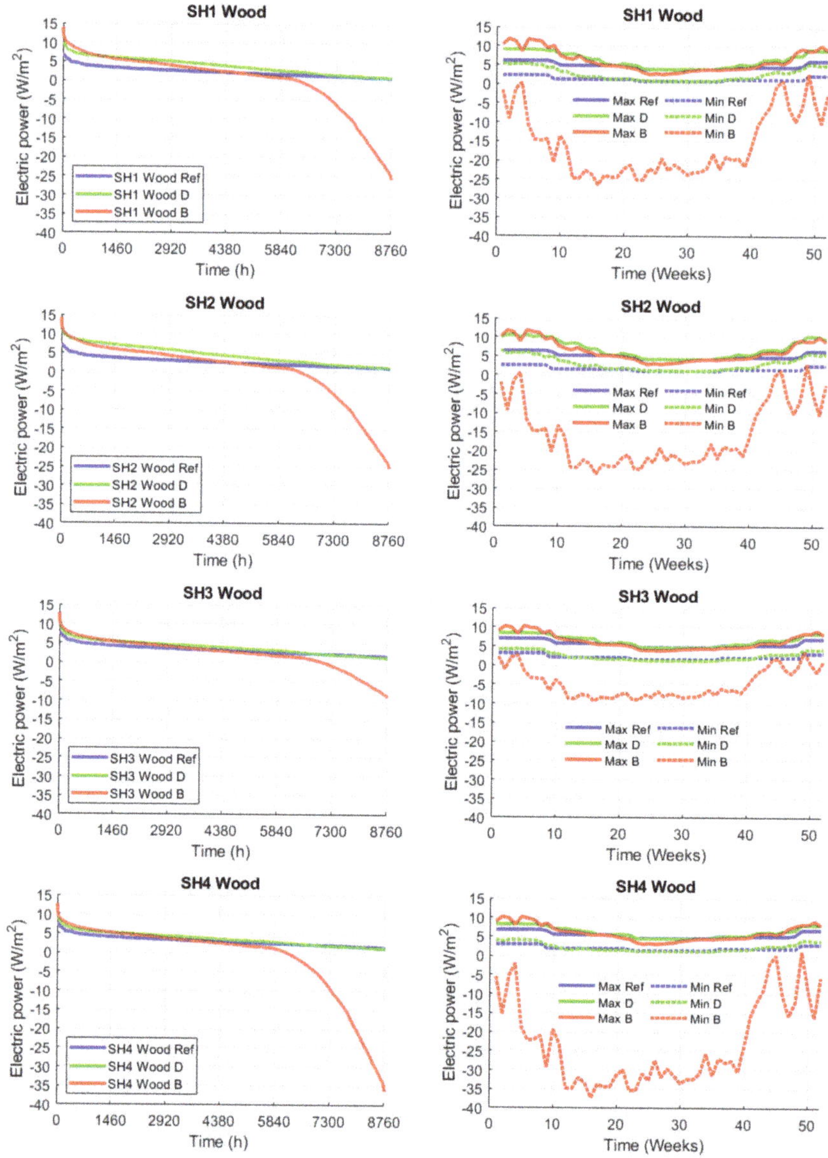

Figure 6. Electric power demand of the wood-heated buildings. Negative values represent exports of excess electricity back Table 5 of demand, and dotted lines depict the weekly bottom 5% of demand.

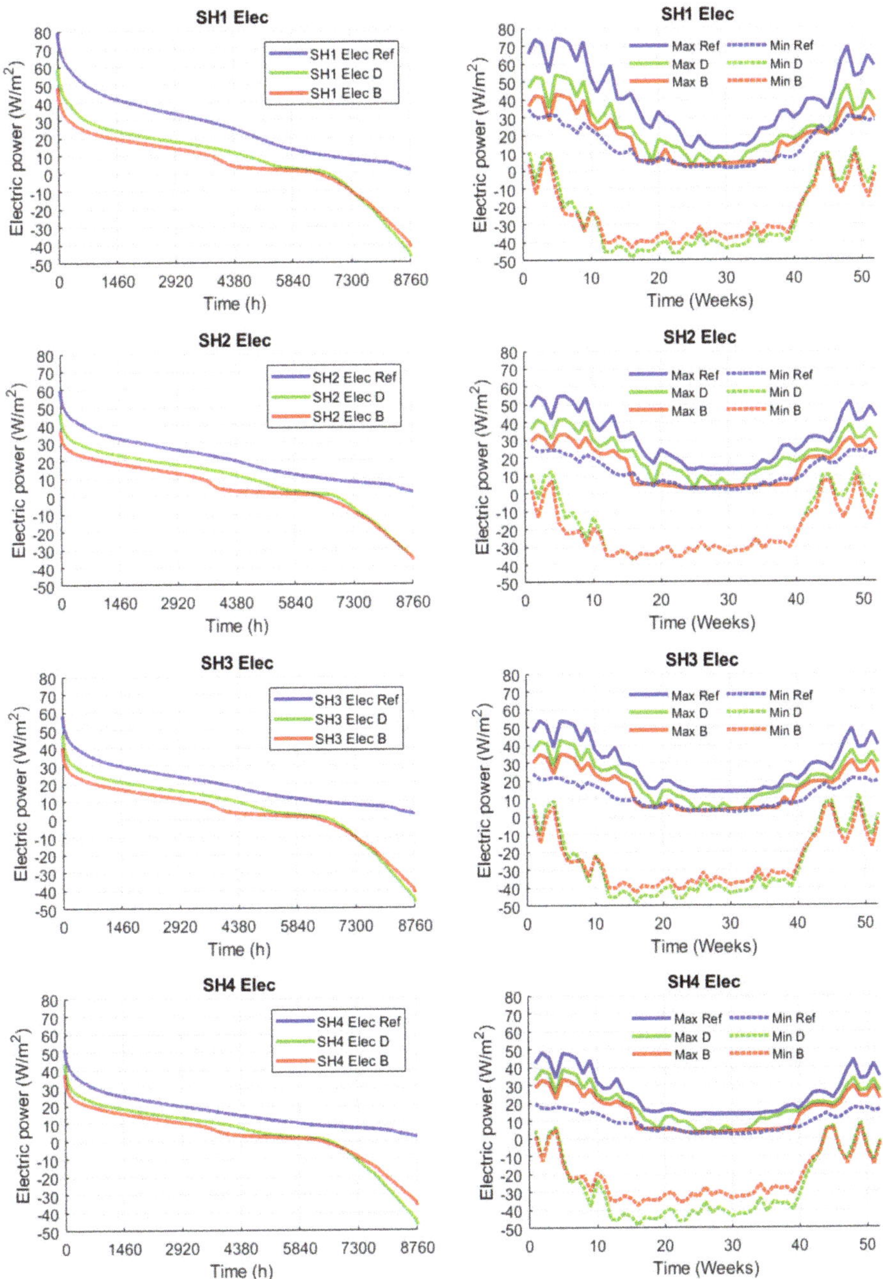

Figure 7. Electric power demand of the buildings with direct electric heating. Negative values represent exports of excess electricity back to the grid. On the left: The hourly duration curve of power demand for the whole year. On the right: the sustained peak/bottom power demand. Solid lines depict the weekly top 5% of demand, and dotted lines depict the weekly bottom 5% of demand.

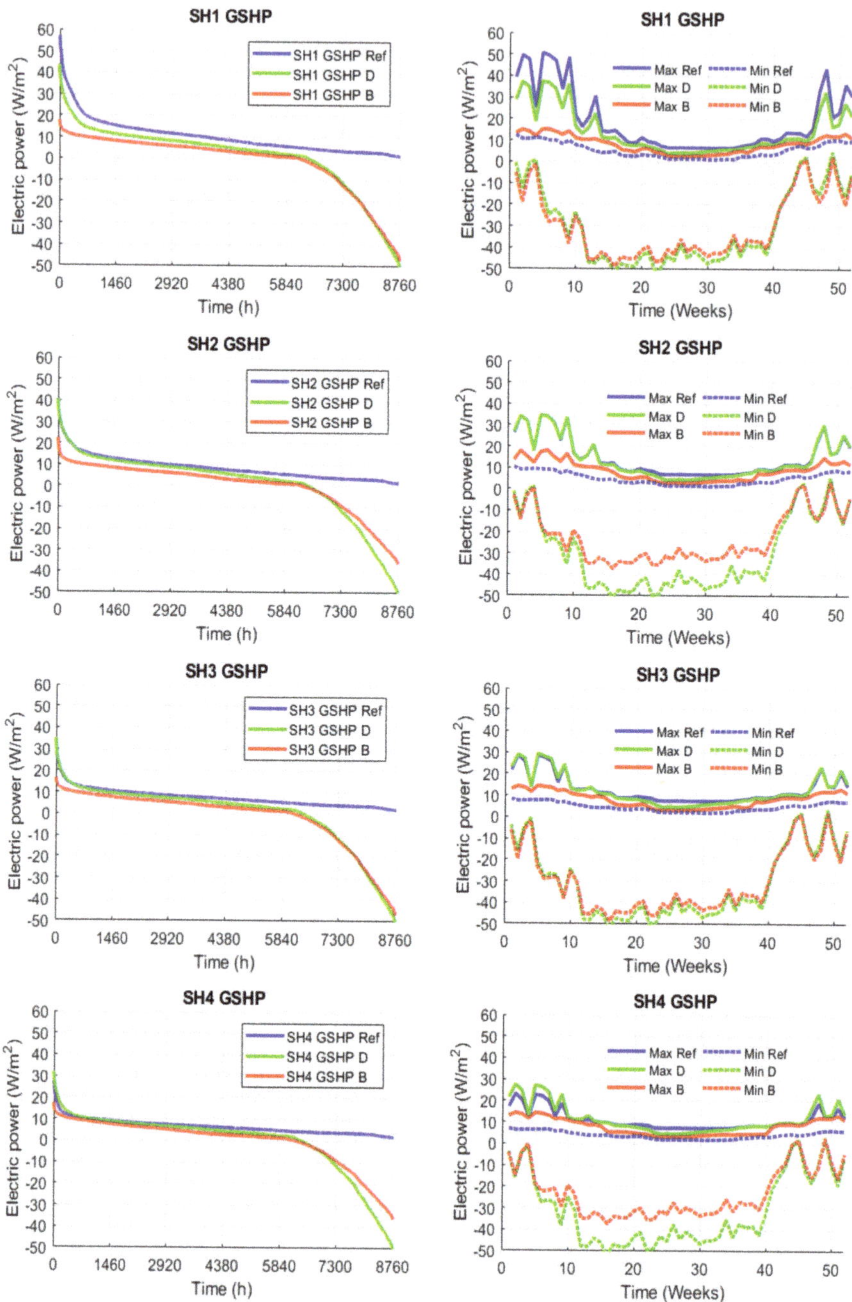

Figure 8. Electric power demand of the buildings with ground-source heat pumps. Negative values represent exports of excess electricity back to the grid. On the left: The hourly duration curve of power demand for the whole year. On the right: Sustained peak/bottom power demand. Solid lines depict the weekly top 5% of demand, and dotted lines depict the weekly bottom 5% of demand.

3.5. Effect on Building Stock

This section shows the effect these energy renovations could have on the whole building stock, namely the electricity consumption levels. Some retrofit actions increase electricity consumption, while others decrease it.

Significant changes in both district heating and electrical power levels were realized with energy retrofits in all the building age categories. In the previous study [13], it was found that switching to electrified heating resulted in significant emission reductions, when monthly average emission factors of Finnish electricity generation were used. However, as the number of retrofitted buildings goes up, the changes in consumption patterns start to influence the national grid. With increased electricity use, the average emission factors may no longer be reasonable. Thus, the changes in DH and electricity demand need to be quantified on the building stock level.

To estimate the potential influence of retrofits on a larger scale, assumptions about the retrofit levels were made in reference [13]. Of buildings that use wood or oil boilers for heating, 50% switch to GSHP and the other 50% switch to or keep using a wood boiler, while also doing other improvements. Buildings already equipped with GSHP are not renovated. In the other buildings, the main heating system remains unchanged while other retrofit actions are performed as described in Tables 2–5 and reference [13]. The buildings are retrofitted to either the minimum cost level D or the costlier but high impact level B.

The total annual electricity demand in each scenario is shown in Figure 9. Buildings with direct electric heating consumed most of the electricity in the base scenario. In both of the retrofit scenarios, the electricity consumption of ground-source heat pumps increased significantly. However, the improved energy efficiency of buildings with direct electric heating more than compensated for the increase in heat pump use and the total electricity consumption went down in both the scenarios D (−11%) and B (−38%). The total PV capacity of the retrofitted building stock was 4400 MW in scenario D and 5600 MW in scenario B. It was assumed that no PV panels were installed in the reference buildings, although at the end of 2018 there was actually a total installed solar electric capacity of 120 MW in Finland [38]. Solar electricity produced in the retrofitted buildings significantly exceeded how much could be used in detached houses without energy storage technologies. Self-consumed solar electricity has been subtracted from the demand values presented, but the surplus amounts are shown separately as negative demand. A small part of the surplus could be utilized in other detached houses (Usable solar), which have unmet electricity demand, but the majority of it is excess energy that needs to be used in some other sector or by other building types (Excess solar).

The electric power levels in the whole building stock are shown in Figure 10. It shows the weekly top and bottom 5% of power as well as the hourly duration curves. Like in the individual building level, the excess solar power production was significant in the whole building stock as well. Notably, in retrofit scenario D, the peak demand during winter grew compared to the reference scenario, even though the total electric energy demand went down (as shown in Figure 9). In retrofit scenario B, the peak demand remained almost the same as in the reference scenario, even though the annual energy demand was significantly lower. This is elaborated in the duration curve of Figure 10, where the power demand of scenario B is lower than the original scenario for every moment after the peak. The largest differences are seen after hour 6570, when the large PV arrays of the retrofitted scenarios result in significant excess power. Solar electricity reduces power demand mainly when the demand is not very high anyway and has no effect on the peak demand. On the other hand, the large PV capacities greatly increased the power flow during the summer. In scenario B, the summertime export power reached 5 GW, while the imported peak power was only 4 GW.

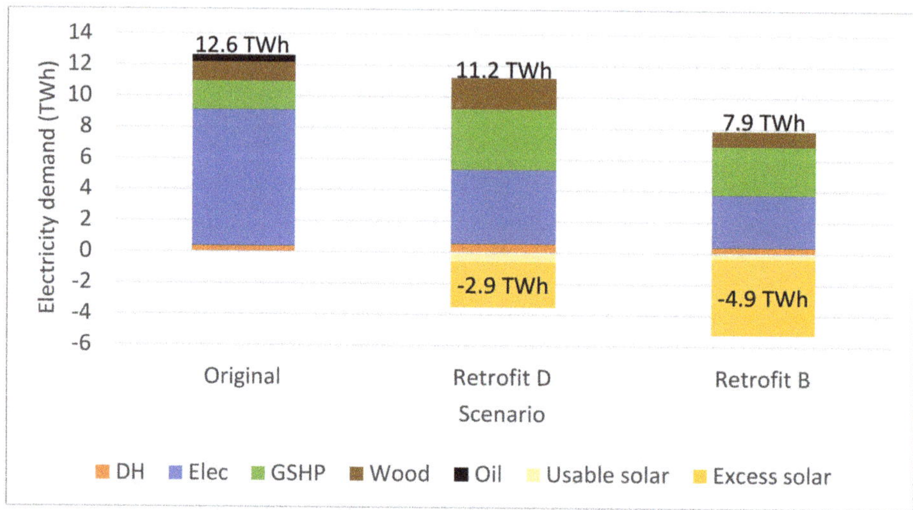

Figure 9. Annual electricity demand for the whole detached house stock. Also shown is the excess solar electricity produced in the buildings. Usable solar is surplus production from some detached houses that could be used in other houses, while Excess solar is surplus that needs to be used in some other sector (values presented).

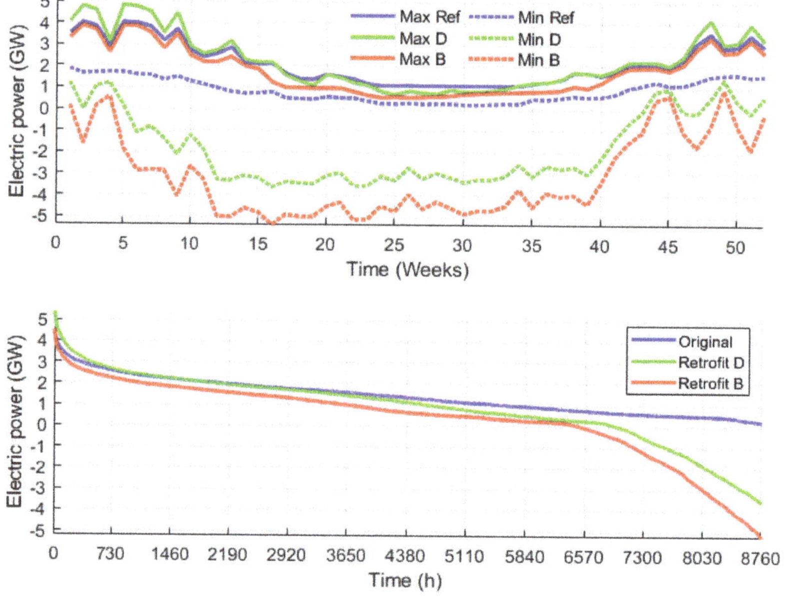

Figure 10. Sustained peak and bottom electric power (5% of weekly hours) in the whole building stock with the original systems and in the retrofit scenarios D and B. Also shown are the hourly duration curves of the same original data.

3.6. Summary

In DH and wood heated cases, the energy retrofit decreased the use of DH and boiler, but increased electricity consumption due to the air-source heat pumps that were included in all optimal solutions.

Solar power did not influence peak power demand. In buildings with a ground-source heat pump or direct electric heating, the electricity consumption went down for both the D and B level retrofits. From the national and regional energy system point of view, it is useful to know the range of power demand of the buildings before and after the retrofits. These results are shown in Table 6. Shown in the table are the maximum and minimum (single hour) power demand for the month of January (high demand and emissions) and July (low demand and emissions). Also shown is the sustained peak/bottom power demand, which is the average power of the 5% of hours in the month (37 h) with the highest/lowest demand. Finally, it shows the median power demand of the whole month. Negative values represent exports of excess solar electricity back into the grid.

Table 6. Electric power demand for all reference and retrofit cases. Results are shown for January (a high emission month) and July (a low emission month). The reported values are the absolute maximum and minimum power demand, the monthly top/bottom 5% of power demand (average power of 37 highest/lowest demand hours) and the median power demand. Negative values represent exports of excel solar energy to the grid.

Case		January, Electric Power (W/m²)					July, Electric Power (W/m²)				
		Max	Top 5%	Median	Bottom 5%	Min	Max	Top 5%	Median	Bottom 5%	Min
SH1	DH Ref	6.5	6.2	3.1	2.3	2.3	4.2	3.9	1.3	0.7	0.7
	DH D	9.7	9.1	6.1	5.3	4.7	4.2	3.9	1.3	0.8	0.7
	DH B	11.2	9.7	6.4	−5.7	−29.6	3.8	3.0	−1.3	−32.8	−35.5
	Elec Ref	75.3	70.1	41.4	30.7	26.8	19.3	13.5	6.8	1.9	1.5
	Elec D	56.0	50.1	23.7	6.3	−36.5	9.4	6.3	−0.7	−42.2	−45.7
	Elec B	45.4	39.7	19.6	0.5	−33.9	4.4	3.6	−1.3	−37.3	−40.4
	GSHP Ref	51.2	45.1	15.1	10.9	9.6	6.7	6.6	3.1	1.3	1.0
	GSHP D	38.8	33.3	11.4	−3.1	−36.1	6.0	4.7	−0.2	−45.8	−49.6
	GSHP B	16.5	13.9	8.2	−7.4	−36.9	3.7	2.9	−2.1	−42.6	−46.0
	Wood Ref	6.5	6.2	3.1	2.3	2.3	4.2	3.9	1.3	0.7	0.7
	Wood D	9.7	9.1	6.1	5.3	4.7	4.2	3.9	1.3	0.8	0.7
	Wood B	13.9	11.0	5.7	−2.8	−20.4	3.9	3.1	−0.5	−23.1	−25.0
SH2	DH Ref	6.8	6.5	3.4	2.6	2.6	4.5	4.2	1.6	1.0	1.0
	DH D	11.5	10.1	7.1	5.6	5.0	4.5	4.2	1.6	1.1	1.0
	DH B	13.5	10.7	5.9	−6.1	−29.3	4.1	3.3	−1.0	−32.5	−35.2
	Elec Ref	56.5	52.0	32.4	24.1	22.4	13.9	13.4	7.1	2.2	1.9
	Elec D	43.3	39.0	23.3	8.0	−26.5	9.6	6.5	−0.1	−32.5	−35.2
	Elec B	36.0	30.8	18.8	−0.9	−29.3	4.1	3.3	−1.0	−32.5	−35.2
	GSHP Ref	35.8	30.8	12.7	9.2	8.5	7.0	6.9	3.4	1.5	1.3
	GSHP D	35.8	30.9	11.9	−3.2	−35.7	6.3	5.0	0.2	−45.5	−49.3
	GSHP B	20.6	15.1	8.4	−4.6	−29.3	4.1	3.3	−1.0	−32.5	−35.2
	Wood Ref	6.8	6.5	3.4	2.6	2.6	4.5	4.2	1.6	1.0	1.0
	Wood D	11.5	10.3	7.3	5.8	5.2	4.5	4.2	1.6	1.1	1.0
	Wood B	13.7	10.9	6.0	−2.9	−20.0	4.2	3.4	−0.2	−22.8	−24.7
SH3	DH Ref	7.4	7.0	4.0	3.2	3.2	5.1	4.8	2.2	1.6	1.6
	DH D	8.9	8.5	5.3	4.2	4.1	5.0	4.7	1.9	1.3	1.2
	DH B	12.5	10.0	5.6	−6.4	−29.3	4.4	3.6	−0.8	−32.5	−35.2
	Elec Ref	56.2	50.5	30.0	21.5	19.1	14.6	14.0	7.5	2.9	2.5
	Elec D	45.4	39.2	21.1	3.2	−36.5	9.5	5.6	−1.0	−42.1	−45.6

Table 6. *Cont.*

Case		January, Electric Power (W/m²)					July, Electric Power (W/m²)				
		Max	Top 5%	Median	Bottom 5%	Min	Max	Top 5%	Median	Bottom 5%	Min
	Elec B	38.0	32.2	17.2	−1.8	−33.9	4.4	3.6	−1.2	−37.3	−40.4
	GSHP Ref	31.5	24.9	10.7	7.7	7.1	7.6	7.5	3.9	2.1	1.9
	GSHP D	32.5	25.8	9.5	−5.9	−37.1	6.8	5.5	0.3	−45.4	−49.2
	GSHP B	16.4	13.4	7.7	−8.2	−38.2	4.3	3.6	−1.6	−42.2	−45.7
	Wood Ref	7.4	7.0	4.0	3.2	3.2	5.1	4.8	2.2	1.6	1.6
	Wood D	8.9	8.5	5.3	4.2	4.1	5.0	4.7	1.9	1.3	1.2
	Wood B	12.0	9.6	5.5	1.5	-6.1	4.7	4.2	1.3	−8.2	−8.9
SH4	DH Ref	7.2	6.9	3.8	3.0	3.0	4.9	4.7	2.0	1.5	1.4
	DH D	8.8	8.3	5.1	4.1	3.4	4.8	4.6	1.8	1.2	1.2
	DH B	11.9	9.5	5.3	−6.8	−29.2	4.4	3.6	−0.8	−32.5	−35.2
	Elec Ref	50.8	45.1	25.7	17.5	15.8	14.3	13.8	7.3	2.5	2.2
	Elec D	41.5	35.5	18.3	0.7	−36.6	9.4	5.2	−1.1	−42.2	−45.7
	Elec B	37.0	30.4	15.9	−1.5	−29.2	4.4	3.6	−0.9	−32.5	−35.2
	GSHP Ref	26.0	20.0	9.2	6.5	5.9	7.4	7.4	3.7	1.9	1.8
	GSHP D	30.3	24.2	8.9	−5.8	−37.3	6.7	5.4	0.5	−45.1	−48.9
	GSHP B	16.6	13.5	7.7	−5.6	−29.3	4.4	3.6	−0.9	−32.5	−35.2
	Wood Ref	7.2	6.9	3.8	3.0	3.0	4.9	4.7	2.0	1.5	1.4
	Wood D	8.8	8.3	5.1	4.1	3.4	4.8	4.6	1.8	1.2	1.2
	Wood B	12.2	9.5	5.3	−6.9	−29.3	4.4	3.6	−0.9	−32.5	−35.2
Building stock	Original	26.4	23.4	13.8	10.4	9.3	6.9	6.8	3.4	1.5	1.3
	Retrofit D	31.3	27.5	13.8	5.4	−14.0	6.7	5.0	1.2	−20.3	−22.0
	Retrofit B	26.4	22.2	11.6	−1.0	−24.1	5.0	4.0	−0.3	−29.3	−31.7

In January, retrofitting district heated or wood heated buildings increased median electric power demand by 33 to 108% or 1.3 to 3.7 W/m^2. The increase was due to the air-source heat pumps included in all the retrofitted cases. Switching from a wood boiler to GSHP increased median power demand by 95 to 272%. This was 5.1 to 8.5 W/m2 for the D level retrofit and 3.7 to 5.2 W/m^2 for the B level retrofit. The increase was smaller for the new buildings SH3 and SH4. The maximum power demand increase in the switch from wood to GSHP was much larger. Level D retrofit increased maximum power by 23 to 32 W/m^2 and level B retrofit increased it by 9 to 14 W/m^2. In July the difference between peak and median demand is smaller. The largest effect is the solar electricity production, which is seen as highly negative minimum and sustained bottom power demand.

In January, the median electricity demand in the electrically heated buildings was reduced by as much as 22 W/m^2, while the maximum power demand was reduced by up to 30 W/m^2. In buildings with GSHP, retrofit B reduced maximum power demand by 9 to 35 W/m^2 (36 to 68%), depending on the building age. Median power demand was reduced by only 2 to 7 W/m^2. The large difference between these changes was caused by electric backup heating, which is only used during peak demand hours. In the GSHP B cases, the heat pump capacity was large enough to meet all loads without backup heating.

The results for the building stock scenarios (as described in Section 3.5.) are shown at the end of Table 6. In the detached house building stock, electrically heated buildings (with heat pumps or direct electricity) make up a significant portion of the houses, especially in the retrofit scenarios (as described in Section 3.5). The maximum combined specific power of the building stock was below that of GSHP buildings only, but above that of wood and DH heated buildings. The maximum power in the Retrofit D scenario was increased vs. the Original case, but remained on the original level in the Retrofit B scenario. The retrofits in electrically heated buildings reduced the power demand enough to compensate for the higher penetration of heat pumps. The median electrical power demand in January was 13.8 W/m^2 for the Original and Retrofit D scenarios and was reduced to 11.6 W/m^2 in the Retrofit B scenario. In July, the power demand never rose above 7 W/m^2.

4. Discussion

Air-to-air or ground-source heat pumps were utilized in all optimal retrofit solutions. Thus, some buildings will increase their electric power demands on the grid. Other studies have therefore assumed that new heat pumps cannot use the average low-emission electricity of the grid and would instead need to utilize the marginal production that is typically high emission coal generation [25]. However, the issue can be bypassed if the existing loads are lowered at the same time as new ones are added. This was the case in this study on the building stock level. New heat pumps increased electricity demand in the retrofitted wood heated or district heated buildings, but this was offset by the same solutions (heat pumps and envelope improvements) reducing electricity demand in the most power intensive buildings that were heated directly with electric radiators or electric boilers.

The optimization of retrofit solutions favored very large PV systems (4400 MW$_p$ and 5600 MW$_p$) for the scenarios aiming for the largest emission reduction. However, the majority of the solar electricity generated by the oversized systems could not be used at the buildings and had to be exported to the grid. The maximum power levels of the exports were several times larger than the peak power demand in building without electric main heating systems. In those cases, the high-power requirement of the solar arrays could be a problem for the distribution grid, if it is not designed to handle such power. However, in electrically heated buildings the peak winter demand was on a similar level as the exports and the grid would presumably be able to handle the loads. However, a study on integration of variable renewables in the Finnish grid estimated that more than 1100 MW$_p$ of solar electricity would decrease wind energy integration potential and significantly increase costs [39]. This shows the need for an additional study that looks at the building stock in more detail, while also including the effects of the national power grid and international transfers through the Nordic electricity market.

With the large amounts of excess solar energy going to the grid, the electricity spot price would likely drop. Solar electricity is produced in all buildings at the same time, so with enough excess power the price could go to zero or even to negative values. This would influence the LCC of the building retrofits, by lowering the lifetime value of solar electricity generation. Thus, if large scale retrofitting was done, the cost-optimal PV array size would go down. To avoid this, more ways to use the electricity are needed. Communities could use seasonal thermal energy storage to shift the use of electricity in summer to meet heating needs in winter. For example, solar electricity combined with borehole thermal energy storage for Finnish conditions was examined in [40]. Typically, demand response and short-term thermal energy storage in water tanks is also useful for increasing the value of solar electricity [41,42], but in the retrofit cases of the current study, solar thermal collectors were also included and handled most of the heating demand in summer. District heating could be produced with heat pumps [43]. Totally new uses for electricity are likely to appear. For example, the number of new electric car registrations in Finland has almost tripled in a year, though the absolute numbers are still low [44]. Other uses for excess electricity are in the energy intensive Finnish industry [45] or synthetic fuel production (also known as power-to-X) [46].

When GSHP was utilized, the annual peak electricity demand was significantly lower for the retrofit B cases than the original or retrofit D cases. This was due to higher heat pump thermal power capacity relative to the heating demand. When the heat pumps were sized to 60% or so of peak demand, electric backup heaters saw more use. Sizing the heat pump to above 90% made the peak electricity demand drop, also making the daily variance in demand smaller. This helps in sizing the electricity distribution infrastructure and designing energy storage systems. It is easier to optimize an energy storage system for power or energy capacity compared to having to maximize both.

The energy demand data for all the buildings were obtained through simulation. While dynamic simulation with IDA-ICE has been shown to be accurate, the results are sensitive to the background assumptions. Different age classes of single-family houses were modelled, but the shape and size of the basic building was the same for every case. The results could be different for smaller houses. In addition, only the southern climate zone of Finland was used for weather input data, creating a southern bias in the data. Further north, heating demand would be higher while solar energy generation would suffer. However, the majority of houses are located in the southern zone. Doing detailed calculations for two more climate zones would have tripled the number of cases and the need for time-consuming optimization. The results were obtained using the Test Reference Year 2012. Since building retrofitting is a long-term investment, climate change can influence the energy demand of the buildings during the lifetime of the buildings, as we move towards the year 2050. Cooling demand was ignored in this study, but it could be that as air-to-air heat pumps become more common, people will start using them for cooling as well, even though the heat pumps were purchased mainly for reducing heating expenses. This would increase the electric loads during summer, though this increase could mostly be mitigated by the increased amount of solar power.

The changes in the building stock were accounted for in a simplistic way, assuming all buildings are immediately retrofitted. In practice, many buildings in regions with declining populations and house values would likely not be retrofitted, due to the resident's unwillingness to do long-term investments. A separate study is needed to calculate more feasible retrofit pathways, taking into account that change happens gradually and that new buildings are added while some old buildings are completely torn down. No flexibility or demand response methods were utilized, which removes the balancing element that appears when a large amount of buildings with different use profiles and energy storage systems are combined. In practice, on the building stock level, the peak power demand could thus be expected to be lower than in the cases presented in this study. The houses were assumed to be oriented south for solar energy purposes. In practice, some buildings are oriented badly, receive a lot of shading or are otherwise not suitable for solar energy installations. Thus, only a fraction of houses would be feasible for solar energy production.

There is uncertainty in the heating systems in use in the current building stock. Building owners do not always report changes to their heating system, such as when replacing an oil boiler with a heat pump. Some wood-heated buildings might actually use wood only as a backup energy source, while others use it as the main heat source. Thus, the real distribution of heating systems is not known. Possible changes to the electricity use of equipment and appliances in the buildings were not considered in the study. On the one hand, old appliances are gradually upgraded into more energy-efficient devices, which should reduce electricity consumption, but on the other hand, people are adding new electricity consuming equipment, which increases power demand. These trends could have an influence on the heating demand of buildings in the future through the excess heat they release.

5. Conclusions

Analyzing the hourly power demand of buildings helps in planning future generation capacity and backup and energy storage investments. The hourly heating and electric power demand of the Finnish detached house building stock was simulated using four different age categories of buildings, four different main heating systems and three levels of energy performance (reference, low cost retrofit D, and high impact retrofit B). Energy retrofits to improve energy efficiency had a significant effect on the peak and average power demand in all examined buildings. The main contribution of this paper was to show the power demand distribution before and after retrofits. Typically retrofit studies only show the effects of retrofits on the annual level, but this study presented the seasonal changes in power demand, to better understand what additional changes to the energy system are needed inside and outside the building sector. Another important contribution was the presented estimate of the net change in power demand in the building stock level if large-scale building energy retrofits are done.

The lower emissions of electricity compared to on-site boilers or district heating favor electrification of heating, through the use air-source heat pumps. This resulted in increased electricity demand in buildings with district heating or on-site wood boilers. At retrofit level B, the peak power demand of these building rose by 60 to 70%, but the absolute impact was low. On the other hand, buildings with direct electric heating significantly lowered their demand through the retrofits (peak demand down by 27 to 40% in retrofit B), as did buildings with ground-source heat pumps (peak demand down by 36 to 68%), with significant absolute impact.

These effects were combined in scenarios where all single-family houses of the whole building stock were retrofitted, which resulted in a net decrease in annual electricity use, −11% for low cost retrofits (scenario D), and −38% for high impact retrofits (scenario B). On the building stock level, peak power demand increased by 19% for low cost retrofits, but remained unchanged for the combined high impact retrofits. However, it is not likely that all buildings could be retrofitted in the same way in practice, due to both social and technical issues related to different conditions in each building.

The optimal solar electricity generation capacity on the individual building level was high. When the individual optima were utilized in the whole building stock, the peak excess power of solar electricity was 3.5 GW for the low cost retrofit scenario and 5 GW for the high impact retrofit. Such high values for unnecessary power generation could be difficult for the grid to handle. Such a scenario is also sensitive to price assumptions and might not be feasible if increasing excess production were to reduce solar energy value. This calls for further research on the optimization of individual building retrofits together with the power system as a whole. Future studies need to combine the changes in buildings and conventional power sector, as well as include new potential ways to use the available renewable energy. Seasonal thermal energy storage could be one way to solve the problem of overproduction, along with electric cars and power-to-X technologies.

Retrofitting old detached houses in Finland can reduce emissions significantly by improving thermal insulation values and by utilizing electrified heating with air-source or ground-source heat pumps. Fears of increasing the marginal electricity demand seem to be unfounded. While the amount of heat pumps is increased, reducing the energy demand in buildings with direct electric heating can prevent both the total electricity demand and peak power demand from rising at the building stock level. This bodes well for major retrofit projects based on electrification of the heating market. However, more accurate modelling of the building stock is needed. A future study should consider how the Finnish building stock could realistically be retrofitted, taking into account both the addition and removal of buildings as well as regional trends in population and economic activity.

Author Contributions: Conceptualization, J.H., J.J. and R.K.; methodology, J.H., J.J. and R.K.; software, J.H.; validation, J.H., J.J. and R.K.; formal analysis, J.H.; investigation, J.H.; resources, J.H., J.J. and R.K.; data curation, J.H. and J.J.; writing—original draft preparation, J.H.; writing—review and editing, J.H., J.J. and R.K.; visualization, J.H.; supervision, J.J. and R.K.; project administration, R.K.; funding acquisition, R.K. All authors have read and agreed to the published version of the manuscript.

Funding: This research was funded by the Academy of Finland, grant number 309064. The study was made as part of the project Optimal transformation pathway towards the 2050 low-carbon target: integrated buildings, grids and national energy system for the case of Finland. The study received further funding from the FINEST Twins project, that is co-funded by the European Union (Horizon 2020 programme, grant number 856602) and the Estonian government.

Conflicts of Interest: The authors declare no conflict of interest. The funders had no role in the design of the study; in the collection, analyses, or interpretation of data; in the writing of the manuscript, or in the decision to publish the results.

References

1. European Commission. EU Climate Action-2050 Long-Term Strategy. 2016. Available online: https://ec.europa.eu/clima/policies/strategies/2050_en (accessed on 4 January 2020).
2. Kavvadias, K.; Jiménez-Navarro, J.P.; Thomassen, G. Decarbonising the EU Heating Sector - Integration of the Power and Heating Sector. *Publications Office of the European Union, EUR 29772 EN*. 2019. Available online: https://publications.jrc.ec.europa.eu/repository/bitstream/JRC114758/kjna29772enn.pdf (accessed on 7 February 2020).
3. European Parliament. *Directive 2010/31/EU of the European Parliament and of the Council of 19 May 2010 on the Energy Performance of Buildings*; European Parliament: Brussels, Belgium, 2010.
4. European Parliament. *Directive (EU) 2018/844 of the European Parliament and of the Council Amending Directive 2010/31/EU on the Energy Performance of Buildings and Directive 2012/27/EU on the Energy Efficiency*; European Parliament: Brussels, Belgium, 2018.
5. Ministry of the Environment. 4/13 Asetus Rakennuksen Energiatehokkuuden Parantamisesta Korjaus-ja Muutostöissä (4/13 Decree on Improving Energy Performance in Renovations). 2013. Available online: http://www.ym.fi/download/noname/%7B924394EF-BED0-42F2-9AD2-5BE3036A6EAD%7D/31396 (accessed on 17 January 2020).
6. Statistics Finland. Number of Buildings by Intended Use and Year of Construction on 31 Dec. 2015. 2016. Available online: http://pxnet2.stat.fi/PXWeb/pxweb/en/StatFin/StatFin__asu__rakke/?tablelist=true (accessed on 10 November 2019).
7. Oberegger, U.F.; Pernetti, R.; Lollini, R.; Ulrich, F.O.; Roberta, P.; Roberto, L. Bottom-up building stock retrofit based on levelized cost of saved energy. *Energy Build.* **2020**, *210*, 109757. [CrossRef]
8. Ekström, T.; Blomsterberg, Å. Renovation of Swedish Single-family Houses to Passive House Standard–Analyses of Energy Savings Potential. *Energy Procedia* **2016**, *96*, 134–145. [CrossRef]
9. Moran, P.; O'Connell, J.; Goggins, J. Sustainable energy efficiency retrofits as residenial buildings move towards nearly zero energy building (NZEB) standards. *Energy Build.* **2020**, *211*, 109816. [CrossRef]
10. Niemelä, T.; Kosonen, R.; Jokisalo, J. Cost-effectiveness of energy performance renovation measures in Finnish brick apartment buildings. *Energy Build.* **2017**, *137*, 60–75. [CrossRef]
11. Hirvonen, J.; Jokisalo, J.; Heljo, V.J.; Kosonen, R. Towards the EU emissions targets of 2050: Optimal energy renovation measures of Finnish apartment buildings. *Int. J. Sustain. Energy* **2018**, *38*, 649–672. [CrossRef]
12. Niemelä, T.; Levy, K.; Kosonen, R.; Jokisalo, J. Cost-optimal renovation solutions to maximize environmental performance, indoor thermal conditions and productivity of office buildings in cold climate. *Sustain. Cities Soc.* **2017**, *32*, 417–434. [CrossRef]
13. Hirvonen, J.; Jokisalo, J.; Heljo, J.; Kosonen, R. Towards the EU Emission Targets of 2050: Cost-Effective Emission Reduction in Finnish Detached Houses. *Energies* **2019**, *12*, 4395. [CrossRef]
14. Paiho, S.; Pulakka, S.; Knuuti, A. Life-cycle cost analyses of heat pump concepts for Finnish new nearly zero energy residential buildings. *Energy Build.* **2017**, *150*, 396–402. [CrossRef]
15. Grimm, V.; Zoettl, G. Investment Incentives and Electricity Spot Market Competition. *J. Econ. Manag. Strat.* **2013**, *22*, 832–851. [CrossRef]
16. Grimm, V.; Martin, A.; Schmidt, M.; Weibelzahl, M.; Zöttl, G. Transmission and generation investment in electricity markets: The effects of market splitting and network fee regimes. *Eur. J. Oper. Res.* **2016**, *254*, 493–509. [CrossRef]
17. Chen, Y.; Peng, X.; Gu, J.; Schmidt, F.P.; Li, W. Measures to improve energy demand flexibility in buildings for demand response (DR): A review. *Energy Build.* **2018**, *177*, 125–139. [CrossRef]
18. Xu, L.; Wang, S.; Xiao, F. An adaptive optimal monthly peak building demand limiting strategy considering load uncertainty. *Appl. Energy* **2019**, *253*, 113582. [CrossRef]
19. Deng, X.; Lv, T. Power system planning with increasing variable renewable energy: A review of optimization models. *J. Clean. Prod.* **2020**, *246*, 118962. [CrossRef]
20. Qu, K.; Chen, X.; Ekambaram, A.; Cui, Y.; Gan, G.; Økland, A.; Riffat, S. A novel holistic EPC related retrofit approach for residential apartment building renovation in Norway. *Sustain. Cities Soc.* **2020**, *54*, 101975. [CrossRef]
21. Garriga, S.M.; Dabbagh, M.; Krarti, M. Optimal carbon-neutral retrofit of residential communities in Barcelona, Spain. *Energy Build.* **2020**, *208*, 109651. [CrossRef]

22. Rakhshan, K.; Friess, W.A. Effectiveness and viability of residential building energy retrofits in Dubai. *J. Build. Eng.* **2017**, *13*, 116–126. [CrossRef]
23. Hirvonen, J.; Jokisalo, J.; Heljo, V.J.; Kosonen, R. Effect of apartment building energy renovation on hourly power demand. *Int. J. Sustain. Energy* **2019**, *38*, 918–936. [CrossRef]
24. Patronen, J.; Kaura, E.; Torvestad, C. *Nordic Heating and Cooling*; Nordic Council of Ministers: Copenhagen, Denmark, 2017.
25. Dodoo, A. Techno-economic and environmental performances of heating systems for single-family code-compliant and passive houses. *E3S Web Conf.* **2019**, *111*, 03039. [CrossRef]
26. EQUA Simulation. IDA ICE-Simulation Software. 2019. Available online: https://www.equa.se/en/ida-ice (accessed on 8 August 2019).
27. EQUA Simulation. *Validation of IDA Indoor Climate and Energy 4.0 with Respect to CEN Standards EN 15255-2007 and EN 15265-2007*; EQUA Simulation: Solna, Sweden, 2010.
28. Loutzenhiser, P.; Manz, H.; Maxwell, G. *Empirical Validations of Shading/Daylighting/Load Interactions in Building Energy Simulation Tools*; IEA-International Energy Agency: Zürich, Switzerland, 2007.
29. Koivuniemi, J. Lämpimän käyttöveden mitoitusvirtaama ja lämpötilakriteerit veden mikrobiologisen laadun kannalta kaukolämmitetyissä taloissa Domestic Hot Water Design Flow and Temperature Criteria as Pertains to Water Microbiological Quality in District Heated Houses. In Finnish. M.Sc. Thesis, Helsinki University of Technology, Espoo, Finland, 2005.
30. Degefa, M.Z. *Electrical Load Disaggregation-A Project Report of SGEM*; SGEM: Espoo, Finland, 2012.
31. Palonen, M.; Hamdy, M.; Hasan, A. MOBO a new software for multi-objective building performance optimization. Presented at the 13th Conference of International Building Performance Simulation Association, Chambéry, France, 26–28 August 2013.
32. Deb, K.; Pratap, A.; Agarwal, S.; Meayarivan, T. A fast and elitist multiobjective genetic algorithm: NSGA-II. *IEEE Trans. Evol. Comput.* **2002**, *6*, 182–197. [CrossRef]
33. Saari, A.; Airaksinen, M. *Energiatehokkuutta koskevien vähimmäisvaatimusten kustannusoptimaalisten tasojen laskenta*; [In Finnish]; EU Commission: Brussels, Belgium, 2012. Available online: https://docplayer.fi/388965-Energiatehokkuutta-koskevien-vahimmaisvaatimusten-kustannusoptimaalisten-tasojen-laskenta-suomi.html (accessed on 7 April 2020).
34. Ministry of the Environment. Directive on Building Energy Certificates, Attachment 1 [In Finnish]. 2017. Available online: https://www.finlex.fi/data/sdliite/liite/6822.pdf (accessed on 12 December 2018).
35. Ministry of the Environment. Ympäristöministeriön asetus uuden rakennuksen energiatehokkuudesta (1010/2017) (Decree of the Ministry of the Environment on the Energy Performance of the New Building (1010/2017) [In Finnish]. 2018. Available online: https://www.finlex.fi/fi/laki/alkup/2017/20171010 (accessed on 12 January 2020).
36. Kalamees, T.; Jylhä, K.; Tietäväinen, H.; Jokisalo, J.; Ilomets, S.; Hyvönen, R.; Saku, S. Development of weighting factors for climate variables for selecting the energy reference year according to the EN ISO 15927-4 standard. *Energy Build.* **2012**, *47*, 53–60. [CrossRef]
37. Finnish Meteorological Institute. Test Reference Years for Energy Calculations [In Finnish]. 19 November 2018. Available online: https://ilmatieteenlaitos.fi/energialaskennan-testivuodet-nyky (accessed on 2 April 2019).
38. Finnish Energy. Aurinkosähkön tuotantokapasiteetti lisääntyi 82% vuodessa. Energiavirasto. Available online: https://energiavirasto.fi/tiedote/-/asset_publisher/aurinkosahkon-tuotantokapasiteetti-lisaantyi-82-vuodessa (accessed on 13 February 2020).
39. Zakeri, B.; Syri, S.; Rinne, S. Higher renewable energy integration into the existing energy system of Finland–Is there any maximum limit? *Energy* **2015**, *92*, 244–259. [CrossRef]
40. Hirvonen, J.; Sirén, K. A novel fully electrified solar heating system with a high renewable fraction—Optimal designs for a high latitude community. *Renew. Energy* **2018**, *127*, 298–309. [CrossRef]
41. Hirvonen, J.; Kayo, G.; Hasan, A.; Sirén, K. Zero energy level and economic potential of small-scale building-integrated PV with different heating systems in Nordic conditions. *Appl. Energy* **2016**, *167*, 255–269. [CrossRef]
42. Schibuola, L.; Scarpa, M.; Tambani, C. Demand response management by means of heat pumps controlled via real time pricing. *Energy Build.* **2015**, *90*, 15–28. [CrossRef]

43. Arabkoohsar, A.; AlSagri, A.S. A new generation of district heating system with neighborhood-scale heat pumps and advanced pipes, a solution for future renewable-based energy systems. *Energy* **2020**, *193*, 116781. [CrossRef]
44. Traficom, Uusien henkilöautojen ensirekisteröinnit laskussa–sähköautojen ja käytettyjen ladattavien hybridien määrät kasvussa, Traficom. Available online: https://www.traficom.fi//fi/ajankohtaista/uusien-henkiloautojen-ensirekisteroinnit-laskussa-sahkoautojen-ja-kaytettyjen (accessed on 13 February 2020).
45. Rehfeldt, M.; Worrell, E.; Eichhammer, W.; Fleiter, T. A review of the emission reduction potential of fuel switch towards biomass and electricity in European basic materials industry until 2030. *Renew. Sustain. Energy Rev.* **2020**, *120*, 109672. [CrossRef]
46. Hänggi, S.; Elbert, P.; Bütler, T.; Cabalzar, U.; Teske, S.; Bach, C.; Onder, C. A review of synthetic fuels for passenger vehicles. *Energy Rep.* **2019**, *5*, 555–569. [CrossRef]

Article

A Smart Hybrid Energy System Grid for Energy Efficiency in Remote Areas for the Army

Umberto Berardi [1,*], Elisa Tomassoni [1,2] and Khaled Khaled [1]

1 Department of Architectural Science, Ryerson University, Toronto, ON M5B 2K3, Canada; e.tomassoni@univpm.it (E.T.); kkhaled@ryerson.ca (K.K.)
2 Dipartimento di Ingegneria Civile, Edile e Architettura DICEA, Universita' Politecnica delle Marche, 60131 Ancona, Italy
* Correspondence: uberardi@ryerson.ca; Tel.: +1-416-979-5000 (ext. 553263)

Received: 14 March 2020; Accepted: 1 May 2020; Published: 5 May 2020

Abstract: The current energy inefficiencies in relocatable temporary camps of the Armed Force troops create logistic challenges associated with fuel supply. The energy needs of these camps are primarily satisfied by diesel engine generators, which imply that a significant amount of fuel needs to be continuously provided to these camps, often built in remote areas. This paper presents an alternative solution, named Smart Hybrid Energy System (SHES), aiming towards significantly reducing the amount of fuel needed and minimizing transportation logistics while meeting camp energy demands. The SHES combines the existing diesel generators with solar power generation, energy storage, and waste heat recovery technologies, all connected to a microgrid, ensuring uninterrupted electricity and hot water supplies. All components are controlled by an energy management system that prioritizes output and switches between different power generators, ensuring operation at optimum efficiencies. The SHES components have been selected to be easily transportable in standard shipping 20 ft containers. The modularity of the solution, scalable from the base camp for 150 people, is designed according to available on-site renewable sources, allowing for energy optimization of different camp sizes in different climates.

Keywords: hybrid energy system; energy efficiency; microgrid; military applications; renewable energy; remote areas

1. Introduction

The Armed Forces operate in remote locations for training and military operations, even under natural disaster conditions or in foreign territories during conflicts, and must be ready to deploy on short notices, in any climate and for prolonged periods. As such, they currently rely on relocatable temporary camps (RTCs) for their deployments through extreme operational and environmental conditions. To sustain operations, as there is no utility grid, RTCs depend on logistics for the continuous supply of fossil fuel (primarily diesel) as the main source of energy. Inefficiencies in current practices lead to vulnerabilities in energy infrastructures, such as shortfalls in power generation and higher requirements for fuel resupply, with the knock-on effect of greatly increasing the transportation logistics during operations. Moreover, RTCs typically use spot generation by connecting loads to a common set of generators, where each generator is oversized to satisfy peak loads, even when these loads are infrequent. Consequently, generators typically are selected at a significantly higher capacity, resulting in an inefficient and costly source of power, increased maintenance, and wet sacking, a condition resulting from poor fuel combustion.

In recent years, military engineers have therefore encountered several operational challenges associated with energy logistic convoys and infrastructure, limited supplies, and climate change. Scientific literature identified a spectrum of approaches and technologies to address energy consumption under these conditions. Few combinations of components have been proposed according to the site-specific characteristics [1–3], however, the definition of further integrated configurations remains rarely investigated, although it is evident that the Armed Forces could benefit from holistically assessing these approaches as integrated systems.

Significant gains in the efficiencies of RTC utility systems (renewable energy systems; improved generators and energy storage or grid efficiency) and energy conservation measures (e.g., insulation of the camp tent fabric, building controls, etc.) would have an overall increasing benefit on the deployed operations. Meanwhile, stand-alone hybrid energy systems have been proposed as valuable means of supplying energy to remote areas, such as isolated rural villages [4–7], and for various other purposes, such as medical clinic practices [8] or military operations [9–11].

Some researchers investigated solutions aiming at reducing the dependency on fossil fuels during prolonged emergencies by proposing self-contained demonstration units that make use of hybrid generation from solar, wind, and biomass and, minimally, fossil sources [12]. Some of these systems have already been introduced to the market, as described below. Besides microgrids, clusters of electricity sources and load operating systems are being used to improve the reliability of electrical grids, manage the addition of distributed clean energy resources like wind and solar photovoltaic generation, reduce fossil fuel emissions, and provide electricity in areas not served by centralized electrical infrastructure [13].

Some models described the components of a microgrid [5,9,14,15], but not much is known about its behavior as a whole system. Some studies aimed to model microgrids at steady-state and study their transient responses to changing inputs [16]. However, researchers have built a full-scale microgrid model, including the power sources, power electronics, and load and mains models [5].

One of the main challenges towards the development of isolated microgrids is the management of various devices and energy flows to optimize their operations, particularly regarding the hourly loads and the availability of power produced by renewable energy systems. Energy management systems could be a solution to tackle these issues [16–18]. Regarding the provision of energy services with modular and transportable systems by making use of microgrid technology, some examples can be found in the market. For example, examples of possible technical solutions include the following:

- Power Box Containers by Out of the Box Energy Solutions (http://www.outofthebox.energy/power-box/powerbox-containers/)
- Energy Containers by Intech Clean Energy (https://www.intechcleanenergy.ca/energy-container/)
- Hybrid Smart Total by Golden Peniel Limited (http://gplnigeria.com/hybrid.html)
- Container Box Systems by Hakai Energy Solutions (https://www.hakaienergysolutions.com/services/container-box-systems/)
- Multi Box Microgrid by BoxPower (https://boxpower.io/products/multi-box-microgrid/)
- PowerPlus Hybrid Power Generator by Firefly (https://www.fireflyhybridpower.com/products/powerplus/)
- ES Box by Schneider Electric (https://solar.schneider-electric.com/product/es-box/)

The Cross-Power unit, e.g., uses modular hybrid wind and solar systems, integrated with battery storage, to produce electricity in remote locations. However, most of the existing solutions use black box intelligent energy management systems to ensure a continuous supply and avoid shortfalls in power generation.

This paper presents a scalable and transportable solution, named Smart Hybrid Energy System (SHES), for providing energy-efficient services to soldiers in protracted displacement situations. The SHES combines the existing diesel generators with solar power generation, energy storage, and waste heat recovery technologies, all connected to a microgrid, ensuring uninterrupted electricity and hot water supplies. The reliable and energy-efficient system helps to manage generator output. By transforming an independently operating system of generators into a demand-managed microgrid, SHES provides power only where and when it is needed, instead of completely relying on fuel-burning generators. The system also provides the Armed Forces with critically needed power surety by utilizing intelligent load management technologies to prevent grid collapse in the event of generator fault, as the SHES prevents a stoppage of energy flow by shifting demand onto supporting generators if one generator fails. The system is designed to manage the energy needs of a 150 to 1500-person RTC, operating in a temperate climate zone and allowing for the occasional deployment to extremely hot or cold climatic zones. Finally, this paper considers the energy savings achievable through technologies that improve the accommodation's insulation, such as a thermoreflective multilayer system developed for emergency architecture, or that provide additional layers of solar protection, reducing the heat transfer through the shelter exterior thus reducing the daily air conditioning loads and reliance on diesel fuel [19,20].

2. Methodology

Regarding the design criteria of the RTCs solution, different technologies for energy production and storage concerning containerized solutions for emergencies were analyzed for the SHES. The equipment was selected from a range of commercial products based on sizing calculations and container space. The travel weight and volume, logistical support, required maintenance, and any hazards associated with the systems were also considered. Finally, for each of the selected technologies, detailed design work was conducted.

The system was proposed to the Canadian Armed Forces, and as such, the annual energy performance reported in this study was analyzed for the temperate climate zone of Brandon (Manitoba, Canada) at 49.85° N. Through dynamic energy and energy management simulations with a combination of software including the DoE Energy Plus and HOMER Pro software, the performance of the SHES system was analyzed.

First, an energy model reproducing the existing baseline 150-person military camp was created using Energy Plus, and data related to geometry, constructions, occupancy, HVAC, lighting, equipment, operation, climate, and energy management system (EMS) was assigned. Furthermore, the energy model was calibrated to match the actual net energy and heating energy consumptions of the RTCs and deriving and collating data from past RTC deployments as a reference for the design process. For this purpose, the net energy consumption was described as the combined energy consumptions of electricity generation and diesel-heating equipment. Second, the Energy Plus-generated electric and thermal hourly load profiles were imported to HOMER Pro, a microgrid design, simulation, and optimization tool, used for the purposes as a design and investment decision support tool for selecting the optimal portfolio, sizing, placement, and dispatch of the multiple energy sources feeding the decentralized energy system and serving the camp loads. HOMER Pro was also used in performing sensitivity analyses to identify the most cost-effective system configuration at various fuel costs and nominal discount rates.

The dynamic studies made it possible to conduct comparisons between different utility systems scenarios, comprising multiple distributed energy resources and energy conservation measures (i.e., advanced insulation materials), over the current base camp practices. Further simulations were subsequently made on the most cost-effective proposed solution to estimate the annual fuel use and energy savings for different climate zones.

In order to evaluate the economic and technical feasibility of the many options and to account for variations in technology costs and energy resource availability, the operation of the different system configurations was simulated in HOMER Pro by performing dynamic energy balance. For each time step and for each system configuration considered, HOMER Pro compared the electric and thermal demand to the energy that the system can supply and calculated the flow of energy to and from each component of the system. In each time step, the analyses focused on how to operate the generators and whether to charge or discharge the batteries and determined whether a configuration is feasible. The study also looked at the system cost calculations in terms of capital, replacement, operation and maintenance, fuel, and interest rates. Furthermore, HOMER Pro used optimization algorithms to search for the most cost-effective system configuration in terms of net present cost (life cycle cost). Details about the calculations are available in the HOMER Pro user manual [21].

It is important to note that the lowest net present cost did not necessarily indicate the lowest energy consumption; however, incorporating the operational costs in the calculation, it was used to find the optimal system design rather than net energy consumptions.

2.1. Case Study

RTCs provide accommodations, administration shelters, ablutions, maintenance, storage, hangar, and kitchen facilities. RTCs generally include tents identical to those deployed by militaries throughout the world (Figure 1). The population supported by RTCs can vary considerably from day to day, depending on operational activities as well as surges due to the rotation of personnel. Deployed personnel are provided a bed space in a tent that holds 4–10 people. The accommodations are located in a condensed area adjacent to the ablutions. Ablutions units are composed of a shower, a toilet, and a sink, each unit serving 10 members of the camp population.

Figure 1. Tent demonstration unit in Toronto, ON, built within Ryerson University (**left**) and full assembly of relocatable temporary camp (RTC) (**right**).

Multiple, independent systems are in place for electrical power generation and distribution, heating and cooling, storage and distribution of fuel and water, and waste disposal. These systems are not designed to promote energy efficiency. The current energy management approach in RTCs relies on diesel-powered generators for electricity production. Electrical energy is provided to the camp via multiple single-speed generator farms that incorporate variants of 300, 350, and 500 kW generators.

To avoid low load operation, load banks are employed to keep the generators running at optimal conditions and efficiency points. Excess electricity not required in the camp is ultimately diverted to a load bank where it is converted to waste heat. In the current "baseline" scenario, diesel-fired space heaters are used for heating. Cooling is provided by electric environmental conditioning units. Heating and cooling units are attached to each tent and are controlled by individual users.

2.2. Energy Modeling

The virtual model reproducing the existing base camp (baseline) was sized for accommodating 150-persons and included 15 accommodations, each hosting 10 persons, 15 ablutions, administration, maintenance, laundry and storage hangar, kitchen, and dining facilities (Table 1). Each accommodation tent was 5.80 m length and 5.10 m width, with a medium internal height of 2.30 m. Data have been obtained in line with the technical sheet of the model by Montana 29 tent, Ferrino (Figure 2). The relative U-values of the envelope reported in Table 2 were estimated assuming that the standard shelter system was a canvas tent with low thermal resistance, in agreement with the literature [19].

Table 1. Modeled zone properties of the military base camp.

Zone	Occupancy	Plug Loads	Exhaust Air Rate	Heating/Cooling Set Point	Target Illuminance
	(m^2/Person)	(W/m^2)	(m^3/s)	(°C)	(Lux)
Accommodation	4	5	–	20/24	300
Kitchen	9	100	0.21	17/21	500
Dining hall	1.4	20	–	20/24	300
Ablution	2.4	2	0.0125	20/24	200
Admin office	20	15	–	20/24	400
Maintenance	20	10	3.15	NA/24	400
Storage	3	-	–	NA/24	200
Laundry	10	70	–	20/24	300

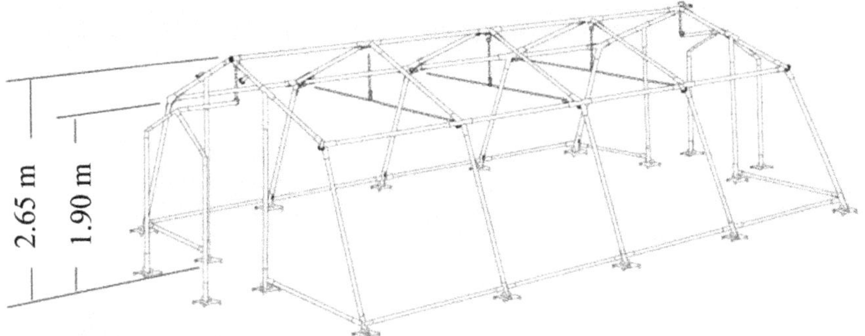

Figure 2. Geometric characteristic of the modeled tent.

Table 2. Modeled envelope construction: thermal properties of the fabric tent of the military shelters.

Component	Construction	Overall U-Value ($W/m^2{\cdot}K$)
Floor	50.8 mm OSB + RSI-0.175 standard insulation	1.32
Window	6 mm clear sheet	5.78
Wall, roof, door	RSI-0.175 standard insulation	5.71

Camp electric power capacity was sized for 1.5 kW/person, considering data related to past RTC deployments. Specific electrical load profiles were also taken into consideration. Generators were sized with a 10% overload and a further 10% expansion capability factor.

Typical fuel consumption for a 1.5 kW/person load provision was approximately 2000 L of diesel per person per year. Besides, it was assumed that 500 L of diesel fuel per person per year was used for direct combustion, which after accounting for an 80% efficiency, provided 15 GJ of energy for space heating.

Figure 3 shows the detailed HVAC system and power generation scheme of the base camp, which, for a 150-person camp size, has a total diesel consumption including heating of 2500 L/person*year, has a heating diesel consumption of 500 L/person*year, and allows a hot water consumption of 30 L/person*day.

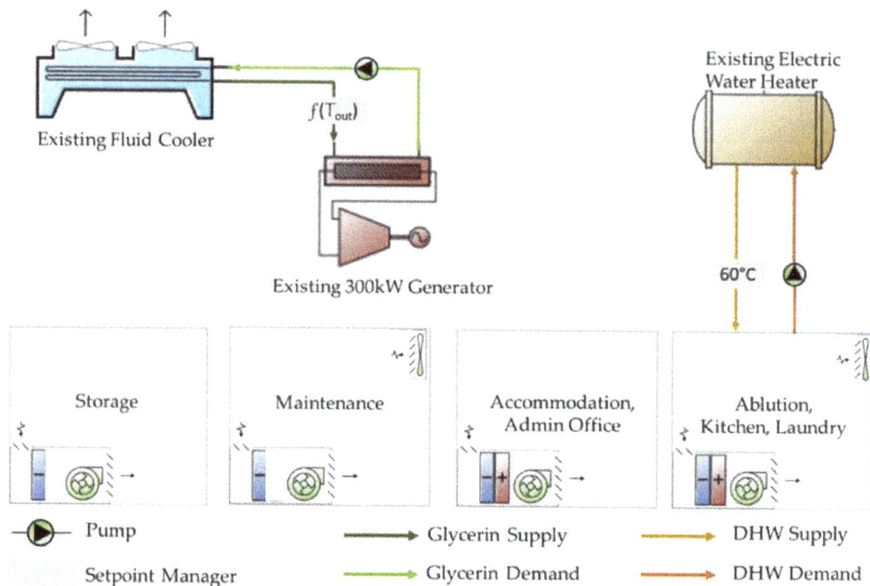

Figure 3. Baseline camp: HVAC system and power generation scheme for a base camp.

Solution designs were also conceived to operate for prolonged periods in extremely hot (up to 50 °C) or cold (down to −40 °C) climates, as well as in all the temperate climatic conditions. Temperature, precipitation, daylight, and wind data of a temperate climate zone, Brandon, Manitoba, Canada, was used for all design calculations as required by the Canadian Defense Department. Further simulations were made to analyze the proposed system performance in different climate zones with severe conditions, including: Vancouver (British Columbia, Canada), Kanoya (Japan), Churchill (Manitoba, Canada), and Changi (Singapore).

3. Results

3.1. Smart Hybrid Energy System (SHES) Design

The SHES is designed to work in stand-alone mode or connected to the local grid. The SHES combines into a single, integrated system of the following technologies: (a) photovoltaic (PV) array, (b) an energy storage system, (c) existing diesel generators, (d) waste-to-heat energy recovery system (WHRU) for space heating, (e) solar hot water (SHW) system for domestic hot water, and (f) energy management system (EMS) that actively monitors and manages base camp equipment and zones.

Figure 4 shows the SHES schematic configuration of the energy vectors, while Figure 5 shows the detailed HVAC system and power generation scheme of the SHES.

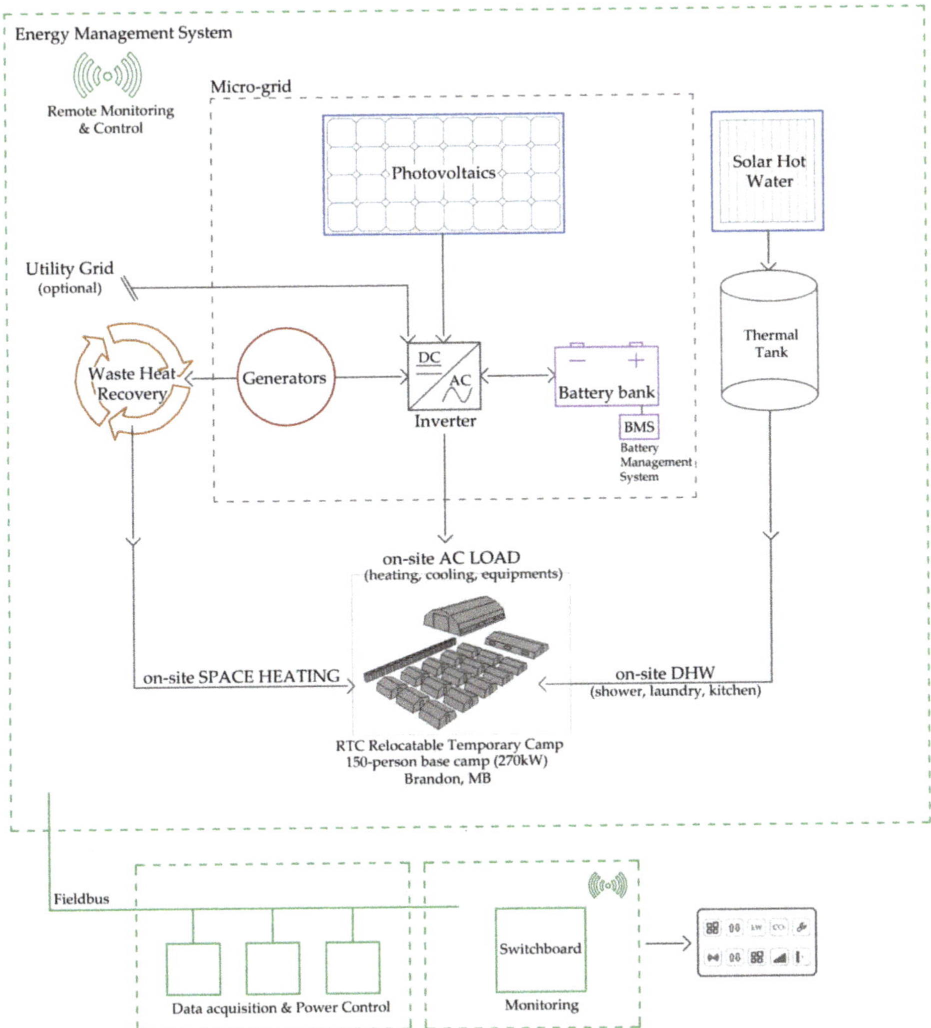

Figure 4. Smart Hybrid Energy System (SHES) schematic configuration of the energy vectors.

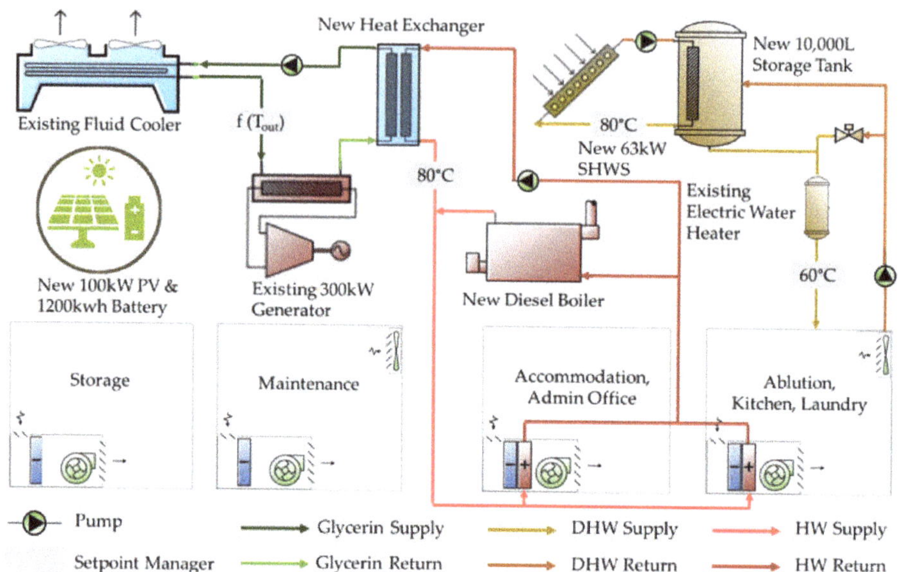

Figure 5. Smart Hybrid Energy System (SHES): detailed HVAC system and power generation scheme.

In particular, the SHES system was composed of the following components:

- PV modules, with a 21% efficiency (SPR-X21-345) and a combined power output of 100 kW with a surface of 446 m^2, are strung together. Each array is south-oriented with an optimized tilt angle, which varies by location, anchored to the ground with a mounting aluminum system. As the base camp loads use alternating current (AC), inverters are used to transform the direct current (DC) produced from PV. It is important to note that to satisfy the entire energy demand of the camp, the PV configuration would have employed a larger system (about four times larger); however, considering the army requirements for continuous transportation and reinstallation in addition to the high initial costs of the system and the spatial limitations, the PV system was sized to sufficiently cover the peak electric loads except for cooling; thus, only systems up to 100 kW were considered.
- Energy storage allows overcoming the intermittent nature of renewable energy sources. When PV systems have a peak above the load, the surplus energy is stored in sodium–sulfur batteries. On the other hand, when there is a high energy demand and the electricity generated by the PVs is insufficient, the batteries are discharged. The cells are monitored and protected by a battery management system (BMS). Various battery models were considered, and the most cost-effective model was selected. In particular, containerized NAS lithium-ion energy storage system by BASF was found to be a cost-effective solution for this project purposes due to its relatively large capacity (1250 kWh), discharge output (286.1 kW max), long duration (4.4 h), long lifespan (20 years or 6,250,000 kWh), reasonable warranty periods (10 years), and physical sizing considerations. A 40% minimum state of charge was assumed, while an 80% state of charge setpoint was found to be the most cost-effective strategy, as discussed later.
- Existing internal combustion diesel generator was integrated into the microgrid to ensure continuous power supply. A 15% minimum part-load ratio was assumed.
- WHRU allows recovering the waste heat from generators to use it for space heating (combined heat and power), which is supplemented by an efficient (89%) diesel boiler. The WHRU consisted

of counter-flow heat exchangers and water-distribution systems, delivering hot water to terminal heating equipment. The terminal heating equipment was modeled as fan–coil units.

- SHW system used has a surface of 90 m^2 and a power of 63 kW that supplies a significant fraction of the base camp domestic hot water (DHW) requirements, while the remaining portion is satisfied by the existing electric water heater. The strategy undertaken allows a 10,000 L storage tank to reach a higher temperature (80 °C) to increase the efficiency of the SHW system.
- EMS controls all components and ensures grid stability, continuously balancing energy generation, and consumption. It maximizes the power output by establishing a hierarchy of sources and prioritizing the use of renewable energy while optimizing the interactions between different components. The EMS is equipped with real-time remote monitoring and controls the base camp parameters, enabling central and informed decisions.

The selection of equipment from a range of commercial products was made based on 20 ft container space (Table 3) in addition to sizing calculations (Table 4) to optimize the annual energy production.

Table 3. Key components dimensions of the Smart Hybrid Energy System (SHES).

Equipment	Example of Possible Component Models	Quantity	Size (mm)*
PV panels	SunPower (SPR-X21-345)	290	1559 × 1046 × 46
Inverter	SUNSYS PCS[2] IM-200kVA-TL	1	805 × 806 × 2150
Battery	NAS	1	6100 × 2400 × 2400
PV/SHW mounting systems	Fast-Rack (GMX Caribou Ground)	2	Various
SHW panels	Ritter Solar (CPC45 Star Azzurro)	20	2426 × 2032 × 121
Insulated storage tank	Niles Steel Tank (JS-36-072)	10	2032 height × 1016 diameter
Heat exchanger	Doucette Industries (CUI 35M5)	1	1346 × 873 × 533

* ISO 6346 [22]: 20 ft container: length = 20 ft (6096 mm); width = 8 ft (2438 mm); height = 8 ft 6″ (2590 mm).

Table 4. Components of the Smart Hybrid Energy System (SHES) for a 150-person relocatable temporary camp (RTC) in Brandon, MB.

Technology	Smart Hybrid Energy System (150-Persons Base Camp)	Sizing Based On
PV panels	Number of collectors = 290	80 kW peak (excluding cooling) electric loads
	Total collectors aperture/Gross area = 446/465 m^2	Spatial limitations
	System size = 100 kW	Logistic requirements
	Optimal collector tilt = 45°, Azimuth =0° S	Spatial limitations
Converter	175 kW (Eff. 96%)	Selected PV size Net present cost
Battery	1250 kWh, Max. discharge output = 286.1 kW Duration =4.4 h	Selected PV size Net present cost
Generator	Existing 300 kW diesel generator	Existing
SHW panels	Number of collectors = 20	30 L/Day. Person
	Total collectors aperture/Gross area= 90/98.8 m^2	60 °C HW temp.
	System size = 63 kW	4 °C supply temp.
	Optimal collector tilt = 45°, Azimuth = 0° S	70% solar fraction for sunniest Month
	Required hot water storage volume = 10,000 L	
	Backed-up by existing electric heater (Eff. 65%)	
Waste heat recovery unit	Counter-flow fluid heat exchangers	1.25 L/s design heat recovery water flow rate

The system is adaptable to different installation and climates, and it was designed such that some existing equipment could be incorporated into the system, thus reducing investment costs for the Army. At the same time, individual components could be integrated without being tied to one manufacturer too. Consequently, the models of the components reported in Table 3 are only indications of possible solutions, as the system could integrate alternative components with comparable performances.

Other renewables can also be optionally integrated into the microgrid according to their on-site availability. The SHES is fully scalable for electric power outputs of 270 kW up to 2.7 MW to meet the varying energy needs from 150 to 1500-persons base camps. Several units of the system, eventually centralized at each tent, can be interconnected to complement the system provided for a 150-person base camp with a larger operational power range. The scaling options include the deployment of additional PV arrays, supplementary battery units, larger power generators, a large-scale WHRU, and a large-scale SHW system.

The SHES design architecture provides redundancy to ensure continuous operation through any subsystem failure, while the microgrid supply power guarantees the longer service and lifespans. The hybrid power generation design prioritizes renewable, followed by battery power, resulting in less generator runtime, thus requiring less maintenance. The EMS allows for identifying operations and maintenance issues before they become problematic, improving problem response time while contributing to the overall system reliability. Furthermore, the central and remote monitoring of the system parameters improves the maintenance supervision, scheduling, and management control. The proposed solution is provided with fire protection and security functions. The SHES incorporates existing technologies and state-of-the-art components; therefore, eventual replacements parts are widely available in the market. The system is prewired, preconfigured, and designed to be rapidly deployed as a plug and play system, following minor on-site assembly.

3.2. Energy Simulation Results

Energy simulations were conducted to quantify the reduced energy demand of the combination of microgrid, renewable resources, and energy recovery measures.

Figure 6 shows the results obtained using different software for the baseline scenario in different climate zones. Maximum differences of 6% and 8% were obtained for the heating and net energy consumptions in Vancouver (BC), respectively. This confirmed the validity of the approaches followed in modeling.

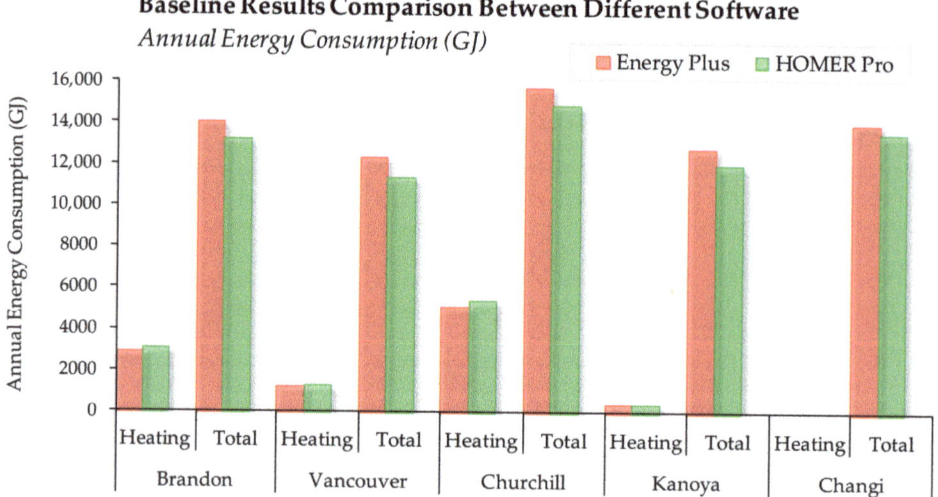

Figure 6. Comparison between Energy Plus and HOMER Pro software for the current base camp practice (Baseline) for a 150-persons base camp in different climate zones.

Table 5 reports the simulation results for the annual energy consumption of a 150-persons RTC in Brandon (MB) according to different scenarios: Baseline, Solar Hot Water, Microgrid: generators + battery, Microgrid with a Waste Heat Recovery (a counter-flow fluid heat exchanger), Microgrid with a PV array (100 kW PV system to cover peak electric loads excluding cooling), and the scenario with all SHES technologies integrated.

Table 5. Simulation results of the annual energy consumption for a 150-persons RTC in Brandon, MB, for the various scenarios with new technology solutions incrementally adopted.

Sc.	Technology	Equipment	Annual Energy Consumptions
1	Baseline	1- Diesel Generator 300 kW + Fluid Cooler 2- Electric Water Heater (Eff. 65%) for DHW 3- Packaged Terminal AC Unit for: A- Electric Space Cooling (COP 2) B- Diesel Space Heating (Eff. 80%)	Net Energy Consumption = 13041 GJ/year Space Heating Energy Consumption = 3034 GJ/year Space Cooling Energy Consumption = 377 GJ/year DHW Energy Consumption = 529 GJ/year
2	Solar Hot Water	1- Diesel Generator 300 kW + Fluid Cooler 2- SHW System + Back-up Existing Electric Water Heater (Eff. 65%) for DHW 3- Packaged Terminal AC Unit for: A- Electric Space Cooling (COP 2) B- Diesel Space Heating (Eff. 80%)	Net Energy Consumption = 13041 GJ/year (−1%) Space Heating Energy Consumption = 3034 GJ/year Space Cooling Energy Consumption = 377 GJ/year DHW Energy Consumption = 99 GJ/year (−81%)
3	Microgrid: generators + battery	1- Diesel Generator 300 kW + Fluid Cooler 2- Electric Water Heater (Eff. 65%) for DHW 3- Energy Storage System (1 NAS Battery) 4- Packaged Terminal AC Unit for: A- Electric Space Cooling (COP 2) B- Diesel Space Heating (Eff. 80%)	Net Energy Consumption = 13139 GJ/year Space Heating Energy Consumption = 3034 GJ/year Space Cooling Energy Consumption = 377 GJ/year DHW Energy Consumption = 529 GJ/year
4	Microgrid: Waste Heat Recovery	1- Diesel Generator 300 kW (0.5 Thermal to Electric Power Ratio) + Fluid Cooler 2- Counter-flow Fluid Heat Exchanger 3- Electric Water Heater (Eff. 65%) for DHW 4- Packaged Terminal AC Unit for: A- Electric Space Cooling (COP 2) B- Space Heating Using Recovered Heat (Eff. 89%)	Net Energy Consumption = 13639 GJ/year (−16%) Space Heating Energy Consumption = 913 GJ/year (−70%) Space Cooling Energy Consumption = 377 GJ/year DHW Energy Consumption = 529 GJ/year
5	Microgrid: PV	1- Diesel Generator 300 kW (Variable Eff.) + Fluid Cooler 2- 100 kW PV System to Cover Peak Electric Loads Except for Cooling + Energy Storage System 3- Electric Water Heater (Eff. 65%) for DHW 4- Packaged Terminal AC Unit for: A- Electric Space Cooling (COP 2) B- Diesel Space Heating (Eff. 80%)	Net Energy Consumption = 11593 GJ/year (−12%) Space Heating Energy Consumption = 2865 GJ/year Space Cooling Energy Consumption = 377 GJ/year DHW Energy Consumption = 516 GJ/year
6	All SHES technologies	1- Diesel Generator 300 kW (0.5 Thermal to Electric Power Ratio + Fluid Cooler 2- Counter-flow Fluid Heat Exchanger 3- 63 kW SHW System & Existing Electric Back-up Water Heater (Eff. 65%) 4- 100 kW PV System to Cover Peak Electric Loads Except for Cooling + Energy Storage System 5- Packaged Terminal AC Unit for: A- Electric Space Cooling (COP 2) B- Space Heating Using Recovered Heat & Diesel Boiler (Eff. 89%)	Net Energy Consumption = 8340 GJ/year (−37%) Space Heating Energy Consumption = 2027 GJ/year (−33%) Space Cooling Energy Consumption = 377 GJ/yearDHW Energy Consumption = 100 GJ/year (−81%)

The simulation results indicated that up to 37% fuel savings over current base camp configurations are achieved when all the SHES technologies are implemented for accommodating 150-person in a temperate climate (Brandon, MB) (Figure 7). Considered individually, the most impactful technology was found to be the Microgrid with a Waste Heat Recovery system (scenario 4), as evident in both Figure 7 and Tables 5 and 6.

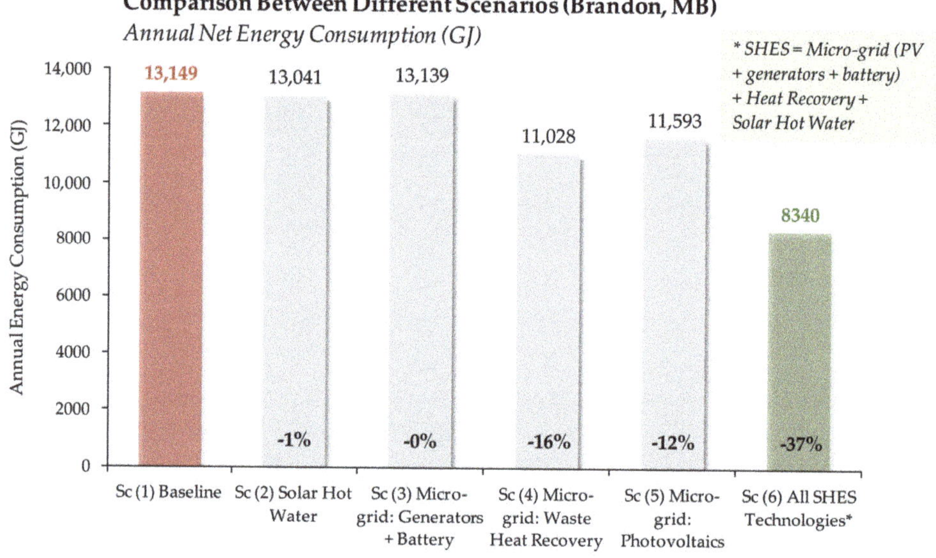

Figure 7. Comparison between different utility systems scenarios: annual energy consumption savings over the current base camp practice (Baseline) for a 150-persons base camp in Brandon, MB.

Table 6. Summary of simulation results for the various scenarios for a 150-person RTC in Brandon, MB.

Scenario	Technology	Consumptions Reduction Compared to the Baseline
Scenario 1	Baseline	Net Energy Consumption = 13985 GJ/year Space Heating Energy Consumption = 2866 GJ/year DHW Energy Consumption = 529 GJ/year
Scenario 2	Solar Hot Water	1% reduction in net energy consumption 81% reduction in DHW energy consumption
Scenario 3	Microgrid: generators + battery	-
Scenario 4	Microgrid: Waste Heat Recovery	16% reduction in net energy consumption 70% reduction in space heating energy consumption
Scenario 5	Microgrid: Photovoltaics	12% reduction in net energy consumption
Scenario 6	All SHES technologies	37% reduction in net energy consumption 33% reduction in space heating energy consumption 81% reduction in DHW energy consumption

The technoeconomic assessment indicated that the implementation of all SHES technologies significantly reduced the net-present cost (life cycle cost), defined as the present value of all installation and operation costs over the project lifetime, excluding the present value of all the revenues earned over the same interval.

Actual components' costs were used in the analysis. The project lifetime was assumed to be 25 years, while an 8% nominal discount rate, defined as the simple interest rate on borrowed capital before factoring the inflation rates in, was assumed. The inflation rate was also assumed to be 2% over the project lifetime, and it was used in combination with the nominal discount rate to calculate the real discount rate, which is used to convert between one-time and annualized costs. Finally, the fuel (diesel) cost was assumed to be $1.00 for each liter.

Results indicated that up to 32% reduction in net present value over current base camp configurations are achieved when all the SHES technologies are implemented for accommodating 150-person in a temperate climate (Brandon, MB) (Figure 8). The levelized cost of energy was also significantly reduced by 25%, as indicated. Figure 8 shows that the SHES acts as the most cost-effective solution; thus, it is the optimal solution that maximizes the use of renewable-generated energy while minimizing the project lifetime capital costs. More importantly, it is evident that the consideration of individual technologies eliminated the benefits gained from the integrated solution, and in fact, most technologies (except for WHRU), although reducing the net-present costs compared to the baseline, produced increased levelized energy costs due to their high initial costs and limited energy savings and utilization of all available energy resources when compared to SHES.

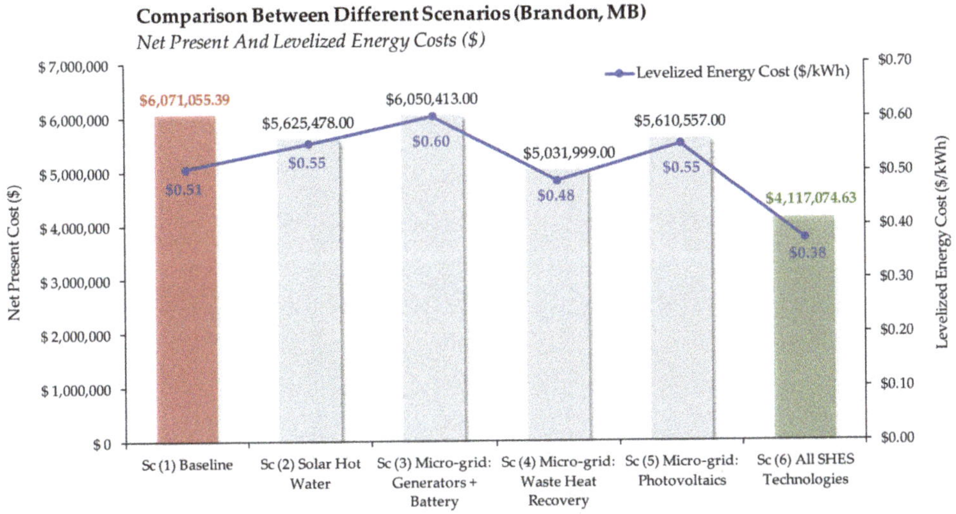

Figure 8. Comparison between different utility systems scenarios: net present cost and levelized energy cost over the current base camp practice (Baseline) for a 150-persons base camp in Brandon, MB.

Diesel costs may vary overtime and by the region it is sold depending on the cost of crude oil. Similarly, interest rates fluctuate overtime, being influenced by the economic growth, fiscal and monetary policies, and inflation rates. Thus, it is critical to design an optimal system in terms of life cycle costs at the expected fuel and interest rates.

A sensitivity analysis was conducted to evaluate the proposed system due to variations in the nominal discount rates and diesel costs (Figure 9). As evident, SHES acts as the optimal system design for the widest range of fuel and simple interest rates, assuming a constant inflation rate of 2%. In particular, the baseline configuration is only considered cost-effective at combined low fuel rates and high simple interest rates. It is also evident that the SHES's selected photovoltaics, energy storage, and converter capacities are optimal at the specified fuel cost and interest rates of $1/L and 8%, respectively. On the other hand, reducing these capacities would act as a more cost-effective solution at combinations of low and high fuel and simple interest rates, respectively.

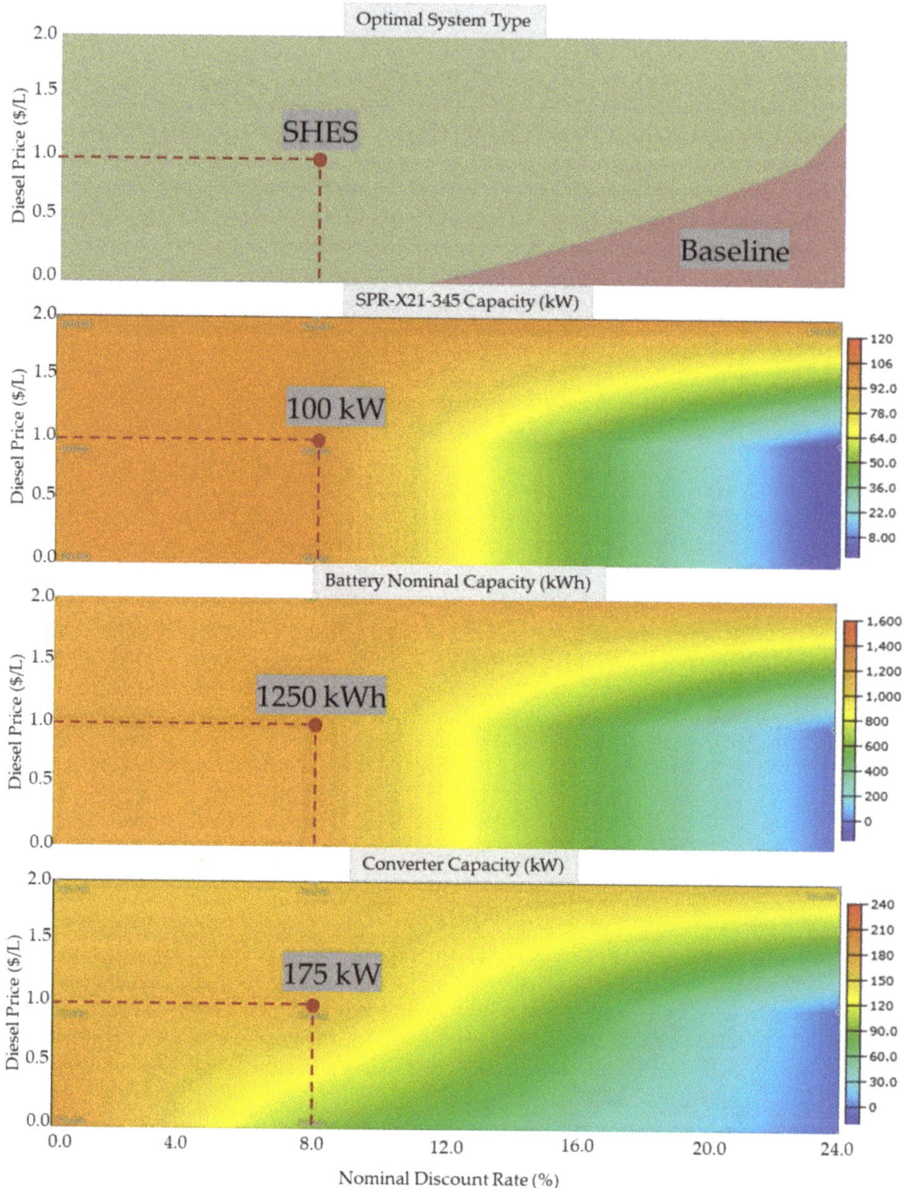

Figure 9. Optimal system design for variations in diesel price and nominal discount rate.

As discussed earlier, only PV capacities up to 100 kW, which sufficiently covers the camp peak electric loads excluding the cooling loads, were considered to reduce initial costs and due to spatial and logistics purposes. Also, various battery and converter models and capacities were considered, and the most cost-effective configurations were selected considering the preselected PV size. Figure 9 shows that the optimal battery capacity is strictly influenced by the selected PV capacity, thus, using a lower PV capacity would imply using a smaller battery capacity except in cases where the interest rates are incredibly high, for which investing in energy storage would not be cost-effective at all.

Similarly, the converter capacity is dependent on the selected PV capacity. However, the rate of capacity reduction is higher than that of the battery, particularly at moderate interest rates (8%–12%), driven by high converter capital costs when compared to energy storage and PVs.

One of the main challenges towards the development of isolated microgrids is the management of the various devices and energy flows to optimize their operations, particularly regarding the hourly loads that must be served, and the availability of power produced by renewable energy systems depending on daily and seasonal variations. The SHES combines the existing diesel generators with solar power generation, energy storage, and waste heat recovery technologies, all connected to a microgrid, ensuring uninterrupted electricity and hot water supplies. The reliable, energy-efficient system helps to manage generator output. By transforming an independently operating system of generators into a demand managed microgrid, SHES provides power only where and when it is needed, instead of completely relying on fuel-burning generators. A critical part of designing SHES was understanding the electric and thermal load and generation profiles to identify the most cost-effective energy management strategy while maximizing the renewable generation, without significantly increasing the initial costs of system while considering army spatial and logistic requirements. It is crucial to identify the parts of the system that carry these loads at different times of the day and different seasons, particularly at peak loads. The peak electric load typically occurs during the warmest period of the year due to increased cooling loads, while thermal loads during the same period would be low due to the absence of heating requirements. For these purposes, various dispatching strategies were considered for energy management purposes of controlling generator and battery operation in periods of insufficient renewable energy to supply the load, including "cycle charging" and "load following" strategies. A cycle charging dispatching strategy was found to be the most cost-effective. The cycle charging strategy implies that the generator runs at its maximum power output when it is needed to serve the electrical loads, while any surplus electrical production is diverted towards charging the battery until the battery setpoint state of charge of 80% is reached. This is accomplished by selecting the optimal combination of power sources, based on fixed and marginal costs, to serve the electric and thermal loads at the minimum cost and excess electricity production, while still satisfying the operating reserve requirements. On the other hand, a load following strategy, which implies that the generator produces enough power only to serve the load while the battery is charged by the renewable sources, would be more cost-effective in situations where the renewable generation is comparable to the magnitude of the served load.

The results of the control strategy can be observed in Figures 10 and 11, which show the electric and thermal and the generation profiles for the summer and winter peak demand days, respectively. The electric load served is initially constant and relatively low during early and late hours of the summer days and in the absence of solar radiation (Figure 10). Thus, this low electric load is satisfied solely by the energy stored in the battery while the generators are off. As the electric load starts to increase, the generators are turned on to satisfy parts of these loads, while the remaining parts are satisfied using the PV-generated power. To reduce the generator runtimes, at several intervals of the day, the reliance of the generator is reduced and eventually eliminated. At the same time, the loads are still being served by the energy stored in the battery during the previous hours. For the winter day, similar trends can be seen. However, due to lower electrical loads and increased PV generation as a result of a lower sun altitude, after the generators are turned on to serve partial loads and charge the battery, they are turned off for extended periods of the day, thus significantly reducing fuel consumption by relying on renewable resources. The generator's runtime was reduced considerably to 4600 h (48% reduction) compared to 8760 h for the baseline configuration. It is also evident that the system could benefit from a greater PV size due to the high availability of solar radiation that is not being taken advantage of in both summer and winter months.

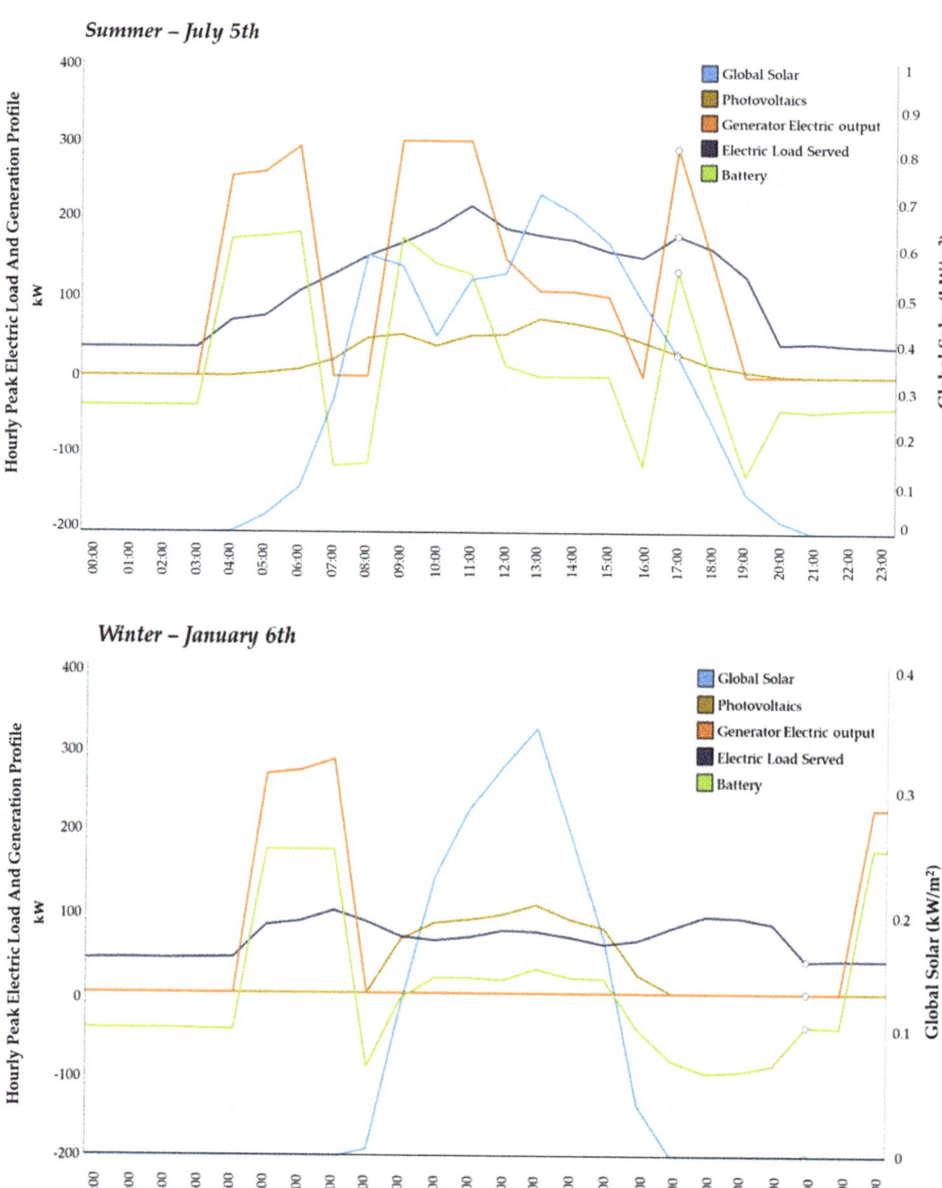

Figure 10. SHES dynamic electric load management for a 150-persons base camp in Brandon, MB, in summer and winter peak days.

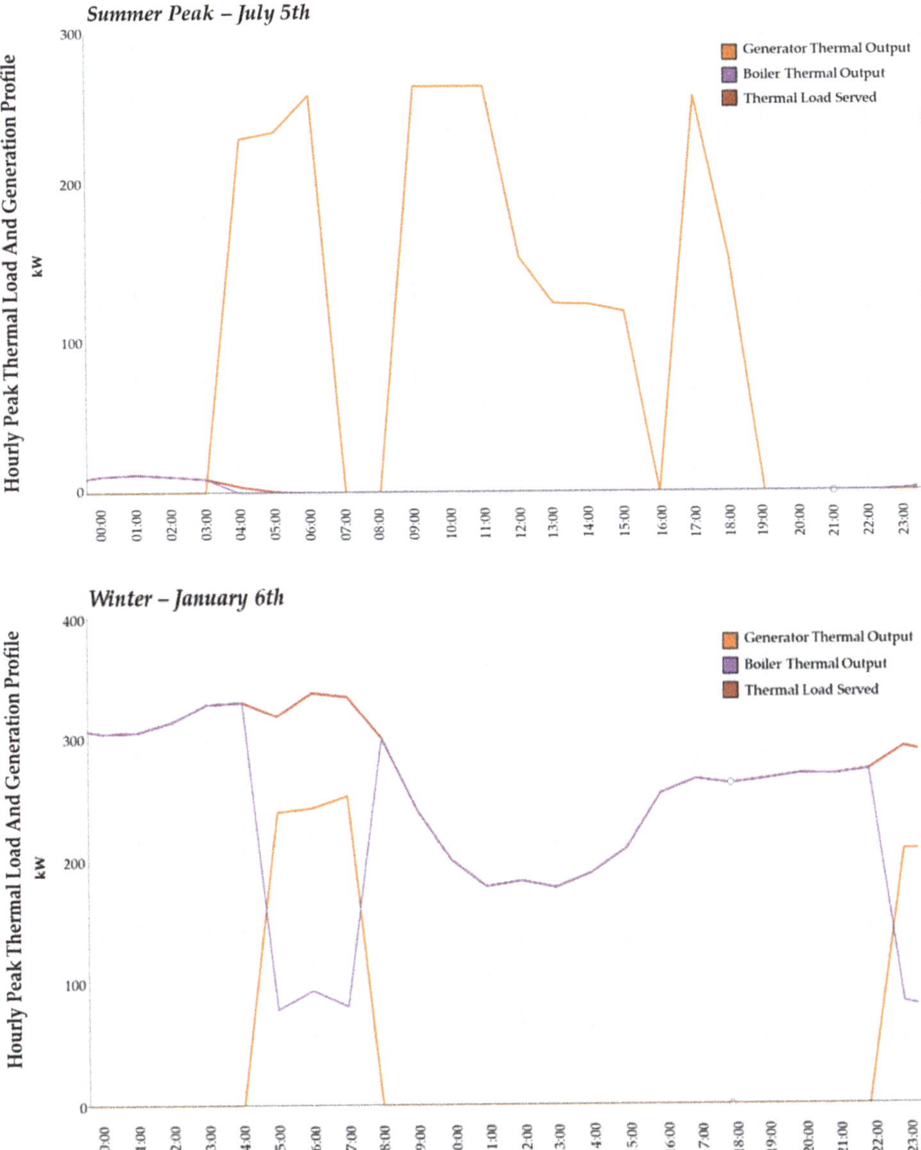

Figure 11. SHES dynamic thermal load management for a 150-persons base camp in Brandon, MB, in summer and winter peak days.

In Figure 11, the thermal loads are very low in summer due to the absence of heating requirements while some heating is only required during night times, which suggested increasing the insulating properties of the tent fabrics. In the winter, the peak thermal heating load is served mainly using the diesel boiler, which is supplemented by the waste-to-heat recovery system that uses the generator's heat to warm up the water delivered to the terminal fan–coil units. This heat is drawn from the generator only when it is running to serve the electrical loads during the winter; therefore, the WHRU system can only serve a part of the daily thermal load, which highlights the importance of considering synergies and differences between the different seasonal loads and designing a cost-optimal system in regards to full-year expected loads and availability of resources. It is also evident that a significant amount of excess heat is wasted in the summer months due to the absence of a simultaneous end use, thus, a heat storage system might be beneficial to store the heat and allowing for its use when needed in the colder months or in cold summer nights.

The SHES solution for the camp was also simulated in different geographic locations, to evaluate its performance, feasibility, and expected energy savings outcomes in different climate zones. In particular, the analysis involved the city of Vancouver (British Columbia, Canada), Kanoya (Japan), Churchill (Manitoba, Canada), and Changi (Singapore). Results indicated that fuel reductions from 21% up to 39% are also achievable for extremely hot and frigid climates when the solar collectors' tilt and orientation are optimized for the specific location (Table 7). It is important to note that the solution was not reoptimized for different locations except for the PV and SHW tilt angles, and component and system sizes were kept constant to satisfy the army requirements of standard sizing.

Table 7. Simulation results for extreme climatic zones when all SHES technologies are implemented.

Climates	Scenario	Technology	Consumptions Reduction Compared to Baseline	Optimal Angle
Vancouver, BC	1	Baseline	Net Energy Consumption = 11297 GJ/year Space Heating Energy Consumption = 7408 GJ/year DHW Energy Consumption = 529 GJ/year	33°
	6	All SHES technologies	34% reduction in net energy consumption 34% reduction in space heating energy consumption 66% reduction in DHW energy consumption	
Kanoya, Japan	1	Baseline	Net Energy Consumption = 11917 GJ/year Space Heating Energy Consumption = 392 GJ/year DHW Energy Consumption = 529 GJ/year	24°
	6	All SHES technologies	27% reduction in net energy consumption 34% reduction in space heating energy consumption 83% reduction in DHW energy consumption	
Churchill, MB	1	Baseline	Net Energy Consumption = 14804 GJ/year Space Heating Energy Consumption = 5355 GJ/year DHW Energy Consumption = 529 GJ/year	55°
	6	All SHES technologies	39% reduction in net energy consumption 29% reduction in space heating energy consumption 66% reduction in DHW energy consumption	
Changi, Singapore*	1	Baseline	Net Energy Consumption = 13461 GJ/year DHW Energy Consumption = 529 GJ/year	0°
	6	All SHES technologies	21% reduction in net energy consumption 98% reduction in DHW energy consumption	

* The heat recovery system is not required for the Changi climate zone due to the extremely low heating demand.

3.3. Environmental Considerations

The SHES reduces the dependency on fossil fuel lowering the environmental footprint of RTCs since the transportation logistics are minimized and the consumption of vehicle-fuel transporting fuel to the base campsite is also reduced. The generators, rather than being fueled by diesel (as currently done in Canadian Armed Forces), could be powered by LPG (Liquefied Petroleum Gas) to produce lower amounts of harmful greenhouse gases. However, the risks due to the transportation of more hazardous materials must be considered.

Table 8 reports significant annual CO_2 emissions savings compared to the baseline when all the SHES technologies are implemented for accommodating 150-person in a temperate climate (Brandon, MB). CO_2 reductions up to 39% are also achievable with the deployment of the system in extreme hot and cold climates (Table 8). The CO_2 emissions were calculated based on the annual diesel consumption to produce power by the generator and thermal energy by the boiler, assuming an emissions factor of 2.4 kg-CO_2e produced as a result of burning each liter of consumed diesel. It was also assumed that diesel has a carbon content of 88%.

Table 8. CO_2 annual emissions for different scenarios of a 150-person RTC operating in different climates.

Climates	Scenario	Technology	Annual CO_2 Emissions kg/Year
Brandon, MB	(1)	Baseline	919,500
	(6)	All SHES technologies	583,270 (−37%)
Vancouver, BC	(1)	Baseline	788,879
	(6)	All SHES technologies	517,335 (−35%)
Kanoya, Japan	(1)	Baseline	831,462
	(6)	All SHES technologies	606,460 (−27%)
Churchill, MB	(1)	Baseline	1036,790
	(6)	All SHES technologies	636,805 (−39%)
Changi, Singapore	(1)	Baseline	938,783
	(6)	All SHES technologies	743,139 (−21%)

The SHES is designed to be easily integrated with other water and waste infrastructure systems, e.g., it could be combined with deployable water purification systems (e.g., Aspen Water) powering them as AC loads or using directly the solar energy of the photovoltaic panels or with deployable waste-to-energy systems (e.g., Energos Technology and Eco Waste Solution), which convert wastes into thermal energy (e.g., DWECX-TEEPS) that can be used locally. Moreover, the waste heat coming from the waste converter exhaust of the waste-to-energy system could be captured and used to produce space heating.

3.4. Enhanced RTC Construction

The existing standard RTCs are constructed with materials (canvas) that suffer from poor thermal performance leading to higher heating and cooling energy consumptions to maintain the thermal comfort of deployed soldiers. Consequently, the low thermal resistance of the existing shelter systems requires the design of bigger HVAC systems. Standard tents are constructed of such materials mainly due to their lightweight and waterproofing abilities, allowing for their easier transportation, installation, and long service life. However, recent developments in innovative materials made it possible to provide such properties combined with improved thermal performance to achieve higher energy savings [23,24]. Notably, 13 mm aerogel-enhanced blankets were proposed to be integrated into the existing construction to make significantly improved RSI-values of 0.625 m^2K/W compared to 0.175 m^2K/W as demonstrated in Table 2 earlier [23]. Such blankets have an extremely low thermal conductivity of 0.015 W/m. K, while having low toxicity, low density of 200 kg/m^3, excellent fire rating, and high recyclability. Figure 12 shows a recently patented aerogel blanket obtained at Ryerson University using an ambient pressure drying (APD) process.

Energy simulations were conducted to assess the energy performance of the camp integrating SHES and high-performance tents incorporating the aerogel blanket produced into the interior of the existing canvas. Table 9 reports the results for different climate zones. The results show that enhancing the thermal resistance of the RTCs has significant impacts on increasing the camp energy efficiency for all climate zones. When compared to Scenario 6 results (SHES), it becomes clear that aerogel-enhanced blankets have their highest energy reduction impacts in extreme climates as Churchill, while still offering significant reductions in the other climate zones.

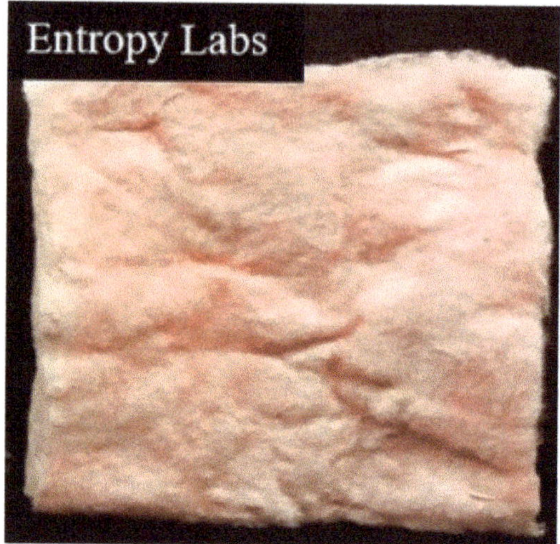

Figure 12. New aerogel-enhanced blanket obtained by the authors and used to enhance the tent material fabric providing higher thermal resistance.

Table 9. Energy consumption and CO_2 annual emissions results for different scenarios (SHES and aerogel blanket are used) of a 150-person RTC operating in different climates.

Climates	Consumption Reduction Compared to the Baseline	Annual CO_2 Emissions, kg/Year
Brandon, MB	48% reduction in net energy consumption 85% reduction in space heating energy consumption 81% reduction in DHW energy consumption	477,291 (−48%)
Vancouver, BC	47% reduction in net energy consumption 90% reduction in space heating energy consumption 66% reduction in DHW energy consumption	486,372 (−47%)
Kanoya, Japan	31% reduction in net energy consumption 89% reduction in space heating energy consumption 83% reduction in DHW energy consumption	570,357 (−31%)
Churchill, MB	57% reduction in net energy consumption 84% reduction in space heating energy consumption 66% reduction in DHW energy consumption	446,767 (−57%)
Changi, Singapore	29% reduction in net energy consumption 26% reduction in space cooling energy consumption 98% reduction in DHW energy consumption	664,548 (−29%)

4. Conclusions

This paper has described a new solution for a military army camp. The new approach shows significant improvements over the existing solutions thanks to the combination of microgrid with energy storage systems, renewable resources, and waste heat recovery technologies to reduce the fuel supply to RTCs significantly. In particular:

- A properly sized SHW system can supply a significant fraction of a base camp water heating requirements using solar energy;
- Modeling indicates that a combination of smart microgrid and renewable energy sources can reduce base camp energy demand and fuel use significantly, in addition to a significant carbon emission reduction. Considered individually, the technology with the lowest energy consumptions (up to 16% reduction) is the microgrid connected waster heat recovery system (scenario 4);

- Smart microgrids with energy storage systems supply power with improved voltage and frequency stability increased grid reliability and longer life of end use equipment;
- The EMS, equipped with real-time monitoring and control of base parameters, enables central and informed decision making. Configurable automatic load distribution provides the potential for reducing camp energy consumption for normal operations and unplanned events;
- The simulation results indicated that up to 37% of fuel savings and up to 37% annual CO_2 emissions savings over current base camp configurations are achieved when all the SHES technologies are implemented in a temperate climate;
- Fuel and CO_2 reductions from 21% up to 39% are also achievable for extremely hot and frigid climates when the solar collectors' tilts are optimized.

Author Contributions: Conceptualization, U.B. and E.T.; methodology U.B., E.T., and K.K.; software, K.K.; data curation, U.B. and E.T.; writing—review and editing, U.B.; and supervision, U.B. All authors have read and agreed to the published version of the manuscript.

Funding: The authors would like to thank the financial support provided by the Government of Canada through the Pop-up City project. They would like to gratefully thank ISSNAF (Italian Scientists and Scholars in North America Foundation) and Italian National Order of Engineers for funding a scholarship for allowing the second author to dedicate time to this project.

Conflicts of Interest: The authors declare no conflicts of interest.

References

1. Cioccolanti, L.; Fonti, A.; Comodi, G. Dynamic Modeling of Thermal and Electrical Microgrid of Multi-Apartment in Different European Locations. In Proceedings of the 17th International Stirling Engine Conference and Exhibition, Northumbria University, Newcastle upon Tyne, UK, 24–26 August 2016.
2. Razmjoo, A.; Shirmohammadi, R.; Davarpanah, A.; Pourfayaz, F.; Aslani, A. Stand-alone hybrid energy systems for remote area power generation. *Energy Rep.* **2019**, *5*, 231–241. [CrossRef]
3. Ismail, M.; Moghavvemi, M.; Mahlia, T.M.I. Techno-economic analysis of an optimized photovoltaic and diesel generator hybrid power system for remote houses in a tropical climate. *Energy Convers. Manag.* **2013**, *69*, 163–173. [CrossRef]
4. Ghasemi, A.; Asrari, A.; Zarif, M.; Abdelwahed, S. Techno-economic analysis of stand-alone hybrid photovoltaic–diesel–battery systems for rural electrification in eastern part of Iran—A step toward sustainable rural development. *Renew. Sustain. Energy Rev.* **2013**, *28*, 456–462. [CrossRef]
5. Puglia, G.; Moroni, M.; Fagnani, R.; Comodi, G. A Design Approach of Off-grid Hybrid Electric Microgrids in Isolated Villages: A Case Study in Uganda. *Energy Procedia* **2017**, *105*, 3089–3094. [CrossRef]
6. Daud, A.-K.; Ismail, M.S. Design of isolated hybrid systems minimizing costs and pollutant emissions. *Renew. Energy* **2012**, *44*, 215–224. [CrossRef]
7. Zhai, H.; Dai, Y.; Wu, J.; Wang, R. Energy and exergy analyses on a novel hybrid solar heating, cooling and power generation system for remote areas. *Appl. Energy* **2009**, *86*, 1395–1404. [CrossRef]
8. Higier, A.; Arbide, A.; Awaad, A.; Eiroa, J.; Miller, J.; Munroe, N.; Ravinet, A.; Redding, B. Design, development and deployment of a hybrid renewable energy powered mobile medical clinic with automated modular control system. *Renew. Energy* **2013**, *50*, 847–857. [CrossRef]
9. Kashem, S.B.A.; De Souza, S.; Iqbal, A.; Ahmed, J. Microgrid in military applications. In Proceedings of the 2018 IEEE 12th International Conference on Compatibility, Power Electronics and Power Engineering (CPE-POWERENG 2018), Doha, Qatar, 10–12 April 2018; Institute of Electrical and Electronics Engineers (IEEE): New York, NY, USA, 2018; pp. 1–5. [CrossRef]
10. Van Broekhoven, S.B.; Judson, N.; Nguyen, S.V.; Ross, W.D. Microgrid Study: Energy Security for DoD Installations. *Lincoln Lab.* **2012**.
11. Skowronska-Kurec, A.G.; Eick, S.T.; Kallio, E.T. Demonstration of Microgrid technology at a military installation. In Proceedings of the 2012 IEEE Power and Energy Society General Meeting, San Diego, CA, USA, 22–26 July 2012; Institute of Electrical and Electronics Engineers (IEEE): New York, NY, USA, 2012; pp. 1–2. [CrossRef]

12. Nerini, F.F.; Valentini, F.; Modi, A.; Upadhyay, G.; Abeysekera, M.; Salehin, S.; Appleyard, E. The Energy and Water Emergency Module; A containerized solution for meeting the energy and water needs in protracted displacement situations. *Energy Convers. Manag.* **2015**, *93*, 205–214. [CrossRef]
13. Hirsch, A.; Parag, Y.; Guerrero, J.M. Microgrids: A review of technologies, key drivers, and outstanding issues. *Renew. Sustain. Energy Rev.* **2018**, *90*, 402–411. [CrossRef]
14. Khmais, A.; Nasir, M.; Mohamed, A.; Shareef, H. Design and Simulation of Small Scale Microgrid Testbed. In Proceedings of the 2011 Third International Conference on Computational Intelligence, Modelling & Simulation, Langkawi, Malaysia, 20–22 September 2011; Institute of Electrical and Electronics Engineers (IEEE): New York, NY, USA, 2011; pp. 288–292. [CrossRef]
15. Van Broekhoven, S.; Judson, N.; Galvin, J.; Marqusee, J. Leading the Charge: Microgrids for Domestic Military Installations. *IEEE Power Energy Mag.* **2013**, *11*, 40–45. [CrossRef]
16. Julian, A.L.; Oriti, G.; Ji, C.; Zanchetta, P. Power Quality Improvement in a Single-Phase Energy Management System Operating in Islanding Mode. In Proceedings of the 2018 IEEE Energy Conversion Congress and Exposition (ECCE), Portland, OR, USA, 23–27 September 2018; Institute of Electrical and Electronics Engineers (IEEE): New York, NY, USA, 2018; pp. 208–214. [CrossRef]
17. Anglani, N.; Oriti, G.; Colombini, M. Optimized Energy Management System to Reduce Fuel Consumption in Remote Military Microgrids. *IEEE Trans. Ind. Appl.* **2017**, *53*, 5777–5785. [CrossRef]
18. Naval Facilities Engineering Command. *Technology Transition Final Report: Smart Power Infrastructure Demonstration for Energy Reliability and Security (SPIDERS)*; U.S. Department of Defense: Arlington, VA, USA, 2015.
19. Salvalai, G.; Imperadori, M.; Scaccabarozzi, D.; Pusceddu, C. Thermal performance measurement and application of a multilayer insulator for emergency architecture. *Appl. Therm. Eng.* **2015**, *82*, 110–119. [CrossRef]
20. Ghanmi, A. Energy management in military operational camps: A cost-benefit analysis. In Proceedings of the 2014 5th International Renewable Energy Congress (IREC), Hammamet, Tunisia, 25–27 March 2014; Institute of Electrical and Electronics Engineers (IEEE): New York, NY, USA, 2014; pp. 1–6. [CrossRef]
21. Software, HOMER Pro Version 3.13. 2020. Available online: https://www.homerenergy.com/products/pro/version-history.html (accessed on 1 May 2020).
22. *ISO 6346:1995 Freight Containers - Coding, Identification and Marking*; International Organization for Standardization: Geneva, Switzerland, 1995.
23. Berardi, U.; Zaidi, M. Characterization of commercial aerogel-enhanced blankets obtained with supercritical drying and of a new ambient pressure drying blanket. *Energy Build.* **2019**, *198*, 542–552. [CrossRef]
24. Zhang, L.; Menga, X.; Liu, F.; Xu, L.; Long, E. Effect of retro-reflective materials on temperature environment in tents. *Case Stud. Thermal Eng.* **2017**, *9*, 122–127. [CrossRef]

Article

Flattening the Electricity Demand Profile of Office Buildings for Future-Proof Smart Grids

Rick Cox [1], Shalika Walker [1,*], Joep van der Velden [2], Phuong Nguyen [1] and Wim Zeiler [1]

1 Department of the Built Environment, Eindhoven University of Technology, PO Box 513, 5600 MB Eindhoven, The Netherlands; rick.j.g.cox@gmail.com (R.C.); P.Nguyen.Hong@tue.nl (P.N.); W.Zeiler@tue.nl (W.Z.)
2 Kropman Installatietechniek, Lagelandseweg 84, 6545 CG Nijmegen, The Netherlands; Joep.van.der.velden@kropman.nl
* Correspondence: S.W.Walker@tue.nl

Received: 11 April 2020; Accepted: 3 May 2020; Published: 8 May 2020

Abstract: The built environment has the potential to contribute to maintaining a reliable grid at the demand side by offering flexibility services to a future Smart Grid. In this study, an office building is used to demonstrate forecast-driven building energy flexibility by operating a Battery Electric Storage System (BESS). The objective of this study is, therefore, to stabilize/flatten a building energy demand profile with the operation of a BESS. First, electricity demand forecasting models are developed and assessed for each individual load group of the building based on their characteristics. For each load group, the prediction models show Coefficient of Variation of the Root Mean Square Error (CVRMSE) values below 30%, which indicates that the prediction models are suitable for use in engineering applications. An operational strategy is developed aiming at meeting the flattened electricity load shape objective. Both the simulation and experimental results show that the flattened load shape objective can be met more than 95% of the time for the evaluation period without compromising the thermal comfort of users. Accurate energy demand forecasting is shown to be pivotal for meeting load shape objectives.

Keywords: electricity; HVAC; demand forecasting; flexibility; office building; Smart Grid

---

1. Introduction

The European Union agrees on drastically lowering CO_2 emissions in order to mitigate the effects of climate change [1,2]. Currently, in the Netherlands, electricity generation is mostly achieved by means of fossil fuels and is responsible for a significant portion of the total emissions [3]. To meet European targets, a transition to more sustainable energy generation is necessary within the country to decarbonize the grid [4,5]. When transitioning towards a low-carbon society, not only sustainable generation, but also energy saving on the demand side is even more important [6,7]. Transport and the built environment account for approximately 24% and 36% of total energy consumption in the Netherlands and are therefore responsible for much of the emissions due to fossil fuels [3,8]. It is evident that an effective transition to a sustainable future also requires technologies on the demand side [9,10] that can be powered by Renewable Energy Sources (RES), such as electric heat pumps [11,12] to fulfill the heating demand of buildings and electric vehicles for transport [13], in the Dutch context [4].

A transition to more sustainable energy generation is expected to bring about a variety of challenges. Firstly, the foreseen large-scale deployment of RESs may seriously affect the stability of energy grids [14,15]. An increase in power grid-connected RESs results in a change in power generation characteristics and grid operation [16]. In contrast to conventional fossil power plants, RESs are often relatively small power generators and are distributed throughout the low- and medium-voltage grid levels [17]. When RESs are integrated into the built environment, buildings will both consume and supply energy to the grid and become active 'prosumers' [18]. This creates multi-directional energy flows on the low- and medium-voltage grid levels [19]. Additionally, through the continuing electrification of space heating by heat pumps and transport by electric vehicles, the pressure on the transmission and distribution grids will increase further, thereby increasing the risk of congestion [20,21].

A specific problem that can be encountered is "overgeneration," with increasing penetration of solar photovoltaics (PVs). The Californian Independent System Operator published a "duck chart" which shows during springtime a significant drop in midday net load as more PVs are added to the system [22]. This introduces a huge problem in ramping up the generation, as PV power production rapidly decreases as the sun sets in the evening. Notably, no research reports, papers, or other documents were found which describe the duck curve or a similar problem explicitly in the Dutch context. However, a quick analysis of the installed PV capacity growth over the years and the grid loads in the Netherlands indicate that the problem increases as the current PV growth trend progresses. Large-scale integration of PV generation could also lead to local problems. The decentralized generation of PV power could lead to overvoltage and congestion in the low-voltage grid level when there is high PV power generation but low demand [23].

All the aforementioned problems call for more intelligent ways of consuming electricity. One possible way is a Smart Grid [24], where both demand and local production in the distribution grid are controlled in order to stabilize the grid [14]. Many definitions of a Smart Grid exist [25]. According to the Institute of Electrical and Electronics Engineers (IEEE) [26], the Smart Grid has come to describe a *next-generation electrical power system that is typified by the increased use of information and communication technology (ICT) in the generation, delivery and consumption of electrical energy*. The future power grid is expected to provide unprecedented flexibility in how energy is generated, distributed and managed [27]. The Dutch branch organization of energy network operators (Netbeheer Nederland) estimates that the total need for flexibility in the Netherlands will double towards 2030 compared to 2015, and increase even further by a factor of three towards 2050 [28].

Power system flexibility can be achieved through a variety of different interventions at both the supply and demand side [29]. The traditional approach is supply-side flexibility, which could be delivered by supply-side energy storage, power plant response, curtailment of variable renewable electricity generators or dedicated power plants such as combined heat and power (CHP) and combined cycle gas turbines [29]. The demand side, which includes the built environment, can also adapt its electricity demand according to grid needs through the adoption of Demand Side Management (DSM) programs. Gellings [30] describes DSM as: *"the planning and implementation of those electric utility activities designed to influence customer uses of electricity in ways that will produce desired changes in the utility's load shape"*. Techniques such as peak shaving and valley filling could be used to accomplish the load shape objective [31], especially with the use of storage systems. The built environment could thereby provide energy flexibility, which is defined by the International Energy Agency (IEA) Annex 67 as *a building's ability to manage its demand and generation according to local climate conditions, user needs and grid requirements* [32].

Building energy flexibility is not just limited to the utilization of storage systems but also energy systems inside the buildings to create a balance with the building-integrated renewable energy production. Of the energy systems present in a commercial office building, Heating Ventilation and Air Conditioning (HVAC) [33] systems account for approximately 63% of the energy requirements (in the Netherlands in 2017) [34]. Therefore, the employment of HVAC systems for realizing Demand-Side Flexibility (DSF) services is interested and driven by the following factors [19]:

- HVAC systems are equipped with automation and control systems that enable implementation of strategies for DSF actuation;
- HVAC systems have significant thermal inertia, thereby they can function as a buffer for the electricity grid for short periods of time by reducing air-handling unit (AHU), chiller or heat pump loads;
- HVAC systems have continuous control systems that allow operational cycle modification for energy advantages, rather than on–off control.

The significant energy demands and the aforementioned control advantages allow HVAC systems and energy storage systems to provide effective energy flexibility and management services for the built environment. Buildings could, therefore, offer DSF by manipulating installations to respond to power system requirements by increasing or reducing electricity consumption patterns while maintaining a comfortable and productive environment for the occupants [19]. Additional DSF could also be delivered through the control of lighting and plug loads [35].

To provide the above-mentioned decision-making requirement, it is necessary to perform accurate short-term and small-scale electricity load forecasting on subsystem levels of individual buildings [36]. The energy behavior of a building is influenced by many factors, such as weather conditions, building construction, the thermal properties of the building, the occupancy, and occupant behavior [37]. Forecasting subsystem-level loads is therefore considered a complex and challenging problem [36]. However, this type of demand prediction could be a valuable contribution to maintaining a reliable electricity grid.

Contributing to solving the mentioned problems, the objective of this research is to identify and implement building energy management opportunities using subsystem-level electricity demand prediction and a Battery Electric Storage System (BESS). The objective of this study is to stabilize/flatten a building energy demand profile to demonstrate energy flexibility for a future Smart Grid without compromising user comfort. The proposed methodology in Section 2 is advantageous to the research community because it discusses the demand prediction of several subcomponents (AHU, HVAC, chiller, lighting, and plug loads) of buildings, which is otherwise rare in the existing literature. Moreover, a major contribution would be the implementation of the discussed methodology in a real-life office building. Next, the quantitative results and qualitative findings have been presented in Sections 3 and 4.

2. Materials and Methods

This section first introduces the case study building and its electrical load groups. Subsequently, three steps are described in the methodology used:

- Step 1: Prediction models are established for each load group and the performance metrics that are used to analyze prediction accuracy are described;
- Step 2: The operational strategy of the Battery Electric Storage System (BESS) is described alongside key performance indicators (KPIs) that are used to quantify the impact of the energy flexibility provided;
- Step 3: The implementation procedure in the real BMS system is described.

2.1. Case Study Building

The case study office building is located in The Netherlands, which is a European country with a temperate oceanic climate (Cfb type) according to the Köppen–Geiger climate classification [38]. The office was built in 1993 and can be described as a traditional office building. The building has an approximate floor area of 1500 m^2 and a practical maximum occupancy count of 35 [19]. The building is provided with a photovoltaic system and a Battery Electric Storage System. An impression of the building is shown in Figure 1. A general overview of the loads of the building is shown in Figure 2.

Figure 1. (**Left**) Impression of the building. (**Right**) A Battery Electric Storage System (BESS) installed inside the building.

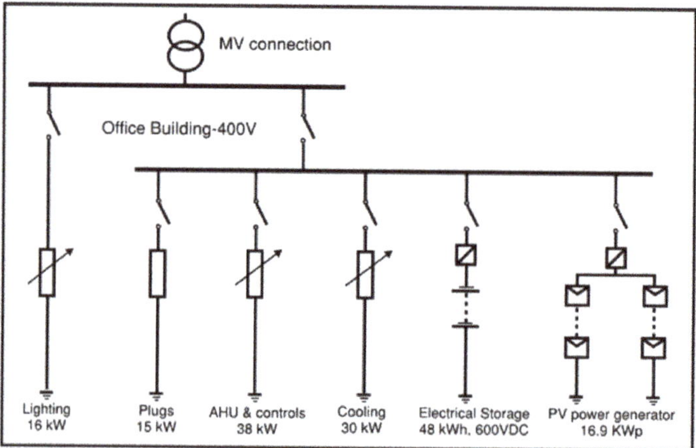

Figure 2. Electrical load diagram of the building.

A detailed description of the subcomponents of the building follows:

(a) **AHU and HVAC control unit:** The ventilation system of the building is designed for two functions: refreshing the air in the building and transporting the desired cooling capacity into the building [39]. This group is composed of a variety of different components including supply and return fans, HVAC controls, a heat recovery wheel, and the pumps for the heating system and cooling system.

(b) **Chiller:** The chiller is an electric, double-stage air-source compression cooling machine. It is an on–off operated machine, which means that it starts and stops multiple times per hour depending on the cooling load. This cooling machine is equipped with a small buffer. The distribution system supplies cold temperature to the rooms utilizing water-to-air aftercoolers.

(c) **Lighting:** The building is largely equipped with fluorescent lighting. One office space has an LED lighting system with motion sensors and the ability to individually set light temperature settings for each workplace.

(d) **Plug loads:** Plug loads consist of all the remaining electricity consuming devices such as computers, printers, coffee machines, etc.

(e) **PV system:** The case study office has 65 solar Photovoltaic (PV) panels on its roof, with 260 Wp per panel, corresponding to a total installed capacity of 16.9 kWp [40], with a 15 kW inverter. Each solar panel is individually optimized to its Max Power Point (MPP) with the use of DC/DC optimizers.

(f) **Battery Electric Storage System:** The building is equipped with a Nilar NiMH Battery Electric Storage System (BESS) with 48 kWh of storage capacity. The power conversion system is formed by the combination of a bi-directional inverter and transformer in a single cabinet. The advantage of this configuration is that the equipment can be disconnected completely. This prevents unnecessary power loss when the battery is not utilized. Table 1 provides an overview of the specifications of the Nilar BESS and energy conversion system.

The method used to ultimately flatten the electricity demand profile for the case study building consists of three steps elaborated in Sections 2.2–2.4.

Table 1. Battery storage and conversion system specifications.

System	Technical Features	Value	Unit
SUNSYS-PCS2-33TR (bi-directional inverter and transformer)	Maximum current (DC)	80	A_{dc}
	- limited in setup to	35	A_{dc}
	Maximum current (AC)	53	A_{rms}
	Rated current (AC)	48	A_{rms}
	Rated power (AC)	33	kW
	Maximum efficiency	97	%
	European efficiency	96	%
Nilar ECI-600V—48 kWh (BESS)	No. of battery packs	40	
	System voltage	600	V
	Rated capacity	80	Ah
	Energy	48	kWh

2.2. Step 1: Establish Prediction Models

Building electricity demand predictions are an essential part of developing a suitable control strategy. The total electricity consumption of the case study building consists of 5 major load groups and a BESS which were extensively monitored. Due to the different behavior of all load groups, different prediction methodologies are proposed depending on the group's characteristics. An advantage of this approach is that the cause of prediction errors is easier to trace back to the load groups which are inaccurately predicted, after which the model could be adjusted or optimized. Another argument for predicting each load group separately is that relatively simple prediction methods could be used which are specifically designed for predicting the loads of a particular load group. This also makes practical implementation of the predictions in the BMS more transparent. A priori knowledge obtained through inspection of the building's individual load groups and their characteristics enabled the construction of the prediction models as presented. On the other hand, large fluctuations in loads such as the chiller make it exceptionally difficult to accurately predict electricity demands intra-hourly. Consequently, predictions in this study are calculated for all load groups on a 1 hour resolution. Figure 3 illustrates the overview of subloads and corresponding day-ahead prediction models. BMS data from 1 January 2017 to 31 December 2018 are used in the establishment of the prediction models except for Solargis® predictions for the outdoor temperature and PV yields. The Solargis dataset contains data from 25 May 2018 to 4 April 2019 (~10 months).

(a) AHU and HVAC control unit:

For this load group, a parametric approach is chosen because of the characteristic S shape of the data (see Figure 4). Parametric modeling techniques involve two steps [41]: first identifying the function form, and then fitting the parameters of the mathematical model. In order to determine a better fit for the mathematical model, the natural logarithm of the dataset is taken, which is also known as a variance-stabilizing transformation.

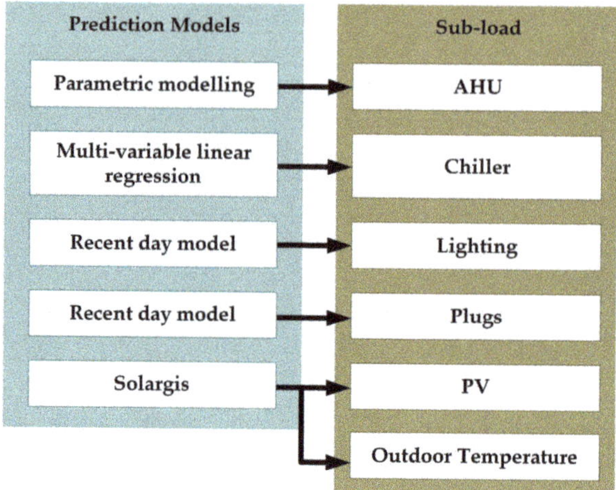

Figure 3. Overview of the subloads and corresponding prediction models.

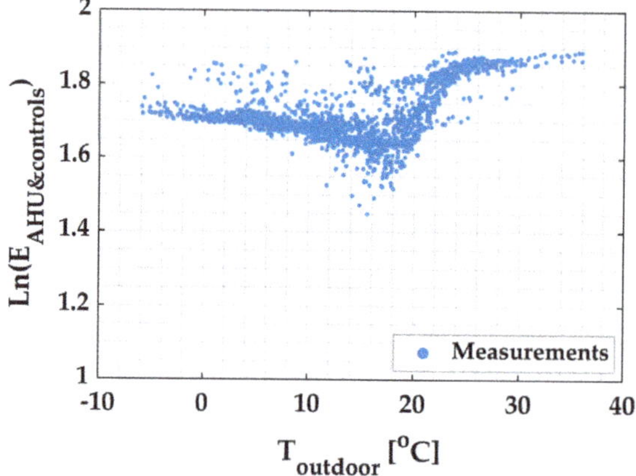

Figure 4. Demand variation in the air-handling unit (AHU) and the Heating Ventilation and Air Conditioning (HVAC) control unit with respect to the outdoor temperature ($T_{outdoor}$).

Data points with an energy demand < 4 kWh.h^{-1} are considered outliers and were removed from the dataset. The data in Figure 4 are plotted in an S shape. This shape can be described mathematically by combining a logistic function and parabolic function. The transformation of this combined equation back to the scale of the original dataset is achieved by exponentiation. The final equation to predict the AHU and HVAC control unit energy demand ($E_{AHU\&controls}$) as a function of the ambient temperature ($T_{outdoor}$) is given by Equation (1).

$$E_{AHU\&controls,t=i} = \exp\Big((A\cdot T_{outdoor}{}^2 + B\cdot T_{outdoor} + C) \cdot\Big[\alpha + \frac{\beta}{1+\exp(-\gamma\cdot(T_{outdoor}-\delta))}\Big]\Big)\left[\text{kWh.h}^{-1}\right] \quad (1)$$

(b) Chiller:

Figure 5 shows a scatter plot of the hourly electricity demand of the chiller as a function of the outdoor temperature. Considering the shape of the data, a linear regression model will be used in this case. Engineering expertise also tells us that measurements at the previous time step (t−1) may have the largest impact on the building cooling load at time t [42]. Therefore, one of the proposed models is a multi-variable linear regression model that uses a time series of $T_{outdoor}$ as an input and provides a prediction for the chiller's energy demand at t = i ($E_{chiller,t = i}$) as the output. The mathematical definition of this model is shown in Equation (2). By fitting the coefficients a_i to the dataset, a single equation is obtained which is capable of providing predictions. The number of terms/variables that should be included in the model is determined through k-fold cross-validation [41] with k = 10. Note that k-fold cross-validation is used for the establishment of the chiller model.

$$E_{chiller,t=i} = a_0 + a_1 \cdot T_{outdoor,t=i} + a_2 \cdot T_{outdoor,t=i-1} + a_3 \cdot T_{outdoor,t=i-2} + \cdots + a_n \cdot T_{outdoor,t=i-(n-1)} \left[\text{kWh.h}^{-1}\right] \quad (2)$$

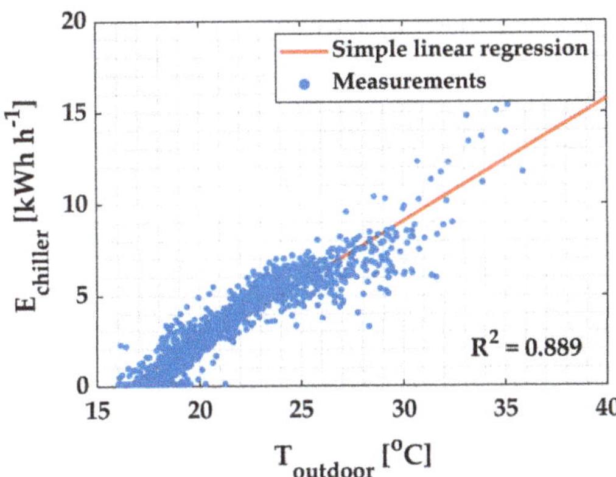

Figure 5. Demand variation of the chiller with respect to the outdoor temperature.

(c) Plug loads and lighting:

Occupancy is known to be related to plug loads [43]. However, because the day-ahead occupancy cannot be predicted accurately, an approach is chosen wherein the future plug load and lighting demand are based on (recent) historic demands. The proposed model makes energy demand predictions for an hour i ($E_{pred,t=i}$), based on the historic demands of the same clock hour. This model predicts hour i based on the power demands at hour i of the N *most recent workdays* when predicting workdays, likewise for weekend days. Equation (3) describes the model. In this equation, 24 describes the number of hours per day. This prediction model is hereby named the "recent day model".

$$E_{pred,t=i} = \frac{\sum_{k=1}^{N} (E_{t=i-24*k})}{N} \left[\text{kWh.h}^{-1}\right] \quad (3)$$

(d) Photovoltaic panels:

Solargis® is a Slovakian company that provides solar, weather, and PV yield forecasts for almost any location on earth. The case study building has been using its services since May 2018. Solargis® provides both temperature and PV yield predictions on an hourly basis for every hour of the day and up to 48 hours ahead.

(e) Outdoor temperature:

Some of the aforementioned models use the outdoor temperature as a predictor variable. Making future predictions with these models, therefore, requires outdoor temperature predictions. Solargis® services are again used to provide predictions for the outdoor temperature. The data obtained are post-processed with elevation correction and bias correction [44].

2.2.1. Performance Evaluation of Predictions

The performance of the prediction models requires quantification. This is achieved by introducing various error metrics [45]. Two types of error metrics are used: those without dimensions (without units) and those with dimensions (with units). Error metrics *without* dimensions are essentially normalized errors which are necessary for comparing results with studies with different sized installations [46].

To normalize the data, error metrics with dimensions use the sum of all measured points or the average of all measured points in their denominators. This denominator has a downside when interpreting these error metrics for parameters such as the temperature or when assessing the seasonal performance of the PV predictions, since there is the possibility that the temperatures of a dataset, for example, may average to approximately 0 °C in wintertime, which in turn gives an unreasonable scaling to the performance metrics. Choosing an error metric with dimensions omits this problem and gives an intuitive value with units.

(a) The Coefficient of Determination (R^2)

R^2 evaluates how much of the variability in the actual values is explained by the model [41]. Generally, R^2 takes a value between 0 and 1, wherein 1 represents the best performance. It should be emphasized that while R^2 is a powerful metric when assessing linear models, it is an inadequate measure when assessing non-linear models [47]. R^2 is therefore only used for the assessment of the chiller model in this research. The mathematical definition of R^2 is given by Equation (4).

$$R^2 = 1 - \frac{\sum_{i=1}^{N} (\hat{x}_i - x_i)^2}{\sum_{i=1}^{N} (x_i - \overline{x})^2} \; [-] \tag{4}$$

where

$\hat{x}_i$	The predicted value for data point i (e.g., power demand),
x_i	The measured (observed) value for data point i, and
$\overline{x}$	The mean of all observed values in the dataset.

(b) The Weighted Average Percentage Error (WAPE)

The WAPE describes the average magnitude of error produced by the model, relative to the measured values. It is widely used as a performance measure in forecasting, since it is easy to interpret and understand [48]. This metric is robust to outliers. Forecasting is best when the *WAPE* is close to 0. Equation (5) shows the mathematical definition of the *WAPE* [48].

$$WAPE = \frac{\sum_{i=1}^{N} \left| \frac{\hat{x}_i - x_i}{x_i} \right| x_i}{\sum_{i=1}^{N} x_i} = \frac{\sum_{i=1}^{N} |x_i - \hat{x}_i|}{\sum_{i=1}^{N} x_i} \; [\%] \quad for\; x_i \geq 0 \tag{5}$$

(c) The Coefficient of Variation of the Root Mean Square Error (CVRMSE)

The *CVRMSE* is a performance metric that penalizes larger errors more than the WAPE [49]. The American Society of Heating, Refrigerating, and Air Conditioning Engineers (ASHRAE) recommends *CVRMSE* values below 30% [50] for hourly predictions and so this standard is also adopted in this research. The mathematical definition is provided in Equation (6) [49].

$$CVRMSE = \frac{\sqrt{\frac{\sum_{i=1}^{N} (x_i - \hat{x}_i)^2}{N}}}{\overline{x}} \; [\%] \tag{6}$$

(d) The Mean Absolute Error (MAE)

The *MAE* is the average of the absolute difference between the predicted values and observed value; see Equation (7) [51]. The closer the value is to 0, the better the prediction performance.

$$MAE = \frac{\sum_{i=1}^{N} |x_i - \hat{x}_i|}{N} \tag{7}$$

(e) The Root Mean Square Error (RMSE)

The *RMSE* is the same as the CVRMSE, except for scaling by the average of all observations; see Equation (8) [52].

$$RMSE = \sqrt{\frac{\sum_{i=1}^{N}(x_i - \hat{x}_i)^2}{N}} \tag{8}$$

(f) The Mean Bias Error (MBE)

The *MBE* indicates whether a forecasting model, in general, tends to overestimate or underestimate in comparison to the actual values [46]. This metric could then be used to correct such systematic deviations. The *MBE* can be calculated according to Equation (9) [52]:

$$MBE = \frac{1}{N}\sum_{i=1}^{N}(\hat{x}_i - x_i) \tag{9}$$

2.2.2. Total Demand Prediction of the Building

Finally, in order to predict the total day-ahead (lead time: 24 h) electricity demand of the building, the established prediction models for all load groups and Solargis® temperature and PV prediction services are integrated into one combined model (see Figure 6). The predictions are performed each day at 00:00 and error metrics are computed. The dataset used in the integrated model consists of Solargis® and historic building energy demands from 25 May to 4 April 2019. MATLAB is used for the integration of the models and assessment of the data. After combining the predicted energy demands of each subcomponent, the resulting total building demand prediction with the hourly resolution data in kWh.h^{-1} is taken as the power demand in kW.

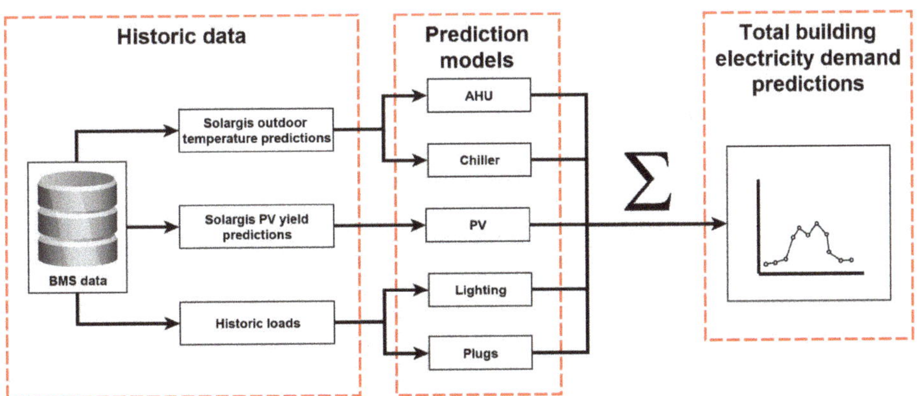

Figure 6. Total demand prediction of the building.

2.3. Step 2: Establish the Operational Strategy and BESS Simulations

The objective of this study is to stabilize/flatten a building energy demand profile during office work hours using a BESS. Peak shaving and valley filling are necessary to meet the load shape objective. A peak refers to a significantly higher power demand than desired, and a valley to a significantly lower power demand than desired. Before peak shaving and valley filling can be considered, a *'desired power demand profile'* of the building should be established. A comparison between the actual building load and the desired load then allows for the identification of peaks and valleys. Instead of the term *'desired power demand'*, henceforth, the term *'baseline'* (BL) is used. An illustrative example of a BL which is set between 07:00 and 17:00 (working hours of the building) is shown in Figure 7. By charging and discharging the BESS, load shape objectives can be met. In principle, this baseline can be developed to reflect the different objectives of the building owners such as maximizing self-energy consumption, minimizing electricity costs, and matching flexible Smart Grid demands.

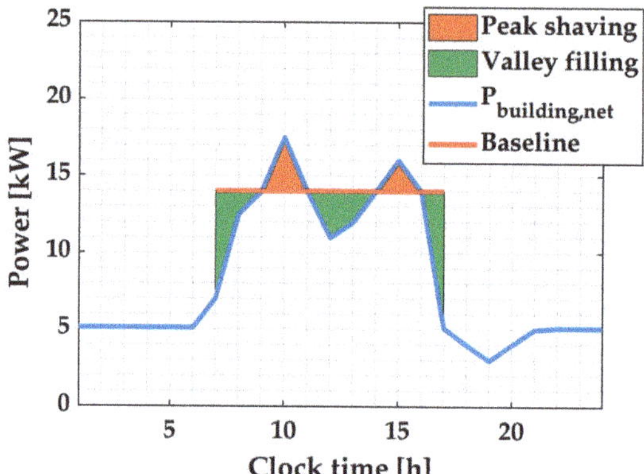

Figure 7. Peak shaving and valley filling depending on the established baseline.

The operation of the BESS relies heavily on the established BL. When the baseline is too high, power demands are unnecessarily high, and the BESS may not be able to fill all valleys. On the other hand, when the BL is too low, the BESS may not be able to deliver the power necessary to shave all the peaks. Another important parameter that is dependent on the baseline (and vice versa) is the initial state of charge (SoC_{ini}) of the BESS before the load balancing period starts. In this case, the load balancing period starts at 07:00. The mutual dependence of BL and SoC_{ini} calls for a strategy to determine the best balance. The steps taken to determine the best balance are described below. For both workdays and weekend days, the predictions are calculated at 00:00 for the upcoming 24 h and then used when determining the BL and SoC_{ini}.

2.3.1. Workdays

The determination of suitable values for the BL and SoC_{ini} is a dynamic process performed by using the established energy predictions for each day. On the other hand, battery operation is constrained to work between a 0.8 SoC and 0.2 SoC margin, which means a capacity of ~28.8 kWh out of the total 48 kWh is available for operation throughout the day. The other operational constraints, e.g., the limited number of charging cycles, which affect the degradation of the battery are not taken into account.

The algorithm starts at 00:00 by receiving the predicted energy demand profile for the upcoming day. Based on the maximum predicted power demand of the prediction profile, a series of test baselines (BL_{test}) are generated as shown in Figure 8a. Next, for different SoC_{ini} values ranging from 20% to 80% (20%, 30%, ... , 80%), the charging and discharging patterns of battery storage are evaluated for each BL_{test} profile as illustrated in Figure 8b. For the evaluation, the charging efficiency is taken as 85.5%, and the discharging efficiency is taken as 95%.

For each case, the cumulative energy which could not be delivered by the BESS to shave the peaks throughout the day, denoted by $X_{discharge}$ (read: 'inability to discharge'), and the cumulative energy which could not be stored by the battery to fill the valleys throughout the day, denoted by X_{charge} (read: 'inability to charge'), are calculated. Performing the simulations for each case results in a complete overview with all different combinations of BL and SoC_{ini}, and corresponding values for $X_{discharge}$ and X_{charge}. This information forms the basis in order to decide which case is expected to perform the best. Then, the chosen BL and SoC_{ini} are used for the operation of the specific workday.

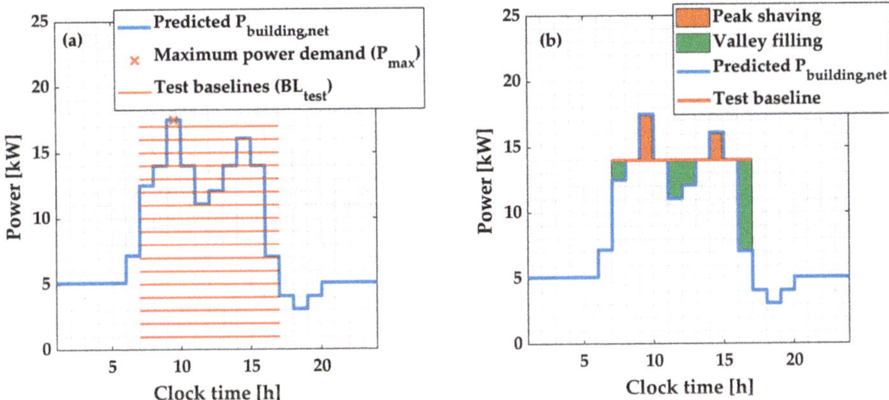

Figure 8. (**a**) Principle of defining test baselines based on the maximum predicted power demand. (**b**) Peak shaving and valley filling depending on the test baseline and building energy demand forecast.

2.3.2. Weekend Days and Holidays

The operational strategy of the weekend/holiday is to maximize PV self-consumption and prevent net power injection into the grid, meaning BL = 0 kW is chosen for weekends. At 00:00, an energy demand profile of the building is generated based on the predictions for the upcoming 24 h. By using these energy predictions and the assumed charging efficiency of the battery, the expected required storage capacity of the battery to prevent net injection between 07:00 and 17:00 is estimated. From this, the required SoC_{ini} is calculated. Figure 9 shows a visual representation of an example weekend day.

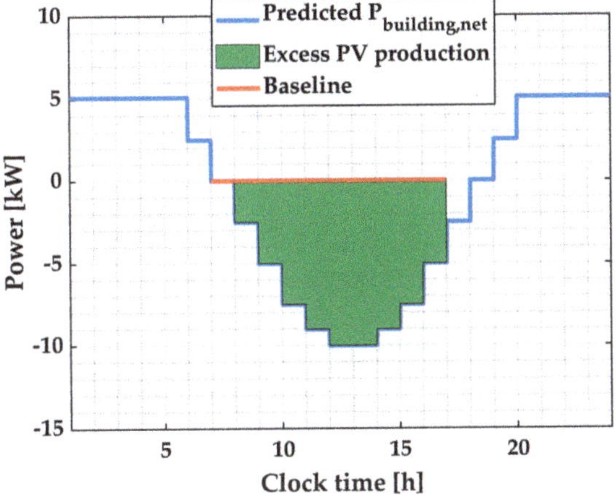

Figure 9. Foreseen excess photovoltaic (PV) production depending on building energy demand forecast on weekend days and holidays.

An illustration of the operational strategy to maintain the flattened demand profile on weekdays and weekends/holidays is shown in Figure 10. Before implementing it in the real building, the prediction models that are established in Section 2.2, the operational strategy and algorithm to determine the BL and SoC_{ini}, as well as a BESS model are implemented in MATLAB.

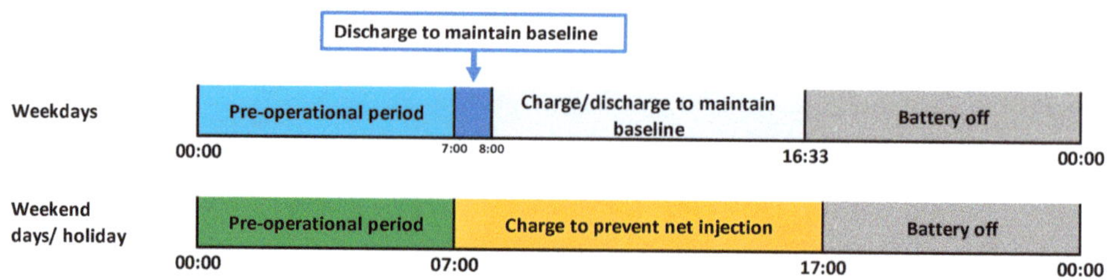

Figure 10. BESS operational schedule.

Even though the demand predictions are carried out with hourly resolution data, after the determination of the required baseline (BL) and SoC_{ini}, operational strategies are simulated with the highest-resolution data available, i.e., 1 min resolution data. This is becausecontrol of the real system occurs on the time scale of seconds rather than hours. Furthermore, battery behavior can only be accurately modeled when simulating with very high-resolution data. After simulation of the power flows in the BESS, the 1 min operation data are averaged to 15 min resolution data. The 15 min resolution is of interest because national electricity grid balancing in the Netherlands is carried out in time blocks of 15 min (clock quarters), also known as the program time unit (PTU) [53]. It is, therefore, reasonable to assess the performance of flexibility efforts at the same resolution.

2.3.3. Assessment of Operational Strategy

Because this research focuses on assessing the building's electrical energy flexibility, key performance indicators (KPIs) are chosen wherein the actual impact of energy flexibility is quantified. Important KPIs that evaluate overall building energy performance and which are used in this research are:

(a) KPI 1: Total energy consumption (excluding PV power generation) [54]. In this paper, this is limited to electricity only;
(b) KPI 2: Exported electricity (feed in from the building's PV system into the AC grid) [54];
(c) KPI 3: Imported electricity (power from the grid) [54];
(d) KPI 4: Battery Electric Storage System (BESS) losses;
(e) KPI 5: Self-consumption [55];
(f) KPI 6: Self-sufficiency [55];
(g) KPI 7: Percentage indicating the proportion of working hours wherein the baseline is successfully maintained.

Furthermore, a qualitative assessment is included using the load duration curves only for working hours and for both working and non-working hours. In addition, similar to the load duration curve, a Baseline Deviation Duration Curve (BDDC) is defined, since the aim of the model is to maintain the power demand as set by the baseline for the building through the operation of the BESS. This curve visualizes how well the system can maintain the baseline during the operational hours. The BDDC can be constructed by calculating the offset between the baseline and electricity consumption from the grid ($P_{building,net}$) for all working hours, as shown in Equation (10).

$$P_{baseline,deviation,t=i} = P_{building,net,\ t=i} - P_{baseline,t=i}\ [kW] \tag{10}$$

Then, the values of $P_{baseline,deviation}$ are sorted in descending order. This leads to a curve that is analogous to the load duration curve. The obtained Baseline Deviation Duration Curve visualizes how well the BESS strategy can maintain the baseline.

2.4. Step 3: Real Building Management System (BMS) Implementation

The final step concerns the practical implementation of the prediction models, algorithms, and BESS control strategies in the InsiteView® Building Management System (BMS). The BMS platform coordinates sensor-based measurements, actuators, and monitoring data at all operational levels in the building and provides an environment where advanced control algorithms can be implemented. After beta testing for ~4 weeks, an experimental phase was conducted for 13 days, from 7 August 2019 to 19 August 2019.

An overview of the general methodological steps that are used to structure Section 2 and the results (Section 3) is provided in Figure 11.

	Section 2: Materials and Methods	Section 3: Results
Step 1: Prediction Models	• Establish prediction models for each load group and implement in MATLAB • Introduce prediction performance evaluation metrics	• Analyze the prediction performance of each load group • Analyze the prediction performance of total electricity demand
Step 2: Operation Strategy and BESS Simulations	• Establish the operation strategy for BESS • Introduce KPIs to quantify the impact • Implement operation strategy with prediction models in MATLAB	• Analyze the building's energy performance according to the established operation strategy
Step 3: Real BMS Implementation	• Implement operational strategies, prediction models and algorithms in BMS	• Analyze the building energy performance when operating with the real BESS according to the operation strategy

Figure 11. Summary of the methodological framework.

3. Results

The results are discussed following the steps described in Sections 2.2–2.4.

3.1. Step 1. Establish Prediction Models

(a) AHU and HVAC control unit: Parametric model

As mentioned in Section 2.2, from the scatter plot obtained for the energy demands of the AHU and HVAC control unit load group ($E_{AHU\&controls}$) against the outdoor temperature ($T_{outdoor}$), a parametric approach was considered for demand prediction. Figure 12 shows the results of the curve fitting, where (a) shows the measurements and fitted curve, and (b) the residuals. The fitted parameters to Equation (1) are provided in Table 2. Thereby, the parametric model is represented by Equation (11).

$$E_{AHU\&controls,t=i} = \exp\Big(\big(0.0075 \cdot T_{Outdoor}{}^{2} - 0.3576 \cdot T_{Outdoor} + 123.8934\big) \cdot \Big[0.01555 + \frac{-0.0016}{1+\exp(-(-1.0342)\cdot(T_{Outdoor}-21.2692))}\Big]\Big) \tag{11}$$

A major factor for the deviations from the function fit above the curve is found to be the opening of the variable air volume (VAV) valve depending on the indoor CO_2 concentration. The error calculation metrics for the fitted curve are a WAPE of 2.81% and a CVRMSE of 4.36%. Since these values are in line with the requirements, the model is considered accurate enough for its predictions.

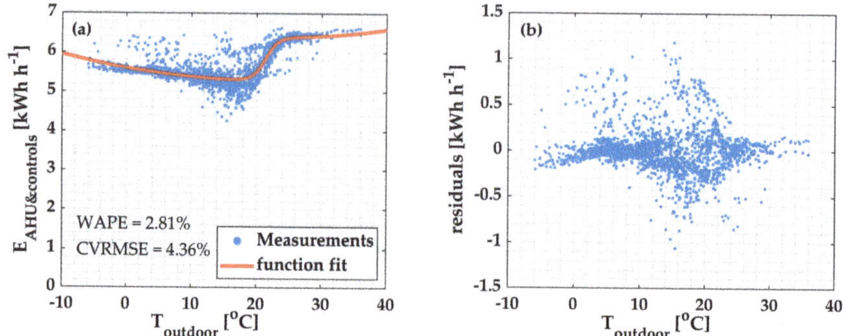

Figure 12. (**a**) Final function fit power demand for the AHU and HVAC control unit during office hours. (**b**) Residuals plot.

Table 2. Numerical values for parameters.

Parameter	Value
A	0.0075
B	−0.3576
C	123.8934
α	0.0155
β	−0.0016
γ	−1.0342
δ	21.2692

(b) Chiller: Multi-variable linear regression

For chiller demand prediction, multi-variable linear regression is used. The results of the k-fold cross-validation for different numbers of variables involved in the regression formula (see Equation (2)) are shown in Figure 13. Overall, the performance is very similar regardless of the number of variables involved. At first, there is a (slightly) increasing performance when using one variable compared to two variables. This behavior is in line with the literature claiming that temperatures at previous time steps (t−1) may have the largest impact on the building cooling load at time t [42]. Including more variables in the regression results in unexpected behavior; at first, the performance decreases, and then it appears to increase again. Since this observed behavior cannot be proven, the sudden increase in performance using ≥ 6 variables is thought to have a mathematical or coincidental origin rather than a physical origin.

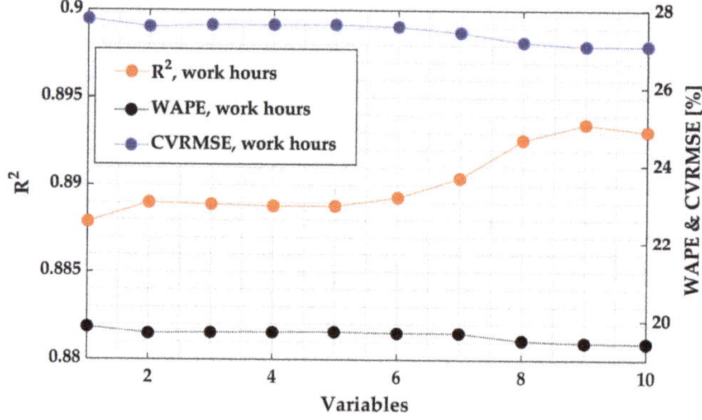

Figure 13. K-fold cross-validation results for multi-variable linear regression.

The results of the k-fold cross-validation for different numbers of variables involved in the regression formula are investigated for up to 10 variables; see Figure 14. As discussed in the previous section, k-fold cross-validation where k = 10 is used—this means that the process of taking the 9 training folds, determining the regression coefficients, and then inputting the validation fold as input is repeated 10 times in total until all folds are used once as a validation set. Each repetition of this process yields a set of regression coefficients. These regression coefficients are plotted in Figure 14. It is observed that the value for a_1 is nearly the same value for all folds, indicating a low uncertainty in the value of this parameter. The results for parameter a_2 show that most values are slightly below zero, indicating a weak negative correlation. For a_3 and beyond, parameter values become extremely uncertain; as shown by the distribution of values around zero, it can be concluded that it is not even clear whether there is a small positive correlation, small negative correlation, or no correlation at all. Therefore, these and subsequent terms were not considered.

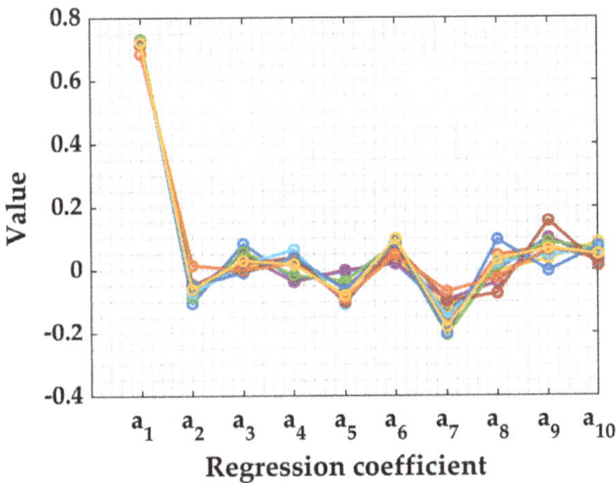

Figure 14. Regression coefficients for k-fold cross-validation using 10 variables.

The regression with two variables reaches a local performance maximum with an R^2 value of 0.89, a WAPE of 19.7% and a CVRMSE of 27.6%. Since the best results are obtained by using two variables, a model of the form as shown in Equation (12) is used and fitted to the dataset. The corresponding coefficients are provided in Table 3.

$$E_{chiller,t=i} = a_0 + a_1 \cdot T_{outdoor,\ t=i} + a_2 \cdot T_{outdoor,t=i-1} \left[\text{kWh.h}^{-1}\right] \quad (12)$$

Table 3. Fitted parameters to the multi-linear regression model.

Parameter	Value
a_0	−10.9190
a_1	0.7902
a_2	−0.1223

(c) Plug loads and lighting: Recent day model

The results of the performance assessment for the lighting and plug load predictions according to the model described by Equation (3) are shown in Figure 15. The performance for all hours (= working + non-working) of the dataset is shown in blue, and the prediction performance for only work hours, is shown in orange.

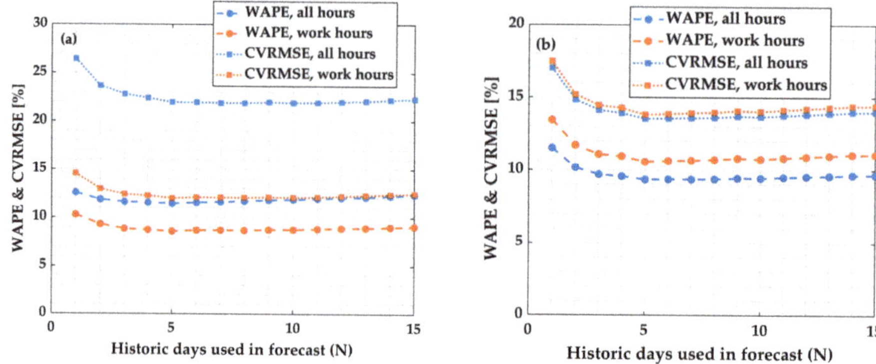

Figure 15. (**a**) Performance assessment of predictions for lighting loads. (**b**) Performance assessment of predictions for plug loads.

The predictions show a sharp increase in performance between $N = 1$ and $N = 2$ for both lighting and plug loads. Using multiple historic data points for the prediction of a future data point is thought to have a stabilizing effect because the outliers which may be present in the historic data are combined with more representative historic values. The predictions, which are made by taking the average of these historic data points, are therefore less affected by outliers. Both the lighting and plug load predictions reach optimum performance when using five historic days ($N = 5$) in the forecast. With $N = 5$ for work hours, the lighting load predictions yield a WAPE of 8.6% and a CVRMSE of 12.0%, and the plug load predictions yield a WAPE of 10.6% and a CVRMSE of 13.8%.

(d) PV power: Solargis®

During the night, there is no sunlight, and PV power predictions for these hours are always zero. These predictions are of course always 100% accurate. These predictions should not be considered in performance evaluation. Specifically, predictions which are done for hours between 17:00 and 07:00 are not included in the Solargis® PV prediction evaluation. Prediction accuracy is determined by comparing Solargis® predictions with the AC-side power measurements of the PV system.

Figure 16 gives an overview of the Solargis® prediction performance as a function of the lead time for the case study building. The overall performance of the predictions shows a rapid decrease in prediction accuracy in the first few lead hours. A peak at 5 lead hours is followed by a sharp decrease in the MAE and the RMSE, after which a long approximately stable period follows. The peak and subsequent decrease mark the transition between satellite-based models and numerical weather prediction models used by Solargis® [46].

The subplots for the different seasons all show similar behavior in terms of the MAE and the RMSE. However, the MBE clearly shows different fluctuations depending on the season and the lead time. As stated earlier, the MBE indicates whether there is systematic overestimation or underestimation in the predictions. In principle, the MBE can be used to easily correct prediction inaccuracy, e.g., by subtracting the value of the MBE in case of overestimation. Due to the various fluctuations, there will be no attempt to improve prediction accuracy through MBE compensation. Such an analysis is beyond the scope of this research.

(e) Outdoor temperature: Solargis®

The accuracy of the temperature predictions is assessed by comparing Solargis® temperature predictions with the temperature measurements which are calculated by the weather station on the roof of the building; see Figure 17.

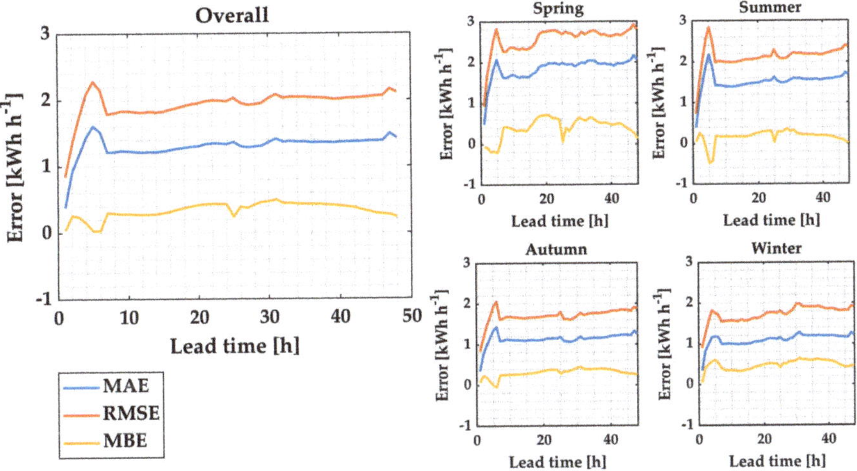

Figure 16. PV yield prediction performance assessment.

Figure 17. Weather station at the case study building.

The results of the analysis of the temperature forecasts for the case study building are shown in Figure 18. The overall performance of all data points is shown as well as subplots for the performance during the different seasons. From the magnitude of the MAE and the RMSE (without MBE correction), it can be seen that temperature predictions are quite accurate overall. The autumn and winter temperatures are predicted best. The errors increase only gradually for longer lead times. For an unknown reason, a small spike at lead time = 25 h is observed. The behavior of the MBE in the overall assessment shows a steady underestimation, with a value of −0.5 °C. Although this value is slightly changed for the different seasons, a systematic underestimation (indicated by the "−" sign) occurs throughout the seasons. MBE correction can be used to improve the predictions. This is achieved by adding a value of 0.5 °C to all the Solargis® temperature predictions of the dataset. The dashed lines in Figure 18 show the modified results. The figure clearly shows that the MBE has shifted upwards to the desired value of ~0 °C. The lower values for the MAE and the RMSE in all plots show that this correction is an appropriate measure to better align the predictions with the measurements.

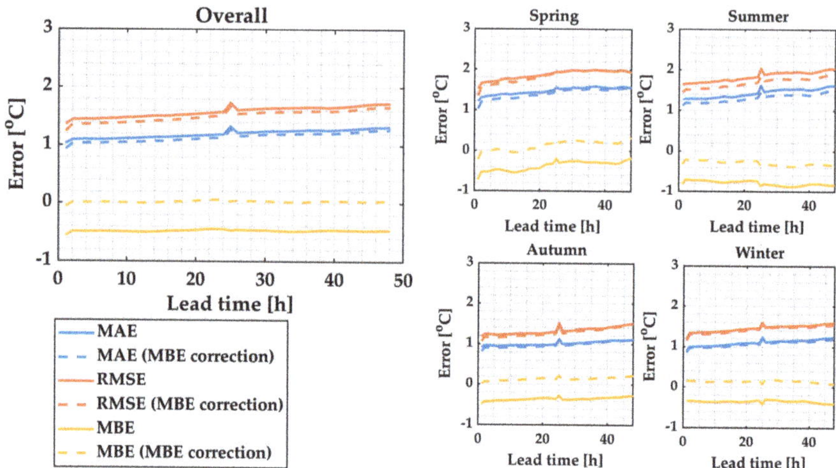

Figure 18. Outdoor temperature prediction performance assessment.

3.1.1. Summary of Subcomponent Prediction Models

The subprediction models and Solargis® services that were demonstrated in Section 2.2 form the building blocks of the complete building energy demand prediction. A summary of the developed models and corresponding performance metrics for each load group is provided in Table 4. For the purpose of this research, the proposed model accuracies are considered sufficient.

Table 4. Summary of the prediction performance of the best performing model for each load group.

Load Group	Model	R^2	WAPE	CVRMSE
AHU and controls	Parametric fitting	N/A	2.8%	4.4%
Chiller	Multi-variable linear regression	0.89	19.2%	27.1%
Lighting	Recent day model	N/A	8.6%	12.0%
Plugs	Recent day model	N/A	10.6%	13.8%

3.1.2. Total Demand Prediction of the Building

The established prediction models for all load groups and Solargis® temperature and PV prediction services are integrated into a combined model, wherein the building's total energy demand is predicted. The dataset used in the integrated model consists of the Solargis® and historic building energy demands from 25 May 2018 to 4 April 2019. The error metrics are computed for the predictions which are calculated at 00:00. Only the predictions calculated for workdays between 07:00 and 17:00 are included in quantifying the error matrices. Figure 19 shows the prediction accuracy of all the models on average for all the months. Predictions for lighting and plug loads are combined and simultaneously assessed for convenience.

From Figure 19b it can be seen that prediction errors are largest during the summer months. Since the chiller operates the most during these months, and with the highest energy demands, the magnitude of error is also larger. During the colder months, the chiller is mostly in standby mode and is thus nearly perfectly predicted because standby power is constant. In Figure 19c, the months January and February show an above-average error magnitude. From the large relative difference between the MAE and the RMSE, it follows that there were a few moments with a relatively large prediction error. These are caused by the opening of the variable air volume (VAV) valve due to CO_2 concentration, which is not a factor that is considered in the predictions.

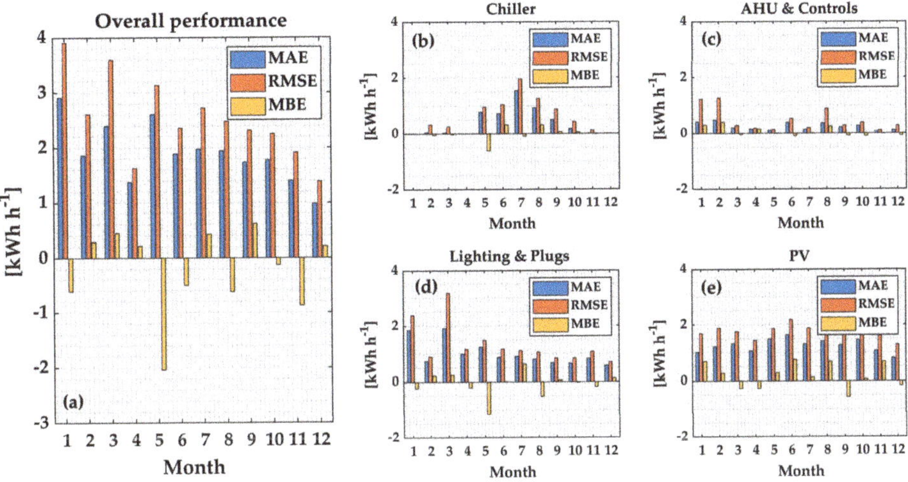

Figure 19. Prediction errors for work hours (07:00 to 17:00).

Lighting and plug load prediction accuracy show above-average larger errors in January and March. As can be seen for the first day in Figure 20a, and in Figure 20b, the building shows completely different behavior compared to expectations. Due to abnormal building operation, which is probably caused by anomalously low occupancy, the predictions are far off, resulting in a large prediction error. Additionally, due to the history-based model used for lighting and plug-load predictions, these abnormal days are still incorporated in the predictions for the next day. Since data from historic days are used to make the forecasts, this again results in the estimation of the demand in the upcoming days. One abnormal day could, therefore, trigger a cascade of prediction accuracy deviation for several days. Nonetheless, overall, prediction errors for this load group are comparatively small and prediction results are satisfactory.

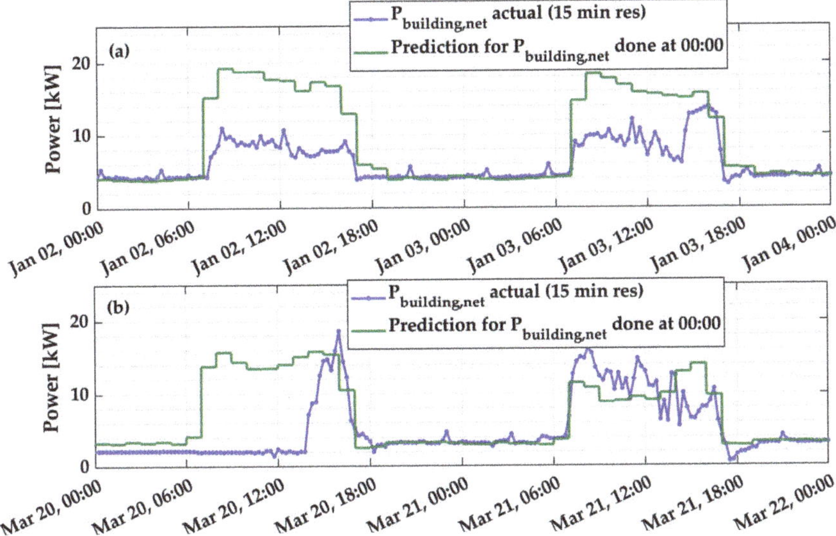

Figure 20. Bad prediction performance in (**a**) January and (**b**) March.

3.2. Step 2: BESS Simulations

In the previous step, multiple prediction models were developed and ultimately combined to predict the total demand of the building. The prediction models are integrated with the proposed operational strategy and simulated in MATLAB. The results of the simulations are evaluated in this section. Table 5 provides an overview of the assessed Key Performance Indicators (KPIs). The overview shows that total energy consumption (KPI 1) has increased by 2.2%, which is caused by conversion loss in the BESS (KPI 4). From a decrease of 60.9% in exported electricity (KPI 2), it follows that the operational strategy has significantly increased self-consumption (KPI 5) from 82.3% to 93.1%. Due to the storage of this excess PV power that would otherwise be exported, the amount of imported electricity (KPI 3) is reduced by 0.4%, as the BESS was capable of providing (some of) the required energy. Overall, self-sufficiency (KPI 6) has increased by 2%, which means that the ratio of self-consumed electricity from PV to total energy consumption (KPI 1) has improved. KPI 7, which quantifies the ability of the system to maintain the baseline, shows that the baseline was successfully maintained for 97.2% of the time on weekdays between 07:00 and 16:33 (see also Figure 10).

Table 5. Assessment summary.

Key Performance Indicator	Without BESS	With BESS Load Balancing	Difference [%]
KPI 1: Total energy consumption (excluding PV power generation) [54] (in this paper, this is limited to electricity only)	55,538 kWh	56,781 kWh	+2.2%
KPI 2: Exported electricity (feed in from the building's PV system into the AC grid) [54]	2309 kWh	902 kWh	−60.9%
KPI 3: Imported electricity (power from the grid) [54]	44,769 kWh	44,604 kWh	−0.4%
KPI 4: Battery Electric Storage System (BESS) losses	0 kWh	1245 kWh	N/A
KPI 5: Self-consumption [55]	82.3%	93.1%	+10.8%
KPI 6: Self-sufficiency [55]	19.4%	21.4%	+2%
KPI 7: Percentage indicating the proportion of working hours wherein the baseline is successfully maintained	N/A	97.2%	N/A

In Section 2.3.3, the Baseline Deviation Duration Curve (BDDC) was defined. This curve provides a visual impression of the ability of the demand curve to maintain the baseline throughout the day. Figure 21 illustrates the baseline deviation.

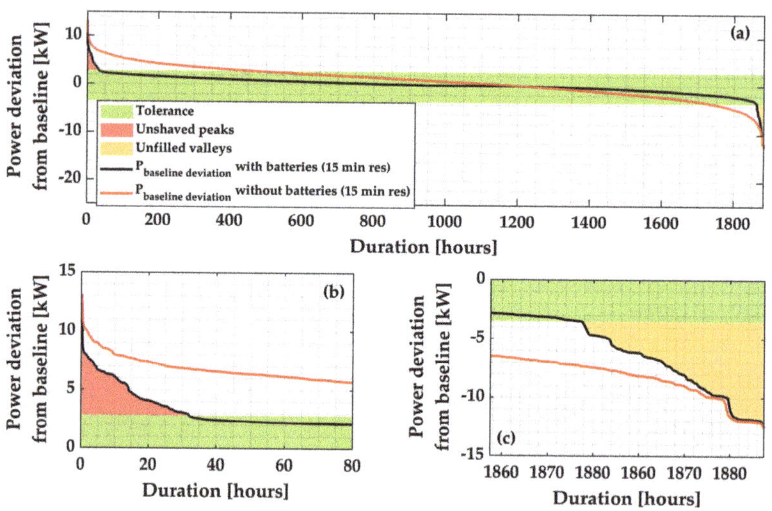

Figure 21. (**a**) Baseline Deviation Duration Curve (BDDC) with 15-minute resolution data. (**b&c**) Close-up of the corners of (**a**).

It is important to realize that the BESS cannot always store/deliver power due to the applied constraints—this means that whenever the difference between $P_{building,net}$ and the baseline is too small, the battery will not deliver or store power. A slight deviation from the baseline (BL) value cannot be prevented, and it is not a problem. This is why the tolerance band, which marks the baseline deviation between which the deviation is considered acceptable, is defined. The green area marks the bandwidth around the zero line of –3.51 kW to 2.85 kW. These values naturally follow when considering the minimum requirement of a 3 kW charging/discharging power constraint before the BESS starts to operate and charging/discharging efficiencies (85.5% and 90% are the charging/discharging efficiencies):

- The BESS is controlled such that it starts charging when there is at least 3 kW/0.855 = 3.51 kW of AC power available. Or in other words, when $P_{building,net}$–BL ≤ –3.51 kW.
- Similarly, the BESS starts discharging when at least 3 kW × 0.95 = 2.81 kW of AC power is required by the building. This means that when $P_{building,net}$–BL ≥ 2.81 kW, then the BESS starts discharging.

Therefore, whenever the baseline deviation ≤ –3.51 kW, the BESS should have charged to fill the valley. Whenever the baseline deviation ≥ 2.85 kW, the BESS should have discharged to shave the peak. Finding baseline deviation values outside of the tolerance means that the BESS was incapable of maintaining the BL and this was not caused by the minimum power constraint.

From the parts of the load duration curve outside of the green area, it can be seen that it was not always possible to discharge/charge to deliver/store the power necessary to maintain the baseline. The peaks which could not be shaved by the BESS are marked by the red area and have a total duration of ~32 hours during the evaluated period. The valleys which could not be filled by the BESS are marked by the yellow area and have a total duration of ~19 hours. Nevertheless, it can be concluded that the system is well capable of actively maintaining the baseline 97.2% of the time (within the green tolerance band). From Figure 22 it can be seen that, overall, there is a decrease in load duration of high positive power and an increase in the duration of low positive power. This is the direct consequence of the load balancing strategies wherein peaks are shaved and valleys are filled.

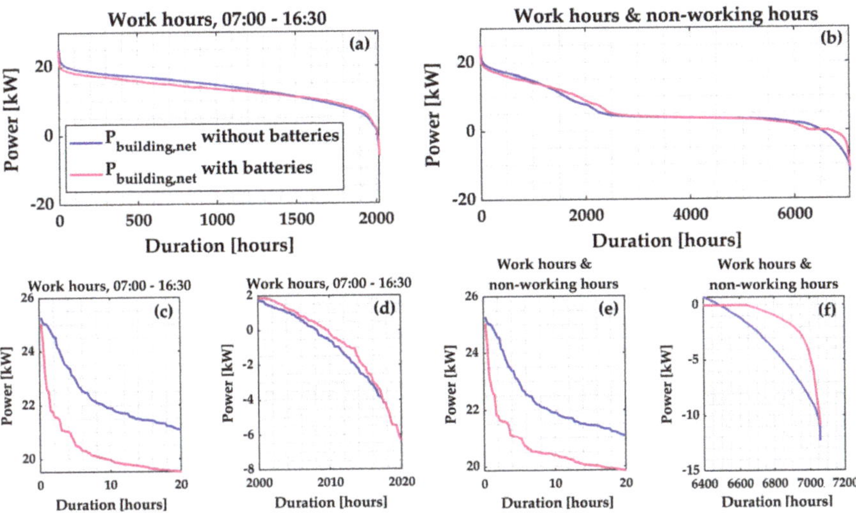

Figure 22. Load duration curves for (**a**) work hours only and (**b**) all hours of the dataset. (**c&d**) Close-ups of (**a**). (**e&f**) Close-ups of (**b**).

3.3. Step 3. BMS Implementation

The KPIs of the building when operating the BESS can readily be calculated from the measurements that are extracted from the Building Management System (BMS) during the experimental period from

7 August 2019 to 19 August 2019. An overview of the resulting KPIs for the experimental period is shown in Table 6.

Table 6. Key performance indicator (KPI) assessment for experimental results.

Key Performance Indicator	Base Case	With BESS Load Balancing	Difference [%]
KPI 1: Total energy consumption (excluding PV power generation) [54] (in this paper, this is limited to electricity only)	2190	2317	+5.8%
KPI 2: Exported electricity (feed in from the building's PV system into the AC grid) [54]	115	70	−39.1%
KPI 3: Imported electricity (power from the grid) [54]	1500	1582	+5.5%
KPI 4: Battery Electric Storage System (BESS) losses	0	127	-
KPI 5: Self-consumption [55]	85.8	91.3	+5.5%
KPI 6: Self-sufficiency [55]	31.5	31.7	+0.2%
KPI 7: Percentage indicating the proportion of working hours wherein the baseline is successfully maintained	N/A	96.2	-

After introducing the BL strategy, total energy consumption increased due to BESS losses. Furthermore, it follows that exported electricity to the national grid is reduced from 115 to 70 kWh, and imported electricity increased from 1500 to 1582 kWh. Self-sufficiency has increased from 31.5% to 31.7% and self-consumption from 85.8% to 91.3%. Finally, during 96.2% of the time, the BESS was able to successfully maintain the BL within the tolerance, thereby, demonstrating that the load shape objectives are most often met. In the future, the value of flexibility can be established if the relevant guidelines and regulations are provided by the energy markets. The load duration curves for the experimental period on the real building are shown in Figure 23.

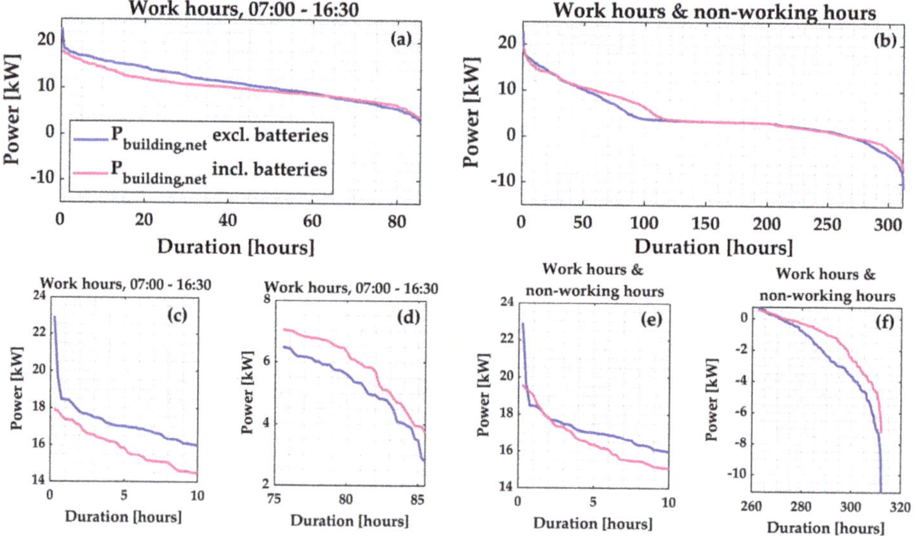

Figure 23. Load duration curves for the experimental results (**a**) Work hours only. (**b**) All hours of the dataset. (**c&d**) Close-ups of (**a**). (**e&f**) Close-ups of (**b**).

The ability of the demand curve to maintain the BL is visualized in Figure 24 using a Baseline Deviation Duration Curve (BDDC). During 96.2% of the time, the BL was maintained within the constraints. There was a total duration of 3 hours wherein peaks were not shaved. However, all valleys were effectively filled during the experimental period.

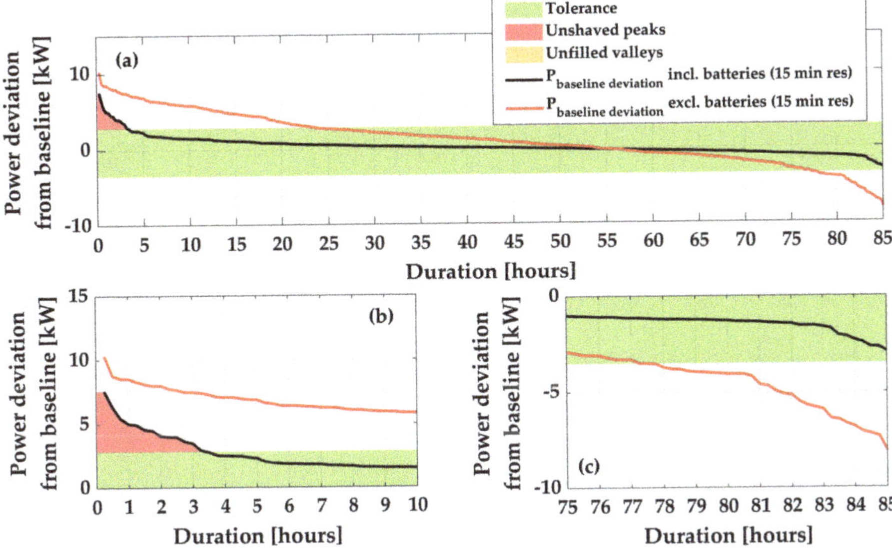

Figure 24. (**a**) Baseline Deviation Duration Curve (BDDC) for the experimental results of strategy 2 with 15-minute resolution data. (**b&c**) Close-up of (**a**).

4. Discussion

The objective of this study was to stabilize/flatten a building energy demand profile during office hours by means of peak shaving and valley filling using a Battery Electric Storage System. This was achieved by defining load shape objectives in the form of a baseline that is determined based on electricity demand forecasts for the building. Before doing so, predicting the electricity demand of the various load groups in the building was achieved through relatively simple models. All individual prediction models of each load group proved to be sufficiently accurate for use in the control strategy of the BESS. Finally, testing the operational strategy with BESS after the predictions resulted in meeting the flattened load shape objectives over 95% of the time in both simulations and practical implementation. The practical implementation was performed without compromising the thermal comfort of the building users. Peak loads, which increase the risk of congestion, were also successfully reduced both in magnitude and duration. Due to BESS losses, total energy consumption is shown to have increased marginally.

Total energy demand forecasting of the building was achieved by combining the separate predictions for each load group. The level of detail required to assess these separate models in order to determine the best performing algorithm makes this approach a labor-intensive process. Even though data-driven machine learning prediction methods are expected to increase prediction accuracy while allowing for higher levels of abstraction, with the current BMS structure of the case study building, the question remains whether that would be practically implementable. The prediction models that were developed in this work were constrained by practical considerations. Nevertheless, the relatively simple prediction models that were developed and optimized proved to be well capable of predicting the building's energy demands with sufficient accuracy within the practical setting.

5. Conclusions

Utilization of flexibility in the future Smart Grid will occur through information and communication technology (ICT) regardless of the exact market design. In the future, it is unknown whether a steady load profile as demonstrated in this paper will be the true load shape objective of a Smart Grid. However, because no operational Smart Grid exists that can be used to define the load shape objective in the Dutch context, and since the method introduced in this article has proven to be successful after implementing in the building, it is safe to conclude that the baseline approach is adequate for demonstrating load flexibility for a future electricity grid setting.

As an extension of this work, in the future, the baseline approach can be extended to a dynamic baseline which varies hourly depending on the grid's needs. BESS flexibility, as well as other sources of flexibility (e.g., HVAC system flexibility), could then be integrated to achieve even greater flexibility. The optimal load shape could be determined based on the combination of the predicted energy demand of the buildings or neighborhoods according to the Smart Grid's needs. However, this work did not include an economic assessment of a Smart Grid using PVs and a BESS. An extension of this study could also be made with an added economic assessment. However, in such a study, a time-of-use tariff structure and the economic value of provided flexibility should be incorporated.

The built environment has the potential to contribute to maintaining a reliable grid by actively participating in future grids. The case study building provided a perfect opportunity for demonstrating and investigating such characteristics.

Author Contributions: R.C., Conceptualization, Methodology, Software, Validation, and Writing—Original Draft Preparation; S.W., Methodology, Supervision, and Writing—Original Draft Preparation; J.v.d.V., Resources, Review, and Supervision; P.N., Review; W.Z., Funding Acquisition, Review, and Supervision. All authors have read and agreed to the published version of the manuscript.

Funding: This research received no external funding.

Acknowledgments: This work was conducted in collaboration with the Netherlands Organization for Scientific Research (NWO) Perspective program TTW Project B (14180) "Interactive energy management systems and lifecycle performance design for energy infrastructures of local communities" (https://ses-be.tue.nl/). We wish to express our gratitude to Kropman Installatietechniek for providing access to necessary data, helping with the data collection process, and providing the building for experiments.

Conflicts of Interest: The authors declare no conflict of interest.

Nomenclature

AHU	air-handling unit	MBE	mean bias error
BDDC	baseline deviations duration curve	PVs	photovoltaics
BESS	battery electric storage system	R^2	coefficient of determination
BL	baseline	RES	renewable energy sources
BMS	building management system	RMSE	root mean square error
CVRMSE	coefficient of variation of the root mean square error	SoC	state of charge
DSF	demand-side flexibility	SoC_{ini}	initial state of charge of the BESS
DSM	demand-side management	t	time
$E_{AHU\&controls}$	predicted energy demand of AHU and HVAC control unit	$T_{outdoor}$	outdoor temperature
$E_{chiller}$	predicted energy demand of the chiller	WAPE	weighted average percentage error
$E_{pred,t=i}$	predicted energy demand at t = i	$\overline{x}$	the mean of all observed values in the dataset
HVAC	heating ventilation and air conditioning	X_{charge}	inability of the BESS to charge
i	index	$X_{discharge}$	inability of the BESS to discharge
KPI	key performance indicators	x_i	the measured (observed) value for data point i
MAE	mean absolute error	$\hat{x}_i$	the predicted value for data point i

References

1. Mancini, F.; Nastasi, B. Energy Retrofitting Effects on the Energy Flexibility of Dwellings. *Energies* **2019**, *12*, 2788. [CrossRef]
2. Rijksoverheid. *Energierapport, Transitie Naar Duurzaam*. 2016. Available online: https://www.rijksoverheid.nl/documenten/rapporten/2016/01/18/energierapport-transitie-naar-duurzaam (accessed on 3 February 2018).
3. CBS StatLine. Available online: https://opendata.cbs.nl/statline/#/CBS/nl/ (accessed on 3 February 2018).
4. PBL Energietransitie. Available online: https://themasites.pbl.nl/energietransitie/ (accessed on 28 March 2020).
5. Koirala, B.; Chaves-Ávila, J.; Gómez, T.; Hakvoort, R.; Herder, P. Local Alternative for Energy Supply: Performance Assessment of Integrated Community Energy Systems. *Energies* **2016**, *9*, 981. [CrossRef]
6. Gvozdenovic, K.; Maassen, W.; Zeiler, W.; Besselink, H. Roadmap to nearly Zero Energy Buildings. *REHVA* **2015**, *2*, 6–10.

7. Arteconi, A.; Polonara, F. Assessing the Demand Side Management Potential and the Energy Flexibility of Heat Pumps in Buildings. *Energies* **2018**, *11*, 1846. [CrossRef]
8. Agentschap, N.L. *Monitor Energiebesparing Gebouwde Omgeving 2018*. Available online: https://www.rvo.nl/sites/default/files/2019/12/monitor-energiebesparing-gebouwde-omgeving-2018.pdf (accessed on 3 February 2018).
9. Walker, S.; Labeodan, T.; Boxem, G.; Maassen, W.; Zeiler, W. An assessment methodology of sustainable energy transition scenarios for realizing energy neutral neighborhoods. *Appl. Energy* **2018**, *228*, 2346–2360. [CrossRef]
10. Javaid, N.; Hussain, S.; Ullah, I.; Noor, M.; Abdul, W.; Almogren, A.; Alamri, A. Demand Side Management in Nearly Zero Energy Buildings Using Heuristic Optimizations. *Energies* **2017**, *10*, 1131. [CrossRef]
11. Fischer, D.; Madani, H. On heat pumps in smart grids: A review. *Renew. Sustain. Energy Rev.* **2017**, *70*, 342–357. [CrossRef]
12. Walker, S.; Katic, K.; Maassen, W.; Zeiler, W. Multi-criteria feasibility assessment of cost-optimized alternatives to comply with heating demand of existing office buildings—A case study. *Energy* **2019**, *187*, 115968. [CrossRef]
13. Qi, F.; Wen, F.; Liu, X.; Salam, M.A. A Residential Energy Hub Model with a Concentrating Solar Power Plant and Electric Vehicles. *Energies* **2017**, *10*, 1159. [CrossRef]
14. Jensen, S.Ø.; Marszal-Pomianowska, A.; Lollini, R.; Pasut, W.; Knotzer, A.; Engelmann, P.; Stafford, A.; Reynders, G. IEA EBC Annex 67 Energy Flexible Buildings. *Energy Build.* **2017**, *155*, 25–34. [CrossRef]
15. Kuiken, D.; Más, H.; Haji Ghasemi, M.; Blaauwbroek, N.; Vo, T.; van der Klauw, T.; Nguyen, P. Energy Flexibility from Large Prosumers to Support Distribution System Operation—A Technical and Legal Case Study on the Amsterdam ArenA Stadium. *Energies* **2018**, *11*, 122. [CrossRef]
16. Manditereza, P.T.; Bansal, R. Renewable distributed generation: The hidden challenges—A review from the protection perspective. *Renew. Sustain. Energy Rev.* **2016**, *58*, 1457–1465. [CrossRef]
17. Coster, E.J.; Myrzik, J.M.A.; Kruimer, B.; Kling, W.L. Integration Issues of Distributed Generation in Distribution Grids. *Proc. IEEE* **2011**, *99*, 28–39. [CrossRef]
18. Salpakari, J.; Lund, P. Optimal and rule-based control strategies for energy flexibility in buildings with PV. *Appl. Energy* **2016**, *161*, 425–436. [CrossRef]
19. Aduda, K.O. Smart Grid-Building Energy Interactions: Demand Side Power Flexibility in Office Buildings. Ph.D. Thesis, Technische Universiteit Eindhoven, Eindhoven, The Netherlands, 2018.
20. Verzijlbergh, R.A.; Vries, L.J. De Renewable Energy Sources and Responsive Demand. Do We Need Congestion Management in the Distribution Grid? *IEEE Trans. Power Syst.* **2014**, *29*, 2119–2128. [CrossRef]
21. Protopapadaki, C.; Saelens, D. Heat pump and PV impact on residential low-voltage distribution grids as a function of building and district properties. *Appl. Energy* **2017**, *192*, 268–281. [CrossRef]
22. Denholm, P.; Connell, M.O.; Brinkman, G.; Jorgenson, J.; Denholm, P.; Connell, M.O.; Brinkman, G.; Jorgenson, J. *Overgeneration from Solar Energy in California: A Field Guide to the Duck Chart*; National Renewable Energy Lab.(NREL): Golden, CO, USA, 2015.
23. Aghahassani, M.; Grillo, S. Voltage regulation by means of storage device in LV feeder using OpenDSS interfacing with MATLAB. Master Thesis, Politecnico Di Milano, Italy, 2017.
24. Shafiullah, D.S.; Vo, T.H.; Nguyen, P.H.; Pemen, A.J.M. Different smart grid frameworks in context of smart neighborhood: A review. In Proceedings of the 2017 52nd International Universities Power Engineering Conference, UPEC 2017, Heraklion, Greece, 28–31 August 2017; Institute of Electrical and Electronics Engineers Inc.: Piscataway, NJ, USA, 2017; Volume 2017, pp. 1–6.
25. Shabanzadeh, M.; Moghaddam, M.P. What is the Smart Grid? Definitions, Perspectives, and Ultimate Goals. In Proceedings of the 28th International Power System Conference (PSC), Tehran, Iran, 4–6 November 2013; p. 5.
26. Slootweg, J.G.; Veldman, E.; Morren, J. Sensing and control challenges for Smart Grids. In Proceedings of the 2011 International Conference on Networking, Sensing and Control, Delft, The Netherlands, 11–13 April 2011; pp. 1–7.
27. Jetcheva, J.G.; Majidpour, M.; Chen, W.P. Neural network model ensembles for building-level electricity load forecasts. *Energy Build.* **2014**, *84*, 214–223. [CrossRef]
28. Sijm, J. *Demand and Supply of Flexibility in the Power System of the Netherlands, 2015–2050*; ECN Policy Studies: Petten, The Netherlands, 2015.

29. Aduda, K.O.; Labeodan, T.; Zeiler, W.; Boxem, G.; Zhao, Y. Demand side flexibility: Potentials and building performance implications. *Sustain. Cities Soc.* **2016**, *22*, 146–163. [CrossRef]
30. Gellings, C.W. The Concept of Demand-Side Management for Electric Utilities. *Proc. IEEE* **1985**, *73*, 1468–1470. [CrossRef]
31. Voerman, M. Grid Connected Active Office Building with Integrated Electrical Storage. Master's Thesis, Eindhoven University of Technology, Eindhoven, The Netherlands, 2017.
32. Jensen, S.Ø.; Madsen, H.; Lopes, R.; Junker, R.G.; Daniel, A. *Annex 67: Energy Flexible Buildings—Energy Flexibility as a Key Asset in a Smart Building Future*. 2017. Available online: http://www.annex67.org/ (accessed on 1 May 2017).
33. Corten, K.; Willems, E.; Walker, S.; Zeiler, W. Energy performance optimization of buildings using data mining techniques. *E3S Web Conf.* **2019**, *111*, 05016. [CrossRef]
34. Agentschap, N.L. Monitor energiebesparing gebouwde omgeving 2017. Available online: https://www.rvo.nl/sites/default/files/2018/12/Monitor%20Energiebesparing%20gebouwde%20omgeving%202017.pdf (accessed on 31 January 2019).
35. Labeodan, T.M. A Multi-Agents and Occupancy Based Strategy for Energy Management and Process Control on the Room-Level. Ph.D. Thesis, Eindhoven University of Technology, Eindhoven, The Netherlands, 2017.
36. Amarasinghe, K.; Marino, D.L.; Manic, M. *Energy Load Forecasting Using Deep Neural Networks*; IEEE: Piscataway, NJ, USA, 2016; pp. 7046–7051.
37. Zhao, H.X.; Magoulès, F. A review on the prediction of building energy consumption. *Renew. Sustain. Energy Rev.* **2012**, *16*, 3586–3592. [CrossRef]
38. World Maps of Köppen-Geiger Climate Classification. Available online: http://koeppen-geiger.vu-wien.ac.at/present.htm (accessed on 28 April 2020).
39. Thomassen, T. Smart Grid Building Energy Management. Master's Thesis, Eindhoven University of Technology, Eindhoven, The Netherlands, 2014.
40. de Bont, K. Developing a Photovoltaic- and Electrical Storage System for Investigation of Demand Side Management Strategies in Office buildings. Master's Thesis, Eindhoven University of Technology, Eindhoven, The Netherlands, 2016.
41. James, G.; Witten, D.; Hastie, T.; Tibshirani, R. *An Introduction to Statistical Learning*; Springer Texts in Statistics; Springer: New York, NY, USA, 2013; Volume 103, ISBN 978-1-4614-7137-0.
42. Ashrae. *ASHRAE HANDBOOK, Heating, Ventilation and Air-Conditioning APPLICATIONS*; ASHRAE: Atlanta, GA, USA, 2015; ISBN 9781936504947.
43. Mahdavi, A.; Tahmasebi, F.; Kayalar, M. Prediction of plug loads in office buildings: Simplified and probabilistic methods. *Energy Build.* **2016**, *129*, 322–329. [CrossRef]
44. Solargis Meteorological Models. Available online: https://solargis.com/ (accessed on 9 July 2019).
45. Walker, S.; Khan, W.; Katic, K.; Maassen, W.; Zeiler, W. Accuracy of different machine learning algorithms and added-value of predicting aggregated-level energy performance of commercial buildings. *Energy Build.* **2019**, *209*, 109705. [CrossRef]
46. Ruiz-Arias, J.A.; Goenka, H. How Solargis is improving accuracy of solar power forecasts. Available online: https://solargis.com/blog/best-practices/improving-accuracy-of-solar-power-forecasts/ (accessed on 9 July 2019).
47. Spiess, A.N.; Neumeyer, N. An evaluation of R2 as an inadequate measure for nonlinear models in pharmacological and biochemical research: A Monte Carlo approach. *BMC Pharmacol.* **2010**, *10*, 1–11. [CrossRef]
48. Louhichi, K.; Jacquet, F.; Butault, J.P. Estimating Input Allocation from Heterogeneous Data Sources: A Comparison of Alternative Estimation Approaches. *Agric. Econ. Rev.* **2012**, *13*, 83–102.
49. Fan, C.; Xiao, F.; Zhao, Y. A short-term building cooling load prediction method using deep learning algorithms. *Appl. Energy* **2017**, *195*, 222–233. [CrossRef]
50. Monfet, D.; Corsi, M.; Choinière, D.; Arkhipova, E. Development of an energy prediction tool for commercial buildings using case-based reasoning. *Energy Build.* **2014**, *81*, 152–160. [CrossRef]
51. Landsman, J. Performance, Prediction and Optimization of Night Ventilation across Different Climates. Master's Thesis, University of California, Berkeley, CA, USA, 2016.
52. Cort, J.W.; Kenji, M. Advantages of the mean absolute error (MAE) over the root mean square error (RMSE) in assessing average model performance. *Clim. Res.* **2005**, *30*, 79–82.

53. Tennet. The Imbalance Pricing System. 2016. Available online: https://www.tennet.eu/fileadmin/user_upload/SO_NL/ALG_imbalance_pricing_system.doc.pdf (accessed on 5 April 2019).
54. Finck, C.; Clauß, J.; Vogler-Finck, P.; Beagon, P.; Zhang, K.; Kazmi, H. *Review of Applied and Tested Control Possibilities for Energy Flexibility in Buildings*. 2018. Available online: http://www.annex67.org/media/1551/review-of-applied-and-tested-control-possibilities-for-energy-flexibility-in-buildings-technical-report-annex67.pdf (accessed on 3 December 2018).
55. Luthander, R.; Widén, J.; Nilsson, D.; Palm, J. Photovoltaic self-consumption in buildings: A review. *Appl. Energy* **2015**, *142*, 80–94. [CrossRef]

Article

Cost-Aware Design and Simulation of Electrical Energy Systems

Yukai Chen [1,*], Sara Vinco [1], Donkyu Baek [2,*], Stefano Quer [1], Enrico Macii [3] and Massimo Poncino [1]

1 Department of Control and Computer Engineering (DAUIN), Politecnico di Torino, 10129 Torino, Italy; sara.vinco@polito.it (S.V.); stefano.quer@polito.it (S.Q.); massimo.poncino@polito.it (M.P.)
2 School of Electronics Engineering, Chungbuk National University, Cheongju 28644, Korea
3 Interuniversity Department of Regional and Urban Studies and Planning (DIST), Politecnico di Torino, 10129 Torino, Italy; enrico.macii@polito.it
* Correspondence: yukai.chen@polito.it (Y.C.); donkyu@cbnu.ac.kr (D.B.)

Received: 21 April 2020; Accepted: 4 June 2020; Published: 8 June 2020

Abstract: One fundamental dimension in the design of an electrical energy system (EES) is the economic analysis of the possible design alternatives, in order to ensure not just the maximization of the energy output but also the return on the investment and the possible profits. Since the energy output and the economic figures of merit are intertwined, for an accurate analysis it is necessary to analyze these two aspects of the problem concurrently, in order to define effective energy management policies. This paper achieves that objective by tracking and measuring the energy efficiency and the cost effectiveness in a single modular framework. The two aspects are modeled separately, through the definition of dedicated simulation layers governed by dedicated virtual buses that elaborate and manage the information and energy flows. Both layers are simulated concurrently within the same simulation infrastructure based on SystemC-AMS, so as to recreate at runtime the mutual influence of the two aspects, while allowing the use of different discrete time scales for the two layers. Thanks to the tight coupling provided by the single simulation engine, our method enables a quick estimation of various cost metrics (net costs, annualized costs, and profits) of any configuration of EES under design, via an informed exploration of the alternatives. To prove the effectiveness of this approach, we apply the proposed strategy to two EES case studies, we explored various management strategies and the presence of different types and numbers of power sources and energy storage devices in the EES. The analysis proved to allow the identification of the optimal profitable solutions, thereby improving the standard design and simulation flow of EES.

Keywords: design-time optimization; cost modeling and simulation; cyber-physical system; electrical energy system; sustainable energy planning; sustainable power planning; design space exploration; SystemC-AMS

---

1. Introduction

In the design of large-scale electrical energy systems (EESs), cost is a dimension at least as important as the energy efficiency of the system: given the initial investment, in fact, users do want an effective solution that can provide a return on the investment in the shortest possible time.

Designing an EES encompasses a number of options, such as the choice of components (which power sources and which storage devices), their sizing, and particularly, their management (how the energy flow is controlled among all the actors), possibly in a way that is aware of the load profiles. The problem of optimizing the cost under an initial investment cost constraint is therefore a complex problem, as it involves both "physical" aspects (i.e., the dynamics of the various devices, their non-idealities, the electrical characteristics of the loads, etc.) as well as "cyber" issues (the algorithms that manage the flow of energy among these devices and the loads).

It is quite evident that accounting for (i) such a set of heterogeneous variables, (ii) numerous significant non-idealities, and (iii) complex inter-dependencies between components can only be handled effectively by the simulation of the EES as a cyber-physical system (CPS). This would allow one to describe accurate (power and cost) models for the components, fed by accurate traces of environmental data for the power sources, and exercised under realistic power demand traces [1,2]; on top of that, management policies modeled in software can evaluate a number of alternative scenarios. Although throwing all these aspects into an optimization problem would be possible [3], this could be done only using average quantities as representative values for the variables of the problem.

The literature presents many solutions for the simulation of these cyber-physical electrical energy systems (CPEES), with different levels of accuracy, complexity, generality, and flexibility [4–8]. Most of these approaches lack one fundamental feature which could be regarded as *modularity*. With this term we mean the possibility of separating the different layers of information to be tracked in the CPEES simulation. For instance, the analysis of the power flow (an "power layer") could be carried out to extract information that could be used for different purposes by another "layer" of simulation that sits on top of that power layer. Such information could be used, for example, to track the reliability (e.g., the mean time to failure, MTTF, or the mean time between failures, MTBF) of the CPEES, using appropriate reliability models that depend on how energy is used and are fed by the power traces obtained by the simulation at the power layer. Alternatively, as done in this work, one could use the power traces to feed *cost* models to assess the overall economic balance of the system. In some cases, a user might want to have both layers (reliability and cost), while in others, one might be interested in only one of them. This degree of modularity requires a specific architecture of the overall simulation framework.

An interesting solution that follows this modular approach was proposed in [9], where the authors design a framework for the concurrent simulation of both functionality and extra-functional properties, yet in a different context. The work refers to smart electronic systems [10–12], which can be seen as small scale CPSs; here, the bottom layer of the simulation is the *functionality*; i.e., what the overall system does and its timing evolution in terms of digital signals. Layers built on top of this baseline layers (called "non-functional") track other quantities (called properties), such as power consumption, temperature, and reliability, stacked in this order. The key for modularity in this work was the definition of a multilayer, bus-centric framework where each layer has a similar structure: each simulated quantity corresponds to a simulation layer, and the bus-centric organization in each layer implies the definition of a virtual bus, which conveys and elaborates quantity-specific information (i.e., power-bus, temperature bus, etc.) to ease synchronization and information exchange.

In this work we adopt the paradigm of [9] to use it to add support for a new "property"; i.e., *cost*. Cost is modeled as a new layer of the framework of the bus-centric approach: component-specific costs are estimated locally to each component, while the bus merges them and keeps track of the power balance and of any operation of the grid; i.e., to buy or sell power. We additionally extend the framework to focus on the simultaneous simulation of cost with the power layer, to reproduce the mutual interactions of the two properties, and to investigate such mutual inter-dependency.

Finally, we apply the extended framework to the design of a custom EES, that is used to highlight and investigate the characteristics of the proposed modeling and simulation approach.

The paper is organized as follows: Section 2 discusses the background, including related work and a brief introduction of the multi-layered framework of [9]. Section 3 illustrates how to build the cost-layer and the information exchanges with the other layers. The implementation of proposed simulation framework is introduced in Section 4. Section 5 exemplifies the overall approach on a reference EES case study to prove the effectiveness of the proposed solution, and Section 6 draws our conclusions.

2. Background

2.1. Cost Estimation for EES Systems

The estimation of the total cost (and the possible resulting economic benefit) of an EES should consider a number of cost items; namely, the initial investment, the runtime operation expenditures (i.e., maintenance and obsolescence), and the cost of energy consumption, which depends on the overall architecture and the load profiles. Once a comprehensive model is available, it will allow one to compare different solutions (mainly the allocation of the energy flow, and the choice and sizing of the components), so as to choose the most profitable one, by taking into account a number of constraints and of optimization goals.

Given its relevance in many domains (industry, residential, large-scale energy generation installations), the literature on the topic is quite rich. The most recent solutions proposed at state of the art are summarized in Table A1 (moved to the Appendix A not to interrupt the natural flow of the paper). The Table highlights the goal of each work, how EES components are modeled, the considered costs, and most importantly, the solution proposed by each work, which mostly fall under two categories: optimization-based approaches and simulation-based approaches.

Optimization-based solutions use analytical or empirical equation-based models of the power characteristics of EES components and the corresponding costs, and formulate the problem into the constrained optimization of a given target; e.g., maximization of power production, or minimization of a cost function [13–31]. Unfortunately such approaches suffer from many limitations. Given that the focus is optimization of some economic parameter, the evolution of EES components is considered only as a byproduct: either as a constraint or an optimization goal [29–31], or as a dimension of the problem that can be reproduced with simplified models or even simple input traces [13–18,24,25,28,32], thereby sacrificing accuracy to simplify computation. When accurate models are adopted, they are restricted to a subset of the components considered of interest (e.g., only batteries or PV modules), while the other EES components are either ignored or modeled with simplistic equations [20–23,26,27]. Few works take into account the power management strategy used to activate power sources and energy storage devices. When this happens, the goal is determining the optimal day-ahead scheduling of energy storage devices, thereby preventing the comparison of different power management strategies [26–31]. As a result, power dynamics are always considered as a minor dimension of the problem (i.e., a constraint, an optimization goal, or an input), and the mutual impact of power and cost is completely lost. Finally, these approaches provide one solution (or a few) and do not easily allow a comparative analysis of various solutions, nor the sensitivity of a solution with respect to some of the parameters.

Simulation-based approaches optimize EES design from an economic perspective based on dynamic simulations of alternative configurations [33–35]. The goal is the evaluation of the impact of electricity pricing, the feasibility of the constructed EES, or the evaluation of the most economical alternative. However, once again the focus is on the economic dimension, thereby adopting gross grain temporal scales (e.g., 1 h time steps) and restricting accurate modeling only to specific classes of EES components of interest (e.g., PV modules [33,34]). Finally, the focus is restricted to only one direction of the mutual impact between the cost dimension and power dynamics of the EES under analysis, which are evaluated strictly sequentially, depending on the causality of interest.

The main limitations of the aforementioned approaches are thus the adoption of abstract and simplistic models of EES components and the gross grain temporal scale: the energy dimension is thus modeled with a very low level of detail, thereby missing important intrinsic dynamics of the EES components, together with the modeling of inefficiencies and of realistic operating conditions. Additionally, cost and power evolution are never considered as mutually interacting dimensions, thereby missing important dimensions of the analysis.

The work proposed in this paper can be viewed as a different perspective; while it is a simulation-based approach, it *leverages results of an independent simulation of power traces*, which can use models with variable accuracy and possibly with different time scales; these power traces are then fed into the various models of different cost items (where they are power-dependent), which can vice versa influence and impact the power dimension. This kind of modular approach is not, however, the result of two distinct simulation environments, but both power and cost information can be derived by concurrently simulating them: this allows one to expose all mutual dependencies between power and cost, and thus to improve the design of EES with a more informed view of all the variables. There is one study [36] that focused on the modeling the different components in the EES by using a different model of computation (MoC) in one simulation framework by using SystemC-AMS. Cost is indeed mentioned in that study, but there is no indication of how the cost models of that work are linked to the power simulation. These are simply obtained by a post-processing of the power traces, as done in most related works. Our proposed simulation framework speeds up the design-time optimization process and has different methods to evaluate the power flow and economic benefit of EES by concurrently simulating the two quantities. The key for such a modular and concurrent simulation approach is described in the following section.

2.2. A Multi-Layered Approach for Functional and Extra-Functional Simulation

The simultaneous simulation of the various aspects of a system requires the construction of frameworks that integrate different views of a system in a single run. The work in [9], which targeted smart electronic systems, proposed an effective methodology that allows the simulation of system functionality together with its power, thermal, and reliability evolution.

The approach proposed in [9] envisions a multi-layer, bus-centric framework (depicted on the left-hand side of Figure 1, adapted from the original paper): *multi-layer* because it has a stacked layers structure; each layer is associated with one simulated characteristic of the system (called *property*). *Bus-centric* means each property is simulated with a virtual bus in each layer, used to update the property-specific status of the system. All simulated properties are thus reduced to this common structure, thereby easing synchronization and information exchange. The goal is indeed to avoid the construction of co-simulation frameworks, and to rather simulate all layers in a single run by falling on the same implementation language; namely, SystemC-AMS.

Each layer is fixed to a single generic underlying *architecture*, of which the layer-specific bus is the central element, used to carry information between components (*layer-specific signals*). Such information is specific of the property under analysis in the layer (e.g., voltage and current for the power layer). Each component in the system corresponds to a model in each layer, to capture the property-specific evolution over time of the component. The *layer-specific bus* aggregates the property-specific information of each component to control the information flow and update the overall property-specific information. Layers can additionally share information (*inter-layer signals*) to mutually influence each other. Notice that signals can go in both directions; i.e., they can be forwarded to upper layers, as signals that are needed to carry out the simulation at that layer (solid lines); but signals can also be fed back to lower layers as information that can make the lower-level simulation more accurate (dotted lines). In the context of [9], for example, the former type of signals could be power consumption signals forwarded upwards to compute the temperature map of the system; the latter type could be the same temperature maps, which could impact the power consumption of some of the components. Most importantly, when the layers run at different time scales, *time converters* can be introduced to convert the signals towards coarser or finer time scales. For instance, while it makes sense to measure the functionality of an electronic system at the nanosecond scale, this is not a required granularity for power simulation, and can be even more relaxed for the analysis of the thermal flow.

Not shown in the figure, there are also *layer-specific description data* that are strictly referring to that layer and are needed as a sort of "context" of the simulation. In the temperature layer of [9], these data could refer either to the physical locations of the components, which affect the thermal flow, or to the characteristics of the materials (type, thickness, etc.) constituting the system itself.

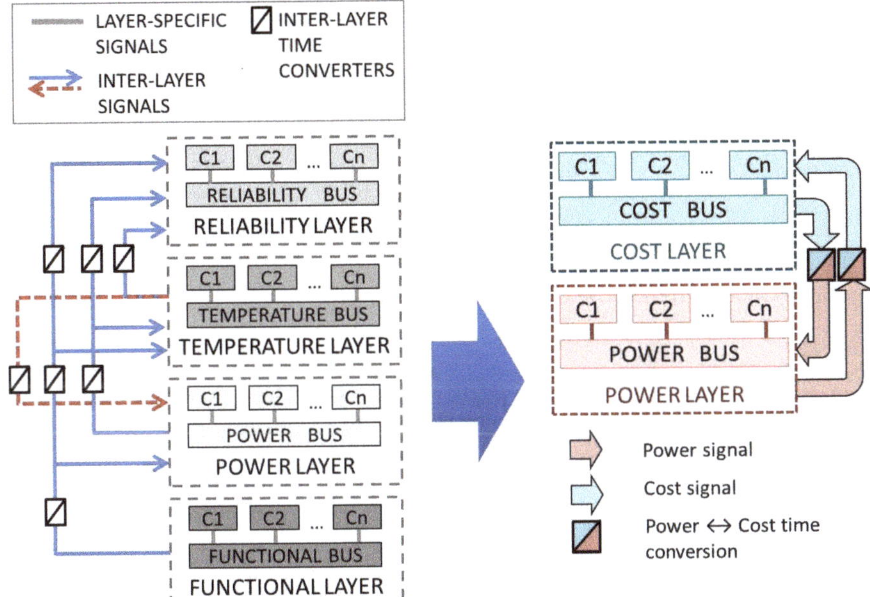

Figure 1. The multiple layered framework for the non-functional properties simulation proposed in [9] (**left**) and application to the simulation of power and cost proposed in this paper (**right**).

This paradigm proved to work very well for the simulation of the different views of a smart system, thanks to its scalability and to the simultaneous evolution of the different system properties, which thus can react simultaneously the one to changes of the other.

For this reason, in this work we tried to fit the above-described paradigm to the *co-simulation of power and cost of an EES*, with the objective of exploiting its two main benefits; namely, modularity and generality. Notice that the contribution of this work is showing that the layered, modular framework of [9] is not a customized architecture, but its paradigm can be extended to other quantities: in this specific case the operational costs of an EES. The work of [9] was applied to small-scale electronic systems, where the non-functional properties were (from bottom to top; see Figure 1 in the manuscript) power, temperature, and reliability. Our work aims at demonstrating the general applicability of the modeling and simulation paradigm of [9] to the context of large-scale EES, and can add different non-functional properties that the properties in this work are power and cost. The right-hand side of Figure 1 shows how the generic bus-based layered architecture maps to our specific context. Since we focus on EES, the first layer of the simulation stack coincides with the power layer. In some sense, this represents the equivalent of "functionality" in the original version; that is, the lowest abstraction level of the semantics. On top of the power layer sits the cost layer, which receives power flow information (used to update the power-dependent cost items, such as electricity cost) and returns cost information to the power layer, which can be used to implement policies in the power bus to decide the optimal power flow among the various components.

Time converters are also expected, since some updates of some of the cost items might generally have different time granularity with respect to power updates, which in the finest granularity are updated every 15 min (as in the most accurate meters).

In the conceptual architecture of Figure 1, the power layer is fundamentally working just as described in [9]; even if components are different in their power scale (100 s of Watts vs. milliwatts), their interfaces and interaction are the same. The cost layer, conversely, is an original layer and its implementation within the constraints of the layered structure demonstrates the claim of modularity of the architecture. In the next section, we will describe the cost models and the technical details to incorporate the cost layer within the template of Figure 1.

3. Modeling of Cost

Calculating the *life-cycle cost* of the EES is an accurate and sound way to estimate the overall cost spent on an asset over the course of its useful life, thereby including the initial capital costs, the projected operating costs, and the maintenance costs, plus possible disposal costs or final residual values of the asset. This Section maps the characteristics of the various models of the life-cycle cost with respect to the layered structure outlined in Section 2.2.

3.1. Main Characteristics of the Cost Layer

The cost of the whole EES is a combination of *component-specific* cost items—that is, that can entirely be determined locally for each component (e.g., initial investment cost, operation, and maintenance cost); and of *global* costs—that is, costs that require an aggregation to be computed; this is essentially the case for electricity cost, whose computation implies the calculation of the balance of the power flow. Therefore, the adoption of a bus-based architecture for the cost layer is relatively natural and straightforward:

- Each component computes its "local" costs over time;
- All individual costs are conveyed to the cost bus;
- The cost bus estimates additional global costs, i.e., due to energy balance with respect to the grid;
- The cost bus determines and keeps track of the total balance over time.

The characteristic information managed in this layer is cost, interpreted as a numeric value in some currency. Cost is thus the only *layer-specific signal* that connects all components to the layer-specific bus, and all components are directly connected to the cost bus. Notice that this architecture may not reflect the actual *physical* organization among the components in the actual EES. For instance, in the real-world EES, a certain component may be connected to the physical power bus through a DC-DC converter; nonetheless, in this *virtual* cost layer, both the component and the DC-DC converter are independently connected to the cost bus. Figure 2a shows a pictorial representation of the general structure of the cost layer.

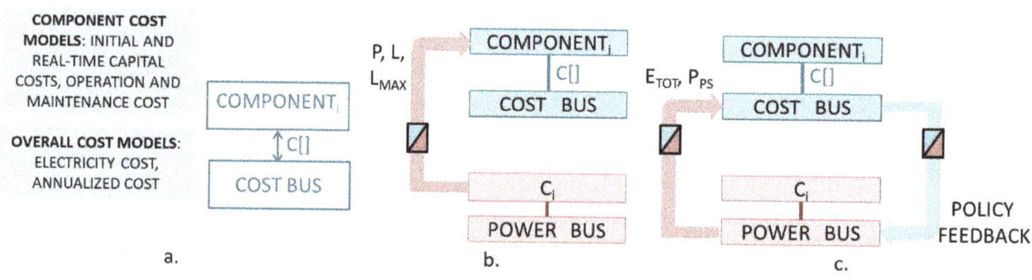

Figure 2. Organization of the cost layer: property-specific signals (**a**), inter-layer signals shared by models of the same component (**b**), and with the power bus (**c**).

3.2. Cost Models

This Section details the various cost components by category, and it maps them to the cost layer template. Notice that the presence of multiple cost models makes the cost signal connecting components to the bus rather an array of cost values, as sketched to Figure 2a, where each array element denotes one cost category listed hereafter. In the following, we will use uppercase letters C to denote cumulative costs; i.e., costs whose definitions encompasses the time intervals over which they are accrued. Lowercase *c* denotes instantaneous costs. The former are expressed in currency × time, e.g., \$ × year, and the latter in currency.

3.2.1. Component-Specific Costs

We consider three main component-specific cost items. Since our focus is on small to medium-scale EESs (such as those in residential installations), we ignore here disposal costs or final residual values as they can be regarded as marginal. These could be, however, easily incorporated without any conceptual difficulty.

Initial Capital Cost

Initial capital cost is the initial investment for the purchase of a component at time zero. This expense is computed for component i by considering its unit cost $C_{unit,i}$ and its cardinality N_i:

$$C_{capital,i} = N_i \cdot C_{unit,i} \tag{1}$$

where C_i is the cost of one unit of component i; N_i is the number of units of i. We decouple the two terms to allow the possibility of tuning the size of components in the exploration. Many components can in fact being seen as modular (e.g., a PV panel consists of a number of PV modules, a battery pack of a number of cells, etc.).

Time-Dependent Capital Cost

Time-dependent capital cost describes the decrease in the value of a component over its useful life as a consequence of obsolescence and/or wearing, and it is in general proportional to its initial capital cost. We consider two types of components, depending on how their loss of value is defined. A first category corresponds to components that have a maximum expected duration; examples are, for instance, PV panels and wind turbines, whose datasheets typically define maximum operational life. For this class, we call this cost *depreciation cost*, defined for component i as:

$$C_{depreciation,i}(t) = C_{capital,i} \cdot CRF \cdot \frac{t}{T_s} \tag{2}$$

where $C_{capital,i}$ is defined as in Equation (1), and T_s serves as a time normalization factor and it denotes the number of simulation samples (ΔT) per year, in order to express the depreciation over each sample time (e.g., if the ΔT is 1s, $T_s = 3600 \times 24 \times 365$). CRF is the *capital recovery factor*, a ratio used for estimating the present value during the lifespan of a system if invested at a particular interest rate [37]. CRF is expressed by:

$$CRF = \frac{R \cdot (1+R)^n}{(1+R)^n - 1} \tag{3}$$

where R is the interest rate, and n denotes the number of years of operations of the system (i.e., component lifetime).

A second class of components has instead a lifetime not defined a priori, but rather determined by their usage characteristics; this is for instance the case of batteries, whose lifetime is defined upon reaching a given value of usable capacity, which depends on several usage-related factors [38]. To distinguish from the former category, we call the time-dependent capital cost for this class as *wear-out cost*, defined for component i as:

$$C_{wout,i}(t) = C_{capital,i} \cdot \frac{L(t)}{L_{max}} \tag{4}$$

where $L(t)$ is the loss of "functionality" over time and L_{max} is the maximum value of the loss after which the component is considered not functional and it needs to be replaced. Thus, when $L(t)$ reaches L_{max}, the entire capital cost has been consumed.

$L(t)$ clearly depends on the type of component and requires an ad-hoc model. Using a battery as an example, $L(t)$ will be the capacity loss due to both calendar and cycle aging, which are affected by several parameters [39], most of which can be attributed to how the power is drawn from the battery, and thus from information derived from the power layer. L_{max} is the maximum capacity loss before the battery is considered as depleted (usually a loss of 20% from the initial capacity). This information must be provided at runtime by power simulation, as will be explained later in this Section.

In summary, the *time-dependent capital cost* depends on the type of component; the term $C_{capital,i}(t)$ will correspond to either Equation (2) or Equation (4) depending on the characteristic of component i.

Operation and Maintenance Cost

Another time-dependent cost is the one due to operation and maintenance [40] that includes both scheduled and major corrective maintenance. This cost is component-specific, as it strictly depends on the characteristics of the component (e.g., it may include periodical cleaning, wiring replacement, and screw and bolt tightening). To this extent, many definitions of this cost are available.

$$C_{operation,i}(t) = C_{operation,yearly,i} \cdot P_i(t) \cdot \frac{t}{T_s} \tag{5}$$

where $C_{operation,yearly,i}$ is expressed in \$/kW/Year; $P_i(t)$ is the instantaneous power flow related to different components i; e.g., the power generation of PV modules and power provided by the battery pack. T_s is the time normalization factor defined as in Equation (2). The $C_{operation,i}$ includes the base cost of element replacement inside each component during its operating lifetime.

3.2.2. Bus Models

The functions of the cost bus are two-fold; firstly, it calculates the costs that depend on the overall power flow of the EES; i.e., the cost of electricity that is bought (and sold) over time from the grid. Secondly, it computes some global cumulative metrics to be used for the exploration of design alternatives or for computing the sensitivity with respect to some parameters of the EES.

Electricity Cost

Electricity cost is the instantaneous cost related to the money paid to (or received from) the utility provider as an effect of the total power balance in the system. This cost can be modeled as:

$$c_{elec}(t) = p_E(t) \cdot \frac{E(t)}{\eta_{conv}} \tag{6}$$

where $E(t) = P(t) \cdot \Delta t$ is the energy to be bought (or sold) at time t; $P(t)$ is the instantaneous power demand (positive or negative), which refers to the total balance of the EES, and its sign determines whether the power is being bought ($P(t) > 0$) or sold ($P(t) < 0$). $p_E(t)$ is the instantaneous electricity price (in currency/kWh); its value depends on the sign of $P(t)$:

$$p_E(t) = \begin{cases} p_{E,buy} & \text{if } P(t) > 0 \\ p_{E,sell} & \text{if } P(t) < 0 \end{cases}$$

where typically $p_{E,sell} < p_{E,buy}$. $\eta_{conv} \leq 1$ is the efficiency of the conversion process; i.e., how the nominally consumed energy $E(t)$ is actually perceived [2].

Given that Equation (6) depends on the overall power flow (production vs. demand) of the EES evolution, this cost can only be computed by the cost bus, that, by collecting the individual signals from the components can have a global perspective on the system.

Net Cost

The first global metric computed by the cost bus is *net cost over time*; i.e., simply the sum of all cost components:

$$C_{net}(t) = \sum_{\tau=0}^{t} \Big(c_{elec}(\tau) \Big) + \sum_{i} \Big(C_{capital,i}(t) + C_{operation,i}(t) \Big) \tag{7}$$

including, thus, time-dependent capital costs, operation and maintenance costs, and the cost of electricity until time t. Notice that the total electricity cost is obtained by integrating (summing) the c_{elec} component over the time interval $[0, t]$, whereas the other components already include the time interval in their definition.

Annualized Cost

The annualized cost of an EES is the cost that, if it were to occur equally in every year in the system lifetime, would give the same net present cost as the actual cash flow sequence associated with the system:

$$C_{annualized} = \frac{C_{net}(t_{MAX})}{\sum_{t=0}^{t_{MAX}} P_{PS}(t)} \tag{8}$$

where t_{MAX} is the maximum system lifetime, and $P_{PS}(t)$ is the power produced by the EES power sources over the same interval. The equation considers all costs from the beginning of simulation to the end of system lifetime, and divides them by the total produced power, with no distinction between whether such power has been used to satisfy load demand or to sell power to the grid.

This cost basically returns the average cost per kWh of consumed energy produced by the system: it is thus a useful metric to compare alternative configurations of the EES and to have a picture of which configuration is more convenient than the others.

Profit

The net cost provides an indication of the total cost over a given time interval. When exploring different design alternatives (EES architectures, policies), a most useful metric is the actual profit of a given configuration. Defining a profit would, however, require to set a baseline to compare against; since the key element that has the most sizable impact on profit is the presence of renewable power sources, our definition of profit focuses on the net balance of the energy provided from power sources to the load and not requested from the grid (at the price $p_{E,buy}$). From this energy we need to subtract the net cost defined in Equation (7) .

$$Profit(t) = \sum_{\tau=0}^{t} \Big(p_{E,buy}(t) \cdot \frac{E_{ps2load}(t)}{\eta_{conv}} \Big) - C_{net}(t) \tag{9}$$

Clearly, the profit is monotonically increasing with t during the lifetime of EES. The positive value of profit illustrates the given configuration can bring real benefit to the users, while the negative value indicates the given configuration is not a profitable one.

3.3. Interaction with the Power Layer

As stated in the previous section, some cost models depend on the power flows in the EES. The concurrent simulation of the cost and power layer is, therefore, essential to get an accurate estimation of the total EES costs. The generic inter-layer interaction depicted in Figure 1 shows generic connection between two layers. However, there are distinct types of interactions among the power and cost layers.

The first one involves *an individual component* in the two layers (Figure 2). The power model of a component C_i sends the following information to its cost model:

- The time-dependent capital cost requires information about the current loss of functionality of the component L, and the maximum accepted loss L_{max} (Equation (4)). Both values are known at the power layer: L_{max} is a configuration value used by policies to determine when the component reached the end of its lifetime; L is a value that is constantly updated at simulation time to correctly estimate the behavior of the component. Using again the example of a battery, this would be approximated by the *full cycle equivalent* [41], i.e., the number of equivalent full charge-discharge cycles, which can be computed by just knowing the nominal battery capacity and the instantaneous power drawn.
- The operation and maintenance cost strictly depends on the power produced or stored by a component over a given time interval (Equation (5)); the power layer naturally keeps track of this quantity during simulation.

The second type of interaction involves the two buses as it concerns *aggregate* information (Figure 2c). The power bus forwards to the cost bus the following information:

- The energy balance over time E_{TOT}, taking into account the difference between the power provided by power sources and the power demand to guarantee the load operations. This quantity allows one to compute the electricity cost over time (Equation (6));
- The total power produced over time by power sources P_{PS}, useful for the estimation of the annualized cost (Equation (8)).

There is, however, also a feedback flow *from* the cost bus *to* the power bus. The former can in fact provide information to the power layer that can be used to design and apply specific power management policies. Examples of such information are:

- The current price of electricity: at times, the price of electricity may be so low that it makes convenient to recharge all energy storage elements (e.g., battery packs), and use the stored energy to power the system when electricity price is higher.
- Data on wear-out and/or depreciation in the form of alarms or warnings that can be used to force the replacement of some components.

The next section will present the implementation details (software infrastructure, timing model, etc.) of the overall intra-layer and inter-layer signal interaction.

4. SystemC-AMS Implementation

In order to replicate the approach proposed in [9], the proposed framework has been implemented in SystemC-AMS. Section 4.1 provides a brief introduction to SystemC-AMS, while Section 4.2 explains the reasons behind this choice and explains how SystemC-AMS has been adopted in this context.

4.1. SystemC-AMS

SystemC-AMS is the extension of SystemC for modeling analog and mixed-signal systems [42]. SystemC-AMS provides three different models of computation (MoC) to cover various domains as indicated in Figure 3.

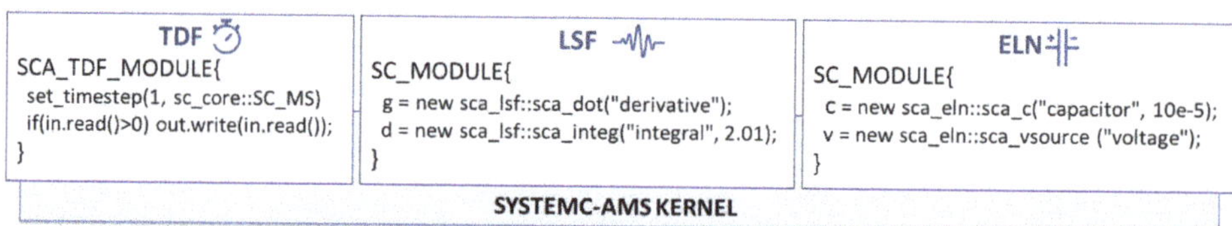

Figure 3. Abstraction levels supported by SystemC-AMS.

Timed data-flow (TDF) models are scheduled statically by considering their producer-consumer dependencies in the discrete time domain. Each TDF module is characterized by a simulation time step, which is used by the TDF solver to insert timed activation events in the standard SystemC event queue. This ensures efficient computation, as it avoids any runtime dynamic event management. Continuous time models can be modeled with two abstraction levels. *Linear signal flow* (LSF) supports the modeling of continuous time through a library of pre-defined non-conservative primitive modules [43] (e.g. in Figure 3 derivative and integrative, respectively). The *electrical linear network* (ELN) MoC models the electrical network by connecting the instantiation of predefined primitives (e.g., in Figure 3, capacitor and voltage source, respectively). All such abstraction levels are handled by the same simulation kernel that derives the system of equations to be solved over time and estimates system evolution.

4.2. SystemC-AMS Implementation of the Proposed Solution

SystemC-AMS is selected as the reference language for heterogeneous modeling for several reasons. The provided multiple abstraction levels unify the modeling work in a wide range of domains by using a single language: models can be built by choosing the most suitable abstraction level, and native converters can be exploited to simulate different abstraction levels simultaneously. SystemC-AMS also has the characteristics of a modular one, in that it divides the definition of interface and implementation, and is a IEEE standard language; thus, it can be easily extended and free from compatibility and reuse issues.

The flexibility of SystemC-AMS allows one to easily integrate the power models and the cost models. Figure 4 shows an example of a component implementation of a battery that is used in the remainder of this section as a reference. Each EES component is instantiated as a SystemC-AMS module (`SC_MODULE`, Figure 4, left) that internally instantiates one SystemC-AMS module for the power model and one for the cost model (line 5). This solution avoids forcing the fact that both models follow the same abstraction level and leaves maximum flexibility in the choice of the implementation style. The interface of the top level module includes the layer-specific signals of power (line 2) and cost (line 3) that will be bound to the corresponding ports of the layer-specific buses.

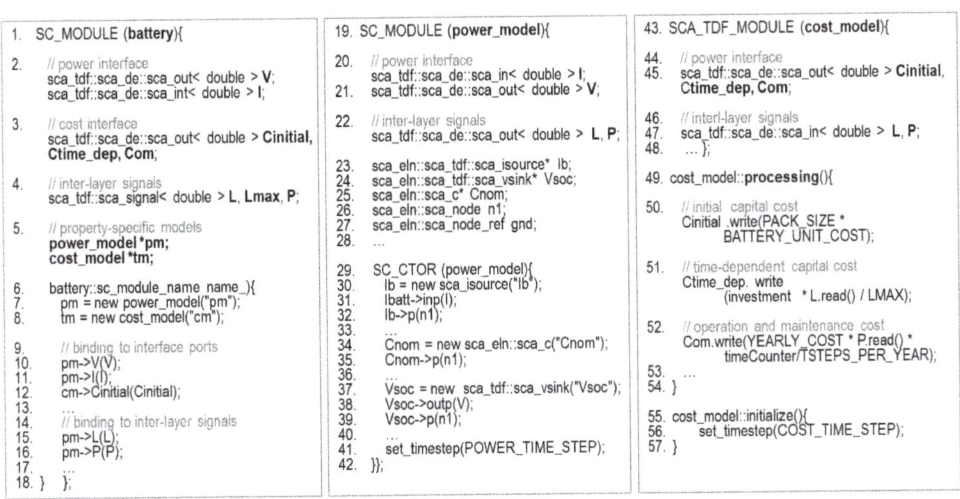

```
1.  SC_MODULE (battery){
2.    // power interface
      sca_tdf::sca_de::sca_out< double > V;
      sca_tdf::sca_de::sca_int< double > I;
3.    // cost interface
      sca_tdf::sca_de::sca_out< double > Cinitial,
      Ctime_dep, Com;
4.    // inter-layer signals
      sca_tdf::sca_signal< double > L, Lmax, P;
5.    // property-specific models
      power_model *pm;
      cost_model *tm;
6.    battery::sc_module_name name_){
7.      pm = new power_model("pm");
8.      tm = new cost_model("cm");
9.      // binding to interface ports
10.     pm->V(V);
11.     pm->I(I);
12.     cm->Cinitial(Cinitial);
13.     ...
14.     // binding to inter-layer signals
15.     pm->L(L);
16.     pm->P(P);
17.     ...
18. }   };
```

```
19. SC_MODULE (power_model){
20.   // power interface
      sca_tdf::sca_de::sca_in< double > I;
21.   sca_tdf::sca_de::sca_out< double > V;
22.   // inter-layer signals
      sca_tdf::sca_de::sca_out< double > L, P;
23.   sca_eln::sca_tdf::sca_isource* Ib;
24.   sca_eln::sca_tdf::sca_vsink* Vsoc;
25.   sca_eln::sca_c* Cnom;
26.   sca_eln::sca_node n1;
27.   sca_eln::sca_node_ref gnd;
28.   ...
29.   SC_CTOR (power_model){
30.     Ib = new sca_isource("Ib");
31.     Ibatt->inp(I);
32.     Ib->p(n1);
33.     ...
34.     Cnom = new sca_eln::sca_c("Cnom");
35.     Cnom->p(n1);
36.     ...
37.     Vsoc = new  sca_tdf::sca_vsink("Vsoc");
38.     Vsoc->outp(V);
39.     Vsoc->p(n1);
40.     ...
41.     set_timestep(POWER_TIME_STEP);
42. }};
```

```
43. SCA_TDF_MODULE (cost_model){
44.   // power interface
45.   sca_tdf::sca_de::sca_out< double > Cinitial,
      Ctime_dep, Com;
46.   // interl-layer signals
47.   sca_tdf::sca_de::sca_in< double > L, P;
48.    ... };
49. cost_model::processing(){
50.   // initial  capital cost
      Cinitial .write(PACK_SIZE *
              BATTERY_UNIT_COST);
51.   // time-dependent  capital cost
      Ctime_dep. write
              (investment  * L.read() / LMAX);
52.   // operation and maintenance cost
      Com.write(YEARLY_COST * P.read() *
              timeCounter/TSTEPS_PER_YEAR);
53.   ...
54. }
55. cost_model::initialize(){
56.       set_timestep(COST_TIME_STEP);
57. }
```

Figure 4. Example of SystemC-AMS of a battery: top level module (**left**), power model, implementing a circuit model in ELN (**center**), and a cost model (**right**).

The interface of the power and cost modules includes the layer-specific signals (e.g., in case of the cost layer, one port per cost computed by the component, line 45) and the inter-layer signals, used to communicate with the other layer (line 22 for the signals propagated from the power module to the cost module). This naturally enables inter-layer communication: thanks to the encapsulation of both power and cost models in a single top level SystemC module, inter-layer communication signals are set as TDF signals, binding the property-specific ports of the modules (lines 4 and 14–16).

The implementation of power models in SystemC-AMS has been discussed in many works at the state of the art that proved that SystemC-AMS can find a good trade off between accuracy and simulation time [36,44,45]. The presence of multiple abstraction levels allows one indeed to adopt for each power model the most suitable solution; e.g., ELN for circuit or circuit-equivalent models, TDF for equations, and LSF for dynamic models. In the case of Figure 4, the power model adopted for the battery is the circuit model proposed in [38] that is implemented as a network of connected ELN primitives (e.g., *Ib* is a current source, lines 23 and 30–32; *Cnom* is a capacitor, lines 25 and 34–35).

Modeling the cost equations detailed in Section 3 is straightforward, as they can be easily mapped on C++ functions and primitives, encapsulated by the TDF semantics of SystemC-AMS. The right-hand side of Figure 4 shows a snapshot of code: the `processing` function of TDF repeatedly evaluates the cost models of the battery over time, in terms of capital cost (line 50), real-time capital cost (line 51), and operation and maintenance cost (line 52).

Note that the separation of the power model and the cost model in two different SystemC-AMS modules allows one to decouple their activation frequency. Power models require a fine grain activation time step (in the order of 1s down to 1ms) to accurately evaluate the internal dynamics of the component. Vice versa, the cost models allow a larger time step, so as to reduce the computation overhead. As a result, the activation time step of the two modules is different (lines 41 and 56, respectively), and the inter-layer signals are handled with a conversion between different time scales.

5. Simulation Results

All simulations reported in this section have been implemented in SystemC-AMS 2.1 and run on a server installed with Intel Xeon 2.40 GHz CPU (16 cores, 2 threads each) and 128GB RAM, with Ubuntu operating system 18.04.1.

5.1. EES Case Study 1

As an example case study, we used a grid-connected EES built upon the prototype of [46], and sketched on the left hand side of Figure 5. The EES includes a wind turbine, a photovoltaic (PV) array, a battery pack, various AC loads, a common DC bus, and the necessary converters. Tables 1 and 2 report the main characteristics of the EES components, in terms of rated power and costs, respectively. The right hand side of Figure 5 shows the mapping of the case study to the proposed two-layer approach. The following subsections will detail the construction of the power and cost layers.

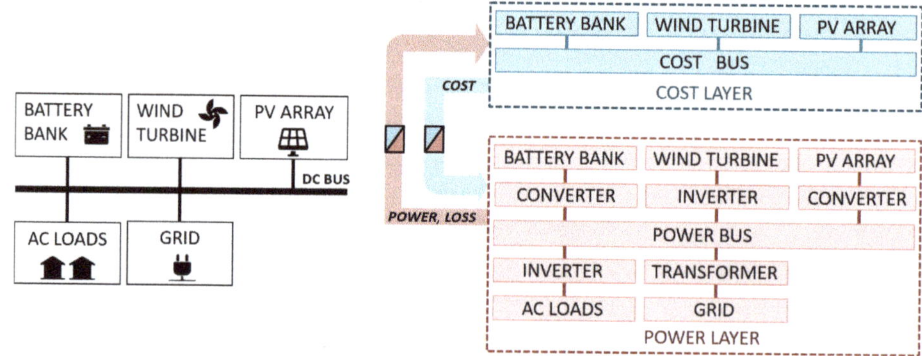

Figure 5. Structure of the electrical energy system (EES) case study 1 (**left**) and result of the application of the proposed approach (**right**).

Table 1. Characteristics of prototype EES components.

Component	Unit Rated Electric Characteristic	Cardinality (#)	Overall Rated Electric Characteristic
Wind Turbine	10 kVA	1	10 kVA
PV Module	57 V 5.49 A	30	10 kW
Battery Cell	3400 mA 3.7 V	40 × 60	200 Ah 144 V

Table 2. Initial capital cost and nominal lifetime of EES components.

Component	Unit Capital Cost ($)	O&M Cost ($/kW/Year)	Nominal Lifetime
Wind Turbine	19,500.00	15	20 Years
PV Module	675.00	15	20 Years
Battery Cell	2.50	10	SOH→ 80%

5.1.1. Power Layer

In the power layer, each EES component is described by a model. More specifically:

- The wind turbine is modeled with a mechanical model, proposed in [47];
- The PV array is modeled with a model of a single PV module, built by adopting the solution in [48], scaled up to the size of the PV array;
- The battery pack is based on the circuit-equivalent model in [38] (which models a single battery cell), scaled up to the size of the pack;
- AC loads reproduce power consumption of a residential community including 15 houses [49];
- Converters are modeled in terms of their conversion efficiency as introduced in [50];
- The grid component is used only to keep track of the power balance between demand and supply, and of any inefficiency introduced by the presence of a transformer between the grid and the power bus.

Table 1 collects the most relevant power characteristics of these components, plus the initial cardinality (# of elements) considered in the initial installation.

The power sources need input traces of irradiance and wind speed, which have been downloaded from the datasets of the National Renewable Energy Laboratory's (NREL's) Measurement and Instrumentation Data Center (MIDC) [51]. The time scale of irradiance and wind speed traces (one sample per minute) is longer that those of the load (one sample per second). We adopted the conversion methodology proposed in [9] to solve the issue of different time resolutions.

The *power bus* implements an initial *non-cost-aware energy management policy* similar to the one proposed in [52]: AC loads are satisfied by the renewable power sources whenever possible, and the battery pack is used to compensate whenever necessary (until its state of charge reaches a minimum of 10%). If the sum of energy stored in the battery pack and the power generated from the power sources cannot satisfy the AC loads, the houses purchase the missing energy from the grid. Otherwise, if the demand of the AC loads is less than the total power generation of the power sources, the unused power is used to charge the battery pack until it reaches 90% SOC, and then it is sold back to the grid.

5.1.2. Cost Layer

In the cost layer, only some components of the EES are relevant from the cost perspective. AC loads are not considered as a "variable," as they are assumed to be given upfront in terms number and type; as such, they are not associated with a cost model, as they do not contribute to the overall economic evolution. Therefore, loads are an input of the system; i.e., a trace of the input power demand over time.

Another difference with respect to the power layer is that all costs connect to the grid; i.e., the costs of buying/selling energy to the grid, are taken into account *inside the cost bus*; it is not, therefore, necessary to include a specific component for the grid.

Thus, the cost layer features three components: the two power sources (i.e., the wind turbine and the PV array) and the energy storage (battery pack). Table 2 reports all cost information for such components. All components model their initial and time-dependent capital costs (Equations (1) and (2) or (4)) and their operation and maintenance cost (Equation (5)). Note that the operation and maintenance cost is expressed in terms of \$/kW/Year, as it considers the annual routine operating and maintenance costs, and not accidental inside parts replacement. The interest rate used in Equation (3) is 7% for all components, and the CRF is thus set as 0.094.

The main difference between the models of the components is in terms of their time-dependent capital cost:

- The wind turbine and the PV array are considered as fixed lifetime components (with lifetime 20 years, as derived from their datasheets); thus, their time-dependent capital is modeled as in Equation (2).
- The battery pack has a variable lifetime, depending on its usage profile; thus, its time-dependent capital cost is modeled as in Equation (4), by considering the loss of functionality L as a aging degradation of battery capacity over time. The state of health (SOH) of the battery pack is represented by $1 - L$.

The *cost bus* estimates the total electricity cost with Equation (6), plus the net cost and the annualized cost, as defined in Equations (7) and (8). In our analysis, electricity price depends on the kind of operation (i.e., buying or selling) and on the time slot of the day, as shown in Table 3. Electricity buying price $p_{E,buy}$ is initially defined in three time slots, with the highest price at peak demand hours of the day. Electricity selling price $p_{E,sell}$ is instead independent of time and is much lower than $p_{E,buy}$.

Table 3. Electricity prices in different time of the day, as defined in [53].

Operation	Rate	Value (\$/kWh)	Time Slot of Day
Buying	F1	0.220	10 a.m.–3 p.m., 6 p.m.–9 p.m.
	F2	0.215	7 a.m.–10 a.m., 3 p.m.–6 p.m., 9 p.m.–11 p.m.
	F3	0.200	11 p.m.–7 a.m.
Selling	-	0.030	all day

5.1.3. One-Month Example Simulation Traces

In order to illustrate the different quantities that can be trace with the proposed simulation framework, we extract one-month simulation results of the prototype EES configured as shown in Table 1. For the simulation, the environmental traces used are relative to the observation site of MIDC at the University of Arizona [51], which has dry and windy weather all year long, with up to 90% sunny days.

Evolution of the Power Layer

Figure 6 depicts the evolution of the prototype EES in the initial 30 days by focusing on the power layer. Plot *A* shows the evolution of the environmental traces, in terms of solar irradiance (blue line) and wind speed (orange line), while plot *B* shows the power production of the corresponding power sources (same colors as in *A*). Plot *C* shows residential load demand over time. Plots *D* and *E* show the results of the application of the power bus policy in terms of state of charge (SOC) over time of the battery (*D*) and power balance in the system that leads to buying or selling energy.

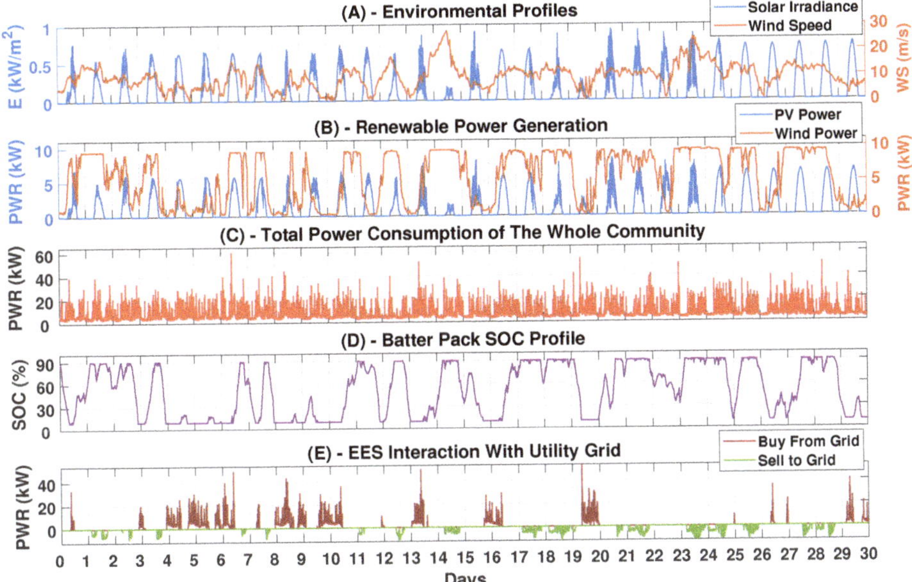

Figure 6. One-month long simulation of the EES in terms of power tracing quantities. The solar irradiance and wind profiles shown in (**A**) illustrate different daily weather conditions; the corresponding power generation by PV array and wind turbine indicated by (**B**); (**C**) reports the load power consumption of the whole residential community; battery SOC profile is shown in (**D**), within its operating range from 10% to 90%; the interactions with utility grid to buy or sell energy due to the energy surplus and deficit are illustrated in (**E**).

Evolution of the Cost Layer

The corresponding cost information evolution is shown in Figure 7, which reports one subplot per cost equation described in Section 3.2.1. The plots refer to the aggregate cost for all components, as determined by the cost bus, to provide a global view of the system rather than focusing on single components.

The plots in *A* and *B* show the global evolution of the time-dependent capital cost and of the operation and maintenance cost over time, respectively, as in Equations (2) and (5). Such graphs linearly grow over time, as the system components' values decrease over time and maintenance is necessary to allow their correct operation.

The graphs in *C*-*E* are used to comment on the instantaneous electricity cost, dropped down into money spent to buy electricity from the grid (*C*), and money earned by selling to the grid (*D*). Such graphs reflect the application of the policy implemented by the power bus. For the sake of readability, we report in *E* the evolution of the battery SOC: from this plot, it is evident that electricity is bought from the grid when the battery is discharged ($SOC < 10\%$) and the loads demand too much power, and that electricity is vice versa sold to the grid when power sources can feed the loads and the battery is fully charged ($SOC > 90\%$).

Plot *F* shows the total net cost, as from Equation (7), that grows almost linearly over time, as a result of the sum of electricity cost with the time-dependent capital cost and operation and maintenance cost of all components.

Plot *H* reports the evolution of profit over time (Equation (9)) that mitigates net cost by considering the intrinsic benefit generated by using the produced green energy (reported in plot *G*), rather than satisfying the entire load demand by buying from the grid. As the graph reports, exploiting renewable power sources generates a positive advantage for the EES: profit tends to grow linearly over time.

The decreasing periods correspond to time slots when it was necessary to buy electricity from the grid, as the battery SOC was equal to 10% and renewable energy could not feed the loads (e.g., in the daytime of 6th and 20th days, or in the nighttime between the 9th and the 10th day).

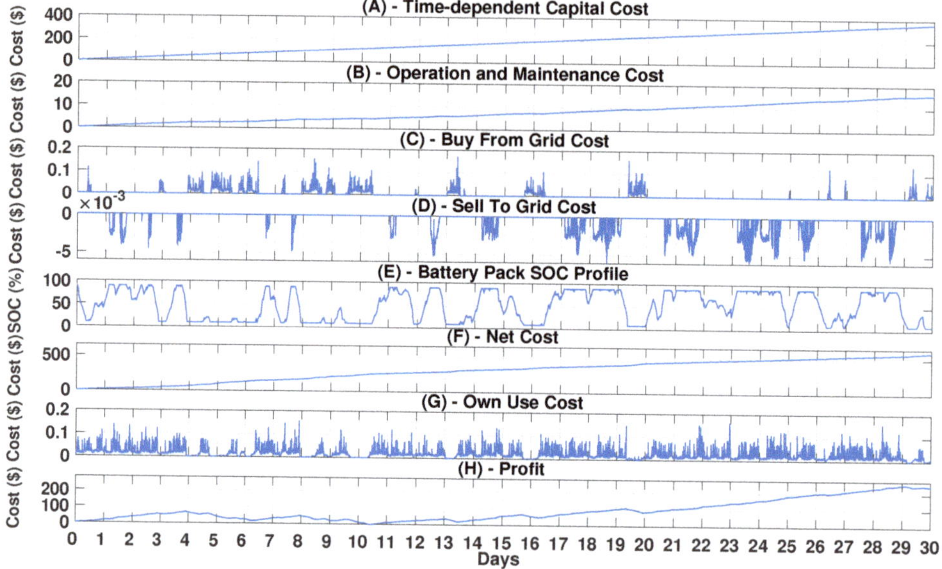

Figure 7. Evolution of the different cost quantities referring to the overall EES system and to the snapshot of simulation reported in Figure 6: time-dependent capital cost (**A**); operation and maintenance cost (**B**); electricity cost, divided into buy cost (**C**) and sell cost (**D**); SOC evolution of the battery (**E**); net cost (**F**); benefit generated by using power sources to feed the loads (**G**); and profit (**H**).

Evolution of Component-Specific Costs

Figure 8 reports the detailed evolution of time-dependent capital cost and *O&M* cost for each component in the EES.

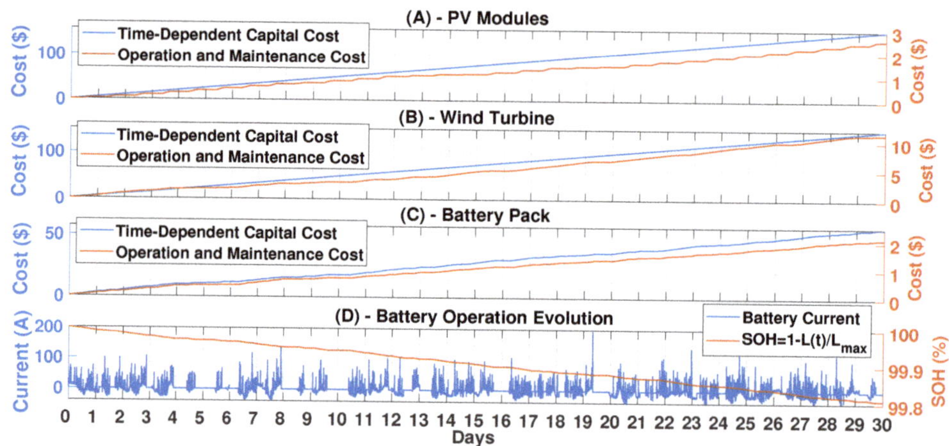

Figure 8. Time-dependent capital cost and *O&M* cost of the PV array (**A**), the wind turbine (**B**), the battery bank (**C**) and SOH and battery current profile (**D**) referring to the snapshot of simulation reported in Figure 6.

The time-dependent capital cost evolves linearly for the PV installation, A, and the wind turbine, B, which have a constant depreciation over time, according to Equation (2). The time-dependent capital cost of the battery (C) reflects the capacity loss over time, as it is calculated with the wear-out Equation (4). The *full cycle equivalent* battery pack aging mode [41] is adopted in our simulation, which correlates (Equation (10)):

$$N_{cyc} = \frac{\int_0^T |I(t)|dt}{2 * C_{nom}} \text{ and } L = \frac{N_{cyc}}{N_{cyc,max}} \tag{10}$$

where *effective number of cycles* N_{cyc} is an amount of charging and discharging energy divided by a nominal battery capacity C_{nom}; N_{cyc} divided by the *maximum charging and discharging cycles of the battery* $N_{cyc,max}$ indicates the lost capacity (L).

We define $1 - L$ as the state of health (SOH) of the battery pack; the real-time available battery pack capacity is thus computed by $SOH \times C_{nom}$. The SOH profile shown in *D* illustrates the available capacity decreases based on the battery charging/discharging current, whereas it keeps stable when there is no current flow in the battery pack (blue line in *D*); e.g., during the night between 5th and 6th days; then results the time-dependent cost also do not change during such period (Orange line in *C*).

The operation and maintenance cost grows almost linearly for all components, in a way that is linearly proportional to the unit operation and maintenance costs listed in Table 2; i.e., 10 $/kW/Year for the battery pack and 15 $/kW/Year for the other components. However, the cost growth is strictly related to the power handled by the component over time. This is clear from the plot in *A*, as the *O&M* cost of the PV module has approximately a stair-case waveform shape. This happens because at night there is no PV power production, and thus no increase in the *O&M* cost. Concerning the *O&M* cost of the battery, it shows itself to be similar to its time-dependent cost due to there being no power value sent to the cost layer during the battery pack idle period.

Comparisons with Previous Works

Concerning the validation of the power simulation accuracy and the comparison with other work, it is not possible to directly compare with other similar methods, as it would imply re-implementing the codes of other authors, since the comparable frameworks are not open-source. Although different in the way the co-simulation of power and cost is carried out, one possible option is to build the same proposed framework in Simulink, the work of [36] has demonstrated that already demonstrated that a SystemC-based homogeneous simulation can conduct the EES power simulation with excellent accuracy compared with Simulink (the average error is smaller than 0.0001%; the maximum error of all different components in the EES is smaller than 0.5%), while achieving a speedup of about 250X on simulation time. In terms of the cost evolution, Simulink it requires additional post-processing of power traces to track the cost metrics, which further supports the benefits of our proposed concurrent simulation, as the post-processing for the evaluation of the cost is proportional to the length of the total simulated interval.

5.1.4. Design Space Exploration through Proposed Simulation Framework

The previous section has shown the type of analysis that our framework can provide; however, its main use is to allow a cost-aware design of EES while simultaneously simulating the power flow of the system.

To demonstrate this feature, this Section provides three possible design space exploration (DSE) experiments to rank different system configurations from the cost perspective. We first compare the *adoption of two possible power policies*: the standard power management policy proposed in Section 5.1.1 and a cost-aware policy, designed for real time pricing rates.

Then, we propose two design space explorations that consider as variables *the amount of power sources and of energy storage*; i.e., the numbers of PV modules, battery cells, and wind turbines included in the system. The reference initial configuration is the one proposed in [46] and listed in Table 1, i.e., one wind turbine, 30 PV modules, and a battery pack of 2400 battery cells. The first experiment carries out an exhaustive exploration of different configurations to determine the one with the highest economic profit as computed by Equation (9); the second one explores the most profitable configuration under the fixed initial capital cost limitation.

Comparison of Different Power Bus Policies

The pricing policies applied by the grid suppliers have an important impact on the profit and the costs connected to EES operation, as evident from the different cost definitions provided in Section 3. When building the power management policy of an EES, it is thus crucial to determine how the energy balance between loads and renewable power sources fits the pricing policy that will be applied by the supplier.

The power management policy presented in Section 5.1.1 is not cost-aware: it assumes a traditional *time of use* (TOU) policy like the one in Section 5.1.2, where electricity price is different at different times of the day (higher/cheaper rates during peak/off-peak hours). However, the smart grid market has started featuring into complex pricing policies [54–56]. Comparing such rates and understanding their economic impact given the power management policy is far from trivial.

The framework proposed in this paper naturally enables this kind of analysis, thanks to the concurrent cost/power co-simulation of the EES on typical environmental traces, with the possibility of evaluating different power management policies implemented by the power bus.

To prove this effectiveness, we compare the adoption of the non-cost-aware policy presented in Section 5.1.1 with a cost-aware policy. In the latter, electricity price is changed dynamically according to a *real time pricing* (RTP) strategy. RTP improves flexibility as electricity price closely reflects the trend of the wholesale market and of the energy demand over time on the grid: prices vary at any time of day, several times per day, and differently on different days (even working ones) of the week; this should encourage users to behave in a flexible manner to reduce demand peaks [56].

As a possible strategy that uses this dynamically changing pricing, we propose the following energy management policy: if the renewable power sources cannot satisfy the load demand, the power bus checks the current price of electricity. If the price is *lower than the daily average buying price* (calculated as the moving average over one week), then electricity is bought from the grid. This allows one to save the energy stored in the battery pack for more expensive time slots. This modification makes the policy cost-aware, as it shifts electricity demand on the grid to cheaper time slots.

Figure 9 analyzes the impact of the two policies (the non-cost-aware and cost-aware ones) by comparing the simulation results on a one-week simulation, from Monday to Sunday. Plot *A* shows the evolution of the total load power consumption of the whole community (blue) and the total power generation from renewables (i.e., PV modules and wind turbine, in red), in order to highlight the power balance in the system. Plot *B* indicates the evolution of electricity prices to buy energy from the grid applied by the cost-aware policy, as derived from [56] (solid) and the moving average used by the policy to implement cost-awareness (dashed); notice that the electricity price exhibits a high heterogeneity not only across the hours of a single day but also across different days of the week. As a concrete comparison between the two policies, plot *C* reports the SOC profile of the battery pack (dashed for the cost-aware, solid for the non-cost-aware).

The main difference is visible on Monday (Day 1), when the total renewable power generation cannot satisfy the load consumption. Since the RTP is lower than the average buying price, the cost-aware policy does not use the battery (whose SOC does not change in this interval), but rather buys power from the grid. Vice versa, the non-cost-aware policy uses the battery to provide, and the SOC curve of non-cost-aware policy (orange color) reaches the minimum threshold of 10%.

Table 4 lists all the costs after one year of operation by applying the two different policies. Net cost and profit are calculated with Equations (7) and (9): net cost is the sum of real-time capital costs, operation and maintenance costs, and the cost of buying/selling power from/to the grid; notice the cost of selling energy (second row) is a negative value since it is treated as a gain of the EES compared to the cost of buying energy; profit is the value gained by using renewable power sources minus the net cost. The year uses the same real load consumption (total AC loads of 15 houses [49]), environmental data collected by MIDC at University of Arizona [51] as the previous simulation, and RTP is extrapolated by [56] repeating the 1-week profile shown in Figure 9B. The cost-aware policy reduces the time-dependent capital cost and operation and maintenance cost of the battery, which is less involved in the EES energy flow. However, the cost to buy from the grid is higher for the cost-aware policy, and it is not compensated by a higher gain to sell to the grid. This is not counter-intuitive: a cost-aware policy is not necessarily improving the overall net cost or profit, as shown in the table, but it is just inclusive of the electricity cost in the decision of the energy flow. While the choice of buying energy when the price is low seems reasonable, it causes in fact a reduction of the energy provided by the EES itself (avoid using the battery and rather buy when electricity price is considered low, thereby reducing the benefit generated by using power sources to feed the loads).

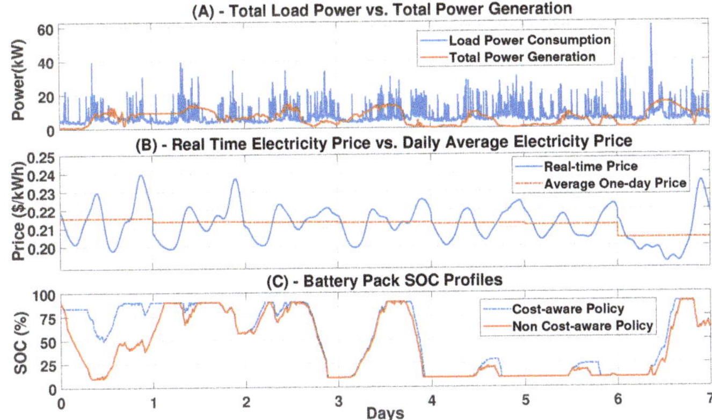

Figure 9. Different scenarios of two power management policies within one example week.

Generally speaking, the design of a smart energy management policy that maximizes the profit is not the target of this work. In this section we just showed that the proposed simulation framework can easily include cost "in the loop" and can efficiently validate the policies over time intervals of practical significance. In this perspective, the result on the cost-aware policy confirms that it is necessary to evaluate the mutual impact of cost and power, to get a complete view of the economic advantage of the EES under design.

Table 4. Different cost values after one year operation by two policies.

	Cost Type	Cost-Aware Policy ($)	Non Cost-Aware Policy ($)
Electricity cost	Buy	2445.34	1855.35
	Sell	−507.34	−416.51
	Own (provided by EES)	9146.64	9736.63
Battery	Time-dependent capital cost	470.86	662.14
	Operation and maintenance cost	17.46	24.73
PV array	Time-dependent capital cost	1903.35	1903.35
	Operation and maintenance	55.88	55.88
Wind turbine	Time-dependent capital cost	1832.79	1832.79
	Operation and maintenance cost	123.57	123.57
Net Cost		6341.97	6041.30
Profit		2804.67	3695.33

Exhaustive Exploration of Different Configurations

The first scenario is an exhaustive DSE to determine the configuration with the highest economic benefit. Two cases are described in here based on the presence of wind turbine. As the capital cost of one wind turbine is approximately equal to the capital cost of 29 PV modules, the ranges of the parameters of power source and energy storage are set as follows:

- The number of PV modules varies from 0 to 100 in steps of 10 when the wind turbine is present, and from 0 to 150 when there is no wind turbine in the EES;
- The number of battery cells varies from 0 to 10,000, in steps of 1000.

The AC loads and input environmental traces are same as the ones introduced in Section 5.1.3. The TOU scheme is adopted as electricity buy price, 0.03 $/kWh as the electricity sell price; then we use the non-cost-aware power management policy in the exploration simulations.

The left-hand side of Figure 10 shows the *net cost* for the cases with (top left) and without (bottom left) wind turbine after 20 years, which is the maximum operating lifetime of wind turbine and PV modules. The right-hand curves show instead the corresponding 20 years *annualized cost* computed by Equation (8). The x-axis and y-axis represent the different numbers of PV modules and battery cells in the pack, and the z-axis represents corresponding cost.

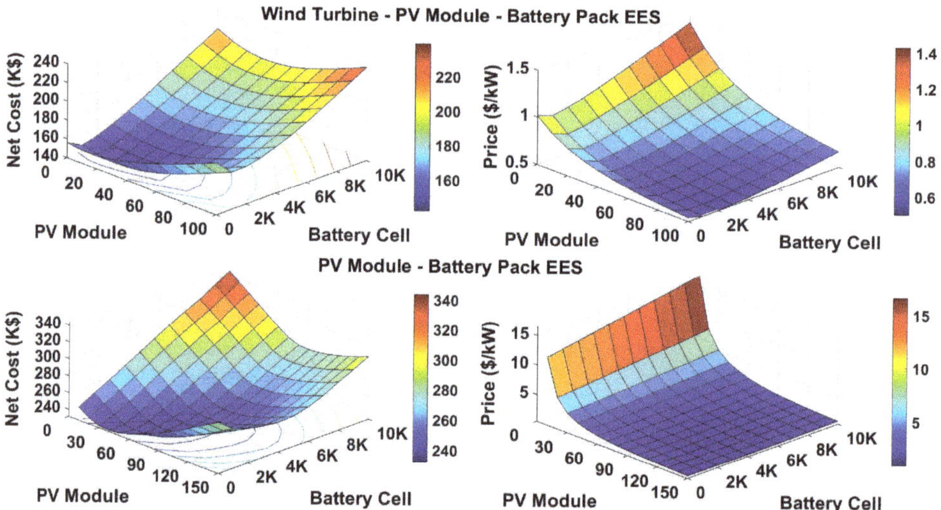

Figure 10. Net cost (**left**) and annualized cost (**right**) of different ESS configurations after 20 years with (**top**) and without (**bottom**) wind turbine.

The results illustrate the advantage of wind turbine in the EES: the annualized cost of those EES including wind turbine is about 10% of the annualized cost of configurations including only PV modules and batteries. Such a big advantage is mainly caused by the complementary effect of the wind turbine with respect to photovoltaic energy generation: the wind turbine mostly generates power at night or on cloudy/rainy days, when the PV modules cannot generate power or can only generate power with low efficiency. Additionally, the wind turbine increases total power generation during the peak hours to reduce the need of buying power from the grid.

The optimal EES configuration without wind turbine is made of 5000 battery cells and 150 PV modules, thereby reaching an annualized cost of 0.9591 $/kW. The optimal configuration when the wind turbine is present has 2000 battery cells and 100 PV modules, with a total 0.4984 $/kW annualized cost. This confirms the intuition that annualized cost can be reduced by increasing total power generation.

However, such configurations may not be optimal from the perspective of *profit*; i.e., when taking into account also the advantage of self-consumption of power generated by the renewable power sources. The corresponding results for the exhaustive exploration results in the perspective of profit are shown in Figures 11 and 12. The x-axis and y-axis represent the different numbers of PV modules and battery cells in the pack. The z-axis represents the total profit of EES computed by Equation (9) for the various configurations, at different points of the lifetime of the EES (i.e., after 5, 10, 15 and 20 years).

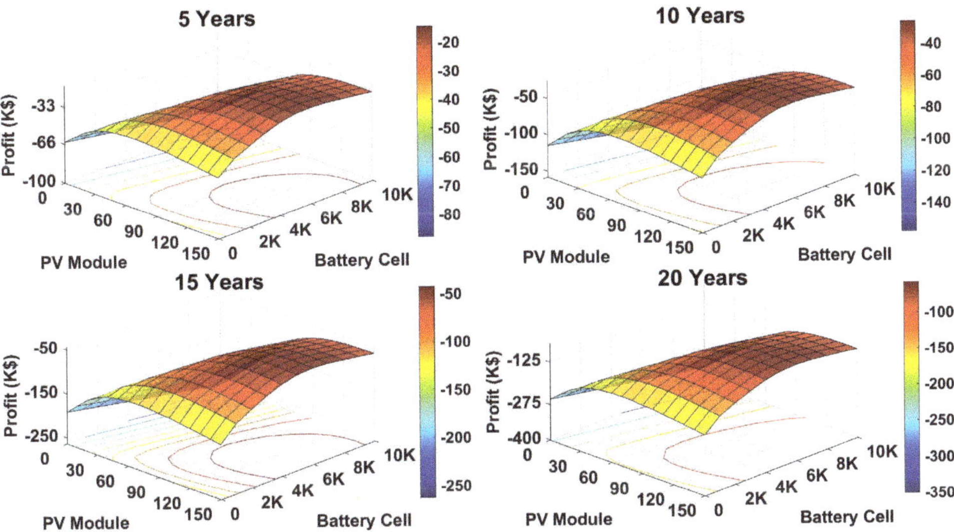

Figure 11. Profit in different years with various configurations for case without wind turbine in EES.

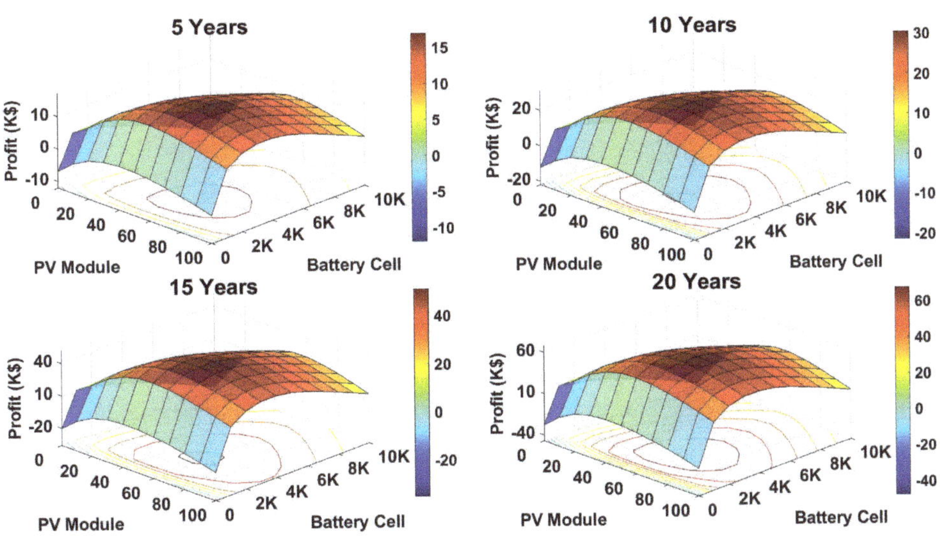

Figure 12. Profit in different years with various configurations for case with wind turbine in EES.

Figure 11 refers to EES configurations without the wind turbine: the results illustrate that **none** of the configurations without a wind turbine results into a positive profit, even after 20 years. The 3-D surface shown in the figures indicates that the more PV modules the higher the benefit, still on the negative profit side. The optimal configurations is different from the previous analysis and always include 6000 battery cells and a number of PV modules that varies from 140 to 150 (due to the varying nature of environmental inputs and load power consumption profiles over the years). Notice that, for longer time horizons, the profit of the EES worsens, e.g., the configuration of 6000 battery cells with 140 PV modules has lost 2805.77 K$ after one year, and this loss is enlarged to 58,080.77 K$ after the EES run for 20 years.

When the EES includes the wind turbine for generating power, there exist several configurations making profit, as revealed in Figure 12. This proves once again the benefit of a wind turbine in the EES.

According to the simulation results, the optimal configuration with highest profit is made of 3000 battery cells with 60 PV modules and one wind turbine. Some years see a higher profit when decreasing PV modules to 50, due to the different weather conditions and load power profiles; however, the increase in profit is minimal; e.g., the configuration with 50 PVs leads to an increase of 86.90 $ after 5 years, while one with 60 PV modules makes a profit 850.42 $ bigger than the 50 PV modules one after 20 years). Thus, the two configurations can be considered comparable, and the user may choose the one he prefers (e.g., the one with lower initial capital cost), knowing that the profit will be comparable.

Exploration with Fixed Initial Capital Cost

A more realistic scenario is the one where the initial capital investment is a fixed constraint, i.e., the compared EES configurations have same initial capital cost. Given the big advantage of the presence of a wind turbine in the EES indicated in the previous exploration, the configurations considered in this analysis always take into account the presence of the wind turbine.

We assume an initial capital cost to 60,000 $: Table 5 lists 20 different configurations with different numbers of PV modules, battery cells and wind turbines, with an initial capital cost as closed as possible to 60,000 $. Note that configurations from 1 to 13 have one wind turbine, and explore the number of PV modules (from 0 to 60) and of battery cells (0 to 16,200). Configurations from 14 to 20 additionally explore the introduction of a second wind turbine, thereby lowering the number of the other EES components to meet the initial capital cost constraint.

Table 5. Different configurations with fixed initial capital cost in the exploration.

Config.	PV Modules (#)	Batteris (#)	Wind Turbines (#)	Config.	PV Modules (#)	Batteries (#)	Wind Turbines (#)
1	0	16,200	1	11	50	2700	1
2	5	14,850	1	12	55	1350	1
3	10	13,500	1	13	60	0	1
4	15	12,150	1	14	0	8400	2
5	20	10,800	1	15	5	7050	2
6	25	9450	1	16	10	5700	2
7	30	8100	1	17	15	4350	2
8	35	6750	1	18	20	3000	2
9	40	5400	1	19	25	1650	2
10	45	4050	1	20	30	300	2

Furthermore, we bring another factor in the exploration to investigate the influence of weather condition. We conduct the exploration with data relative to two locations from the database of NREL's MIDC [51] that have significantly different climatic characteristics: one is located in Eugene, Oregon (cloudy and wet climate), the other in Tucson, Arizona (dry and windy climate).

The exploration results in perspective of profit in both locations show the same finding as previous explorations, the highest profitable configuration is the same one in each year.

Figure 13 shows the profit results after 20 years for the two locations. As expected, the profits in Arizona (right) are always higher than in Oregon, due to its better environmental conditions. Table 6 lists the profit made by highest profitable configuration at both locations in different years: the difference of the profit keeps about 150%, due to the their climatic characteristics. For example, the average annual energy generated by PV modules for configuration 10 in Arizona is about 30,000 kWh, and the wind turbine produces about 50,000 kWh every year; while the same numbers in Oregon become about 20,000 kWh and 25,000 kWh, respectively.

Overall, the highest profit configuration in Eugene is number 19 in Table 5, and the highest one is number 18 for Tucson (peaks of the 3-D surfaces across all years). Note that both optimal configurations have two wind turbines, thereby proving that wind power generation is the critical factor among three modifiable parameters in the EES. The valleys in both 3-D surfaces correspond to 13, which features no battery: this indicates that the battery pack also plays an important role to maximize the profit, as it reduces the need for buying energy from the grid.

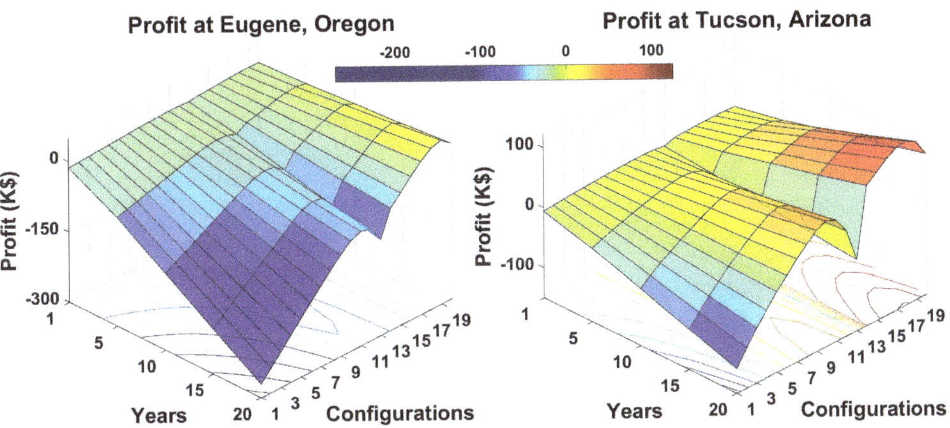

Figure 13. Profit of different configurations have same initial capital cost at two locations.

Table 6. Highest profit at both locations in different years.

Year	Eugene, Oregon		Tucson, Arizona		Absolute	Relative
	Config.	Profit ($)	Config.	Profit ($)	Difference ($)	Difference (%)
1	19	2402.92	18	6062.42	3659.50	152.29%
5	19	11,997.35	18	30,282.70	18,285.35	152.41%
10	19	23,971.38	18	60,679.42	36,708.04	153.13%
15	19	35,432.53	18	90,469.58	55,037.05	155.32%
20	19	49,118.32	18	122,508.33	73,390.01	149.41%

5.2. EES Case Study 2

In order to show the high flexibility of our proposed simulation framework, we built another EES case study as described in Figure 14; the EES is composed of a PV array, an electric vehicle (EV), an AC load, a common DC bus, and the relative converters. The left-hand side of Figure 14 draws the conceptual graph of the new EES case study, the right-hand side displays the corresponding modules of the EES in the proposed simulation framework.

The construction of the power layer and cost layer in the proposed simulation framework has been done similarly to the previous example of Sections 5.1.1 and 5.1.2.

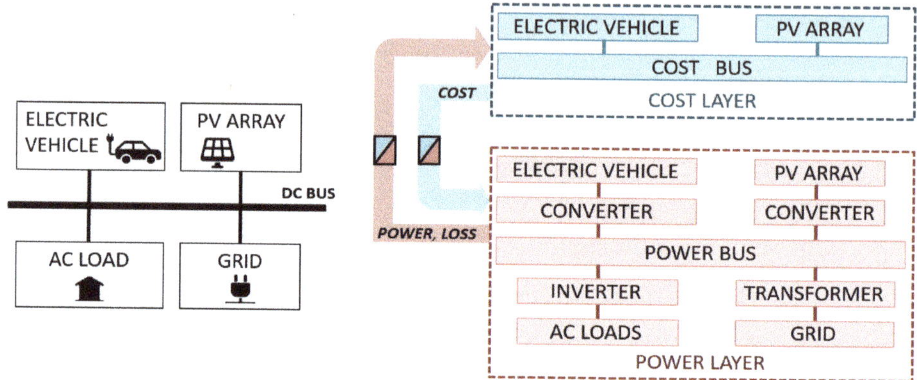

Figure 14. Structure of the EES case study 2 (**left**) and result of the application of the proposed approach (**right**).

Concerning the power layer, the PV array has been modeled by starting from a single PV module model [48], which has then been scaled up to the size of the final array; namely, 10 PV modules to mimic a small-size residential PV installation. In this case study the size of the array is fixed. The AC load represents the power consumption of one single house from dataset [49]; converters are modeled as already described above [50]. The EV consists of two sub-modules; namely, the battery pack module and EV motor module; their models are built by the methods provided in [50]; the grid is used to keep track of the power balance between house power consumption and PV array power generation and energy storage of EV.

The input traces of solar irradiance from the dataset provided by the National Renewable Energy Laboratory's (NREL's) Measurement and Instrumentation Data Center (MIDC) [51]. Concerning the driving profile, we assume the EV operates according to a daily commute routine: the driver leaves at 7:30 a.m. and drives half an hour to arrive at the destination; then he/she drives another half an hour and comes back home at 7 p.m. We assume the EV consumes the identical energy every day in the following simulations to remove from the analysis the influence from the EV driving situations. This scheme can obviously be changed should one be interested in analyzing the fluctuations due to specific driving patterns.

We envision two main scenarios in the EES daily operation; the first one is with the EV plugged, the second one is when PV acts as as the only power source connected to the EES. The power bus module implements a cost-aware energy management policy: (1) when the EV is not connected, the PV array provides its generated power to satisfy load demand of the house; excess power will be sold to the grid while power deficit will be bought from the grid; (2) when the EV is plugged, we set a threshold electricity price to decide whether the power consumption of the house is provided by the EV (price $<$ threshold), or bought from the grid (price $>$ threshold). However, the EV can provide power only until its battery SOC reaches to 10%, then the house will have to buy from the grid the required power, and the EV will start to charge the battery until the electricity price goes below the threshold price. As already discussed for the previous uses case, this is just an example of policy and it has no claim of optimality; our objective is to show the flexibility of the framework and not to provide optimized policies.

Concerning the cost layer, we adopted the three-time slots electricity price as indicated in Table 3 in this case study and we set the threshold electricity price is 0.21 $/kWh; therefore, it means EV plays the power source role in the EES during the electricity price in the F1 and F2 periods if it has residual energy.

The cost items referred to the battery and the PV modules are the same as in Tables 1 and 2. We selected several EVs with different battery pack sizes in the market used in the following simulations.

Notice that the cost analysis does not consider the items involved in the EV operations outside the house EES, such as possible intermediate charging costs or different driving distances.

5.2.1. One-Week Example Simulation Traces

We extract a five-day simulation (one week of working days) results of the EES with EV case study to show the different power and cost quantities that can be tracked in our proposed simulation framework. For the battery pack size of the EV used in this example, the 21 kWh battery pack with $40sX40p$ configuration is adopted. The selected solar irradiance trace is the location at the University of Arizona [51]. The load consumption profile is extracted from the number 1 house in the dataset [49].

Evolution of the Power Layer

Figure 15 shows the EES main quantities evolution of the power layer. Plot A shows the PV array generation power evolution, it illustrates the EES does not have a renewable power source during the night. Plot C shows the result of the power bus policy that leads to buy or sell energy with the grid. Plot D shows the SOC over time of EV battery pack and plot E shows the corresponding battery current profile, the positive value means the discharge current, the negative values means the charge current.

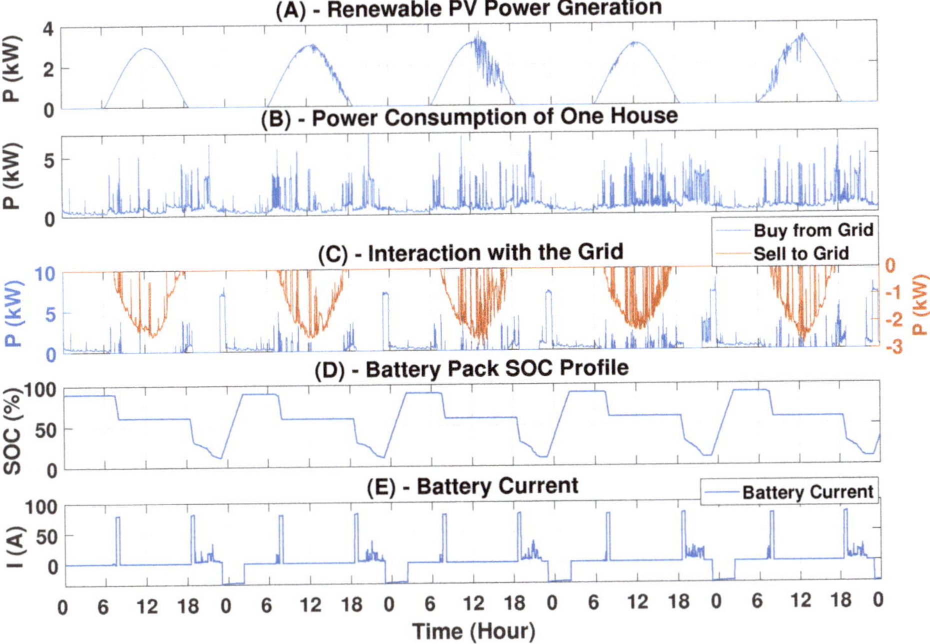

Figure 15. Five-Day simulation of the EES in terms of power tracing quantities. The PV power generation profile is shown in (**A**) illustrate different daily weather conditions; (**B**) reports the load power consumption of one house; the interactions with utility grid to buy or sell energy due to the energy surplus and deficit are illustrated in (**C**); battery SOC profile is shown in (**D**), within its operating range from 10% to 90%; the corresponding battery current is indicated in (**E**).

In order to show how the policy executed in the power bus and how we consider the time-dependent capital and O&M cost of the battery pack in EV in our simulations, we extract one day from Figure 15 to give more details as shown in Figure 16.

Because EV leaves from the house at 7:30 a.m. and comes back home at 7 p.m., there are two periods after 7 a.m. and before 7 p.m. in the plot C indicate the SOC of battery decreases (as estimated) due to the driving. In the period when the EV is plugged in the house, the battery pack in the EV provides the power to the house if the electricity price is higher than the threshold price 0.21 $/kWh, therefore, plot E shows there are discharge currents between 7 a.m. and 7:30 a.m. and from 7 p.m. to 11 p.m. However, the SOC of battery pack reaches its bottom operation limitation 10% during the period from 7 p.m. to 11 p.m. as indicated in plot C, so the house has to buy power from the grid starting from about 9:50 p.m. as shown in plot B.

As discussed above, the cost generated by the EV daily driving and charging does not take into account in our analysis, the reason is that the EV is independent energy storage or power source component to the house, there is no direct relation between load demand of the house and the EV, we only consider the power involved between the EV and the house when the EV plays a role of the power source. Therefore, the power consumed by the daily driving and the charging power to backfill such consumed power is ignored, but the charged power for compensating the power provided to the house should be taken into account. This point is illustrated by plots B and C, the EV starts to provide power to the house when it comes back house at 19:00 and the SOC is about 40% at that time as indicated by red arrow A; then the house start to buy power from the grid after the SOC decreases to 10% at around 21:45 as shown by red arrow B; the electric price is lower than the threshold price after 23:00; thus, the EV starts to charge the battery pack; when the SOC increases to 40% around 24:00 as indicated by red arrow C, the bought power for charging the battery is stopped since it reaches the SOC when the EV arrives at house, only left the bought power for the load demand of house as shown in plot B. We remove the influence of the EV driving conditions from our analysis in this way, only consider the period between arrow A and C which is the period of EV involved with the House.

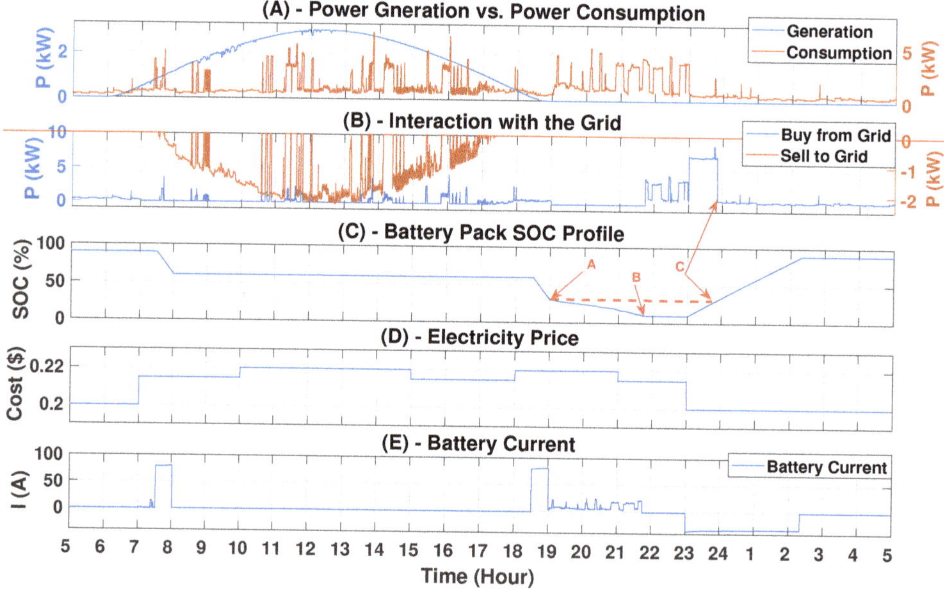

Figure 16. One-day long simulation with power quantities of the EES extracted from Figure 15.

Evolution of the Cost Layer

The corresponding cost quantities traces are shown in Figure 17, each subplot is related to one cost equation formalized in Section 3.2.1.

Plots A and B show the time-dependent capital cost and operation and maintenance cost of PV modules and battery pack in EV; the time-dependent capital cost of PV modules is updated according to Equation (2), so it increases as time elapses. Conversely, the time-dependent capital cost of the battery pack is given by Equation (4), so it only increases when the EV connects with the house to provide the power or to charge the battery pack to backfill the provided power; both actions use Equation (5) to compute the operation and maintenance cost; therefore the curve of PV is related to the power generation profile that keeps stable during the night and increases during the daytime, whereas the curve of the battery pack has the similar trend as time-dependent capital cost.

Plot D shows the cost related to the grid as per Equation (6). It indicates that the surplus PV power is sold to the grid and the house needs to buy the power if PV power cannot satisfy the power demand when the EV is disconnected with the house; it also indicates that the house always buy the power from the grid from 11 p.m. to 7 a.m. due to the low electricity price. The blue line in plot F shows the net cost formalized by Equation (7), which grows over time since it is a result of the sum of the described previous cost. Plot E shows the evolution of intrinsic benefit generated by using the PV power and battery pack of EV, instead of buying power from the grid to satisfy the load demand. The orange line in plot F illustrates the profit profile of the EES computed by Equation (9), the negative values tell this EES cannot bring profit finally, but notice that the EES still can bring the benefit, a more comprehensive comparison is introduced in the following section to illustrate the benefit of EES.

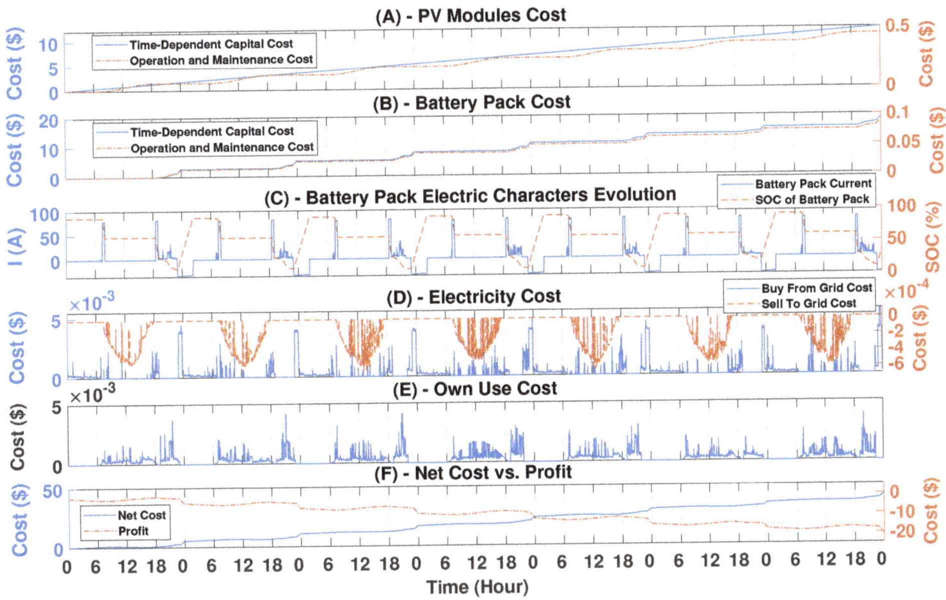

Figure 17. Five-day simulation with cost quantities of the EES corresponding to Figure 15.

5.2.2. Comparison of Different EVs in the EES

As an example of design-space exploration, we investigated the impact of different EVs involved in the EES, in terms of different battery sizes. We select several EVs (with different battery pack sizes) in the following simulation. Table 7 lists the corresponding configurations, chosen from a set the popular common EVs in the market [57]. We also added two configurations without battery pack involved in the EES as a reference. The first one shows a scenario in which the house always buys the power from utility grid since there is no any other power sources; the second one indicates the house can also get the power from PV modules instead of only buying power from the grid, while still having no storage (EV).

Table 7. Different configurations of EES for comparing the cost quantities.

EES No.	PV Modules Number	Electric Vehicle	
		EV Model	**Battery Pack Size (kWh)**
1	0	NaN	0
2	10	NaN	0
3	10	Mitsubishi MiEV	16
4	10	GM Spark	21
5	10	Nissan Leaf	30
6	10	BMW i3 (2019)	42
7	10	Tesla S 60	30

Table 8 shows the one-year long simulation results of the different configurations list in Table 7. The first row indicates the situation when all the load demand needs to be satisfied by the grid, so there is only electricity buying cost, the net cost and profit is computed based on the Equations (7) and (9), respectively; notice that the negative sign in the last column means an absolute cost for the household. The PV array cost columns are constant but in the first case, as it generates the same power. The "buy" electricity cost column indicates that the cost of buy electricity decreases if the battery pack size in the EV increases due to the residual energy of battery pack increases when the EV comes back house, the last three cases show that the "buy" electricity cost are same because of the maximum energy can be provided to the load by the battery pack is reached, it means increasing the battery pack size becomes useless.

The "own" electricity cost behaves similarly to the buy cost; it first increases for increasing battery sizes, then it stabilizes since the maximum energy provided by the EV is reached. The battery pack cost columns indicate that the O&M cost is positively correlated to the power provided from the battery pack, while the time-dependent capital cost column tends to decrease for larger sizes because the time-dependent capital cost (Equations (4) and (10)) states that the aging degradation is reduced when the battery pack size increases.

The last column in the Table 8 illustrates that all the different EVs involved in the EES cannot bring profit for the house, which is not like the results in the previous case study, for example, 3000 battery cells with 60 PV modules and 1 wind turbine combination of previous EES case study can generate positive profit as shown in Figure 12. However, involving EV in the EES can bring economic benefit compared to the situation when this is no EV in the EES; for example, involving an EV with a 16 kWh battery pack (third row) can bring 16.36 $ economic benefits compared to the EES without EV involved (second row). Finally, the profit-optimal battery size is the one in the third case; it provides about a 1104.33 $ benefits compared to the case without PV modules and EV involved in the EES.

Table 8. Cost quantities simulation results of one-year period with different battery packs in the EV.

No.	Electricity Cost ($)			PV Array Cost ($)		Battery Pack Cost ($)		Net Cost ($)	Profit ($)
	Buy	Own	Sell	Time-Dependent Capital Cost	O&M	Time-Dependent Capital Cost	O&M		
1	1513.08	0.00	0.00	0.00	0.00	0.00	0.00	1513.08	−1513.08
2	833.48	672.17	385.03	52.15	1.98	0.00	0.00	1097.27	−425.11
3	812.71	689.20	385.03	52.15	1.98	20.58	0.10	1097.95	−408.75
4	796.10	1051.30	385.03	52.15	1.98	445.34	2.09	1508.09	−456.79
5	793.08	1087.52	385.03	52.15	1.98	477.59	2.30	1537.52	−450.00
6	793.08	1087.52	385.03	52.15	1.98	471.80	2.30	1531.82	−444.30
7	793.08	1087.52	385.03	52.15	1.98	467.80	2.30	1527.73	−440.21

6. Conclusions

This paper proposed a simulation framework for the economic optimization of EESs that relies on the concurrent simulation of energy flows and economic estimations based on a single simulation kernel, namely, SystemC and its extensions. We showed that our framework allows an effective and efficient design exploration of the EES under design and enables decisions, such as the identification of (1) the optimal management policy based on joint energy and economic constraints; (2) the cost-optimal or profit-optimal configurations of the EES in terms of number and type of renewable energy elements used (power sources and energy storage devices).

Author Contributions: Conceptualization and Methodology, Y.C., S.V. and M.P.; Experiment, S.V. and Y.C.; Writing—original draft, S.V., Y.C. and M.P.; and Writing—review and editing, D.B., S.Q. and E.M. All authors have read and agreed to the published version of the manuscript.

Funding: This research received no external funding.

Conflicts of Interest: The authors declare no conflict of interest.

Appendix A

Table A1. Overview of recent representative related works on cost estimation of EES Systems.

Ref.	Goal	Energy Models	Cost Models	Proposed Solution
[13]	Optimize energy storage configuration to minimize fuel costs	Simple linear models, e.g., PV power as function of area and rated power	Operation and maintenance, replacement, capital	Comparison of alternative configurations
[14]	Find sizing of EES components that minimizes levelized cost of electricity	Based on linear equations of rated power and capacity	Capital, operation and maintenance, replacement	HOMER [58] and artificial bee colony optimization
[15]	Optimal configuration while minimizing total net present cost	Simplified models of EES components (e.g., battery as a function of efficiency and depth of discharge)	Capital, replacement, operation and maintenance	Hybrid optimization genetic algorithm
[16]	Optimization of levelized energy cost and payback time, emission reduction	Simple linear models	Annual investment and replacement, emission reduction benefit	Particle swarm optimization to identify optimal component sizing
[17]	Minimize cost and improve reliability	Simple linear models, e.g., power sources as linear function of environmental quantity	Investment, replacement, operation and maintenance, electricity	Constrained optimization problem to determine capacity of power sources and storage
[18]	Optimization of annualized cost through dimensioning of EES components	Power sources as simple linear function of environmental quantities	Operation and maintenance, capital, replacement, electricity	Mixed integer linear programming applied to different scenarios

Table A1. *Cont.*

Ref.	Goal	Energy Models	Cost Models	Proposed Solution
[19]	Optimal sizing of EES components to minimize levelized cost of energy	Accurate models, fixed power management policy	Capital, operation and maintenance, replacement	Genetic algorithm
[20]	Maximize electricity bill savings with optimal sizing of energy storage	Accurate model only of energy storage devices	Capital, electricity, replacement	Non-convex optimization problem plus exhaustive-search solution
[21]	Optimal sizing of EES components given pricing policies	Focus only on power source capacity and battery state of charge	Capital, operation and maintenance, electricity	Cataclysm genetic algorithm
[22]	Reduce cost of microgrid expansion considering impact of battery dynamics	Focuses on batteries, simple model of power sources	Capital, electricity, annualized cost	Mixed integer linear programming
[23]	Optimize battery sizing given battery bank degradation cost	Models only battery aging, fixed power management policy	Electricity, battery degradation	Linear optimization method
[24]	Optimize battery size and scheduling taking into account costs and loads	Simple linear models, input traces for power sources	Electricity, battery capital cost	Convex programming method
[25]	Find sizing of EES components plus operating strategy to minimize costs	Simple models, based on rated power, efficiency coefficients, area occupied by PV modules, etc.	Annualized system cost	Mixed integer linear programming model solved with CPLEX
[26]	Optimization of battery management to minimize electricity cost and carbon dioxide emissions	Focuses on battery state of charge, no model of other EES components	Electricity, emissions equivalent cost	Multi-objective optimization (energy cost and emissions)
[27]	Co-scheduling problem of HVAC control and battery management	Detailed model only of batteries (power sources as input traces, converters as efficiency coefficients)	Electricity, battery degradation	Minimize total cost with the convex optimization tool CVX
[28]	Utility scheduling to minimize total operational cost	Storage devices modeld only as their lifetime, power sources as input traces	Electricity, degradation, pricing schemes	Nonlinear mixed integer optimization to determine schedule of batteries and supercapacitors
[29]	Determine optimal power management policy to reduce operation and emission costs	EES components as min-max constraints for the optimization	Operation and maintenance, electricity	Particle swarm optimization
[30]	Balance total power generation w.r.t. load and minimize total operation cost	Capacity and scheduling as goal of optimization	Capital, operation and maintenance, electricity	Integer linear programming with CPLEX
[31]	Construction of optimal scheduling to optimize usage of storage and grid	EES capacity as constraints and object of optimization	Levelized cost of energy	Predictive optimization given a graph of alternatives
[32]	Techno-economic model to determine optimal capacity of PV system and battery storage	Simple models, e.g., PV as linear function of irradiance and temperature	Capital, operation, payback time	Particle swarm optimization
[33]	Optimal sizing of EES components to maximize profit	Accurate model only of PV modules	Capital, operation and maintenance, electricity	Dynamic simulation to evaluate impact of electricity tariffs
[34]	Feasibility study of wind/PV hybrid system	Accurate model only for PV modules	Capital, operation and maintenance, replacement	Simulation based on MATLAB and HOMER [58]
[35]	Operation cost minimization with different pricing mechanisms	Simple linear models	Annualized cost of system components	Based on HOMER [58]
This	Estimate mutual impact of power dynamics and costs to reach effective design of EES (Sections 3.3 and 5)	Level of detail can be tuned by user, allows one to have high level of accuracy of component dynamics (Sections 2.2 and 4)	Capital, profit, depreciation, operation and maintenance, electricity, net and annualized cost (Section 3)	Simulation-based, allows efficient construction and evaluation of alternative configurations (Section 5)

References

1. Ahadi, A.; Liang, X. A stand-alone hybrid renewable energy system assessment using cost optimization method. In Proceedings of the IEEE International Conference on Industrial Technology (ICIT), Toronto, ON, Canada, 22–25 March 2017; pp. 376–381.

2. Zhu, D.; Yue, S.; Park, S.; Wang, Y.; Chang, N.; Pedram, M. Cost-effective design of a hybrid electrical energy storage system for electric vehicles. In Proceedings of the International Conference on Hardware/Software Codesign and System Synthesis (CODES-ISS), New Delhi, India, 12–17 October 2014; pp. 1–8.
3. Nelson, D.; Nehrir, M.; Wang, C. Unit sizing of stand-alone hybrid wind/PV/fuel cell power generation systems. In Proceedings of the IEEE Power Engineering Society General Meeting, San Francisco, CA, USA, 16 June 2005; pp. 2116–2122.
4. Shi, X.; Li, Y.; Cao, Y.; Tan, Y. Cyber-physical electrical energy systems: Challenges and issues. *CSEE J. Power Energy Syst.* **2015**, *1*, 36–42. [CrossRef]
5. Ilic, M.D.; Xie, L.; Khan, U.A.; Moura, J.M. Modeling future cyber-physical energy systems. In Proceedings of the IEEE Power and Energy Society General Meeting, Pittsburgh, PA, USA, 20–24 July 2008; pp. 1–9.
6. Al Faruque, M.A.; Ahourai, F. A model-based design of cyber-physical energy systems. In Proceedings of the IEEE Asia and South Pacific Design Automation Conference (ASP-DAC), Singapore, 20–23 January 2014; pp. 97–104.
7. Kim, Y.; Shin, D.; Petricca, M.; Park, S.; Poncino, M.; Chang, N. Computer-aided design of electrical energy systems. In Proceedings of the IEEE/ACM International Conference on Computer-Aided Design (ICCAD), San Jose, CA, USA, 18–21 November 2013; pp. 194–201.
8. Molina, J.M.; Pan, X.; Grimm, C.; Damm, M. A framework for model-based design of embedded systems for energy management. In Proceedings of the IEEE Workshop on Modeling and Simulation of Cyber-Physical Energy Systems (MSCPES), Berkeley, CA, USA, 20–20 May 2013; pp. 1–6.
9. Vinco, S.; Chen, Y.; Fummi, F.; Macii, E.; Poncino, M. A layered methodology for the simulation of extra-functional properties in smart systems. *IEEE Trans. Comput. Aided Des. Integr. Circuits Syst.* **2017**, *36*, 1702–1715. [CrossRef]
10. Song, W.J.; Mukhopadhyay, S.; Yalamanchili, S. KitFox: Multiphysics Libraries for Integrated Power, Thermal, and Reliability Simulations of Multicore Microarchitecture. *IEEE Trans. Compon. Packag. Manuf. Technol.* **2015**, *5*, 1590–1601. [CrossRef]
11. Fummi, F.; Lora, M.; Stefanni, F.; Trachanis, D.; Vanhese, J.; Vinco, S. Moving from co-simulation to simulation for effective smart systems design. In Proceedings of the Design, Automation Test in Europe Conference Exhibition (DATE), Dresden, Germany, 24–28 March 2014; pp. 1–4.
12. Görgen, R.; Grüttner, K.; Herrera, F.; Peñil, P.; Medina, J.; Villar, E.; Palermo, G.; Fornaciari, W.; Brandolese, C.; Gadioli, D.; et al. Contrex: Design of embedded mixed-criticality control systems under consideration of extra-functional properties. In Proceedings of the 2016 Euromicro Conference on Digital System Design (DSD), Limassol, Cyprus, 31 August–2 September 2016; pp. 286–293.
13. O.Gbadegesin, A.; Sun, Y.; Nwulu, N.I. Techno-economic analysis of storage degradation effect on levelised cost of hybrid energy storage systems. *Sustain. Energy Technol. Assess.* **2019**, *36*, 100536. [CrossRef]
14. Singh, S.; Chauhan, P.; Aftab, M.; Ali, I.; Hussain, S.; Ustun, T. Cost Optimization of a Stand-Alone Hybrid Energy System with Fuel Cell and PV. *Energies* **2020**, *13*, 1295. [CrossRef]
15. Fulzele, J.B.; Daigavane, M. Design and Optimization of Hybrid PV-Wind Renewable Energy System. *Mater. Today Proc.* **2018**, *5*, 810–818. [CrossRef]
16. Mao, M.; Jin, P.; Chang, L.; Xu, H. Economic Analysis and Optimal Design on Microgrids With SS-PVs for Industries. *IEEE Trans. Sustain. Energy* **2014**, *5*, 1328–1336. [CrossRef]
17. Akram, U.; Khalid, M.; Shafiq, S. An Innovative Hybrid Wind-Solar and Battery-Supercapacitor Microgrid System—Development and Optimization. *IEEE Access* **2017**, *5*, 25897–25912. [CrossRef]
18. Atia, R.; Yamada, N. Sizing and Analysis of Renewable Energy and Battery Systems in Residential Microgrids. *IEEE Trans. Smart Grid* **2016**, *7*, 1204–1213. [CrossRef]
19. Alramlawi, M.; Souidi, Y.; Li, P. Optimal design of PV-Battery Microgrid Incorporating Lead-acid Battery Aging Model. In Proceedings of the IEEE EEEIC/ICPS, Genova, Italy, 11–14 June 2019; pp. 1–6.
20. Zhu, D.; Yue, S.; Chang, N.; Pedram, M. Toward a Profitable Grid-Connected Hybrid Electrical Energy Storage System for Residential Use. *IEEE Trans. Comput. Aided Des. Integr. Circuits Syst.* **2016**, *35*, 1151–1164. [CrossRef]
21. Zhou, L.; Zhang, Y.; Lin, X.; Li, C.; Cai, Z.; Yang, P. Optimal Sizing of PV and BESS for a Smart Household Considering Different Price Mechanisms. *IEEE Access* **2018**, *6*, 41050–41059. [CrossRef]
22. Alsaidan, I.; Khodaei, A.; Gao, W. A Comprehensive Battery Energy Storage Optimal Sizing Model for Microgrid Applications. *IEEE Trans. Power Syst.* **2018**, *33*, 3968–3980. [CrossRef]

23. Hesse, H.; Martins, R.; Musilek, P.; Naumann, M.; Truong, C.; Jossen, A. Economic optimization of component sizing for residential battery storage systems. *Energies* **2017**, *10*, 835. [CrossRef]
24. Wu, X.; Hu, X.; Yin, X.; Zhang, C.; Qian, S. Optimal battery sizing of smart home via convex programming. *Energy* **2017**, *140*, 444–453. [CrossRef]
25. Luo, X.; Liu, Y.; Liu, J.; Liu, X. Optimal design and cost allocation of a distributed energy resource (DER) system with district energy networks: A case study of an isolated island in the South China Sea. *Sustain. Cities Soc.* **2019**, *51*, 101726. [CrossRef]
26. Olivieri, Z.T.; McConky, K. Optimization of residential battery energy storage system scheduling for cost and emissions reductions. *Energy Build.* **2020**, *210*, 109787. [CrossRef]
27. Cui, T.; Chen, S.; Wang, Y.; Zhu, Q.; Nazarian, S.; Pedram, M. Optimal co-scheduling of HVAC control and battery management for energy-efficient buildings considering state-of-health degradation. In Proceedings of the IEEE Asia and South Pacific Design Automation Conference (ASP-DAC), Macau, China, 25–28 January 2016; pp. 775–780.
28. Ju, C.; Wang, P.; Goel, L.; Xu, Y. A Two-Layer Energy Management System for Microgrids With Hybrid Energy Storage Considering Degradation Costs. *IEEE Trans. Smart Grid* **2018**, *9*, 6047–6057. [CrossRef]
29. Esfahanian, H.M.M.; Abtahi, A.; Zilouchian, A. Modeling a Hybrid Microgrid Using Probabilistic Reconfiguration under System Uncertainties. *Energies* **2017**, *10*, 1430.
30. Lotfi, H.; Khodaei, A. Hybrid AC/DC microgrid planning. *Energy* **2017**, *118*, 37–46. [CrossRef]
31. Gupta, N.; Francis, G.; Ospina, J.; Newaz, A.; Collins, E.G.; Faruque, O.; Meeker, R.; Harper, M. Cost Optimal Control of Microgrids Having Solar Power and Energy Storage. In Proceedings of the IEEE/PES Transmission and Distribution, Denver, CO, USA, 16–19 April 2018; pp. 1–9.
32. Bandyopadhyay, S.; Chandra Mouli, G.R.; Qin, Z.; Elizondo, L.R.; Bauer, P. Techno-economical Model based Optimal Sizing of PV-Battery Systems for Microgrids. *IEEE Trans. Sustain. Energy* **2019**, 1. [CrossRef]
33. Buonomano, A.; Calise, F.; d'Accadia, M.D.; Vicidomini, M. A hybrid renewable system based on wind and solar energy coupled with an electrical storage: Dynamic simulation and economic assessment. *Energy* **2018**, *155*, 174–189. [CrossRef]
34. Ramli, M.A.; Hiendro, A.; Al-Turki, Y.A. Techno-economic energy analysis of wind/solar hybrid system: Case study for western coastal area of Saudi Arabia. *Renew. Energy* **2016**, *91*, 374–385. [CrossRef]
35. Khormali, S.; Niknam, E. Operation Cost Minimization of Domestic Microgrid under the Time of Use Pricing Using HOMER. In Proceedings of the Electric Power Engineering (EPE), Kouty nad Desnou, Czech Republic, 15–17 May 2019; pp. 1–6.
36. Chen, Y.; Vinco, S.; Pagliari, D.J.; Montuschi, P.; Macii, E.; Poncino, M. Modeling and Simulation of Cyber-Physical Electrical Energy Systems with SystemC-AMS. *IEEE Trans. Sustain. Comput.* **2020**, 1. [CrossRef]
37. Capital Recovery Factor. 2018 Available online: https://www.homerenergy.com/products/pro/docs/latest/capital_recovery_factor.html (accessed on 6 June 2020).
38. Chen, Y.; Macii, E.; Poncino, M. A circuit-equivalent battery model accounting for the dependency on load frequency. In Proceedings of the Design, Automation Test in Europe Conference (DATE), Lausanne, Switzerland, 27–31 March 2017; pp. 1177–1182.
39. Bocca, A.; Chen, Y.; Macii, A.; Macii, E.; Poncino, M. Aging and Cost Optimal Residential Charging for Plug-In EVs. *IEEE Des. Test* **2017**, *35*, 16–24. [CrossRef]
40. Lambert, T.; Gilman, P.; Lilienthal, P. Micropower System Modeling With Homer. In *Integration of Alternative Sources of Energy*; John Wiley & Sons: Hoboken, NJ, USA, 2006.
41. Baek, D.; Chen, Y.; Chang, N.; Macii, E.; Poncino, M. Optimal Battery Sizing for Electric Truck Delivery. *Energies* **2020**, *13*, 709. [CrossRef]
42. IEEE. *Standard for Standard SystemC Analog/Mixed-Signal Extensions Language Reference Manual*; Std 1666.1-2016; IEEE: Piscataway, NJ, USA, 2016; pp. 1–236.
43. Chen, Y.; Vinco, S.; Macii, E.; Poncino, M. SystemC-AMS thermal modeling for the co-simulation of functional and extra-functional properties. *ACM Trans. Des. Autom. Electron. Syst. TODAES* **2018**, *24*, 1–26. [CrossRef]
44. Unterrieder, C.; Huemer, M.; Marsili, S. SystemC-AMS-based design of a battery model for single and multi cell applications. In Proceedings of the PRIME Conference on Ph.D. Research in Microelectronics Electronics, Aachen, Germany, 12–15 June 2012; pp. 1–4.

45. Vinco, S.; Sassone, A.; Fummi, F.; Macii, E.; Poncino, M. An open-source framework for formal specification and simulation of electrical energy systems. In Proceedings of the IEEE/ACM International Symposium on Low Power Electronics and Design (ISLPED), La Jolla, CA, USA, 11–13 August 2014; pp. 287–290.
46. Kim, S.K.; Jeon, J.H.; Cho, C.H.; Ahn, J.B.; Kwon, S.H. Dynamic modeling and control of a grid-connected hybrid generation system with versatile power transfer. *IEEE Trans. Ind. Electron.* **2008**, *55*, 1677–1688. [CrossRef]
47. Ohyama, K.; Nakashima, T. Wind turbine emulator using wind turbine model based on blade element momentum theory. In Proceedings of the IEEE Power Electronics, Electrical Drives, Automation and Motion (SPEEDAM), Pisa, Italy, 14–16 June 2010; pp. 762–765.
48. Vinco, S.; Chen, Y.; Macii, E.; Poncino, M. A unified model of power sources for the simulation of electrical energy systems. In Proceedings of the Great Lakes Symposium on VLSI (GLSVLSI), Boston, MA, USA, 18–20 May 2016; pp. 281–286.
49. Kelly, J.; Knottenbelt, W. The UK-DALE dataset, domestic appliance-level electricity demand and whole-house demand from five UK homes. *Sci. Data* **2015**, *2*, 1–14. [CrossRef]
50. Chen, Y.; Baek, D.; Kim, J.; Di Cataldo, S.; Chang, N.; Macii, E.; Vinco, S.; Poncino, M. A systemc-ams framework for the design and simulation of energy management in electric vehicles. *IEEE Access* **2019**, *7*, 25779–25791. [CrossRef]
51. National Renewable Energy Laboratory (NREL). Measurement and Instrumentation Data Center. Available online: https://midcdmz.nrel.gov/ (accessed on 6 June 2020).
52. Liu, N.; Yu, X.; Wang, C.; Li, C.; Ma, L.; Lei, J. Energy-Sharing Model With Price-Based Demand Response for Microgrids of Peer-to-Peer Prosumers. *IEEE Trans. Power Syst.* **2017**, *32*, 3569–3583. [CrossRef]
53. Magnani, S.; Pezzola, L.; Danti, P. Design optimization of a heat thermal storage coupled with a micro-CHP for a residential case study. *Energy Procedia* **2016**, *101*, 830–837. [CrossRef]
54. Jabir, H.; Ishak, D.; Abunima, H. Impacts of Demand-Side Management on Electrical Power Systems: A Review. *Energies* **2018**, *11*, 1050. [CrossRef]
55. Eid, C.; Koliou, E.; Valles, M.; Reneses, J.; Hakvoort, R. Time-based pricing and electricity demand response: Existing barriers and next steps. *Util. Policy* **2016**, *40*, 15–25. [CrossRef]
56. Telaretti, E.; Dusonchet, L.; Massaro, F.; Mineo, L.; Pecoraro, G.; Milazzo, F. A simple operation strategy of battery storage systems under dynamic electricity pricing: An Italian case study for a medium-scale public facility. In Proceedings of the Renewable Power Generation Conference (RPG), Naples, Italy, 24–25 September 2014; pp. 1–7.
57. Electric Vehicle (EV). 2019. Available online: https://batteryuniversity.com/learn/article/electric_vehicle_ev/ (accessed on 6 June 2020).
58. Efficient, Informed Decisions about Distributed Generation and Distributed Energy Resources. 2018. Available online: https://www.homerenergy.com/ (accessed on 6 June 2020).

Article

Using Residential and Office Building Archetypes for Energy Efficiency Building Solutions in an Urban Scale: A China Case Study

Chao Ding * and Nan Zhou

Lawrence Berkeley National Laboratory, Berkeley, CA 94709, USA; nzhou@lbl.gov
* Correspondence: chaoding@lbl.gov; Tel.: +1-510-486-6835

Received: 8 May 2020; Accepted: 16 June 2020; Published: 20 June 2020

Abstract: Building energy consumption accounts for 36% of the overall energy end use worldwide and is growing rapidly as developing countries continue to urbanize. Understanding the energy use at urban scale will lay the foundation for identification of energy efficiency opportunities to be deployed at speed. China has almost half of global new constructions and plays an important role in building suitability. However, an open source national building energy consumption database is not available in China. To provide data support for building energy consumptions, this paper used a simulation method to develop an urban building energy consumption database for a pilot city in Wuhan, China. First, residential, small, and large office building archetype energy models were created in EnergyPlus to represent typical building energy consumption in Wuhan. The baseline reference model simulation results were further validated using survey data from the literature. Second, stochastic simulations were conducted to consider different design parameters and occupants' energy usage intensity scenarios, such as thermal properties of the building envelope, lighting power density, equipment power density, HVAC (heating, ventilation and air conditioning) schedule, etc. A building energy consumption database was generated for typical building archetypes. Third, data-driven regression analysis was conducted to support quick building energy consumption prediction using key high-level building information inputs. Finally, a web-based urban energy platform and an interface were developed to support further third-party application development. The research is expected to provide fast energy efficiency building design solutions for urban planning, new constructions as well as building retrofits.

Keywords: urban scale; building energy simulation; EnergyPlus; regression; building archetypes

---

1. Introduction

By 2050, 66% of the world's population will live in urban areas [1], making urbanization one of the critical themes and challenges in this century. This is the case especially for some Asian countries, such as China, where city boundaries are expanding with numerous new constructions every year. China has contributed to approximately 50% of the world's new constructions since 2010 [2]. Rapid global urbanization has resulted in significant increases in energy consumption, greenhouse gas emissions, pollutant emissions, and widespread environmental degradation. Urban areas account for 67–76% of global energy use and 71–76% of CO_2 emissions [3]. Cities around the world are searching for strategies to reduce energy consumption and to become green and low-carbon cities, and enhance their resilience in a changing climate.

Building energy consumption accounts for 36% of the global final energy use in 2017, and this number is much higher in urban areas [4,5]. In the U.S., national level building energy consumption databases have been developed and regularly updated to represent actual building energy usage levels. For example, the Residential Energy Consumption Survey (RECS) and Commercial Buildings

Energy Consumption Survey (CBECS) collect energy-related building characteristics and energy usage information [6,7]. However, this kind of open source national building energy consumption database is not available in China.

To better understand building energy consumption in urban areas, besides survey and measurement, urban datasets and urban-scale building energy consumption platforms have been developed based on urban-scale building energy simulations. Urban-scale building energy simulation can play an essential role in sustainable urbanization, allowing planners and policy makers to develop planning strategies using the lens of energy performance.

A research group from the college of Architecture at Georgia Institute of Technology developed a GIS-based urban building energy modeling system, called Urban-EPC. It includes four main models: the Data Preparation Model, the Pre-Simulation Model, the Main Simulation Model and the Visualization and Analysis Model. This Urban-EPC tool also uses physics models and calculates the hourly heat balance of the whole building. It contains three categories of building vintage (based on the construction year), each of which includes 16 building types representing most of the commercial buildings across 16 US climate zones. The development team also conducted a case study for Manhattan. They obtained the building footprint data from New York city planning database with references to Google Earth 3D building [8].

The sustainable design lab at Massachusetts Institute of Technology (MIT) also developed an Urban Building Energy Model (UBEM) for Boston to estimate citywide hourly energy demands at the building level. In this project, the geometric input for Boston was also extracted from GIS shapefiles into the Rhinoceros 3D V5 CAD environment, and a total of 76 different building archetypes were then assigned to individual buildings based on land use and building age. Bayesian calibration was applied to update the probability distributions of uncertain parameters in archetype descriptions using monthly and annual measured energy usage data. EnergyPlus was used to simulate the energy consumption results of individual building models. The urban energy use pattern of different times of the day is visualized and overlaid with the Boston map. The tool can help local communities to evaluate energy related decisions and building retrofit strategies to reduce building energy use. They also predicated future scenarios, including solar photovoltaic (PV) penetration, and demand response strategy implantation [9,10].

Lawrence Berkeley National Laboratory (LBNL) developed and released a web-based urban-scale building stock simulation platform, called City Building Energy Saver (CityBES). It is designed to support building retrofit analysis. CityBES uses an open standard, CityGML, to represent the 3D city models, and then it categorizes buildings into different types, including small/medium/large offices, hotels, schools, and hospitals. For each type of these buildings, CityBES generates baseline EnergyPlus simulation models based on the cities' building datasets and user-selected energy conservation measures (ECMs). There are three main layers: the data layer, the simulation algorithms and software tools layer, and the use-cases layer. The neighborhood buildings in CityBES are modeled as shading surfaces in EnergyPlus to consider the shading interactions between buildings. Simulation results, such as energy use intensity (EUI), can be color-coded and mapped to the 3D buildings with the GIS database. A case study using CityBES for San Francisco shows a potential retrofit site energy saving of 23–38% per building [11].

In addition, the Oak Ridge National Laboratory and National Renewable Energy Laboratory have also developed urban scale simulation tools, called AutoBEM and URBANopt, respectively [12,13]. They used similar approaches: generate baseline building energy models for each building type as a template, categorize buildings in the area of interest into corresponding archetype and link to the template results, map the simulation results to a GIS platform for visualization. This method can provide quick design support for large scale energy decision making based on archetype data, without running detailed building energy simulation.

However, the above case studies are based mainly on simulation results. It is important to validate the numerical simulations using ground truth building energy survey data and consider occupants' energy usage behavior. Furthermore, the case studies are for large cities in the US, where rapid urbanization has almost been completed. Due to rapid urbanization, China has a large percentage of new constructions. Meanwhile, old buildings with different years of building exist in the same urban region. The building age variation could be as high as several decades. As they were subject to different building design standards/codes, the same type of building, if built in different years, could show very different building consumption profiles. Therefore, building vintage is a key parameter to consider. However, open-source building energy models for typical archetypes have not been well developed in China. It is important to develop an updated urban-scale building energy consumption platform for China, to understand both the spatial and temporal urban energy system.

This paper shows our efforts on building archetype development for the urban energy simulation platform. Three main archetype buildings (residential building, small office building, and large office building) are created and demonstrated in this paper with the following innovations.

- Develop reference building energy models for residential, small office, and large office building types in Wuhan, China, considering different vintages and unique HVAC usage patterns
- Create an application programming interface (API) for Wuhan to support urban building energy platform development.

2. Reference Building Models

Currently, there are no open sourced well-developed reference building models in China. Residential and office building are selected for prototype development in this paper because these two types of buildings are the top two largest building stocks in China, with a percentage of 73% and 9%, respectively [14].

Working with local project partners, the most popular configurations and geometries for residential buildings as well as office buildings were collected through a survey. Figure 1 shows the geometry of a typical residential building. It is a 10-story apartment building with a total building area of 7836 m^2. According to the different orientation, each floor is divided into nine thermal zones: eight apartment units, and one corridor. The floor area of each apartment is about 88 m^2.

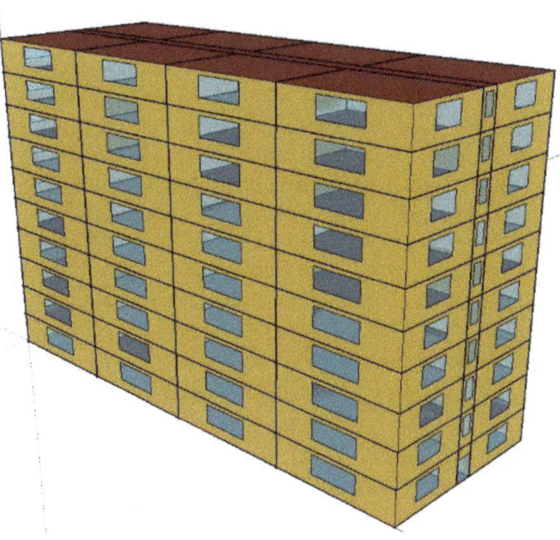

Figure 1. Geometry of the residential reference building.

Figure 2 shows the most popular geometries of a typical small office building (left) and a large office building (right). The small office building has three floors and the large office has eighteen floors. Each floor has four external zones and one core zone. The total building areas are 8176 m^2 for the small office and 26,142 m^2 for the large office, respectively.

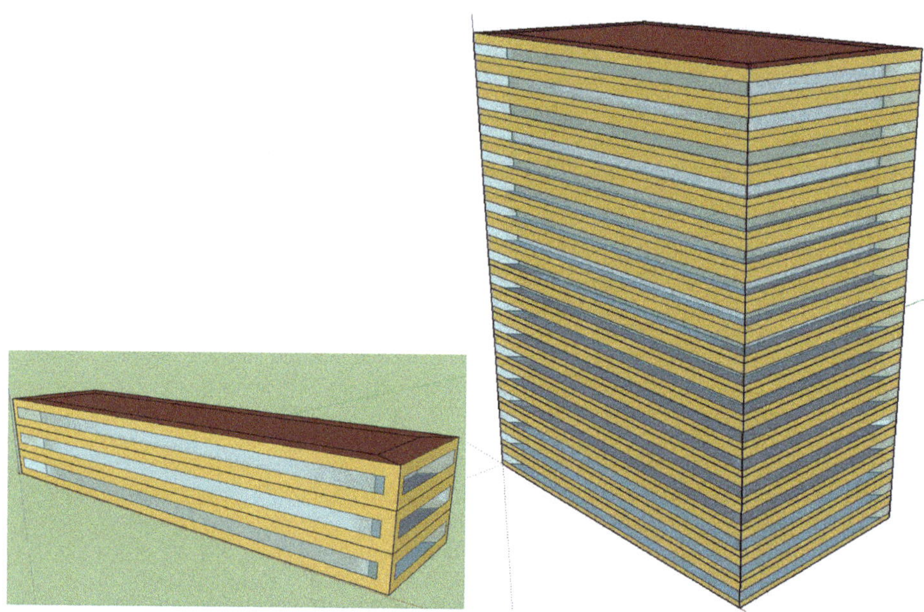

Figure 2. Geometries of reference office buildings (left: small office, right: large office).

Key building design parameters, such as the building envelope's thermal properties, lighting power density, equipment load density, HVAC system and schedules, were defined based on the corresponding residential and commercial building design standards [15,16].

Based on the survey, the HVAC systems for these three building types are different. Residential and small office buildings use ductless mini-split heat pumps for heating and cooling, while large office buildings use chillers and cooling towers for cooling and boilers for heating.

To capture the average energy usage level, the China Residential Energy Consumption Survey (CRECS) data were used to determine the heating and cooling schedules for residential and small office building. The CRECS was conducted by Renming University in 2012. The CRECS2012 includes residential appliance usage and electricity consumption data from 1450 residential buildings across 26 provinces in China [17]. Valid instances numbering 218 from the hot summer and cold winter climate zone (where Wuhan is located) were used to calibrate the baseline residential model. Figure 3 shows the number of heating days in winter and the number of cooling days in summer, respectively. It can be observed that most people only use heating for less than one month in winter, and use cooling for one to two months in summer. Compared with cooling, the residents seem to be more tolerant of heating. Figure 4 shows the daily distribution of heating and cooling hours. It shows that most people use heating or cooling for less than 5 h per day. The heating and cooling schedules (days/year and hours/day) as well as the temperature setpoints of the reference buildings were adjusted to be the average values according to the survey data.

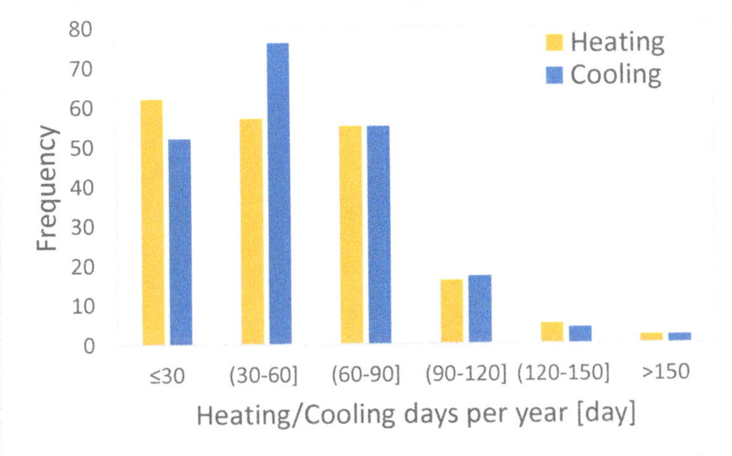

Figure 3. Distribution of heating/cooling days per year in winter/summer.

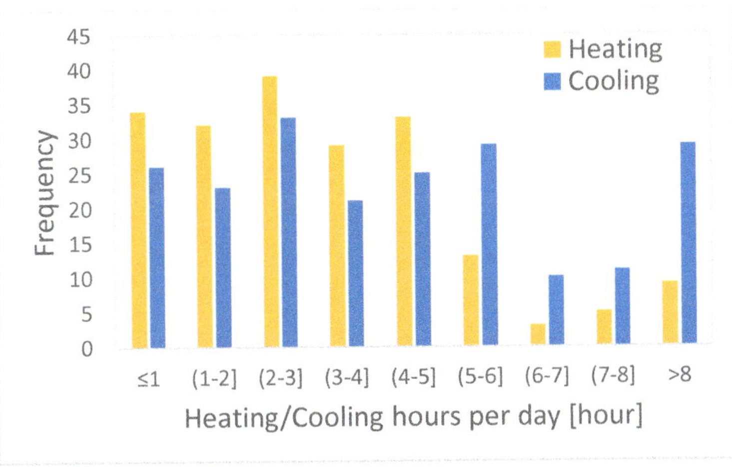

Figure 4. Distribution of heating/cooling hours per day in winter/summer.

The baseline reference buildings were developed using EnergyPlus software. EnergyPlus is an open source simulation engine for whole building energy consumption [18]. It was developed and is supported by the U.S. Department of Energy. EnergyPlus has been widely used and validated by researchers and designers. It is a console-based program, not a user interface. Some graphical interfaces for EnergyPlus, such as DesignBuilder and OpenStudio, are also available. Since the inputs and outputs for EnergyPlus are all text-based, users can easily edit the information to develop a customized system and run parametric simulations using scripts. The detail settings of each model are summarized in Table 1.

Table 1. Baseline EnergyPlus model settings.

Input Parameters	Unit	Small Office	Large Office	Residential Building	Reference
External wall insulation	W/m²·K	0.597	0.597	0.88	Residential building: DB42T-559-2013 [10]; Small and large office buildings: GB50189-2015 [11]
Roof insulation	W/m²·K	0.399	0.399	0.447	
Ground floor insulation	W/m²·K	0.253	0.253	1.2	
External Windows	W/m²·K	2.6	2.6	2.7	
Infiltration rate	ACH	1	1	1	
Lighting power dentistry	W/m²	9	9	Apartment: 4.2, Corridor: 1.8	
Equipment power density	W/m²	15	15	Plug load: 2, Kitchen: 5	
Occupancy density	m²/person	10	10	2	
Infiltration rate	1/h	1.0	1.0	1.0	
HVAC system	-	Mini-split air conditioner	Chiller + Natural gas boiler	Mini-split air conditioner	Survey data [12]
Heating/Cooling setpoints	°C	20.5/23.5	20.5/23.5	18/26	
Heating Schedule	-	9:00–12:00, 1/1–2/14	8:00–15:00, 1/1–2/14	19:00–22:00, 1/1–2/14	
Cooling Schedule	-	12:00–16:00, 7/18–8/31	10:00–17:00, 7/18–8/31	18:00–22:00, 7/18–8/31	

Wuhan's hourly weather data were used to simulate annual building energy consumption [19]. Figure 5 shows the simulation results of a baseline residential building. The simulated total building electricity consumption is 27.8 kWh/m^2. It can be observed that heating and cooling energy consumptions only account for approximately 30% of the total annual electricity consumption. People tend to use the heat pumps only when the weather is too cold or too hot, to reduce their electricity bills. The occupants' behavioral energy saving patterns can be found.

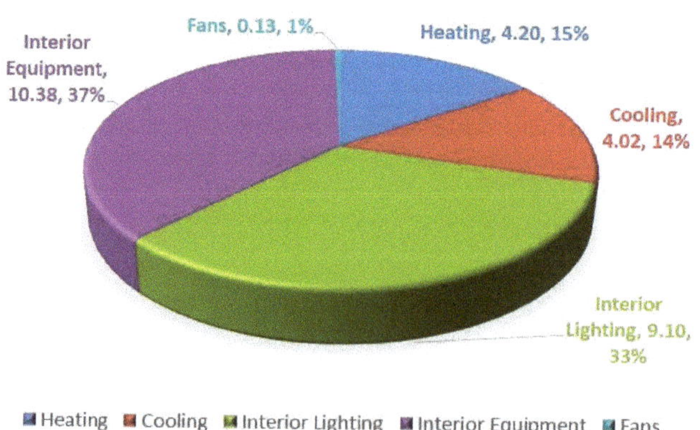

Figure 5. Reference residential building electricity consumption breakdown [kWh/m^2].

Figure 6 shows the distribution of electricity consumption in the hot summer and cold winter climate zone from the CRECS survey. The mean value (25.8 kWh/m^2) matches well with the EnergyPlus simulation result, which further validates the reference building model. It is of note that the electricity consumption is expected to be 35.3 kWh/m^2 (37% higher than the actual mean value) in the Guideline for Energy Consumption Quota of Civil Buildings in Wuhan [20]. Therefore, it is critical to consider occupants' energy use behavior and reflect the actual energy usage when making regional building energy consumption standards.

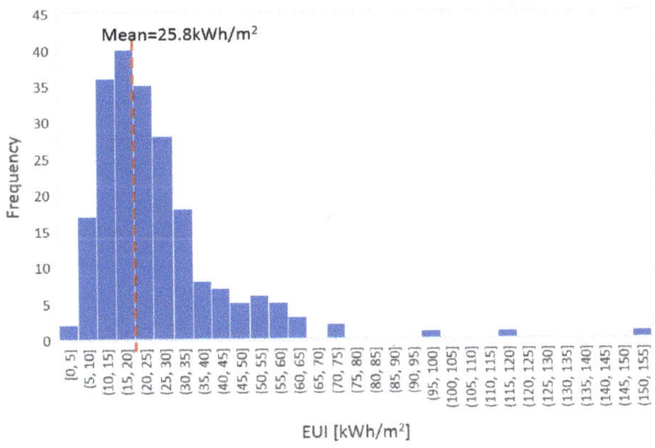

Figure 6. Distribution of electricity consumption from CRECS data.

Similarly, the HVAC schedules of small office and large office buildings were calibrated using the survey data. The annual energy consumptions were simulated in EnergyPlus. Figure 7 shows the simulation results. The total electricity consumption is 61.0 kWh/m^2 for the small office and 130.9 kWh/m^2 for the large office.

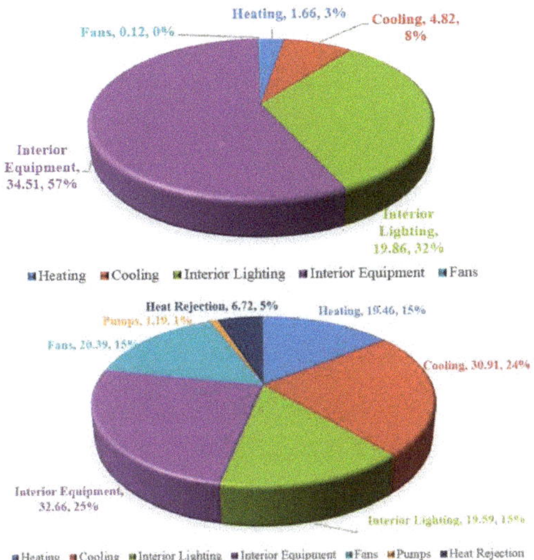

Figure 7. Reference office building electricity consumptions [kWh/m^2] (top: small office, bottom: large office).

3. Stochastic Simulations

After the reference building models were validated, stochastic simulations were conducted to consider different building design variations for the urban building energy consumption database development. Eight different design parameters, such as the building envelope's thermal properties, infiltration, heating and cooling schedules, lighting power density, and equipment power density, were considered to cover different constructions, building design scenarios and the occupant's energy usage patterns.

To differentiate building vintage, three levels (high, medium, and low) of building envelopes were studied by grouping U factors of different parts (external wall, slab, roof, and glass). The building geometry was kept constant to represent the most common configuration in Wuhan.

To reflect the actual energy saving behavior of the occupants and better capture different energy usage patterns, thirteen heating and cooling schedules were proposed. The schedule information was derived based on statistical analysis of the actual building energy data from CRECS. The data from the CRECS energy consumption survey was ranked from low to high. Level of 5% means the top 5% from the ranking. It represents the most efficient energy usage, in terms of heating and cooling hours per day and days per year. Level of 95% represents the least efficient energy usage (bottom 5% from the ranking). It is assumed that the lighting and plug/equipment loads are coupled with heating/cooling schedules, since people's energy saving behavior is consistent. Other energy usage profiles can be interpolated.

Figure 8 shows the stochastic cases of the residential building. In total, 117 design scenarios were considered. The combination of residential building baseline model is highlighted in yellow. In a similar way, 117 small office and 117 large office EnergyPlus models were generated to cover different energy use intensity scenarios for office buildings.

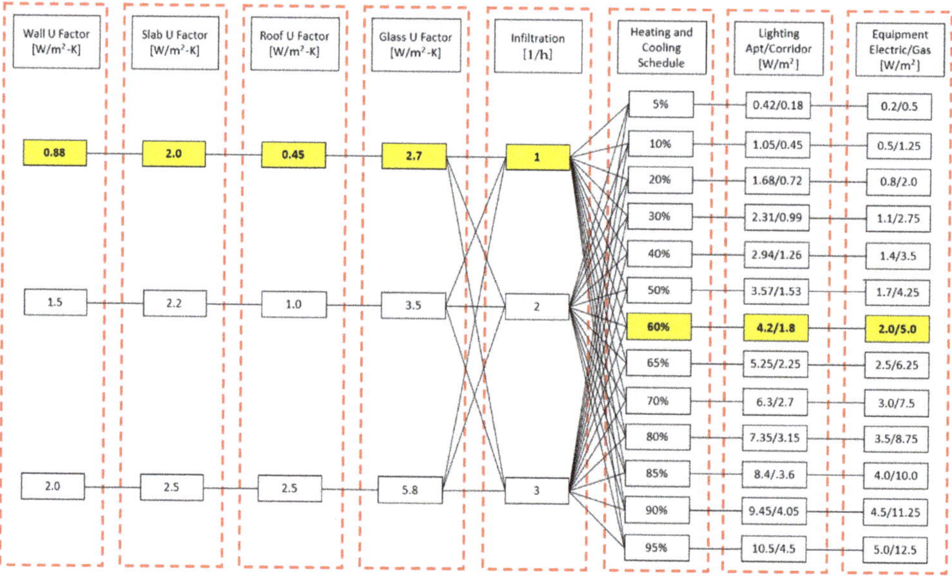

Figure 8. 117 stochastic scenarios of the residential building model.

4. Results and Discussion

Stochastic simulations were performed using EnergyPlus. Figure 9 shows the annual energy simulation results of residential buildings. It is of note that the energy consumptions are based on pure stochastic simulations defined in Section 3, assuming a uniform distribution of the 117 parametric design scenarios of each building type without any additional weighting factor. In reality, there may be less people in the very low (left) and very high (right) energy consumption ranges. To get a more realistic energy consumption distribution, we collected Wuhan's housing price (for residential building) and rent (for office building) information and adjusted the energy distribution accordingly. Figure 10 shows the housing price distribution of 18,864 residential buildings in Wuhan.

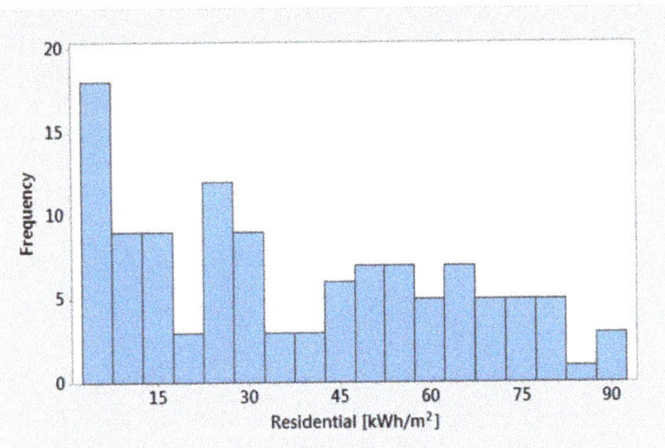

Figure 9. Annual residential building electricity consumption distribution of the stochastic simulations.

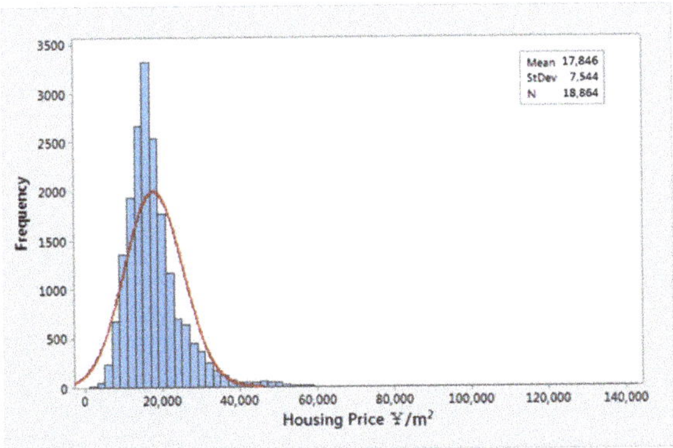

Figure 10. Housing price distribution of residential buildings in Wuhan.

Similarly, the annual electricity consumption distributions for small and large office buildings are shown in Figure 11.

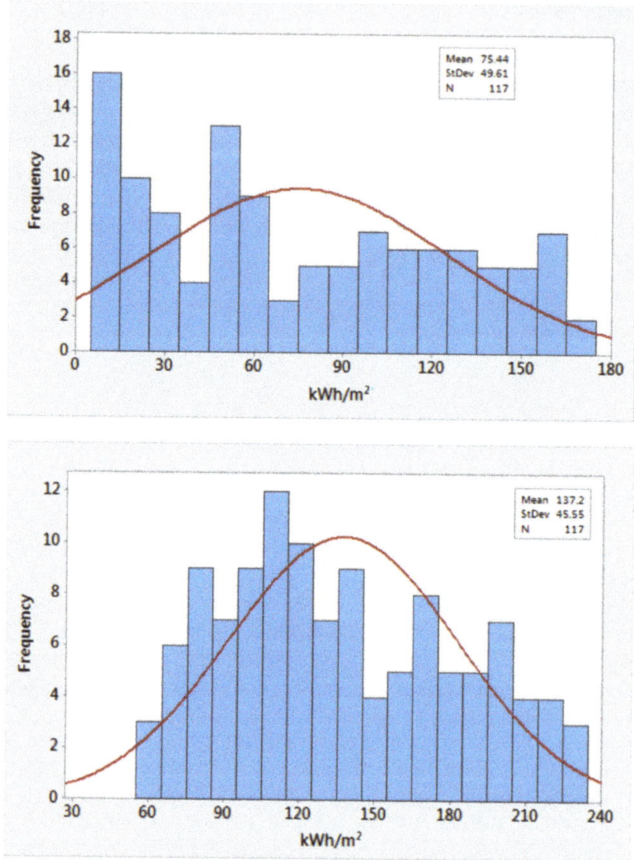

Figure 11. Annual office building electricity consumption distribution of the stochastic simulations (top: small office. Bottom: large office).

Furthermore, a data-driven regression model was developed to predict building energy consumption. Suggested by local urban planners and energy policy makers, building's price/rent and vintage were chosen to be the two key independent variables for the regression model.

Various machine learning algorithms were applied to the dataset to compare prediction accuracy. However, due to the very limited number of inputs, more complex algorithms did not show much advantage. Finally, linear regression models were selected, because of their robustness and high prediction accuracy. Tables 2–4 show the regression functions for residential, small office and large office buildings, respectively.

Table 2. Regression functions for residential buildings.

Item	Regression Function [kWh/m²]	R^2
total electricity	$(0.0034 \times \text{price} - 29.2658) \times (1 + F)$	0.93
heat electricity	$(0.0005 \times \text{price} - 3.4929) \times (1 + F)$	0.77
cool electricity	$(0.0006 \times \text{price} - 5.4794) \times (1 + F)$	0.92
light electricity	$(0.001 \times \text{price} - 9.406) \times (1 + F)$	0.95
equip electricity	$(0.0013 \times \text{price} - 10.7288) \times (1 + F)$	0.95
fan electricity	$(0.00005 \times \text{price} - 0.1587) \times (1 + F)$	0.89
equip gas	$(0.0012 \times \text{price} - 9.941)$	0.95

where

$$F = \begin{cases} 0, \; if \; year = 2010 \\ 0.0245, \; if \; year = 2000 \\ 0.0603, \; if \; year = 1990 \end{cases}$$

Table 3. Regression functions for small office buildings.

Item	Regression Function [kWh/m²]	R^2
total electricity	$(2.2161 \times \text{rent} - 62.4217) \times (1 + F)$	0.97
heat electricity	$\left(1.9 \times 10^{-5} \times \text{rent}^3 - 3.82 \times 10^{-5} \times \text{rent}^2 + 0.37 \times \text{rent} - 7.21\right) \times (1 + F)$	0.81
cool electricity	$(0.2944 \times \text{rent} - 10.2555) \times (1 + F)$	0.95
light electricity	$(0.7 \times \text{rent} - 19.4252) \times (1 + F)$	0.97
equip electricity	$(1.2163 \times \text{rent} - 33.7546) \times (1 + F)$	0.97
fan electricity	$(0.0057 \times \text{rent} - 0.1856) \times (1 + F)$	0.96

where

$$F = \begin{cases} 0, \; if \; year = 2010 \\ 0.0119, \; if \; year = 2000 \\ 0.0228, \; if \; year = 1990 \end{cases}$$

Table 4. Regression functions for large office buildings.

Item	Regression Function [kWh/m²]	R^2
total electricity	$(1.4117 \times \text{rent} + 21.5392) \times (1 + F_{ele})$	0.86
heat electricity	$\left(10^{-6} \times \text{rent} - 7 \times 10^{-5}\right) \times (1 + F_{ele})$	0.88
cool electricity	$(0.0811 \times \text{rent} + 26.5387) \times (1 + F_{ele})$	0.83
light electricity	$(0.4711 \times \text{rent} - 11.3829) \times (1 + F_{ele})$	0.84
equip electricity	$(0.7851 \times \text{rent} - 18.9715) \times (1 + F_{ele})$	0.84
fan electricity	$(0.0549 \times \text{rent} + 18.6239) \times (1 + F_{ele})$	0.65
pump electricity	$(0.0027 \times \text{rent} + 1.0361) \times (1 + F_{ele})$	0.84
heatRej electricity	$(0.0169 \times \text{rent} + 5.6949) \times (1 + F_{ele})$	0.82
heat gas	$(0.0692 \times \text{rent} + 16.2071) \times \left(1 + F_{gas}\right)$	0.80

where

$$F_{ele} = \begin{cases} 0, \; if \; year = 2010 \\ 0.0204, \; if \; year = 2000 \\ 0.0245, \; if \; year = 1990 \end{cases}, \; F_{gas} = \begin{cases} 0, \; if \; year = 2010 \\ 0.0544, \; if \; year = 2000 \\ 0.0847, \; if \; year = 1990 \end{cases}$$

To better illustrate our methods and make it easy and friendly to use, we developed a building simulation platform based on JavaEE technologies [8,9,21,22]. Figure 12 shows the system architecture. The platform consists of two parts. The first part is the service consumer (Application layer). The consumer here refers to the end users or any other third-party applications. The end user can utilize the service which results directly by opening a given service endpoint URL through the browser. Our service can also be incorporated into other external systems easily. The second part is the service provider. It generally includes three main layers: data layer, core algorithms implementation layer, and RESTful WebService layer. The data layer is responsible for providing enough data to make the platform work securely, such as the building information, system data, and stochastic simulation data.

Figure 13 shows how the simulation data is stored in the database. From the E-R diagram, we can see that hourly building energy consumptions can be simulated for the main types of buildings, such as large office, small office, and residential, in different scenarios. The core algorithm layer implements the core algorithms to simulate the building energy consumption. This layer mainly includes regression analysis and interpolation algorithms. To support third-party applications, our platform was designed to be a Service Oriented Architecture (SOA) based program [23]. Specifically, we chose the widely used RESTful WebService to wrap the core simulation APIs, so that everyone would be able to use our platform by just calling these standard WebServices [24]. For instance, users can use the API directly through their browsers by typing into the service endpoint as shown in Figure 14. In addition, third-party applications written in any programming languages can incorporate the APIs easily as these APIs are developed using the standard WebService.

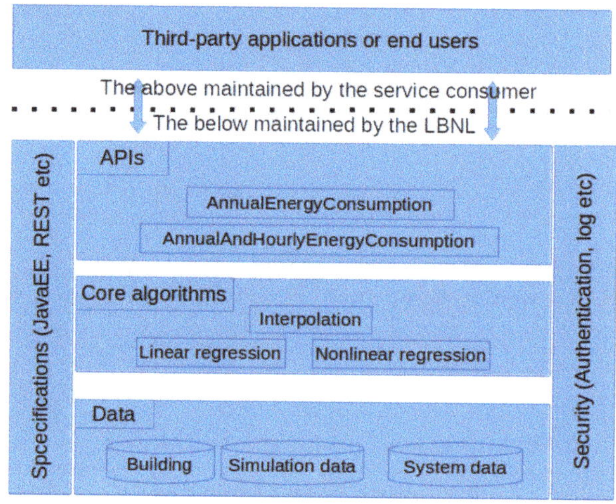

Figure 12. System architecture.

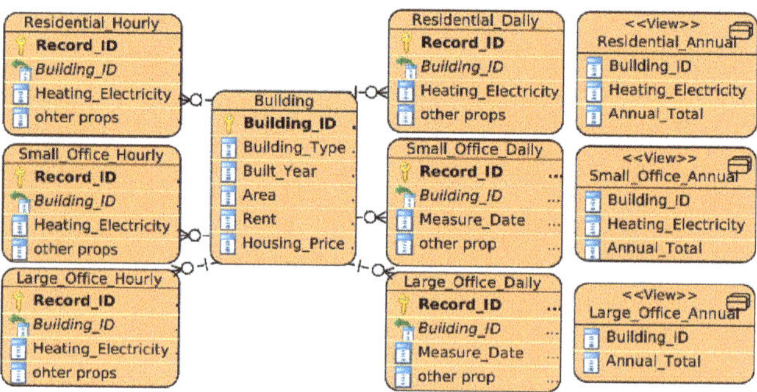

Figure 13. E-R Diagram.

```
{
    status: 200,
    buildingType: "smallOffice",
    description: "OK!",
  - annualEUI: {
        status: 200,
        buildingType: "smallOffice",
        description: "OK!",
        totalEleEUI: 168.91255,
        heatEleEUI: 0.52181,
        coolEleEUI: 20.47633,
        lightEleEUI: 53.6464,
        equipEleEUI: 93.21252,
        fanEleEUI: 0.40941
    },
  + firstDayHourly: […],
  + secondDayHourly: […]
}
```


Figure 14. Output of a RESTful WebService.

The building energy prediction models and the APIs created in this paper can be used to support further third-party urban energy application development. For example, Figure 15 shows an example of an urban building energy prediction platform developed by one of our research partners. Monthly energy consumptions (in EUI) of different building types are color-coded and mapped to individual buildings in a GIS database. Dark red represent a high EUI, while light red represents a lower EUI value. To support HVAC system design and equipment selection, high fidelity hourly EUIs are also provided for typical design days. By clicking any individual buildings from the web-based platform, the annual EUI of the selected building is shown with other building characteristics information, including building height, floor area, year of building, and housing price/rent. If the user toggles the year bar in the bottom, the platform can also visualize energy information in the past and predict future scenarios. This urban-scale 3D platform is currently used by the local government. It provides spatial and temporal building energy assessment and visualization to support design decision makings for city managers and urban planners.

Figure 15. An urban building energy prediction platform developed based on our API [25].

5. Limitations and Future Work

This paper demonstrates the development of residential and office building archetypes using a case study in Wuhan, China. Data-driven regression models were developed based on stochastic simulations. A web-based urban energy platform and an interface were created to support further third-party application development. Future work can be improved based on the following limitations.

A uniform distribution was assumed to generate different design variations for stochastic simulations. The actual distribution was adjusted through post processing to match the distribution of the survey data. In future work, we will apply a Bayesian calibration to consider the probability distribution of key uncertain variables. Due to the limited number of inputs for regression model development, the advantages of more complex non-linear machine learning algorithms, such as support vector machine or gradient boosting, cannot be reflected. In the next step, we will collaborate with our colleagues and partners and collect more available input data to improve our models. In addition, the platform will be fully verified using real-world data from our partners. Furthermore, it is usually straightforward to model building energy consumption for each single building using the traditional physics-based energy simulation methods, but it does not work well for modelling multiple building at community or city level [26–28], hence we are trying to use deep learning to discover the hidden and complex dynamics between multiple buildings so as to make our model more accurate while simulating the city scale energy consumption.

6. Conclusions

Urban-scale building energy consumption data are important for city managers or urban planners. However, an open source national building energy consumption database is not available in China. Instead of an energy consumption survey or measurement, urban scale building energy simulation can play an essential role in sustainable development during the urbanization process. It can enable high resolution analysis to estimate city level energy and track dynamic change. The requirement for citywide dynamic energy consumption information is urgent for city planning and energy policy making. Urban planners and policy makers can use the urban energy simulation platform to support urban-scale spatial and temporal decision-making on energy.

To develop such an urban-scale building energy platform, this paper demonstrates our work on generating a representative building energy consumption database for typical residential building, small office building, and large office building. The reference residential building, small and large office building energy models for Wuhan China were developed in EnergyPlus. The baseline residential reference building was calibrated using China's CRECS2012 building energy survey data to consider different building characteristics and occupants' unique HVAC usage patterns. Stochastic simulations were conducted to generate the numerical building energy consumption database. Three different construction levels were considered to reflect building vintages. Energy consumption distributions were adjusted using Wuhan's housing price and rent data.

Urban-scale building energy simulation requires engineering knowledge and computational resources, which creates a barrier for fast decision-making support. To solve this challenge, the building energy consumption database was further used to develop statistical regression models. To better illustrate our methods and make it easy and friendly to use, we developed a building simulation platform based on JavaEE technologies and standard WebServices. The platform and APIs are expected to provide design support for new constructions as well as for building retrofit. Combined with GIS database, the API can be easily used to develop a 3D urban energy prediction platform. With the support of data visualization, city managers and urban planners can check the spatial and temporal building energy distributions in a city area and assemble fast polices regarding building efficiency and sustainability.

Author Contributions: Conceptualization, C.D. and N.Z.; methodology, C.D.; software, C.D.; validation, C.D.; formal analysis, C.D.; investigation, C.D.; resources, C.D. and N.Z.; data curation, C.D.; writing—original draft preparation, C.D.; writing—review and editing, C.D. and N.Z.; visualization, C.D.; supervision, N.Z.; project administration, N.Z.; funding acquisition, N.Z. All authors have read and agreed to the published version of the manuscript.

Funding: This research was funded by the United States Department of Energy under Contract No. DE-AC02-05CH11231 and Energy Foundation China.

Acknowledgments: The authors would like to thank Chinese researchers in many research institutes by providing materials, guidance, and advice.

Conflicts of Interest: The authors declare no conflict of interest. The funders had no role in the design of the study; in the collection, analyses, or interpretation of data; in the writing of the manuscript, or in the decision to publish the results.

References

1. United Nations, Department of Economic and Social Affairs, Population Division. *World Urbanization Prospects: The 2014 Revision*; United Nations Publications: New York, NY, USA, 2014.
2. Zhou, N.; Khanna, N.; Feng, W.; Ke, J.; Levine, M. Scenarios of energy efficiency and CO2 emissions reduction potential in the buildings sector in China to year 2050. *Nat. Energy* **2018**, *3*, 978–984. [CrossRef]
3. Seto, K.; Dhakal, S.; Bigio, A.; Blanco, H.; Delgado, G.C.; Dewar, D.; Huang, L.; Inaba, A.; Kansal, A.; Lwasa, S.; et al. Human settlements, infrastructure and spatial planning. In *Climate Change 2014: Mitigation of Climate Change, Contribution of Working Group III to the Fifth Assessment Report of the Intergovernmental Panel on Climate Change*; Cambridge University Press: Cambridge, UK; New York, NY, USA, 2014; pp. 923–1000.
4. International Energy Agency; United Nations Environment Programme. *2018 Global Status Report: Towards a Zero-Emission, Efficient and Resilient Buildings and Construction Sector*; International Energy Agency: Paris, France; United Nations Environment Programme: Nairobi, Kenya, 2018.
5. U.S. Energy Information Administration (EIA). Energy Consumption by Sector. 2018. Available online: https://www.eia.gov/totalenergy/data/monthly/pdf/sec2_3.pdf (accessed on 3 June 2020).
6. EIA. Residential Energy Consumption Survey (RECS). Available online: https://www.eia.gov/consumption/residential/ (accessed on 3 June 2020).
7. EIA. Commercial Buildings Energy Consumption Survey (CBECS). Available online: https://www.eia.gov/consumption/commercial/ (accessed on 3 June 2020).
8. Quan, S.J.; Li, Q.; Augenbroe, G.; Brown, J.; Yang, P.P.J. Urban Data and Building Energy Modeling: A GIS-based Urban Building Energy Modeling System Using the Urban-EPC Engine. In *Planning Support Systems and Smart Cities*; Springer International Publishing: Cham, Switzerland, 2015; pp. 447–469.
9. Davila, C.C.; Reinhart, C.F.; Bemis, J.L. Modeling Boston: A work flow for the efficient generation and maintenance of urban building energy models from existing geospatial datasets. *Energy* **2016**, *117*, 237–250. [CrossRef]
10. Reinhart, C.F.; Davila, C.C. Urban building energy modeling–a review of a nascent field. *Build. Environ.* **2016**, *97*, 196–202. [CrossRef]
11. Chen, Y.; Hong, T.; Piette, M.A. Automatic generation and simulation of urban building energy models based on city datasets for city-scale building retrofit analysis. *Appl. Energy* **2017**, *205*, 323–335. [CrossRef]
12. New, J.; Omitaomu, O.A.; Yuan, J.; Yang, H.; Carvalhaes, T.; Sylvester, L.; Adams, M. AutoBEM: Automatic Detection and Creation of Individual Building Energy Models for Each Building in an Area of Interest. In Proceedings of the 2nd International Energy and Environment Summit, Dubai, UAE, 18–20 November 2017.
13. NREL. UrbanOpt. Available online: https://www.nrel.gov/buildings/urbanopt.html (accessed on 7 May 2020).
14. Zhou, N.; Price, L.; Yande, D.; Creyts, J.; Khanna, N.; Fridley, D.; Lu, H.; Feng, W.; Liu, X.; Hasanbeigi, A.; et al. A roadmap for China to peak carbon dioxide emissions and achieve a 20% share of non-fossil fuels in primary energy by 2030. *Appl. Energy* **2019**, *239*, 793–819. [CrossRef]
15. *Design Standard for Residential Buildings of Low Energy Consumption*; DB 42/T 559-2013; Department of Housing and Urban-Rural Development of Hubei Province: Hubei, China, 2013.
16. Ministry of Housing and Urban-Rural Development of the People's Republic of China. *Design Standard for Energy Efficiency of Public Buildings*; GB 50189-2015; China Architecture and Building Press: Beijing, China, 2015.
17. *China Residential Energy Consumption Survey*; School of Economics, Renmin University: Beijing, China, 2012.

18. Crawley, D.B.; Lawrie, L.K.; Winkelmann, F.C.; Buhl, W.F.; Huang, Y.J.; Pedersen, C.O.; Strand, R.K.; Liesen, R.J.; Fisher, D.E.; Witte, M.J.; et al. EnergyPlus: Creating a new-generation building energy simulation program. *Energy Build.* **2001**, *33*, 319–331. [CrossRef]
19. EnergyPlus. Weather Data by Location. Available online: https://energyplus.net/weather-location/asia_wmo_region_2/CHN//CHN_Hubei.Wuhan.574940_CSWD (accessed on 2 December 2019).
20. Wuhan Municipal Development and Reform Commission. *Guideline for Energy Consumption Quota of Civil Buildings in Wuhan-2014 Edition*; Wuhan Municipal Development and Reform Commission: Wuhan, China, 2014.
21. Cheng, J.C.; Das, M. A BIM-based web service framework for green building energy simulation and code checking. *J. Inf. Technol. Constr.* **2014**, *19*, 150–168.
22. Ding, C.; Feng, W.; Tian, Q. Residential Building Archetype and API Development for Urban-scale Building Energy Consumption Platform: A Case Study for Wuhan. In Proceedings of the 16th IBPSA International Conference and Exhibition, Rome, Italy, 2–4 September 2019; pp. 3779–3785.
23. Newcomer, E.; Lomow, G. *Understanding SOA with Web Services*; Addison-Wesley: Reading, MA, USA, 2005.
24. Masse, M. *REST API Design Rulebook: Designing Consistent RESTful Web Service Interfaces*; O'Reilly Media, Inc.: Sebastopol, CA, USA, 2011.
25. CityDNA. Urban Growth Simulation and Scenario Planning. Available online: https://www.citydnatech.com/workCon/3&language=en (accessed on 3 June 2020).
26. Jain, R.K.; Smith, K.M.; Culligan, P.J.; Taylor, J.E. Forecasting energy consumption of multi-family residential buildings using support vector regression: Investigating the impact of temporal and spatial monitoring granularity on performance accuracy. *Appl. Energy* **2014**, *123*, 168–178. [CrossRef]
27. Nutkiewicz, A.; Yang, Z.; Jain, R.K. Data-driven Urban Energy Simulation (DUE-S): A framework for integrating engineering simulation and machine learning methods in a multi-scale urban energy modeling workflow. *Appl. Energy* **2018**, *225*, 1176–1189. [CrossRef]
28. Pisello, A.L.; Taylor, J.E.; Xu, X.; Cotana, F. Inter-building effect: Simulating the impact of a network of buildings on the accuracy of building energy performance predictions. *Build. Environ.* **2012**, *58*, 37–45. [CrossRef]

Article

Determination of the Energy Performance of a Solar Low Energy House with Regard to Aspects of Energy Efficiency and Smartness of the House

Dorota Chwieduk * and Michał Chwieduk

Institute of Heat Engineering, Faculty of Power and Aeronautical Engineering, Warsaw University of Technology, 00665 Warsaw, Poland; michal.chwieduk@itc.pw.edu.pl
* Correspondence: dorota.chwieduk@itc.pw.edu.pl

Received: 15 May 2020; Accepted: 17 June 2020; Published: 22 June 2020

Abstract: The paper shows how difficult it is to prove technically that a building really is both low energy and smart, and that all aspects of energy efficiency have been treated equally. Regulations connected to the determination of the energy performance of residential buildings take into account only space and hot water heating energy consumption and define the indices of maximal primary energy consumption, but not energy needs based on the architecture of the building. A single family house designed and constructed as a low energy solar house in Warsaw's suburbs is considered. Availability of solar energy and its influence on the architecture of the house is analyzed. A specific solar passive architectural concept with solar southern and cold northern buffer spaces incorporated into the interior of the house is presented. Parameters of the building's structure, construction materials, as well as operation parameters of equipment and heating systems based on active use of solar energy, ground energy (via a heat pump) and waste heat from a ventilation system are described. Results of calculations give values of final and primary energy consumption index levels of 11.58 kWh/m^2 and 25.77 kWh/m^2, respectively. However, the official methodology for determination of energy performance does not allow for presenting how energy efficient and smart the building really is.

Keywords: energy performance of buildings; solar passive systems; low energy buildings; energy efficiency; smart buildings

---

1. Introduction

This paper deals with the analysis of the energy performance of a solar low-energy house, which can also be considered a smart house. The smartness of the house is based on its design and construction with the focus on using the ambient surrounding and energy sources, mainly solar energy, in a passive way to reduce energy demand for space heating and cooling, and then to provide the energy demand in an effective way through the well-planned operation of its energy systems. In other words, to achieve real smartness in a house it is necessary first to create the architectural concept, which from the beginning takes into account energy aspects as well as aesthetics, and assures passive and active utilization of renewable energy and waste heat available to the house. Then the well-designed integration of different energy devices and installations assures their complementarity thanks to effective automatic control of the multisource energy system. It can be said that smartness is achieved through coupling passive and active methods of energy conservation, joining architecture, civil engineering and technical energy aspects. Avoiding energy demand is the best way to save energy. Thanks to the low energy demand concept of a house it is easier to consume less energy living there. The paper presents how difficult it is to technically prove, using standard methods of determination of the energy performance of buildings, that a building is really both low energy and smart, and that all aspects of energy efficiency have been treated equally.

In most countries in the world there are specific standards and regulations put into place to assure energy efficiency in buildings. In EU countries the formal background for the development of energy efficiency in buildings has been set out by the directive of the European Parliament and of the Council on the energy performance of buildings (Directive 2010/31/EU) [1]. The directive entered into force in 2010 and was a recast of the previous EU Directive on the Energy Performance of Buildings—EPBD, published in 2002. Further amendments have been made in 2018 (2018/844/EU) [2]. Since the beginning of the new millennium the EU Commission has been fostering the development of low energy buildings and the application of energy saving technologies, including utilization of renewable energy. Nowadays, thanks to Directive 2010/31/EU, measures to reduce the energy consumption in buildings have led to the evident improvement of thermal comfort in buildings and a reduction in environmental pollution. According to Article 9 of the directive, from 2021 all new buildings will have to be nearly zero-energy buildings, and in the case of public utility buildings the rule already came into force at the beginning of 2019. Thus, due to the official EU legislative policy to adopt the rules of nearly zero-energy use in the construction and use of buildings, the idea of energy efficiency in buildings is not only a concept of implementing general energy conservation principles and minimizing their environmental footprint, but it has become a national duty and legal necessity. The EU directive promotes the idea of nearly zero-energy for new buildings and stimulates the transformation of existing buildings to be refurbished into nearly zero-energy ones. Energy conservation in buildings has been supported by other EU regulatory frameworks, mainly by the EU directive on energy efficiency [3].

Nearly zero-energy buildings (NZEBs) should have very high energy performance. The low amount of energy which these buildings require for their effective use comes mostly from the use of renewable energy sources. The idea of zero energy buildings and net zero energy buildings was discussed before the EPBD directive came into force [4]. The EU energy legislative policy (through directives) compels national member state regulations to introduce limits for final and primary energy consumption for buildings with regard to their typology. These indices are different in different EU countries depending on climate and energy mix in those countries. The EU policy leaves it open for the member states to make a decision on the quantification of the boundary indices for energy consumption, i.e., what is the maximum quantity of energy to be used in buildings with regard to their final and primary energy.

Thanks to the implemented regulations the energy needs of buildings have been much reduced in recent times, mainly due to the well-designed envelopes and structures of buildings (i.e., shape of the building, types of construction, one layer or multilayered, type of construction materials and insulation applied with the focus on their thermal parameters). Recently, focus has been put on the thermal quality of buildings, mainly on application of thick thermal insulation of very low conductivity, and on windows characterized by low heat loss (transmission) coefficients (U-values). However, it can be mentioned here, that using insulation in a warm climate can reduce the heat losses necessary in summer to limit overheating. Heat losses are necessary during summer nights to release the excess heat gained during the day, because of high solar irradiation and to keep the indoor air temperature at the required level. Therefore, application of insulation must always be adapted to the given climate conditions. In addition, highly efficient and reliable equipment and energy installations, including ventilation systems with heat recovery, have been implemented. Such measures have caused significant drops in final energy demand. Application of renewable energy systems have also reduced the primary energy consumption based on fossil fuels.

Buildings are responsible for approximately 40% of EU energy consumption and 36% of the CO_2 emissions [5]. For many years high energy consumption in buildings was caused by high heat losses through buildings' envelopes [6]. The space heating demand used to be the highest demand component of energy consumption in buildings and is still responsible on average for about 65% of the total energy needs of buildings in the residential sector [7]. Of course, the share of space heating demand was higher in high latitude countries than those of low latitudes. Energy consumed by space heating and domestic hot water systems now accounts on average for 80% of the total (according to EUROSTATS). Therefore, it is not surprising that measures undertaken to reduce energy consumption in buildings have been mainly focused on decreasing the space heating demand.

In many European countries, as in Poland, new regulations connected with the determination of energy performance of residential buildings take into account only heating energy demand for space heating and domestic hot water (DHW). Cooling energy demand and electricity consumption by lighting units and systems, and electrical appliances are not limited by any official regulations. As electricity consumption was relatively small in residential buildings, it was believed that there was no reason to set limits for electricity consumption in houses. In the case of cooling energy it is arbitrarily assumed, that residential buildings in Poland do not require cooling, because of the relatively cold climate. Nowadays, it turns out that supplying cooling energy to residential buildings is sometimes necessary to maintain thermal comfort. Summer cooling demand can be seen especially for south and west facing rooms with large windows [8]. It turns out that cooling demand becomes a challenging issue for new buildings in moderate climates [9].

In Poland, according to existing regulations [10] since 2021 the indices of primary energy consumption for space heating and domestic hot water of all newly constructed residential buildings, called nearly zero energy buildings (NZEB), cannot exceed 70 kWh/m^{2}a or 65 kWh/m^{2}a for single family houses or multi-family apartment buildings, respectively. In addition, the heat transfer coefficients for external walls must not be higher than 0.2 W/(m^{2}K). As a result, external walls have thick insulation of high thermal quality (in the 1970s the recommended thickness of thermal insulation was 6–8 cm, at present it is 20–25 cm). What is more, the heat transfer coefficients for windows will soon not be allowed to be larger than 0.9 W/(m^{2}K) and currently they cannot be higher than 1.3 W/(m^{2}K) (in the 1970s it was 3.2 W/(m^{2}K)). Existing regulations on energy performance of buildings define the indices of maximal primary energy consumption considering only technical issues. They put the focus on energy efficiency, which results in reduction of final energy consumption and gives support for renewable energy sources, which utilize much less primary energy. Unfortunately, they do not show how crucial for energy consumption is the architectural concept of a building. Its shape, structure, location, the sizing of different elements of the building envelope and their orientation to specific directions of the world, and surroundings are of great importance. Without analysis of all the architectural and local settlement conditions, it is like being halfway to the finish line, but with a slow first half and no chance of winning. When a building is designed and constructed without a real vision of maintaining low energy consumption throughout the whole year, then it will not be possible to reduce the final and primary energy consumption to the set limits relying only on the energy efficiency of devices and installations applied. Therefore, such a building will not be a smart low-energy building. To get a real reduction in energy consumption of any building it is necessary to have a global interdisciplinary approach and look at the process of building design, construction and use in a holistic way.

Many energy simulation programs have been developed to determine building energy performance and they are used in many different countries. Comparison of the features and capabilities of twenty major simulation programs determining the energy performance of buildings can be found in the literature [11]. The authors showed how contrasting the building energy performance approaches can be. So it is not surprising that it is difficult to find such a method of determining energy performance of buildings, which would take into account all aspects of the actual energy efficiency of buildings such as those described in this paper.

This paper presents the problem of how the existing regulations supporting the reduction of energy consumption in buildings through an engineering determination of energy performance cannot fully present a true and reliable assessment and evaluate the real impact on energy consumption. The paper shows how difficult it is to prove technically that a building is really low energy and smart, and that all aspects of energy efficiency mentioned above have been treated equally. An example of a single family house designed and constructed as a low energy solar house located in Warsaw's suburbs has been considered. Section 2 describes a general idea of a low energy building and a smart building and shows how some of their main features are similar to each other, whilst others are different. Section 3 presents solar radiation conditions in Poland and a concept of solar passive architecture, which should be taken into account when a low energy smart solar house has to be designed and constructed. Section 4 describes the heating energy demand as well as final and primary energy consumption of the house under consideration. At the end, both the officially calculated and real energy performances of the building are discussed and general conclusions are formulated.

2. Low Energy and Smart Buildings

Low energy buildings are not usually equated with smart buildings. However, it seems that if a building is low energy it must be smart. However, the smartness of a building can be defined by many parameters that are distinct from those usually applied to low energy buildings. If we would like to consider a modern low energy building as a smart building, we should define what both 'low-energy' and 'smart' mean and whether a building can be both. A low energy building can be defined as a building which needs and consumes a small amount of energy during its life-time. It is a building needing a small amount of energy for space heating or cooling, thanks to its architectural and civil engineering design and construction. Low energy needs result from the specific concept of the building envelope, materials used, specific location of different partitions such as opaque walls and transparent glazing in the structure of the building, as well as specific location of rooms of different functions inside the building. It also needs the correct utilization of solar radiation for gaining energy in the winter and protecting from excessive solar gains in the summer. Of course the aesthetic values of the building envelope cannot be forgotten. A low energy building also consumes a small amount of energy because of the use of energy efficient devices and systems applied to fulfill all its energy needs: space and water heating, cooling, ventilation, air conditioning, lighting, electricity for electrical appliances, etc. Moreover, renewable energy sources are used to reduce primary energy consumption of fossil fuels.

Different means of construction and operation of low energy buildings have been developed in recent decades, with the best achievements in the last decade. Kivimaa and Martiskainen [12] conducted a systematic review of case studies on low energy innovations in the European residential building sector from the beginning of this century. They analyzed drivers important for systemic and architectural innovation in low energy buildings and pointed out how different the low energy buildings can be in their main purpose. There are some key words used to describe buildings of low energy consumption, such as: energy efficient, low energy, zero carbon, passive houses, etc. All these keywords express the main features of low energy buildings, i.e., energy, efficiency, environment and architecture; however, according to the names of those buildings the focus can be put on different aspects.

When we search through the Internet trying to find a definition of the smart building, the most common one is as follows: A smart building is any structure that uses automated processes to automatically control the building's operations including heating, ventilation, air conditioning, lighting, security and other systems [13]. This idea is certainly connected to the energy efficiency aspect of low energy buildings, but it seems to be much more closely connected to energy efficiency measures introduced (mainly to office buildings) and known as BMS—building management system [14]. BMS systems are also known as building automation systems (BASs). Such a system is a computer-based control system used in buildings to monitor and control the building's energy systems and other systems such as fire and security. Smart buildings use sensors, data monitoring and collecting systems, which give the information needed for effective operation of different buildings' systems, including energy systems. Smart buildings use IT—Information Technology and IoT—Internet of Things. Such technology is usually used for office buildings, hospitals, health care and educational facilities, sport centers and sometimes for public buildings, but very rarely for residential houses. There are no standards for smart buildings. Low energy architecture is not one of determinants of the smartness of such buildings. Therefore it can be said, that in many smart buildings like office buildings, the important element of energy efficiency required to reduce energy needs is usually missed. However, without low energy or energy efficient architecture of a building it is really difficult to call any building a smart one.

One more aspect not analyzed in detail in the paper is very important, namely the smartness of building users. Energy consumption in buildings depends on user behavior. It can be said that the inhabitants of residential houses basically want energy savings because they relate to the user costs and directly affect them. The problem with these unthinking building users' behavior is particularly evident in office buildings [15]. Therefore, it should be stated that a smart building also requires smart users.

Taking into account what features should be common to both low energy buildings and smart buildings, it can be seen that energy efficiency is essential. However, it seems that the definition of a smart building should be much wider, especially in the case of residential buildings. The next section presents the concept of a smart low energy building realized at the micro scale, i.e., in a single family house in Polish climatic conditions.

3. Influence of Solar Energy Availability on Architecture of a Building

3.1. Solar Radiation Conditions in Poland

The relation between climate, and especially solar radiation conditions, and architecture of a building should be obvious [16,17]. Unfortunately, nowadays it is quite often forgotten. Any building is under the influence of solar radiation, but the solar building must pay very special attention to solar radiation conditions. In Poland the climate is moderate with the influence of continental climate. The annual average ambient air temperature, depending on the region can be around 8 °C to 11 °C. However, there are relatively large differences in ambient air temperature during the year and especially between summer and winter. Thus in summer, during the daytime, the temperature can be +30 °C or more, as has happened quite often recently. In winter, ambient air temperature can drop to −30 °C, however such a low temperature was last recorded almost 10 years ago. The annual global solar irradiation varies from 900 kWh/m^2 to 1200 kWh/m^2. Annual solar hours are on average equal to 1600. Climate is characterized by relatively large differences in solar irradiation throughout the year. For example, in Warsaw in June, the average monthly solar irradiation is about 160–180 kWh/m^2, but in December only 11–12 kWh/m^2. What is also typical for the climate is the high share of diffuse radiation. The annual share of diffuse radiation usually accounts for 54–56% of the global and in winter this share is especially high and accounts for 70–80%. Only in summer is the share of direct radiation higher and can be on average equal to 60% of global radiation [18]. Figure 1 presents the averaged distribution of the average hourly global solar irradiance on averaged days of the all months of the average year. Figure 2 shows the distribution of average hourly ambient air temperature for averaged days of all months of the average year for Warsaw.

Relatively large differences in solar irradiation and ambient air temperature in summer and winter can easily be seen in Figure 1. In such climatic conditions not every solar passive system can be used in an effective way. Very uneven distribution of solar radiation during the whole year means that specific passive architectural solutions should be recommended [19].

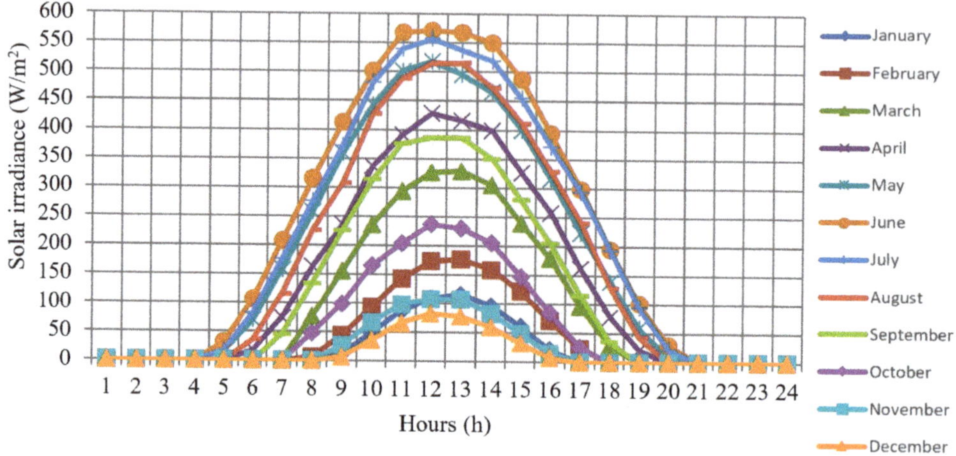

Figure 1. Distribution of average global solar irradiance for every hour on averaged days of different months of the average year in Warsaw.

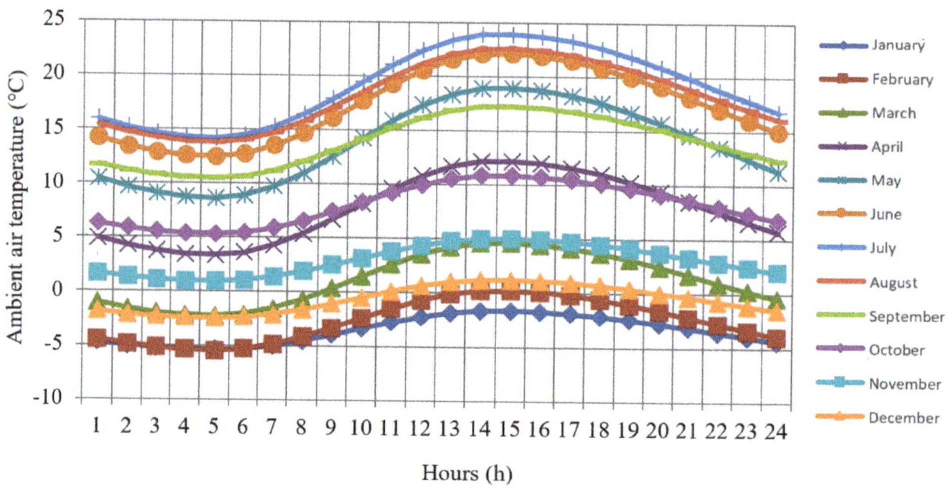

Figure 2. Distribution of average hourly ambient air temperature for averaged days of different months of the average year for Warsaw.

3.2. Buffer Space Incorporated Into Interior of a House as a Specific Solar Passive Architectural Concept

The single family solar house presented here has been designed and constructed with particular attention to passive and active utilization of renewable energy, mainly solar. Availability of solar energy with regard to the climatic conditions and specific location of the building has been considered. It can be mentioned here, that even if the climate is the same, the conditions in a city center are different than in the suburbs and in the country side in the vicinity. The considered house is located in the suburbs of Warsaw. The location of the house was specially selected so that in winter the southern facade of the house is fully exposed to solar radiation. Deciduous trees were planted on the south-east and south-west sides creating shade on these sides in the summer. The south is completely open, as it is beneficial in winter. Elements of building architecture provide shading in summer, as is described in the next section. A plan of the first floor of the house is presented in Figure 3. The main living space area is marked with the red lines.

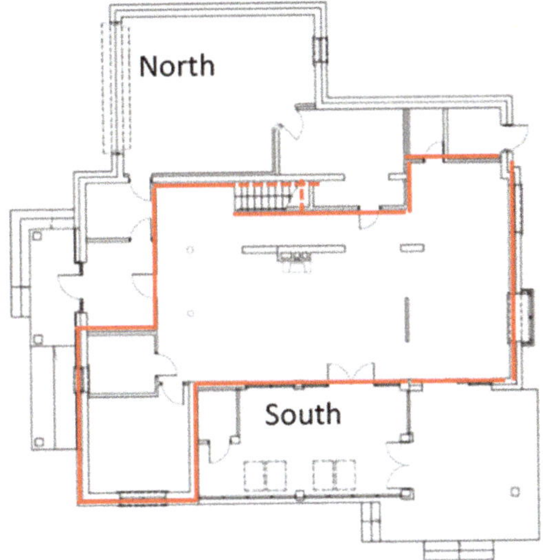

Figure 3. A plan of the first floor of the considered house, main living space area is delimited by red lines.

The south transparent solar buffer space of a special design and north opaque buffer space are described below. Buffer spaces, as the name indicates, create a kind of a buffer between the outdoor and indoor climate. The southern transparent glazed solar buffer space allows solar radiation to penetrate the interior of the house in a planned way. Solar radiation can be fully utilized in winter and significantly reduce the space heating demand, whilst in summer, to reduce solar energy gains, the special architectural form of the southern buffer space is needed. A glazed solar buffer space is incorporated into the interior of the building. This specific architectural concept of a solar passive system is shown in Figure 4. It can be seen that the buffer space contains two theoretical cuboid sub-spaces. The external one is higher and the internal one lower. There is no partition between them. The solar buffer space has external and internal glazed vertical surfaces. External, four meter high glazed partitions are in direct contact with the ambient surroundings on one side, and with the interior of the buffer space, on the other. Internal glazed partitions (regular windows) are at one side in direct contact with the interior of the buffer space on the one side and with the interior of the main living space on the other.

The key architectural concept of the southern glazed façade of the house is to design and plan two overhangs at the south side of the building. The first external overhang is just a regular one being a part of the roof (marked with a symbol E). The internal overhang (marked with a symbol I) is a part of the internal construction of the building. The main point is to properly design (place and size) the internal overhang in accordance with the sizing of the external overhang to protect the interior of a building against too much solar energy gains in summer and to allow solar radiation to penetrate the interior of the house without any obstacles in winter. Of course the size of the external south glazed facade is taken into account. The external overhang of the roof (E) has been designed to shade the buffer space for a few noon hours (between 10 a.m. and 4 p.m.) in the warmest part of the year, i.e., from May to the end of August, but not to block the access of solar radiation in the rest of the year. The internal overhang (I) is formed by part of the floor on the second floor being, at the same time, a part of the ceiling of the first floor (over the lower part of the buffer space, as can be seen in Figure 2). The internal overhang has been designed to allow direct solar radiation to enter fully into the interior space of the house in winter, exactly from November to the end of February, and to fully block the direct solar radiation penetration from May to the end of August. In the remaining months of the year, solar radiation reaches the interior directly in the morning and afternoon. As has been mentioned, the buffer space has full access to solar energy from October to the end of April and partly (mornings and late afternoons) from May to the end of August.

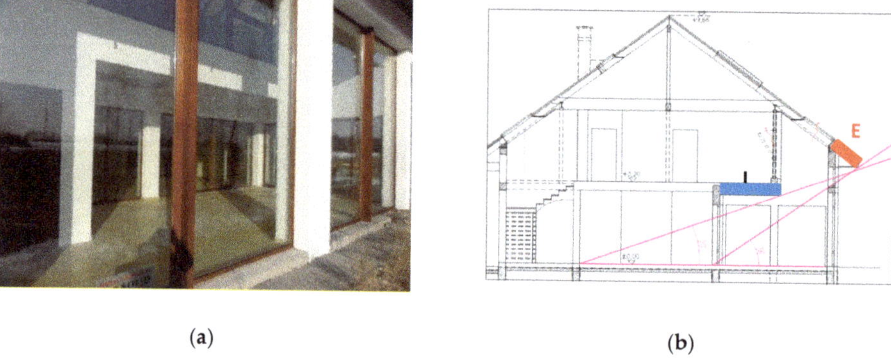

(a) (b)

Figure 4. Architectural solar passive system: (**a**) A view of the south glazed buffer space incorporated into the house; (**b**) and a cross-section of a house along the south–north direction with apparent shading planes.

To have planned such access to solar radiation, the size of the buffer space, the size and position of different partitions of the buffer space, including the size of the external and internal overhangs, have been determined on the basis of the astronomical relationship between position of the Sun and the different partitions of the buffer space (glazed and opaque) [20]. Calculations of solar energy availability have been conducted with a time step equal to one hour throughout the whole year. The anisotropic diffuse solar radiation HDKR (Hay – Davis – Klucher – Raindl) model has been applied to determine solar irradiation of surfaces of different inclination and orientation [21].

3.3. Northern Buffer Space

Analysis of solar energy availability in buildings also requires designers to pay attention to that part of the building envelope, which is not exposed to solar radiation, especially in winter. It is the northern part of the building enclosure which requires this thoughtful approach. In the considered climate, with cold winters and warm (or even hot) summers, a highly opaque insulated buffer zone should be created at the northern part of the building to reduce the effect of the severe climatic conditions in winter. The external north partitions must be characterized by very high thermal resistance (0.5 m thick walls, which include 0.25 m thick insulation) and no windows or any other transparent elements.

Figure 5 presents such a north façade, which is a façade of the considered low energy solar house.

The northern walls and roof are elements of the buffer space. Another important feature is to create usable spaces which do not require heating energy, because they are not for permanent residence of the inhabitants. The air temperature in that space can be, or even is required to be, lower than in the living space of the house. It is possible to plan a cold store, pantry, wardrobe, garage or boiler room there. Northern fully opaque and highly insulated buffer zones can significantly reduce the influence of severe climatic conditions in winter and have a positive impact on the energy balance of the building.

Planning cold and solar buffer spaces in a building allows introduction of temperature zones into the building in a natural way. Due to the architecture of the building, natural passive control of thermal comfort is created. In this way the following air temperature zones are designed:

- A cold northern zone with a seasonally variable indoor air temperature, daily changes of temperature are very small;

- A warm internal zone corresponding to the living space of residents with a requirement of relatively constant indoor air temperature (throughout the year), where a low temperature space heating system is applied;
- A southern solar zone with a variable air temperature on a daily scale (warm day/cold night) and on a seasonal basis depending on the intensity of solar radiation and the influence of the ambient air temperature.

Figure 5. A view of the northern buffer space of the house.

Thus, the architecture of the building creates a smart concept for building structure focused on the planned use of the environment to reduce the need for heating and cooling energy for the living spaces. The temperature zones go from the north, always with the lowest indoor air temperature, to the south, with most variable indoor air temperature during daylight hours, the highest in summer, high or moderate in spring and autumn and high or low in winter. The thermal state of the southern buffer space is directly impacted by the solar radiation and by ambient air temperature. However, the radiation is the dominant component. In such a way a smart solar low energy concept of a house can be created. When the energy needs of a house are accomplished through the smart and energy efficient design of a building then highly energy efficient devices and installations can be introduced.

4. Building Energy Needs and Final and Primary Energy Consumption

4.1. Space Heating Energy Needs of the Considered House

Usually, to reduce building energy consumption, two main measures are applied. The first is focused on improvement of the thermal performance of the building envelope by adding insulation and reducing infiltration rates, and the second is on improvement of energy efficiency of the devices and installations used in buildings, including lighting, heating, ventilation and air-conditioning [22]. However, to reduce the energy consumption significantly the design of the building cannot be based only on reduction of heat and mass flow through the building envelope (through improvement of insulation and reduction of infiltration). The architecture of a building is crucial and it should take into account specific climatic and environmental conditions with a focus on solar energy availability specific for the given climate and location of the building. The single family house presented in this paper was designed using this wide holistic approach to reduce energy needs, as has been described in the previous section. In this section the results of calculations of the energy balance of the building and the final and primary energy consumption are presented. Simulation studies have been performed using our own simulation code. The availability of solar radiation and its impact on the energy needs of the building have been determined. Modeling of the space heating and DHW needs have been based on the methodology on determination of energy performance of buildings [23].

The floor area of the heated space of the house is 350 m^2. In addition to two floors, the building has an unheated attic (there is no basement). The roof directed to the south and north is inclined at 30 degrees. The northern slope is larger than the southern one, required by the specially designed north unheated buffer space. The south façade of the building is transparent in 70% of its total surface area, which in turn is related to the design of the south solar buffer space. There are two types of windows. Large windows form the main part of the façade and their heat transfer coefficient U is equal to 1.2 W/(m^2K). The U value of the other regular windows and internal windows of the south buffer space is 1.4 W/(m^2K). The main idea of the building design was to use standard construction materials, typical nowadays for such a climate, to make the building envelope energy efficient and reduce heat transfer with ambient surroundings. External walls are 0.5 m thick (two layers: mineral wool from outside and bricks from inside). The heat transfer coefficient U is equal to 0.14 W/(m^2K). The U value for the ceiling over the second floor, under the unheated attic, is lower and accounts for 0.12 W/(m^2K), but the heat transfer coefficient for the floor on the ground is larger and is equal to 0.17 W/(m^2K).

The energy balance of the building was formulated and space heating energy needs were calculated. The results are shown in Figure 6. This figure presents monthly space heating energy demand with all components of the energy balance of the building. Thus all heat losses through the building envelope and ventilation (positive values), as well as heat gains: internal and solar (negative values), are presented. It can be noticed that ventilation and heat losses through windows dominate among all others, but it is also evident how large the solar gains are and their impact on the energy balance of the building.

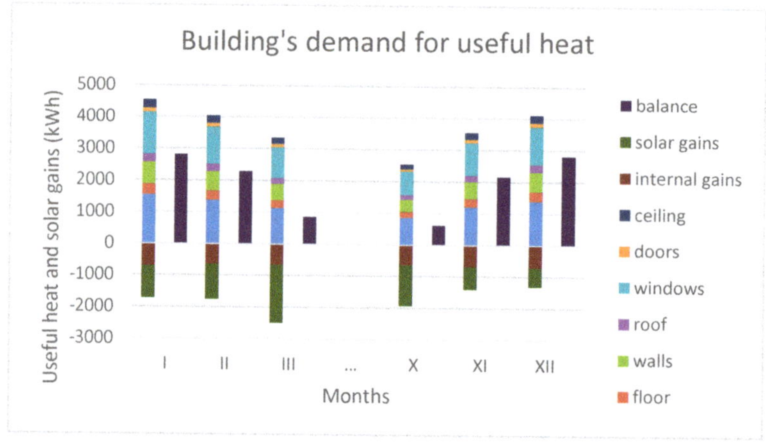

Figure 6. Monthly space heating demands with energy balance components: heat losses and gains.

The main observations on heat losses from Figure 6 are confirmed by the diagrams in Figure 7 which presents the seasonal share of different heat loss components in the total heat loss of the building. As could be expected for a low energy house, the largest losses occur through ventilation and then through windows, and they account for 34% and 29% of the total losses, respectively. In the third place there are heat losses through walls (15%) and all the others take nearly the same share (6–7%) (heat losses through doors are lower at 3%).

It can be noticed that the heat energy demand for the ventilation system results from the natural necessity to exchange air in the building, including exchange for hygienic purposes in rooms such as kitchens and bathrooms. In the building under consideration, the heat demand is significantly reduced due to the use of heat recuperation. In the calculations of energy demand the use of a recuperative unit was taken into account when determining the final energy consumption. Heat recuperation requires a forced ventilation system to be used. The design of the forced ventilation system is usually taken into account at the design stage of the building and its interior, and this was the case with the building under consideration. Ventilation ducts supplying fresh air, which is preheated in a recovery unit, as well as ducts discharging used air outside through the recovery unit, were planned at the time of creating the architectural design of the building. It can be mentioned that the heat recovery ventilation unit is not used all the time, but only when inhabitants are at home. For the remaining months outside the heating season only short-term morning and evening ventilation is used.

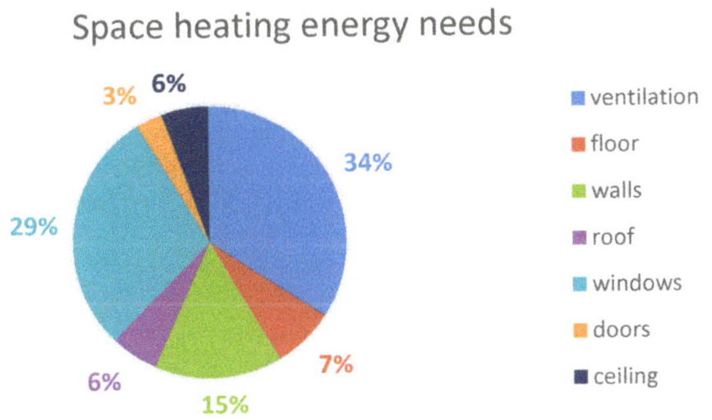

Figure 7. The seasonal share of heat losses in the total amount of heat losses of the building.

The calculated index of the annual space heating energy needs amounts to 36.16 kWh/m^2 of the heated floor area. The low space heating energy demand results from the architectural concept of the building, the introduction of temperature zones resulting from the existence of buffer spaces at the northern and southern sides of the building, including the passive use of solar radiation energy. Low energy demand also results from the use of appropriate building materials of high thermal insulation and thermal capacity. The high thermal capacity is demonstrated by a building time constant, equal to 222 h.

According to existing regulations [23], when the energy performance of a residential building is calculated the energy consumption for domestic hot water (DHW) heating is also taken into account. For the considered house, the annual total heat demand index (for DHW and space heating, which is only seasonal) is equal to 45.6 kWh/m^2. The annual DHW head demand is equal to 3011 kWh (heat demand index for DHW is equal to 9.44 kWh//m^2). Up to now there have not been any official regulations introduced to limit these energy-need indices of buildings; this is a problem which does not help in significantly reducing the energy consumption of buildings, as the authors try to present in this paper. There are limits only for the heat loss coefficients of walls, e.g., for walls U was equal to 0.3 W/(m^{2}K), now it is 0.25 W/(m^{2}K). It can be mentioned, that 10 years ago when the house was constructed according to the obligatory regulations, buildings (of similar compact shape) required a maximum of 90 kWh/m^2 of final energy and 69 kWh/m^2 of primary energy consumption (regulation [10], before amendments in 2013). Nowadays, these indices are even higher, because the official limits stated for primary energy consumption for a single family house is equal to 95 kWh/m^2. Since the beginning of the 2021, even if a new house is to be called "nearly zero energy" the index for primary energy consumption will be at a level of 69 kWh/m^2.

4.2. Final Heating Energy Consumption of the Building

In order to determine the final energy consumption, it is necessary to take into account the energy efficiency and effectiveness of energy devices and installations used to cover heating needs, as well as their time of operation. In addition, the work of auxiliary devices necessary for the operation of heating systems, such as circulation pumps in liquid circuits and fans in air circuits also have to be taken into account.

Operation of heating systems in the low energy house is of course based on using energy efficient devices and installations. Heating demand is met by a ground source heat pump with vertical heat exchangers coupled to a solar thermal systems with flat plate solar collectors via a buffer storage tank. The buffer storage tank has a smaller DHW tank inside. In this way water in the main volume of the buffer tank not only serves as a storage medium, but also as thermal insulation for the water in the internal DHW tank. DHW is preliminarily heated in the internal tank and then it flows to another DHW tank with an auxiliary heat source (electric heater). Solar collectors are integrated into part of the south roof surface. A low temperature underfloor space heating system is used. The flow is forced by a pump into every loop of the system. As has been mentioned, the heat recuperation unit is also applied. Fans are used to force the flow of fresh and used air through the ventilation ducts. The main characteristics of the heating system are presented in Table 1.

Table 1. Main (elements) devices of the heating system of the house and their main parameters.

Devices and Their Parameters	Value	Units
Ground source heat pump		
Ground U-tube heat exchangers: number of boreholes	4	–
Total length of boreholes	200	m
Heat pump heating capacity	8.1	kW
Seasonal COP (Coefficient of Performance)	4.9	–
Solar thermal flat plate collectors		
Gross / Aperture surface area	12/10.92	m^2
Annual average solar energy gains	380	kWh/m^2
Water storage system		
Buffer storage tank - total volume	700	kg
Internal DHW tank volume	100	kg
Auxiliary DHW tank	50	kg
Ventilation heat recovery unit		
Power of fans	280	W
Volumetric air flow	600	m^3/h

The operation of the solar thermal collectors and the heat pump is not directly connected. Both devices operate in parallel. They can operate at different times of the day, but they can also supply heat at the same time. Flat plate solar collectors supply heat to the main buffer storage tank (via a heat exchanger). An antifreeze mixture circulates in a solar collector loop. There is another loop with an antifreeze mixture circulating in vertical ground heat exchangers, which are coupled with the evaporator of the heat pump. Heat can be sent to the buffer storage. It is also possible to supply heat directly from the heat pump to the underfloor heating system or DHW tank (without charging the storage tank). The buffer storage tank with water as a storage medium contains a small tank inside. The small tank is used as a buffer tank for the DHW system. Cold water is supplied to the small tank and when the water is heated up it flows out of the tank and is transferred to the other tank, which is the main DHW storage tank (50 l volume) equipped with an auxiliary electric heater. The ground source heat pump is used only during heating season. Domestic hot water out of the space heating system is accomplished via the solar thermal system operation, which operates very effectively in Polish conditions [18,24]. From May until the end of September the thermal solar energy system can cover all DHW heating demand. In March, April and October, solar energy provides about 60–70% of demand, in winter the share of solar energy is very small and it does not exceed 10% for the DHW and space heating. Figure 8 presents the print screen of a display showing the operation of the ground heat pump system coupled with the solar heating system for space heating and DHW heating.

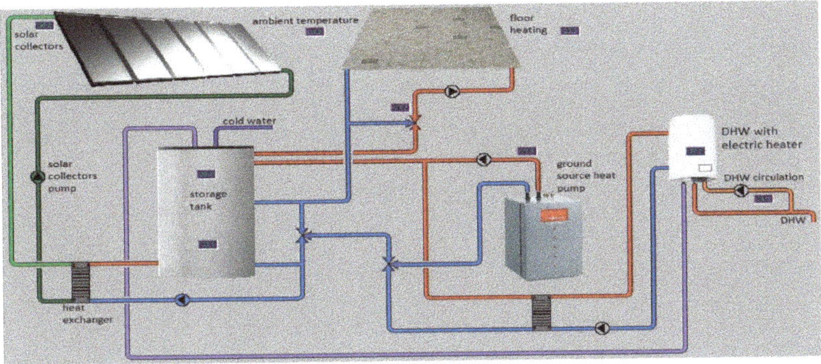

Figure 8. The print screen displaying the configuration and operation of the ground heat pump system coupled with solar heating system.

A micro-scale energy management system is used in the house. Operation of the heating systems is controlled by a central system that continuously monitors the operation of all heating devices and systems. Several variants of operation are possible depending on the availability of the given renewable energy source in time and its adherence with the energy demand in that time. The system is equipped with a number of sensors enabling on-line observation of the system's operation as well as its control. Through a dedicated computer application, it is possible to remotely change the temperature settings in the rooms. It is also possible to change the parameters of the system operation, mainly temperatures and flows, and even completely turn off the operation of individual devices or installations. Operation priorities are set according to the efficiency of energy conversion from a given energy source and the effectiveness of using that energy at a given time. The automatic control system helps in a smart way to ensure the highest energy efficiency in gaining the available renewable energy and consuming it in an effective way.

Figure 9 presents the distribution of monthly space heating final and primary energy consumption for comparing the distribution of monthly space heating needs. It can be noticed that monthly space heating needs are presented in two graphical forms. The highest bars show total energy demand with standard ventilation needs, as shown in Figure 4 (where ventilation is a dominant factor of energy demand). The lower bars (colored red also show the total energy needs, but the demand is much reduced due to application of the heat recuperation ventilation system. The smallest bars represent final energy consumption. It is evident that final energy consumption is really very low thanks to the highly energy efficient energy systems and mainly because of using a heat pump that has been well selected for the given operating conditions and operates with high energy performance (SCOP (Seasonal Coefficient of Performance) nearly equal to 5 after nearly 10 years of operation).

The seasonal index of the final energy demand for space heating accounts for 4.61 kWh/m^2, which is very low. As has been mentioned, determination of the energy performance of any residential building requires taking into account only the heat consumption of the building for space heating and DHW. The so called annual index of energy consumption includes the annual DHW heating and space heating, while the space heating takes place only during the heating season, and for the considered house it lasts only four months. Thus the final annual energy demand index for space heating and domestic hot water is 11.58 kWh/m^2, which is still a very low value even if the electric energy consumption by the auxiliary devices of the heating loops (like pumps and fans) is included. It can be mentioned here, that in Poland a building can be classified as a low energy building, when its final annual energy consumption (for all heating needs) amounts to 30–60 kWh/m^2. Such a range of indices was proposed in 2007 [25] and is still used [26].

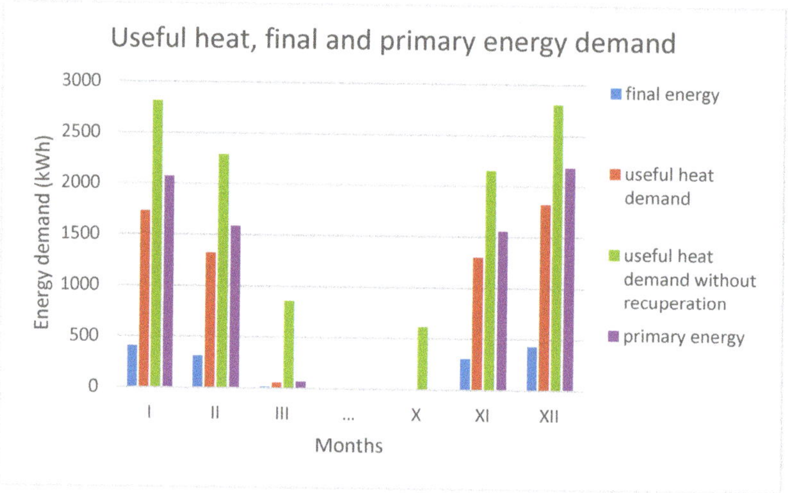

Figure 9. Monthly final and primary energy consumption and energy demand for space heating (bars for ventilation with heat recuperation and without are shown).

4.3. Primary Energy Consumption

The annual index of primary energy consumption for space heating and domestic hot water for the considered low energy solar house is equal to 25.77 kWh/m^2. The annual index for domestic hot water accounts for 4.11 kWh/m^2. These are really very small values, far below the existing and future (since 2021) limits for such indices required by the present regulations. The primary energy consumption is based on renewable energy sources: solar, solar thermal collectors and ground, ground source heat pump; on waste heat, i.e., recovery of heat from the ventilation system and on conventional electrical energy taken from the grid. In Poland, more than 90% of electricity is produced in power plants fired by coal (hard and brown). Thus, taking electricity from the grid means using primary energy based on fossil fuels. The calculated value of annual CO_2 emissions is equal to 596.8 kg per year. To estimate that value the official Polish energy mix with referred emissions for fossil fuel used was taken into account [27].

5. Conclusions

Carefully analyzing the results obtained it can be noted that the building under consideration is characterized by the final space heating energy consumption being lower than the space heating energy needs and lower than the primary energy consumption. At the same time, primary energy consumption is less than the heating energy needs (if heat recuperation from ventilation systems is included in the calculations of the final energy consumption). This low consumption of final energy results from the use of a heat pump for which a seasonal coefficient of thermal performance (SCOP) is used instead of standard efficiency or effectiveness of devices and systems. So the use of a heat pump should always be recommended to achieve a small final energy consumption.

In standard buildings with standard thermal energy systems supplied by fossil fuels the primary energy consumption is always the highest, then the final energy (which is lower than primary energy) and the lowest being the energy heating needs (in this case for the space heating). In energy efficient buildings the application of a heat pump reduces the final energy consumption, which is lower than energy needs (what is also the case in the energy system described in this paper). Such a lowering of final energy can be also achieved and fostered through application of a ventilation heat recovery system (real operational effectiveness of the recuperative heat exchangers is 70% in the considered system, theoretical one accounts for 85%). However, it can be noted why the heat recovery of the ventilation system should be considered as final energy consumption, when the decision to use such a system is made during elaboration of the architectural concept of the building (ventilation ducts for forced airflow must be considered at this stage). So when the space heating of the low energy solar house is considered in this paper and if we assume the heat recovery of the ventilation system at the stage of determining the space heating energy needs, then comparing all three forms of energy, i.e., energy needs, final and primary energy consumption, the final energy consumption is the lowest, and the energy needs are lower than the primary energy consumption. Next, when application of renewables is taken into account then the primary energy consumption is reduced. How much this energy is reduced depends on the share of renewable energy sources used to fulfill the energy requirements. All these considerations show how difficult it is to determine the energy performance of a building. The problem is that in the official requirements for determination of energy performance of a building there are no limits (indices) for the heating energy needs of the building, so perhaps if a heat pump and the heat recovery ventilation system are to be applied, then the space heating energy consumption can just be at the limit of official regulated indices, even if not much has been done to reduce the energy needs.

Another problem is connected to official national regulations, which do not require (or even do not allow) inclusion of the electricity use by electrical lighting and appliances, when determining the energy performance of a residential building. Consequently, when any renewable energy system, like photovoltaic or wind energy, is used, then it can be considered if such a system supplies energy to drive the heating device, like a heat pump or electric heater. This means that any energy produced by such renewable energy systems for electrical appliances during the whole year cannot be included in calculations for determination of the energy performance of buildings. Thus the air heat pump used all year round to supply heat for DHW and seasonally for space heating can turn out to be a much better solution than using solar collectors for heating energy demand, mainly for DHW demand. This paper describes a ground source heat pump operated during the space heating season. Such a heat pump operates in a very efficient way, because out of heating season the ground can recover and come back to its natural thermal state very quickly. As a result the thermal conditions of using the ground source throughout the space heating season are very good (SCOP is quite high). In a case of an air heat pump the SCOP is much lower, because of using an ambient air source during winter.

It turns out, that determination of the energy performance of a solar low energy house based on official regulation does not allow for showing how energy efficient and smart the building really is. The main problem is that there are no limits on the energy demand of the building.

The basis for ensuring the energy efficiency of the building is primarily ensuring significantly reduced energy needs for the energy used in the building, i.e., the heating energy (energy for DHW, for space heating or cooling) and electricity. The biggest energy saving is the lack of demand for it. In buildings, significant savings can be obtained through the appropriate architectural concept of the building, designing the compact shape of the building, opening the southern side of the building to the impact of solar radiation and closing it tightly from the north to limit the impact of the external environment, especially in winter. A suitable concept for the interior of the building, e.g., as described in this paper, is the concept of using buffer zones, including the introduction of a southern buffer zone into the interior of the building, allowing the use of energy from the environment, including primarily solar radiation, and thus significantly reducing heat demand for heating. The architectural concept should ensure a natural temperature zoning of the interior of the building, including high thermal living comfort zones for permanent residence of people, zones for periodic residence (e.g., in transition seasons like spring and autumn) and non-residential zones (e.g., for cold or hot auxiliary facilities). Such natural zoning ensures a significant reduction in final energy consumption, and determines the smartness of the building. One could say it is an innate, inborn intelligence because the smartness is achieved naturally, passively, and not through the use of complex energy management systems in the building.

As has been presented in this paper, a low energy solar building is smart through its architecture, construction, energy efficient devices and systems applied, and through the well managed operation of all components based of utilization of renewable energies. It can therefore be concluded that the building should have the built-in (embodied) smartness of reducing energy demand and consumption, achieved naturally, passively, as well as through the use of energy-efficient devices and installations, planning appropriate operating priorities and the use of efficient low-carbon energy sources, preferably renewables.

Author Contributions: Conceptualization, D.C.; methodology, D.C. and M.C.; software, M.C.; validation, D.C. and M.C.; formal analysis, M.C.; investigation, D.C. and M.C.; resources, D.C. and M.C.; data curation, M.C.; writing—original draft preparation, D.C.; writing—review and editing, D.C. and M.C.; visualization, M.C.; supervision, D.C.; project administration, D.C. All authors have read and agreed to the published version of the manuscript.

Funding: This research received no external funding.

Acknowledgments: Open Access and Article Processing Charge are covered by the Institute of Heat Engineering, Faculty of Power and Aeronautical Engineering of Warsaw University of Technology.

Conflicts of Interest: The authors declare no conflict of interest.

References

1. *Directive 2010/31/EU of the European Parliament and of the Council of 19 May 2010 on the Energy Performance of Buildings*; The Energy Performance of Buildings Directive (EPBD), 2010. Available online: https://eur-lex.europa.eu/eli/dir/2010/31/oj (accessed on 10 May 2020).
2. *Directive 2018/844 of The European Parliament and of The Council of 30 May 2018 Amending Directive 2010/31/EU on The Energy Performance of Buildings and Directive 2012/27/EU on Energy Efficiency*; The Energy Performance of Buildings Directive (EPBD), 2012. Available online: http://data.europa.eu/eli/dir/2018/844/oj (accessed on 10 May 2020).
3. *The Energy Efficiency Directive (2012/27/EU) of the European Parliament and of the Council of 25 October 2012 on Energy Efficiency*, 2012. Available online: http://data.europa.eu/eli/dir/2012/27/oj (accessed on 10 May 2020).
4. Marszal, A.J.; Heiselberg, P.; Bourrelle, J.S.; Musall, E.; Voss, K.; Sartori, N.A. Zero Energy Building—A review of definitions and calculation methodologies. *Energy Build.* **2011**, *43*, 971–979. [CrossRef]
5. Available online: https://ec.europa.eu/eurostat/web/environmental-data-centre-on-natural-resources/resource-efficiency-indicators/resource-efficiency-scoreboard/thematic-indicators/key-areas/improving-buildings (accessed on 10 May 2020).
6. Bragança, L.; Wetzel, C.; Buhagiar, V.; Verhoef, L.G.W. (Eds.) *Cost C16 Improving the Quality of Existing Urban Building Envelopes*; Facades and Roofs, Research Architectural Engineering Series; IOS Press BV: Amsterdam, The Netherlands, 2007; Volume 5.
7. Available online: https://ec.europa.eu/energy/topics/energy-efficiency/energy-efficient-buildings/energy-performance-buildings-directive_en (accessed on 10 May 2020).
8. Chwieduk, D. Impact of solar energy on the energy balance of attic rooms in high latitude countries. *Appl. Therm. Eng.* **2018**, *136*, 548–559. [CrossRef]
9. Chwieduk, D. Cooling demand: A challenging issue for new buildings in moderate climates. In Proceedings of the ICR 2019 IIR the 25th International Congress of Refrigeration, Montreal, QC, Canada, 24–30 August 2019; Refrigeration Science and Technology Proceedings: 2019. pp. 3948–3956, ISBN 978-2-36215-035-7.
10. Rozporządzenie Ministra Infrastruktury w Sprawie Warunków Technicznych, Jakim Powinny Odpowiadać Budynki i ich Usytuowanie Warszawa. National Regulation on Technical Conditions for Buildings and Their Location with Amendments. *Dziennik Ustaw*, Dnia 13 Sierpnia 2013, r. Poz. 926. (In Polish). Available online: https://sip.lex.pl/akty-prawne/dzu-dziennik-ustaw/warunki-techniczne-jakim-powinny-odpowiadac-budynki-i-ich-usytuowanie-16964625 (accessed on 10 May 2020).
11. Drury, B.; Crawley, D.B.; Hand, J.W.; Kummert, M.; Griffith, B.T. Contrasting the capabilities of building energy performance simulation programs. *Build. Environ.* **2008**, *43*, 661–673.
12. Kivimaa, P.; Martiskainen, M. Innovation, low energy buildings and intermediaries in Europe: Systematic case study review. *Energy Effic.* **2018**, *11*, 31–51. [CrossRef]

13. Available online: https://www.rcrwireless.com/20160725/business/smart-building-tag31-tag99 (accessed on 10 May 2020).
14. Wang, S.; Xie, J. Integrating Building Management System and facilities Management on the Internet. *Autom. Constr.* **2002**, *11*, 707–715. [CrossRef]
15. Masoso, O.T.; Grobler, L.J. The dark side of occupants' behavior on building energy use. *Energy Build.* **2010**, *42*, 173–177. [CrossRef]
16. Anderson, B. *Solar Energy: Fundamentals in Building Design*; Total Environmental Action, Inc.: Harrisville, NH, USA, 1975.
17. Kolokotsa, D.; Santamouris, A.; Karlessi, T. Passive Solar Architecture. In *Comprehensive Renewable Energy*; Kalogirou, S.A., Saigh, A., Eds.; Elsevier: Amsterdam, The Netherlands, 2014; Volume 3, pp. 637–665.
18. Chwieduk, D. Solar energy use for thermal application in Poland. *Pol. J. Environ. Stud.* **2010**, *19*, 473–477.
19. Chwieduk, D. Achievement of Low-Energy Buildings in High-Latitude Countries Through Passive Solar Systems. In *Renewable Energy and Sustainable Buildings*; Sayigh, A., Ed.; Innovative Renewable Energy; Springer: Berlin/Heidelberg, Germany, 2019; pp. 955–961.
20. Duffie, J.A.; Beckman, W. *Solar Engineering of Thermal Processes*, 4th ed.; John Wiley and Sons: Hoboken, NJ, USA, 2013.
21. Reindl, D.T.; Beckman, W.A.; Duffie, J.A. Evaluation of Hourly Tilted Surface Radiation Models. *Sol. Energy* **1990**, *45*, 9–17.
22. Ahmad, N.; Ghiaus, C.H.; Thiery, T. Influence of Initial and Boundary Conditions on the Accuracy of the QUB Method to Determine the Overall Heat Loss Coefficient of a Building. *Energies* **2020**, *13*, 284. [CrossRef]
23. Rozporządzenie Ministra Infrastruktury i Rozwoju z dnia 27 lutego 2015 r. w Sprawie Metodologii Wyznaczania Charakterystyki Energetycznej Budynku oraz świadectw Charakterystyki Energetycznej. National Regulation on Methodology of Determination on Energy Performance of Buildings and Energy Performance Certificates. *Dziennik Ustaw*, 2015, poz. 376. (In Polish). Available online: http://isap.sejm.gov.pl/isap.nsf/DocDetails.xsp?id=WDU20170000022 (accessed on 10 May 2020).
24. Sawicka-Chudy, P.; Rybak-Wilusz, E.; Cholewa, M. Thermal efficiency of a solar power system in a collective residential structure based on performance tests. *J. Renew. Sustain. Energy* **2016**, *8*. [CrossRef]
25. Stachowicz, A.; Fedorczuk, C.M. Low energy buildings—Analysis of energy consumption of the all life cycle building. *Czasopismo Techniczne* **2007**, *Z-1B*, 133–141. (In Polish)
26. Matuszko, L.; Parzych, J.; Hozer, J. The low-Energy building—New trends of the construction industry. *Studia i Prace WNEiZ US* **2018**, *54*, 21–31. (In Polish) [CrossRef]
27. Available online: www.kobize.pl (accessed on 10 May 2020).

Article

Extending the Application of the Smart Readiness Indicator–A Methodology for the Quantitative Assessment of the Load Shifting Potential of Smart Districts

Thomas Märzinger [1] and Doris Österreicher [2,*]

1 Department of Material Sciences and Process Engineering, Institute for Chemical and Energy Engineering, University of Natural Resources and Life Sciences, 1190 Vienna, Austria; thomas.maerzinger@boku.ac.at

2 Department of Landscape, Spatial and Infrastructure Sciences, Institute of Spatial Planning, Environmental Planning and Land Rearrangement, University of Natural Resources and Life Sciences, 1190 Vienna; Austria

* Correspondence: doris.oesterreicher@boku.ac.at; Tel.: +43-1-47654-85515

Received: 2 June 2020; Accepted: 3 July 2020; Published: 7 July 2020

Abstract: In 2018, the revised Energy Performance of Buildings Directive (EPBD) included for the first time the application of a smart readiness indicator (SRI). Based on the fact that load shifting in and across buildings plays an increasingly important role to improve efficiency and alleviate the integration of renewable energy systems, the SRI is also aimed at providing an indication of how well buildings can interact with the energy grids. With the clustering of buildings into larger entities, synergies related to the integration of renewable energy and load shifting can be efficiently exploited. However, current proposals for the SRI focus mainly on qualitative appraisals of the smartness of buildings and do not include the wider context of the districts. Quantitative approaches that can be easily applied at an early planning stage are still mostly missing. To optimize infrastructure decisions on a larger scale, a quantifiable perspective beyond the building level is necessary to evaluate and leverage the larger load shifting capacities. This article builds on a previously published methodology for smart buildings with the aim to provide a numerical model-based approach on the assessment of whole districts based on their overall energy storage capacity, load shifting potential and their ability to actively interact with the energy grids. It also delivers the equivalent CO_2 savings potential compared to a non-interactive system. The methodology is applied to theoretical use cases for validation. The results highlight that the proposed quantitative model can provide a meaningful and objective assessment of the load shifting potentials of smart districts.

Keywords: smart buildings; smart districts; smart grids; smart readiness indicator; energy efficiency; energy performance of buildings directive; energy flexibility; load shifting; demand response

---

1. Introduction

At the end of 2019, the newly elected President of the European Commission published the European Green Deal [1], a roadmap for transforming the EU's economy towards sustainability. The goal is for the EU to be climate neutral in 2050 by boosting the efficiency and use of resources, moving to a clean and circular economy and restoring biodiversity and cutting pollution. The de-carbonization of the energy sector, as well as ensuring that buildings become more energy efficient, is amongst the key actions in this long-term strategy. The subsequent proposal for the first European Climate Law [2] aims to ensure that all EU policies contribute to the European Green Deal and that all sectors of the economy and society will play their part. With over 40% of the global energy consumption being attributed to the construction and use of buildings [3,4] and the European building sector being responsible for an estimated 39% of final energy consumption [5] it is evident, that any future oriented government must include buildings and their associated infrastructure at the core of its roadmaps.

Currently EU legislation related to energy use in buildings is based on the Energy Efficiency Directive [6], the Renewable Energy Directive [7], and Energy Performance of Buildings Directive (EPBD) [8], with each directive providing the framework conditions for the national regulations and standards. The EPBD, which came into force for the first time in 2002 has since then be twice revised in 2010 and 2018 [9], with each new version imposing yet stricter regulations on energy efficiency in buildings. In the last amendment the EPBD included for the first time the development and application of a so-called smart readiness indicator (SRI), which should describe how well the building can interact with the grid, manage and optimize itself, and relate information to and from its occupants. Since the SRI has not been fully elaborated in the regulative document, a study has subsequently been commissioned by the EC in order to provide guidance and a coherent framework for the member states [10,11]. The study provides a calculation framework for the assessment of the SRI that is based on qualitative indicators in a matrix approach that covers a series of impact criteria, domains and domain services. The framework focuses mostly on a qualitative assessment that is dependent on certified assessors, thus adding the danger of subjectivity to the process. Since its publication the consortium has carried out several tests to validate the process. An independent study also concluded that the approach shows limitations, particularly when applied to colder climates [12]. Another recently published analysis highlights the inherent subjectivity of the proposed solution as it documents how two independent research groups carrying out the assessment on the same buildings came to highly diverging results [13]. It is expected that more appraisals will follow to assess whether the initial proposal poses a viable way forward. The member states of the European Union will finally have to jointly or individually decide on the specific process to be implemented in their countries.

In a position paper of the "Annex 67: Energy Flexible Buildings" of the International Energy Agency (IEA) the authors argue, that there is a need for a quantitative analysis of buildings' energy flexibility [14]. Whilst a market model is proposed as an assessment, this is not considered entirely future proof, as costs and markets are subject to change [15]. One of the key objectives of the SRI is the assessment of the load shifting potentials of buildings, however this aspect is not explicitly quantified within the above-discussed study [10,11]. Also, the district is not considered, even though the clustering of larger entities becomes more important as load shifting capabilities increase. In a previous publication [16], the authors of this article have already proposed a methodology to integrate a quantitative assessment of the load shifting potential of buildings in order to support an objective judgment and subsequent implementation of the SRI. Based on the definition of the SRI in that paper, conclusions can be drawn on the load shifting potential of buildings. Following the publication, the authors have consulted with relevant stakeholders to gather feedback on the proposed methodology and to identify relevant research gaps.

This article consequently builds on the previously published methodology with the aim to provide a coherent assessment to support the optimization of infrastructure decisions on a larger scale based on the hypothesis that a perspective beyond the building level is necessary to leverage potential load shifting capacities of the built environment. The underlying hypothesis is, that the methodology for the SRI can also be expanded to larger entities, such as districts or cities and that it can provide an adequate approximation for the potential CO_2 savings. The subsequent research questions follow this hypothesis and can be summarized as follows: (1) How can the definition of the SRI be extended in terms of an efficiency limit? (2) Can the assessments for buildings be meaningfully extended to groups of buildings and larger districts? (3) Can the equivalent CO_2 savings potential be derived from the methodology?

As a result, the objective of this study is the adaptation and enlargement of the methodology to also include larger entities as well as infrastructure and CO_2 assessments on a district scale. The aim is to provide a numerical model-based approach on the assessment of whole districts based on their overall energy storage capacity, load shifting potential and their ability to actively interact with the energy grids. In addition to the district SRI and the district load shifting potential it also provides an estimation of the equivalent CO_2 savings compared to a system that does not include the building's load shifting potential. Comparably to the first publication, the approach is applied to theoretical use cases for validation. It shows that a comprehensive quantitative approach can provide meaningful result also on a district level, thus delivering important answers to the question of how much buildings can contribute to actively store and dispatch energy within a district or larger urban quarter.

The following Section 2 highlights the current state of the art and subsequent research gap this publication is addressing in the context of the assessments of smart districts. The regulative framework conditions are briefly outlined, followed by an account of state-of-the-art research related to the load shifting potential in buildings in combination with the increased use of renewable energy systems (RES). The particular focus on smart districts is seen as the logical intermediate step between smart buildings and smart cities. Section 3 describes the overall methodology, respective equations and derivations for the assessment. In Section 4 the approach is tested on a theoretical use case on a small representative district in the City of Vienna. The discussion in Section 5 finally provides a review of this extended methodology, its limitations, as well as potential for a wider application with the goal that the member states include an objective and quantitative assessment within their new regulations related to the Smart Readiness Indicator.

2. Background

Buildings play undoubtedly a crucial role within a sustainable and fossil-free energy system. Whilst the focus on the single entity is hugely relevant in order to develop highly efficient materials, structures and system, the enlargement of the perspective to bigger entities can be crucial to leverage the full potential of connected systems. Especially in dense urban environments, buildings cannot only be viewed as detached elements, but must be perceived within a wider neighborhood in their urban morphological and societal context. Resilient urban development thus sets a particular focus on concepts for sustainable, efficient, and green districts [17]. The de-carbonization of the energy systems will heavily rely on the widespread integration of RES. But since demand and supply can be deeply asynchronous, demand response management and storage potentials must be implemented to match the scale of renewables. Energy grids can provide the required transfer for electrical and thermal energy. Whilst on the building level infrastructure considerations are mostly dependent on the already existing infrastructure on a particular building site, planning on a district scale offers a broader range of options. In addition to larger urban or regional networks, small-scale infrastructure, such as district heating or cooling networks, can be included at this scale.

2.1. Regulative Background and Current Developments on the SRI

Within the latest revision of the EPBD the regulators also foresee a Smart Readiness Indicator (SRI) that rates a building to use information and communication technology (ICT) to adapt the operation of the building to the needs of the occupants and the grid [9]. As a support mechanism, the European Commission has funded a study to provide a coherent methodology for the assessment of the SRI for the member states [10]. After the publication of the original findings in 2018, the consortium subsequently started a stakeholder consultation process to review the applicability of their proposal. This process included the review of a series of topics, including cost and cost-benefits, climatic specificities, scoring system and testing. They also implemented two expert topical stakeholder working groups focused on SRI value proposition and implementation as well as SRI calculation methodology. The findings of the process and adaptations have been summarized in the interim report of the Second Technical Support Study on the Smart Readiness Indicator for Buildings [11]. Related to their Task 1 on the technical support for the consolidation of the definition and the calculation methodology of the SRI, the study concludes that the proposed SRI methodology builds on assessing the smart readiness service in a building. These services improve the performance of the building in regard to energy efficiency, responds to user requirements and support the interaction with the grid. The proposal includes both a simplified and detailed assessment method and the overall methodology has also been tested on 112 test cases [11]. Compared to the initial study, the revised version after the consultation process has not significantly changed other than refinement of the indicators as outlined above. The methodology still relies heavily on qualified assessors and thus on a subjective and quantitative approach. The latest report states that the reliability of and trust in the experts to deliver the scheme will be a key success factor and that high-quality training will be required [11]. It should also be noted that the methodology relates mostly to the electricity demand (as outlined under point 3 in the reference stated above) and does not equally consider flexibility in thermal demand.

Whilst the implementation of the EPBD is up to the individual member states, the Concerted Action on Energy Performance of Buildings Directive (CA EPBD), which is funded under the European Unions' Horizon 2020 program, aims at exchanging knowledge and best practices in the field of energy efficiency amongst the European member countries [18]. Subsequently the SRI and potential methodologies associated with its integration into national building codes will also most likely be discussed within this working group. As the CA EPBD also publishes country reports on the status of the implementation of the EPBD in the member states, it remains to be seen how the SRI methodology as proposed in the above study will be applied throughout Europe.

Nevertheless, it is clearly understood, that the Energy Performance certificates (EPCs), which are an inherent part of the EPBD play a crucial role in transforming the building market and that the directive as such already has been shown to be an effective policy [19,20]. Education and training as well as interdisciplinarity are essential cornerstones in driving the EPBD forward to improve the performance of buildings [21]. The EPCs should ideally provide easily accessible data on building performance and can support the identification and subsequent refurbishment of underperforming buildings [22]. The recently added SRI can also serve as a useful source of information to enhance public awareness on the smartness of a building, however similarly to the EPC, it is key that the indicator is easy to use, transparent and based on reliable data.

The discussion of the regulatory background shows, that policy related to the assessments of buildings are both highly relevant, but there is an evident need for easily applicable and reliable tools that provide an objective assessment. Currently, this aspect is still mostly missing within the context of the SRI.

2.2. Smart Districts

The terminology "smart" has in recent years been extensively used and elaborated on. There are several definitions, when it comes to the labelling of smart buildings or smart districts. Wiggington and Harris conclude that there exist more than 30 separate definitions on the term intelligence in relation to buildings [23]. Other literature on the subject states that intelligent buildings are clearly multi-faceted and whilst they can be summarized by a series of characteristics that include aspects ranging from user safety and comfort to resources, a universal description is challenging [24]. Whilst smartness has been mainly used as a label for buildings and cities, the intermediate scale of the district gains in importance as local energy solutions emerge.

Within the contact of this approach, the smartness mostly related to the efficient use of resources and energy. From an architectural perspective, the planning of a building is mostly limited and defined by the actual plot of the construction. Nevertheless, urban planning considerations as well as scale, morphology and societal setting amongst others heavily influence the design. Energy efficiency is also largely dependent on the immediate context relating to climate, resources, and infrastructure. Incorporating a systemic view beyond the building's edge towards the district can provide added value, as concepts for sustainable neighborhoods play a fundamental role in the development of resilient cities [17].

Following the logic of resource and energy efficient design on a buildings scale [25], the district offers the benefit of the systemic perspective and can subsequently deliver optimization beyond the single entity. Especially when it comes to building renovation, clustering buildings with different thermal qualities and connected energy generation, supply and storage systems can achieve significant primary energy savings with minimal physical intervention compared to the renovation of single entities alone [26].

The steps as outlined in Figure 1 describe the methodology for the development of energy concepts at district scale. In step 1, the passive aspects, defined by the architecture (shape, form, envelope, mass) are the first key measures to reduce the actual energy demand for heating, cooling, lighting and ventilation. At the district scale the influencing factors include the orientation of the building blocks, the density and the functional mix of the various entities. The 2nd step relates to the energy systems and focus on the energy networks, the use of waste heat potential and the efficient control management across buildings. In step 3 the adequate selection of RES and respective energy storage solutions is key to move towards zero-carbon district solutions. The overall load management as well as small scale district heating or cooling grids are relevant aspects in step 4, where the overall district in connection to the larger urban entity or region signifies the move towards smart district and subsequently smart city solutions.

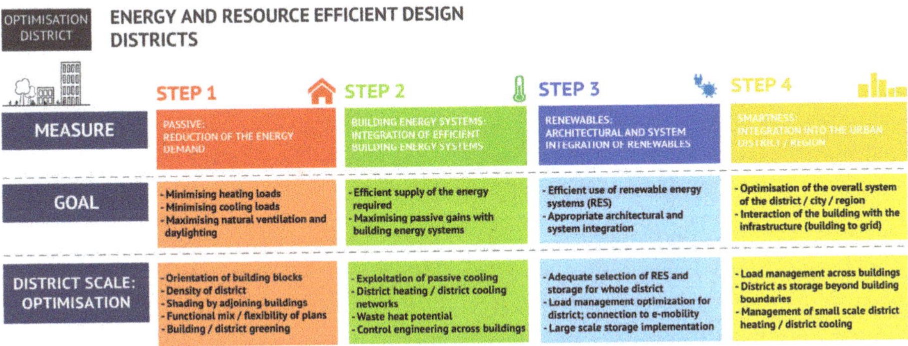

Figure 1. Methodology for resource and energy efficient design at district scale, author's graphic adapted from [25].

There is however still the question how the domains of the smart buildings can be connected to the district and subsequently to the city to ensure interoperability, especially when it comes to digital planning tools but also for operational purposes. In a recent publication where the interoperability of smart buildings into smart city platforms was evaluated, the authors concluded that the five aspects of smart energy, smart mobility, smart life, smart environment and smart data were the key domains related to the interconnectivity between the scales. Subsequently, smart building integration into a smart city has been defined to set out the framework for the various integration levels [27]. On a building scale there are already several assessments in place that are aimed at quantifying the intelligence or smartness of a building, several of which have been analyzed within the context of the named publication. The building intelligent quotient (BIQ) program developed by the BIQ Consortium consisting of CABA (Continental Automated Building Association) members, is a program aimed at evaluating building intelligence. Whilst it is mainly focused on building automation and control, it should function by its own definition as an evaluation tool for a buildings' smartness [28]. The Honeywell Smart Building Score (HSBS) provides a rating on 15 technology asset groups that make a building green, safe, and productive, and similarly offers a broad approach on the rating of devices, software, and control mechanisms within a building [29]. Both ratings encompass a rather complex and elaborate process and mainly focus on intelligent control mechanism and appliances rather than on load management, which provides an entirely different approach as outlined in this paper. In addition, they are also focused on a technology and device-oriented approach, which would need to be adapted as new technologies emerge.

Whilst the assessment of energy flows in buildings is already mostly considered standard practice with more or less detail, the district scale requires a different set of models as other aspects, such as mobility, networks and other infrastructure (e.g., waste heat from industrial processes) need to be factored in. In order to identify optimal strategies at the district level, methodologies should include qualitative and quantitative evaluation procedures based on reciprocal impacts [30]. However, there is also a need for district and energy models that go beyond the scientific community to be applied in practical design developments. Focusing especially on the planning community, the CityCalc tool has been developed to provide a quick assessment for urban planning competitions and initial planning phases [31,32]. Spatial-temporal modeling and thus dynamic assessments on a district scale can also be carried out by the CEA (city energy analyst) [33], a free open-source GIS-(geographic information systems) integrated system that has been conceived as an urban building simulation platform for the analysis and scenario comparison of energy demand, associated CO_2 emissions, financial benefits and production optimization for districts [34,35]. GIS based data analysis coupled with energy workflow modelling can be of particular importance for integrated urban platforms, that aim at modelling a diverse range of CO_2 emission related domains such as energy, resources and mobility. Such multi-domain tools can help to identify, e.g., high impact districts in need of modernization within the different urban sectors [36].

In a recent study, several urban energy-planning tools have been assessed based on their overall user friendliness related to spatial scale, output time and energy services with a few having been considered suitable for widespread use [37]. District energy modeling also supports the development of adequate technical and economical solutions for the existing building stock, as energy efficiency and renewable energy measures can be potentially more cost-effectively integrated at a multi-building scale. IEA Task 75 is specifically dedicated to the development of solutions for existing urban districts [38]. These developments show, that there is a noticeable shift towards the system perspective, which becomes ever more important, as with the exponential increase in information and communication technology, buildings act as distributed consumers, producers and storage of energy and thus develop into active players in the energy system. Similarly, the clustering of buildings to larger entities becomes more relevant as synergies related to energy efficiency and renewable energy sources can be exploited [14,15].

In his recently published book on the Green New Deal, the economist Jeremy Rifkin argues that a factor in his so called third industrial revolution towards a de-carbonized energy system lies in the digitalization of the energy networks. He stipulates that paradigm changes can occur when new communication technology converges with new energy sources and new forms of mobility. Thus he concludes that the world is on verge of a third industrial revolution as the Internet is connected to the energy system and to the mass transport systems of e-mobility [39]. Consequently, the assessment of the flexibility of buildings is significant in this context.

2.3. The Potential of Load Shifting in Buildings for the Integration of RES

The transition towards a sustainable energy system relies heavily both on the lowering of the overall demand and the provision of the required energy by renewable sources. The transformation from a centralized market to an intelligent smart grid requires a fundamental change in how we conceive the production, distribution, storage, and supply of energy [40]. As outlined above, in this context, buildings play a crucial role as they are significant consumers of energy but can at the same time provide surface areas for the integration of de-centralized solar energy systems and storage potential by means of their thermal mass and building services systems.

Aggregating buildings for cooperative energy management can yield substantial energy savings by exploiting their load shifting capabilities and utilizing shared energy systems. Aggregating buildings to clusters allow the exploitation of the variation in energy demand in different building types [15]. Determining efficient control strategies to allow a data driven and robust optimization strategy are necessary to use the potentials at a larger scale [41]. For electrical smart grids (SGs) the integration of renewable energy (RE) generation also depends largely on efficient demand response (DR). Increasing the share of RES implies that both storage systems and DR have to be jointly considered as with an increasingly higher share of RES the flexibility in the grids decrease without adequate management of demand and supply. Studies undertaken in specific micro-grids analyzing the effects of high renewable energy penetration highlight that adequate methods must be applied for an effective demand response management [42]. In order to provide de-centralized storage devices in buildings that increase the participation of end-users in the operation of the grids, micro-storage solutions must be properly planned and managed in order to provide optimized results [43]. Peer to peer (P2P) energy trading can also enable direct energy trading between energy consumers and prosumers. This reduces the exchange between the microgrid and the large-scale utility grids and can subsequently support the wide-ranging penetration of renewable energy into the power grid [44].

Whilst the electrical load shifting undoubtedly dominates the discussions related to smart grids, the thermal integration and consideration of intelligent thermal grids must not be neglected. Integrating PV (photovoltaic) systems in buildings, the so-called building integrated PVs (BIPVs) can, from an architectural and building technical point of view, more easily be achieved compared to solar thermal systems. There is nevertheless still a demand to also incorporate thermal renewables into the urban fabric. Solar thermal collectors, heat pumps, systems based on biomass or waste heat from auxiliary sources can all provide low emission alternatives to fossil-based systems. In a forecast scenario for the European heating and cooling fuel deployment an increase in the share of renewable energy from 16.7% in 2012 to 25.9% in 2030 is possible under the current policy scenario. This is driven mostly by an increased deployment of RES and a simultaneously falling final energy demand due to stricter building efficiency [45].

There is a growing awareness, that district heating networks should also react to the de-centralization of the energy market and subsequently allow the integration of small-scale supply, mostly based on RES, into their systems [46]. This would also allow the use of so-called waste heat (usually low temperature heat) from industrial processes, wastewater or reject heat from cooling systems. Several studies suggest that there is a significant potential to exploit these yet untapped resources [47,48]. Especially data centers, with their large demand in power and cooling energy represent both a potential for waste heat as well as renewable energy integration. A recent study has found that regional climate studies can provide an effective way of improving the efficiency of data centers in both the upstream renewable energy supply and the downstream waste heat reuse [48].

Although supply temperature from so called prosumers (customers that consume as well as produce energy) is usually lower than typical supply temperature, the thermal networks need to effectively manage and control their system to increase the share of decentralized renewable integration [49]. A thorough analysis on the exact scale and potential of the renewable input is however crucial to determine the feasibility of the de-centralized option. Defining a model that combines prosumers, central supply as well as market and emissions aspects can be accomplished by applying stochastic optimization algorithms. However, achieving a fair distribution of economic benefits between a central heat plant and multiple consumers remains a challenging task [50].

A study comparing several scenarios from the (classical) central heat-plant setup, to an agent-based approach and prosumer centric solutions comes to the conclusions that no approach has emerged as superior to the others and that each solution is justified under certain circumstances. It stresses subsequently that mathematical optimization is crucial in determining the best way forward [51]. While economic benefits are achieved in most scenarios, it is a non-trivial task to construct a market model that distributes these benefits in a fair way between the central heat plant and the prosumers.

Focusing on exergy with the aim to use energy efficiently and reduce carbon emissions presents yet another modelling approach. By considering the match between the grade of energy on the demand and supply side analytical models can provide useful decision support for the planning of low- or zero energy districts [52]. Combining heating and power models by providing modeling solutions for the design of co-generation is an essential cornerstone on the development of decision support mechanisms at the district scale. Multi-criteria optimization allows the focus not just on minimization of operation and maintenance system costs, but also taking into account time-varying loads, tariffs, and ambient conditions [53]. The intelligent coupling of heat and power demand and supply and subsequent co-generation is of particular importance on that scale as thermal and electrical loads can more efficiently be balanced on multiple and different building types with varying demands. Integrating renewable energy systems thus requires a multi-objective approach that considers both economical as well as environmental functions [54]. Quantifying relevant characteristics regarding the generation, distribution, and storage of energy in districts consequently represents a highly relevant aspect in the increased integration of RES into the urban environment. Current assessments and simulation tools on a district scale address the energy related aspects, however load shifting is still mostly considered from the perspective of the utility provider and thus mostly neglected in appraisals focusing on the characteristics of smart buildings and districts.

3. Methodology

Following the initial concept for the development of a quantitative approach on a building scale, the below outlined methodology aims at upscaling the concept to a bigger dimension. The numerical model-based approach provides an assessment of whole districts (or even larger entities) based on their overall energy storage capacity, load shifting potential and their ability to actively interact with the energy grids. With this methodology, districts or larger conglomerates of buildings can be rated and subsequently categorized with a single indicator per energy type. In addition, the resulting CO_2 savings potential compared to an equivalent non-interactive system can be defined, which in turn highlights the probable benefits of the load shifting. This last aspect is of particular importance for future funding schemes or other incentives tied to CO_2 emission savings. Since load shifting increases the efficiency of the overall system and subsequently the efficient use of renewable energy, the equivalent savings should be calculated and highlighted. Subsequently the proposed methodology aims at providing an answer to the following questions:

> "What is the potential of the district to take energy from the grid, store it over a certain period of time and again dispatch it back to the grid? What are the potential CO_2 emission savings associated with the load shifting potential of the district?"

In a first step, the previously published methodology is improved based on stakeholder feedback. Based on the adapted equations, the approach is enlarged from the single building to multiple buildings thus allowing the application on a whole district or any bigger logically connected series of buildings. The last sub-section finally provides an estimation for the equivalent CO_2 savings, which might be of particular importance for the communication of the benefits of increasing the load shifting capabilities in buildings.

3.1. Adaptation of the Previously Published Methodology

Following the publication of the initial methodology on the quantitative assessment of the load shifting potentials in buildings [16], a series of discussions were held with relevant stakeholders to gain insight related to the usefulness and potential application of the methodology. Whilst the overall approach to provide a simplified numerical assessment has been positively acknowledged, it has been critically reviewed, that the proposed calculation does not require minimum efficiency standards related to the storage system. It was noted that this could essentially mean that a series of highly inefficient (and potentially environmentally adverse) storage technologies could result in an equally good SRI as highly efficient (and less ecologically detrimental) systems. This could be of particular importance if the SRI is used in future application for any funding mechanisms. Thus, in order not to favor cheap and inefficient storage technologies via the definition of the SRI, the original approach has been extended by a simple extension to include a barrier in regard to minimal efficiency related to the storage type and system.

The preceding equation with the variables as outlined below has been originally published [16] and reads as follows:

$$SRI = \frac{AC}{\left(1 + e^{-6((\frac{SC}{ED} * \eta_{SC} * (1 - \zeta_{SC})) - 1)}\right)} \tag{1}$$

where ED refers to the energy demand of the building per energy source for the selected time period τ, SC, the storage capacity of the respective storage in the building, and η_{SC}, the efficiency factor of the storage capacity (here the efficiency for loading as well as unloading the storage must be considered). $\eta_{SC} = \eta_C * \eta_D$. η_C denotes the efficiency factor of the storage capacity for charge. η_D refers to the efficiency factor of the storage capacity for discharge, ζ_{SC}, the storage loss during the selected period in full storage (e.g., through self-discharge or associated heat losses), and AC the activity coefficient for the building.

Depending on the activity of the building, four different activity coefficients have previously been distinguished: (1) no grid available n/a; (2) no interaction with the grid, the activity coefficient is 0; (3) passive interaction with the grid, the activity coefficient is 1; (4) active interaction with the grid the activity coefficient is 2. ("... In this context "no interaction with the grid" means that no storage or load shifting potential is available, the building is a simple consumer. A "passive interaction with the grid" requires the building to offer storage and/or load shifting potential to the grid. The load shifting is however only one-directional from the grid to the building. The "active interaction with the grid" stands for an energy flexible building that provides storage and/or load shifting capabilities and offers bi-directional load shifting from the grid to the building as well as from the building to the grid. This building would be able to produce as well as consume energy and consequently be a prosumer ... ") [16].

From this equation the required characteristics for the SRI methodology can be achieved. An initial validation of the methodology has also been described in the previous publication [16]. Following the comments for improvements as outlined above, a function has been added to regulate the SRI regarding the storage efficiency. The following variables are necessary for the amendment of the equation:

AF: Attenuation Factor to regulate the SRI related to storage efficiency.
EP_{min}: Definition of the required minimal efficiency of the storage system. Substitute for the definition: $EP_{min} := \eta_{min} \cdot (1 - \zeta_{max})$ with η_{min} the minimal required efficiency factor and ζ_{max} the maximal required losses.
Λ: Definition point of the minimal efficiency $\lambda = 1/EP_{min}$. At this point the Attenuation Factor (AF) is always 0.63.
k: defines how fast the SRI veers with a low efficiency towards 0. With a low k the SRI is slowly reduced. With a high k the AF (SRI will be cut off) is rapidly reduced with minimal efficiency from 1 to 0.
EP: Energy Performance, i.e., the efficiency of the system $EP := \eta_{SC} \cdot (1 - \zeta_{SC})$.

$$AF = 1 - e^{-(\lambda \cdot EP)^k} \tag{2}$$

As shown in Figure 2 the pinch-off characteristics of k influence the overall energy performance as there is a vast difference if $k = 5$ or $k = 100$ as displayed in the figure. The graphic shows the main properties of the function used to calculate the Attenuation Factor (AF). The x shows the definition point for the minimum acceptable efficiency of the storage system. Based on this limit, the parameter k can be used to define how quickly the SRI approaches zero when the EP decreases. A version for a slow decline is shown for the curve $k = 5$. The curve with $k = 100$ shows a rapid decline of the SRI for a decreasing EP. That means with a large k a cut off of the SRI, by a continuous function, is achieved at the definition limit for the minimal EP. In this case, if the EP is greater than EP_{min}, the SRI remains the same as in the first definition and consequently the basic properties are retained. In Figure 2, Equation (2) has been applied.

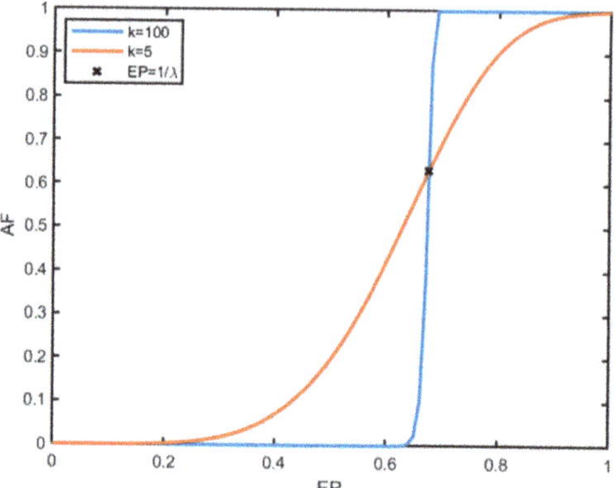

Figure 2. Pinch-off characteristics relating to k.

Following this logic, the *SRI* including the above outlined function for the regulation of the storage efficiency can be derived as follows based on Equations (1) and (2):

$$SRI = \frac{AC \cdot e^{-\left(\frac{\eta_{SC}\cdot(1-\zeta_{SC})}{\eta_{min}\cdot(1-\zeta_{max})}\right)^{k}}}{\left(1+e^{-6\left(\left(\frac{SC}{ED}\cdot\eta_{SC}\cdot(1-\zeta_{SC})\right)-1\right)}\right)} \tag{3}$$

Based on this adapted equation the required characteristics for the *SRI* methodology can be achieved. The new function now considers the regulation of the storage efficiency and thus avoids the use of potentially inefficient storage technologies.

Figure 3 below depicts the SRI curves based on the modified Equation (3) for the various activity coefficients. Graph (a) shows the SRI curves with the activity coefficient 1 and (b) shows the SRI curves with the activity coefficient 2. Both SRIs are calculated with a $k = 100$.

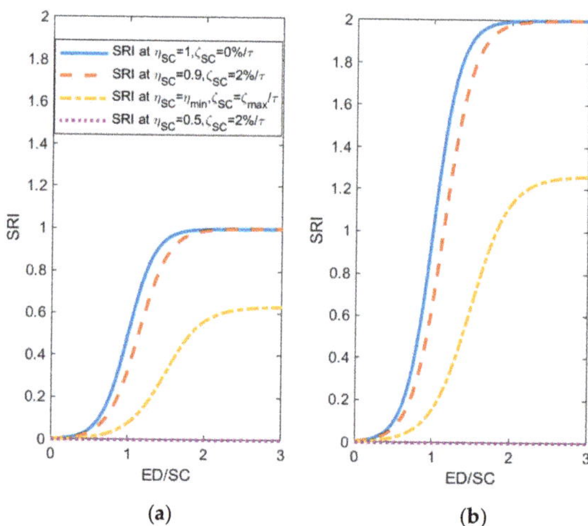

Figure 3. Pinch-off characteristics relating to k with Activity Coefficient 1 (**a**) and Activity Coefficient 2 (**b**).

It is shown that, with a high k, the curve with the EP_{min} (definition of the minimal efficiency and maximum losses, yellow curve) reaches a maximum exactly at 0.63 with the $AC = 1$ and 1.26 with the $AC = 2$. However, with the same k, an EP lower than EP_{min} results in a $SRI = 0$. Consequently, a bad storage efficiency cannot be compensated with a high storage capacity. In Figure 3, Equation (3) has been applied.

3.2. Enlargement to the District Scale

In order to enlarge the methodology to the district scale, the following variables are added:

N Number of buildings
ED_i Energy demand of the building i per energy source for the selected time period τ.
SRI_i SRI for the building i.
ED_{Dist} Energy demand for the whole district.
SRI_{Dist} SRI for the whole district.

$$SRI_{Dist} := \sum_{i=1}^{N} \omega_i \cdot SRI_i \tag{4}$$

With the weighting factor as follows:

$$\omega_i := \frac{ED_i}{ED_{Dist}} \tag{5}$$

The weighting factor has been included in order to ensure a fair comparability of districts with different characteristics. Therefore, this does not represent an average of the district's SRI, but rather a weighted average based on the respective share of energy consumption in relation to the total energy consumption of the district. For example, if a commercial entity has a low SRI with a very high ED, then a single building with a high SRI and low ED does not compensate for this.

The ED_{Dist} can subsequently defined as follows:

$$ED_{Dist} := \sum_{i=1}^{N} ED_i \tag{6}$$

Out of the above equations, the load shifting potential for the whole district can be derived. Based on the calculation as outlined in the previous publication [16], the equation to estimate the storage potential reads as follows:

$$SP = \min\left(\frac{5}{2}, \max\left(0, -\frac{\ln\left(\frac{2}{SRI} - 1\right)}{6} + 1\right)\right) \tag{7}$$

The estimation of the storage potential of a building is derived from the SRI. The reverse function is based on the following assumption: The losses and efficiencies were already taken into account when calculating the SRI. For this reason, and the fact that the SRI is monotonically increasing in terms of storage efficiency a higher efficiency subsequently implies a higher SRI. The efficiency for the inverse was chosen with 1, thus the equation is defined as:

$$\eta_{SC} \cdot (1 - \zeta_{SC}) = 1 \tag{8}$$

Subsequently the energy that can be taken from storage is calculated based on the time τ. Also, the storage potential of the building is defined with:

$$SP := \frac{SC}{ED} \tag{9}$$

In order to maintain the properties of the SRI with regard to the building as consumer (one-directional) or prosumer (bi-directional) the activity coefficient for the inverse is set to 2 (AC = 2). This is following the assumption that a storage with an $AC = 1$ is expected to be less active in shifting loads than a storage with an $AC = 2$. Based on the definition of the SRI and the above assumptions:

$$SRI = \frac{AC}{\left(1 + e^{-6((\frac{SC}{ED} \cdot \eta_{SC} \cdot (1-\zeta_{SC}))-1)}\right)} = \frac{2}{1 + e^{-6(SP-1)}} \tag{10}$$

Following these equations, the SP can be derived as follows:

$$SP = 1 - \frac{\ln\left(\frac{2}{SRI} - 1\right)}{6} \tag{11}$$

In the extreme areas of $SRI = 0$ and $SRI = 2$ the estimate obtained needs to be reasonably limited. This means that since the approach function is defined on $\mathbb{R}$, it must still be restricted to $\mathbb{R}_0^+$. Furthermore, due to rounding errors in the range of $SRI \approx 2$, errors could occur which should be limited by an upper bound. As a suggestion 2.5 was chosen as the upper bound for this study. Based on these assumptions, the storage potential for buildings can be defined as follows:

$$SP = \min\left(\frac{5}{2}, \max\left(0, 1 - \frac{\ln\left(\frac{2}{SRI} - 1\right)}{6}\right)\right) \tag{12}$$

With the equation to calculate the storage potential for the whole district:

$$LP_{Dist} := \sum_{i=1}^{N} SP_i \cdot ED_i \tag{13}$$

With the individual coefficients as follows:

SP Storage potential of the building (Relationship of SC/ED; dimensionless).
LP_{Dist} Load shift potential for the whole district.
B_i Building i.
SP_i Storage potential for building i.
ED_i Energy demand for building i.
N Number of Buildings.

The load shifting potential for the whole district serves as an approximation as derived from the SRI, the SP and the ED as outlined in the above equations and displayed in Figure 4 In this figure, in field 2 Equation (3) has been applied, in field 3 Equation (12) has been applied, and in field 4 Equation (13) has been applied.

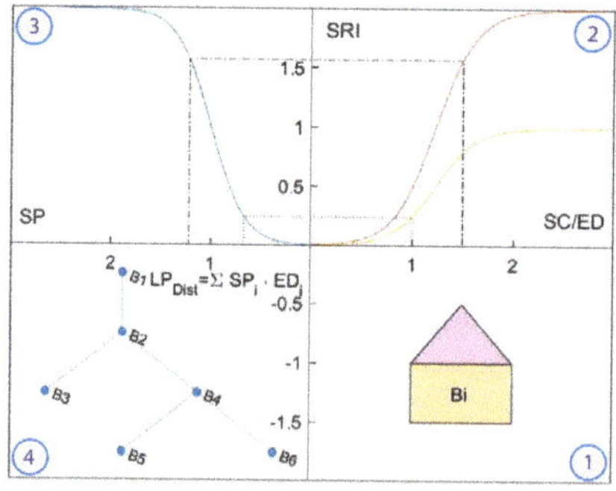

Figure 4. Assessment of the load shifting potential of the whole district.

The general assumption is that only buildings with an activity coefficient of 2 (active interaction) can fully and actively contribute to the load shifting potential in grids. An activity coefficient of 1 (passive interaction) could only contribute to peak shaving but cannot be considered to be fully contributing to the load shifting potential of the district.

- Step 1: At this first step the level of the individual i-number of building includes the data from the energy performance certificate.
- Step 2: At this second step, the SRI has been calculated for the i-number of buildings with the interaction between the energy grid and the building considered separately at this point.
- Step 3: In the third step an a-priori assessment of the storage potential is created from the SRI, which is based on the storage capacity (SC) and energy demand (ED) of the i-number of buildings. At this step the proportion of the storage potential (SP) of the building that is available for a load shifting for the grid is calculated. A building that cannot feed any energy back into the grid is related to a low SP. Thus, a building that cannot shift loads bi-directionally has to consume the stored energy itself, which is less beneficial to the network.
- Step 4: In this fourth step, the product of the storage potential (SP) and energy demand (ED) over all buildings within the district delivers an a priori assessment of the load shifting potential (LP_{Dist}) for a whole district.

As outlined above, the SRI and its subsequent calculations are solely derived from planning data and does not include any monitoring or real time measured data. It gives decision making support in the planning phase and provides answers to the question of how much energy can be theoretically shifted from the grids to the building in addition to the district's own consumption in the time span τ. The calculation provides an approximate order to magnitude for planning purposes only. Thus, this assessment is intended to be applied for a preliminary load shifting analysis of whole districts in regard to their various networks (electrical, thermal, gas).

3.3. Approximation of CO_2 Savings Potential

Whilst the load sifting potential (LP) provides a relevant number for the assessment of whole districts and cities, it will remain closely linked to infrastructure planning decisions. Arguably, load shifting alone does not necessarily increase efficiency of the system. However, if it is linked with the potential to store and dispatch renewable energy, emissions related savings can evidently be made. Especially wind and solar energy is heavily dependent on current and regionally localized weather occurrences. Thus, at times there can either be too much or not enough renewable energy in the system, which would result in wind or solar system being switched off to prevent an overload for the former or fossil-based systems to substitute the remaining demand for the latter. As outlined in the background section, in this context, storage plays an existential role.

Subsequently the assumption is, that a widespread expansion of energy storage enables a higher proportion of renewable energy to be efficiently used. In the area of thermal energy sources, it is also possible to use waste heat (i.e., low temperature heat from buildings or processes) through sufficient storage distribution in the network. Centered on this logic, the following procedure is recommended as a basis for presenting a possible CO_2, saving potential based on the SRI. From the estimate of the load shift potential (LP), an estimate for the potential CO_2 savings can be derived:

$$CO2_a = (CO2_{Curr} - CO2_{renew}) \cdot LP_{Dist} \cdot \frac{year}{\tau} \tag{14}$$

With the individual coefficients as follows:

$CO2_{Curr}$ Actual CO_2 emissions per kWh.
$CO2_{renew}$ CO_2 emissions per kWh from renewable energy sources.
$CO2_a$ Potential total CO_2 savings per year.

The calculation follows the postulation that currently renewable energy production cannot be fully exploited due to limited storage capacities. Thus e.g., wind turbines must be switched off in times of energy overload in the grid. On the other hand, when there is no wind, the energy must be produced from conventional, mostly fossil-based sources. With a SRI > 0, the district can store renewable energy in the amount of LP_{Dist} if the grid cannot take up any more energy. If subsequently energy is in demand again, it can be re-loaded back from the districts to the grid. The extent of the potential total CO_2 savings per year can thus be calculated based on the difference between the CO_2 amount of the type of energy produced multiplied by the amount of energy that can be additionally produced due to the storage capacities of the district.

It should be noted that, since the methodology is focused on load shifting, the resulting CO_2 savings are equivalent to potential savings only. That means, without a corresponding renewable energy generation, there are obviously no CO_2 savings in this context.

4. Application of the Methodology in Theoretical District Use Case

Following the testing of the previously published methodology on the building scale with the use of different building types, the extended methodology has subsequently been applied to a whole district. The selected theoretical use case should include multiple buildings of different size, type and age. For this purpose, a building block situated in the 12th district of the City of Vienna has been used as the SRI has also previously been tested on these buildings [55]. The small district includes nine buildings, differing in type and age, ranging from an erection date of the 1900s to a building constructed in 2010. One of the buildings is for office use while the others are multi-family residential buildings. To simplify the assessment the inner courtyard buildings have been omitted (as no specific use could be determined for these edifices) and the retail areas on the ground floor have been left out. Figure 5 shows a partial land use plan of the City of Vienna, highlighting the selected district as well as an aerial view of the building block.

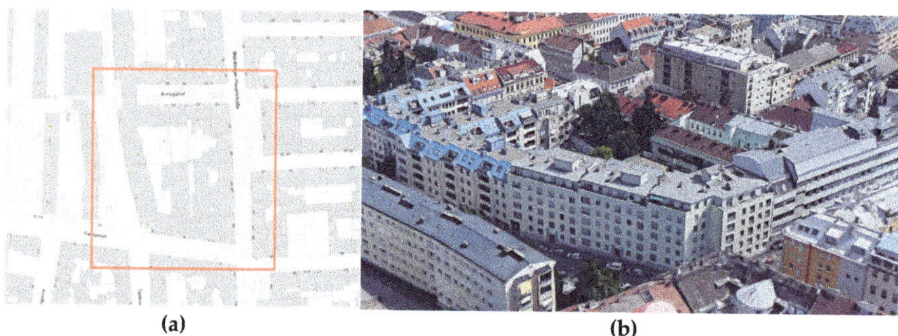

Figure 5. Partial land use map of the City of Vienna with the selected district highlighted in red (**a**) and aerial view of the selected district (**b**) [56,57].

4.1. Description of Theorectical District Use Case

The building data for this district has been derived from a series of publicly available data, such as the land use plan of the City of Vienna [56] and imagery derived from Google maps [57]. Data on typical energy usage depending on the type and age of the building as well as the assumptions on typical heating, cooling and ventilation systems have been based on the Tabula database [58]. It should be noted that, for planning purposes, actual building data derived from, e.g., the Energy Performance Certificate (EPC) or monitored data, should be given preference to proxy data derived from generic databases. However, for the purpose of this study, the accuracy of the energy figures for the base case are of lesser importance, as the aim is solely to assess, whether the proposed methodology can be applied for different types of scenarios for whole districts.

For the assessment of the load shifting potential of the selected district, various scenarios have been defined to cover a wide range of possibilities. The selected options follow a previously carried out study on the SRI validation and are expanded in regard to the scenarios as well as the calculation for the whole district assessment as described in Section 3 above.

- Base Case: For the base case the buildings have been assessed according to the generic data as outlined above. For this case it is assumed that the buildings cannot store, actively load or unload energy to and from the grid. The activity coefficient (AC) is subsequently assumed to be either not available (n/a) in case of e.g., a thermal energy network or (0) where there is no active interaction with the grid, e.g., relating to power or gas.
- Scenario 1: For the first new scenario, a moderate refurbishment of the building envelope is assumed with a 50% improvement compared to the base case. In addition, the gas connection is substituted with a one-directional thermal grid connection and electrical batteries are considered for residential as well as office buildings.
- Scenario 2: In this scenario, the building shell is improved by 90% compared to the base case, thus a high-performance building shell has been implemented. Similar to scenario 1, the gas connection has been severed and the buildings are connected to a low temperature bi-directional district heating system. RES and batteries are included in all buildings.
- Scenario 3: For this scenario the district is doubled in size (18 buildings compared to 9 buildings of the above scenarios) and constitutes a mix of the Base Case and Scenario 1. It is assumed that half of the buildings remain as described in the base case (i.e., un-refurbished) and the other half is considered with a moderate refurbishment as described in Scenario 1.
- Scenario 4: In this scenario, the district is tripled in size (27 buildings) and constitutes a mix of the Base Case, Scenario 1 and Scenario 2. It is assumed that one third of the buildings remains as described in the base case (i.e., un-refurbished), one third is refurbished and follows the characteristics of Scenario 1 and the remaining third follows Scenario 2.

For Scenarios 3 and 4 a double and triple size of the original Base Scenario has been defined. This has been done in order to demonstrate the overall weighting of the SRI across the district. As outlined in Equations (4)–(6) above, a specific weighting factor has been integrated to ensure that districts with different characteristics can be fairly compared. As the SRI_{Dist} is a single number it needed to be avoided that the SRI represents a simple average across all buildings but rather a weighted average based on the respective share of energy consumption in relation to the total energy consumption of the district.

In Table 1, a description of the energy related properties for the base case as well as subsequent scenarios is outlined.

Table 1. Description of energy related properties for base case and scenarios, extended from [55].

Scenario	No. of Buildings	Building Envelope	Electrical Storage/Grid	Thermal Storage/Grid	Gas Storage/Grid
Base Case	9	Un-refurbished	No active storage; one directional connection	No active storage; no thermal grid	No active storage; one directional connection
Scenario 1	9	Improved by 50%	Active storage bi-directional connection	No active storage; one-directional connection (thermal grid)	No connection
Scenario 2	9	Improved by 90%	Active storage bi-directional connection	Active storage bi-directional connection (thermal grid)	No connection
Scenario 3	18	Half of the district (9 buildings) as per Base Case/other half of the district (9 buildings) as per Scenario 1			
Scenario 4	27	One third of the district (9 buildings) as per Base Case/one third of the district (9 buildings) as per Scenario 1/one third of the district (9 buildings) as per Scenario 2			

4.2. Results of Theoretical District Use Case

The results of the theoretical use case comprising of the base case and four scenarios are displayed in the following figures. In Figure 6, the SRI is shown per building for each scenario. It can be seen that, whilst the base case, Scenario 1 and 2, is comprised of nine buildings in the district ($N = 9$), in Scenario 3, the district is doubled in size with 18 buildings ($N = 18$), and in Scenario 4, tripled with 27 buildings ($N = 27$). The SRI_i for the base case is "0" as the activity coefficient is also "0" (i.e., no interaction with the grid, buildings are simple consumers). The SRI for all other buildings varies dependent on type, size, interaction potential and storage capacity. In Figure 6, Equation (3) has been applied.

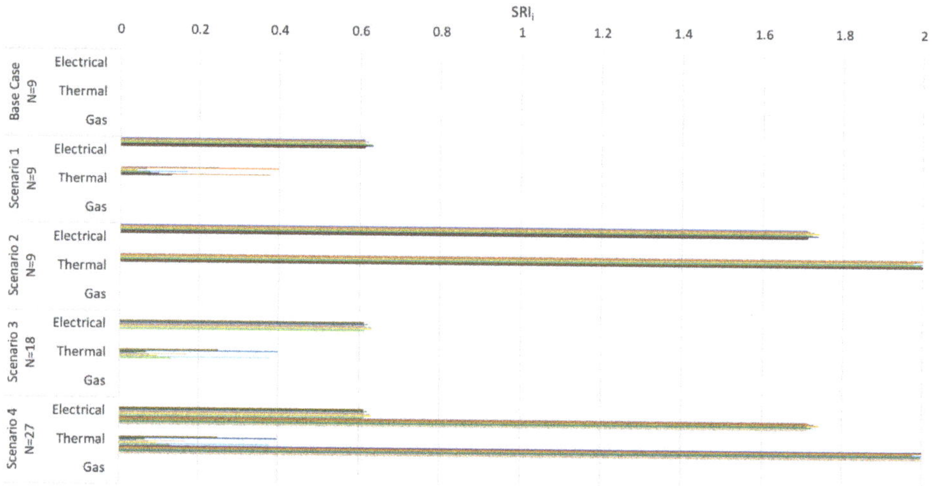

Figure 6. Smart Readiness Indicator for the individual buildings (SRI_i) per district for the Base Case and Scenarios 1–4.

In Figure 7 the SRI_{Dist} for the various scenarios is displayed on a district scale. Since the base case has an activity coefficient of "0" the SRI for the base case for the district is equally "0". Scenario 1 rates already better, however Scenario 2, with the optimized refurbishment option and active storage potential for both electrical and thermal provides the best result. The mix of base case and Scenario 1 which is displayed in Scenario 3 is obviously less good than Scenario 1 as the SRI is weighted across the district. With half the buildings in Scenario 3 being un-refurbished and offering no load shifting potential, the results are poorer. Scenario 4, which is a mix of the Base Case, Scenario 1 and Scenario 2 rates second best, as the base case takes up only a third of the overall scenario and the SRI is weighted across the district. The results highlight that the above described weighting factor does ensure an objective comparability across the various scenarios. In Figure 7, Equation (4) after Equation (5) after Equation (6) after Equation (3) have been applied.

The load shifting potential for the overall district is displayed in Figure 8. The LP_{Dist} represents, different to the SRI_{Dist}, a total figure. The results show, that Scenario 4 can provide with a total number of 27 buildings based on a mix of the base case, Scenario 1 and 2 the highest amount of load shifting in kWh. The base case obviously has no load shifting potential, as there is no interaction with the grid and the SRI_{Dist} as shown above is consequently "0". Scenario 1 and Scenario 3 have exactly the same load shifting potential, as they have the same amount of buildings that offer the same storage capacities, as the base case buildings in Scenario 3 do not contribute at all to the load shifting. Scenario 4 offers the best results due to the highest number of buildings in the district ($N = 27$) and the relatively high storage capacity due to the mix of Scenario 1 and 2 (the base case buildings in this scenario equally do not contribute to the LP_{Dist}). In Figure 8, Equation (13) after Equation (12) after Equation (3) have been applied.

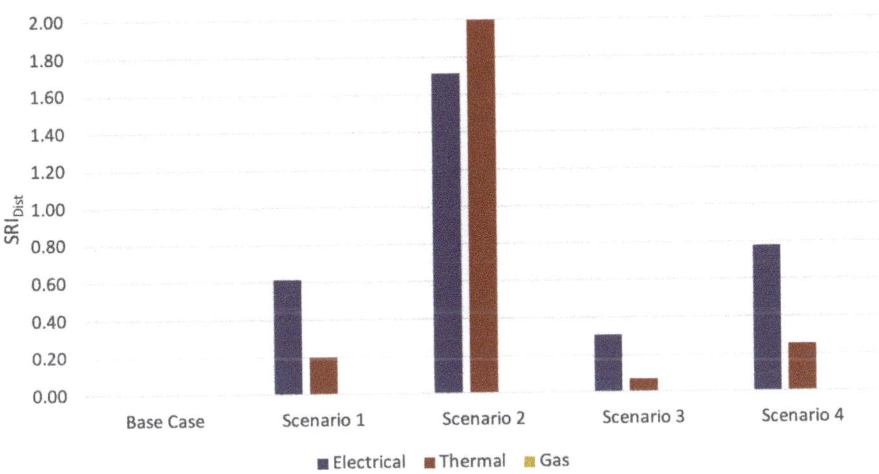

Figure 7. Smart Readiness Indicator District (SRI_{Dist}) for the Base Case and Scenarios 1–4.

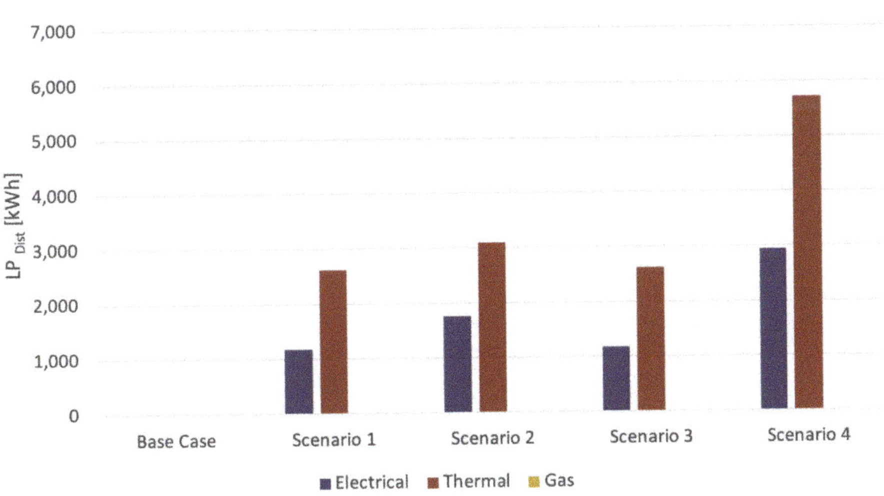

Figure 8. Load Shifting Potential (LP_{Dist}) for the Base Case and Scenarios 1–4.

In Figure 9, the potential CO_2 savings for the district are displayed. The results follow in tendency the results for LP_{Dist} as a higher load shifting potential results in increased CO_2 savings. Similar to the above results, the Base Case does not save any emissions due to the lack of load shifting potential. Scenario 1 and Scenario 3 show the same results (as is also the case with LP_{Dist}) as they both feature the same load shifting potential and subsequently the same potential for emission savings. Scenario 2 and 4 also feature higher equivalent savings due to their higher storage capacity and improved activity (bi-directional connection).

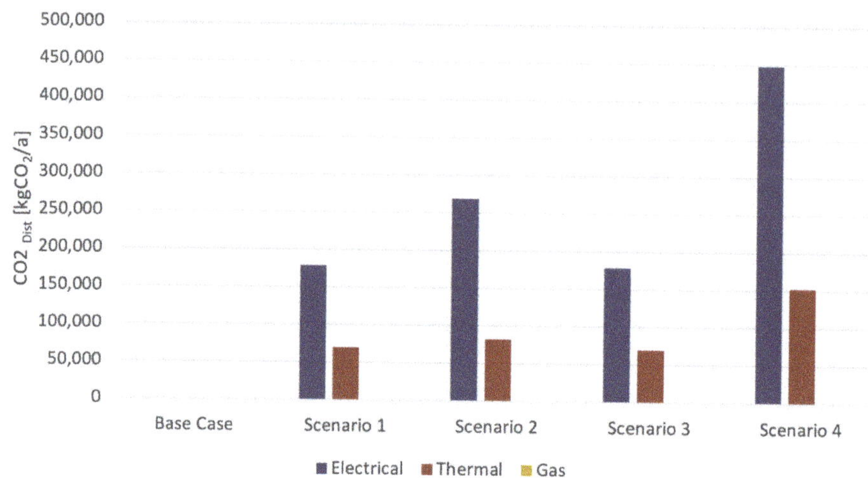

Figure 9. CO_2 equivalent savings ($CO2_{\text{Dist}}$) for the Base Case and Scenarios 1–4.

Evidently, these figures vary, depending on the actual emissions of the energy system ($CO2_{Curr}$, actual CO_2 emissions per kWh) and the emissions from renewables in the energy system ($CO2_{renew}$, CO_2 emissions per kWh from renewable energy sources) as outlined in Equation (14). Since this theoretical use case is based in Vienna, the figures for electrical and thermal energy have been taken from the Austrian standards [59] with a noted power mix of 417 g/kWh CO_2 and district heating from highly efficient CHP (combined heat and power) with 73 g/kWh CO_2. Emissions from renewables have been calculated with 0 g/kWh CO_2. Since the emissions from the current power mix are substantially higher than the emissions from the current thermal sources, the savings on the electrical side subsequently exceed the savings from the thermal side, even though the thermal load shifting potential as outlined in Figure 8 is higher in all four scenarios than the electrical load shifting potential. In Figure 9, Equation (14) after Equation (13) after Equation (12) after Equation (3) have been applied.

5. Discussion

In this study, an expanded methodology aims at providing a coherent quantitative assessment of whole districts based on their overall energy storage capacity, load shifting potential, their ability to actively interact with the energy grids and the resulting CO_2 emission savings compared to a non-interactive system. As the previously published methodology has been discussed with selected stakeholders, an improvement has been undertaken in order to better adjust the SRI to efficiency standards. In the new version a cut-off for the efficiency of the storage system has been introduced so that large scale, but potentially inefficient storage systems cannot easily substitute smaller, but efficient ones. This is of particular importance in order to avoid that outdated and potentially environmentally harmful technologies are rewarded with a good result. Thus, the efficiency limit, as stated as one of the research questions, has been addressed.

The subsequent enlargement towards a district assessment demonstrates that the methodology can be adapted to bigger and spatially logical entities. The application of the methodology in a theoretical case study shows that the approach is suitable to define relevant indications of the load shifting potential by means of SRI_{Dist} and LP_{Dist} as both indicators deliver meaningful results in this context. A weighting factor that has been introduced for the SRI_{Dist} ensures that districts with different characteristics in terms of size, type and quality can be fairly compared. With this measure, an entity that has a low SRI can be avoided, but high energy demand cannot at the same time within a district be easily compensated with a single entity that has a high SRI but low energy demand. Consequently, the methodology has shown to be meaningfully extended to groups of buildings, thus answering to the second research question.

The SRI_{Dist} can also function as a benchmark as whole districts can be assessed based on their possibility for further development without relying on measured or monitored data from the energy providers. As outlined in Figures 7 and 8, for example Scenario 1 and Scenario 3 can be compared based on their potential for expansion. Whilst Scenario 3 features twice as many buildings as Scenario 1, the LP_{Dist} is the same for both scenarios as the load shifting potential is the same across both districts as only half of the buildings in Scenario 3 provide storage capacity. However, from the SRI_{Dist} it can be clearly seen, that in Scenario 3 there is a considerable higher potential for further load shifting as the SRI_{Dist} is considerably lower in Scenario 3 than in Scenario 1. This shows that the SRI_{Dist} can provide a meaningful assessment and comparison of districts that usually greatly differ in size, type, as well as quality and number of buildings.

The approach is however solely focused on the load shifting potential and ability of the district to actively interact with the grid. This differentiates the methodology from other more interdisciplinary approaches such as those outlined, for example, by Garcia-Ayllon [60] and Sharifi [61] where many other factors such as, e.g., resources and governance, have been taken into account. Also, other schemes such as the BIQ [28] and the HSBS [29] rating, that have been comprehensively compared to the SRI by Apanaviciene et al. [27], deliver a much larger, complex set of indicators that are rather dependent on system and device assessment. Our observation offers information targeted on load shifting, thus providing a very focused assessment. The approach outlined in this paper is also not technology specific, but is based on certain qualities, that need to be achieved. This is crucial, as it decouples the methodology from the technology used and makes the approach future proof, as any new technology or device can be similarly represented within this novel approach. The methodology in this paper might therefore serve as one of a series of other quantitative indicators that can be integrated into broader assessments.

In relation to the possible CO_2 savings as stated in the last research question, the methodology also provides a workable approach. When considering the systems of energy generation, energy transmission and consumers in relation to the same three systems and adding energy storage, a potential for CO_2 savings can be generated. The assumption is that energy which comes from renewable sources cannot be produced on demand, but when it can be stored, it can be subsequently released on demand. It follows that there is the possibility that the magnitude of the LP energy from CO_2 neutral but not demand-driven energy sources can be used. This results in a reduction in CO_2-related energy and thus an indirect reduction in CO_2 emissions from energy production. This definition provides a quantitative assessment of the proportional CO_2 savings in relation to the load shifting potential.

Whilst the proposed indicators can be applied for different queries, the following key questions can be answered with the proposed approach: In which area does a high potential for load shifting measures in buildings exist? Which areas already provide enough smart buildings and renewable energy capacity in order to improve the network infrastructure? Is there enough load shifting potential from buildings near, e.g., a wind farm, in order to actively use it as storage? What are the prospective CO_2 savings associated with the potential load shifting capacity of the district? For those and similar questions, the proposed initial assessment can provide meaningful results.

There are however also certain limitations to this approach. For one, the assessment of the LP_{Dist} is not based on actual load profiles and should therefore not be used as a substitute for an exact analysis. Whilst the methodology serves very well for an initial and quick assessment of the load shifting potential within a defined district it is neither intended nor suitable for detailed capacity sizing of energy infrastructure. Care should also be taken related to the choice and extent of the area (system boundaries of the district) under consideration. The SRI_{Dist} evaluates only the load shifting potential of buildings and does not take any information regarding the respective electrical, thermal or gas grids into account. Thus, any bottlenecks in the network infrastructure are not recognized. This means that just because the buildings are able to move a certain amount of energy it is not necessarily the case that the networks are also able to accommodate this. It is subsequently evident that the proposed indicators are centered around the assessment of buildings (or multiple buildings) rather than the energy infrastructure.

For further research, a verification of the indicator could be done with a comprehensive dynamic building simulation for the assessment of load shifting capacities, which is based on monitoring data from a wide range of different buildings types. Based on this assessment, the discrepancies of the monitored and subsequently simulated data with the results from the SRI methodology could be calculated. With this, the functions of the model could be calibrated in order to match the actual requirements. In addition, monitoring data derived from buildings varying in typology and energy profiles could be used to assess how well the calculated approximation matches the actual building data. A validation of that kind is planned for the future, once the SRI has to be implemented for Building Regulations purposes. Additionally, a similar indicator that assesses and highlights the free capacities in different network areas or network nodes could be considered. Conversely, one could also estimate for an area which SRI_{Dist_Max} or LP_{Dist_Max} the network infrastructure can endure and thus monitor the development of load shifting potentials and renewable integration in the area in order to be vigilant of potentially critical conditions related to the energy infrastructure.

6. Conclusions

This paper proposes a novel quantitative approach for the rating of the load shifting potential of smart districts and discusses qualitative and quantitative assessments related to energy load shifting at building and district level. The topic emerged from the new regulations in the Energy Performance of Buildings Directive (EPBD) where a smart readiness indicator (SRI) has been defined in order to rate buildings according to their ability to operate and communicate efficiently with energy grids. Current proposal on the rating of the SRI are mainly focused on qualitative approaches based on the assessments of experts. Also, they mainly focus on buildings without taking larger entities into account. As previous studies have shown, there is a clear need for a quantitative approach to allow objectivity and comparability of the results. The methodology proposed in this publication addresses this research gap. It builds on a previously published quantitative approach for smart buildings and extends the application to bigger entities. It also includes findings from stakeholder consultation and provides an improved version, taking efficiency standards related to energy storage systems into account. The novel SRI_{Dist} can be used to assess whole districts based on their overall energy storage capacity, load shifting potential and their ability to actively interact with the energy grids. In addition, it provides an approximation for CO_2 savings in relation to a non-interactive system. The key aspect of the methodology is an application that integrates the use of building and energy data, that is relatively easily available without the need for monitored data that is either difficult to access or not available at all at an early planning stage. The methodology also does not rely on specific systems, but rather on qualities of a system, thus making the approach suitable for future technologies. The application in a theoretical district use case shows a logical distribution of SRI results across the different scenarios and supports the meaningfulness of an approximation of complex data within a comparable indicator. With a simple rating, load shifting potentials in districts can be more easily assessed, thus leading to a potentially higher integration of renewable energy sources with a volatile generation capacity. The research subsequently contributes with a theoretical framework to the increased exploitation of load shifting capacities in building and districts.

Author Contributions: Author D.Ö. contributed mainly to the conceptualization, methodology, writing of the original draft, review, and editing. Author T.M. contributed mainly to the methodology, the formal analysis, the validation, and the writing of the original draft. All authors have read and agreed to the published version of the manuscript.

Funding: This research is partly funded under the framework of the project "UrbanEnerPlan-Impact of urban planning related framework conditions on energy and resource-efficiency of new residential building developments in Vienna–Quantification of energy and urban planning related influencing factors" under the City of Vienna Jubilee Funds for the University of Natural Resources and Life Sciences, Vienna.

Acknowledgments: Supported by the University of Natural Resources and Life Sciences, Vienna Open Access Publishing Fund.

Conflicts of Interest: The authors declare no conflict of interest.

References

1. European Commission. *The European Green Deal*; COM (2019) 640 Final; European Commission: Brussels, Belgium, 2019.
2. European Climate Law. Available online: https://ec.europa.eu/clima/policies/eu-climate-action/law_en (accessed on 14 April 2020).
3. World Business Council for Sustainable Development (WBCSD). *Energy Efficiency in Buildings Transforming the Market*; WBCSD: Geneva, Switzerland, 2009.
4. Cao, X.D.; Dai, X.L.; Liu, J.J. Building energy-consumption status worldwide and the state-of-the-art technologies for zero-energy buildings during the past decade. *Energy Build.* **2016**, *128*, 198–213. [CrossRef]
5. European Commission. *Technology Assessment*; COM (2013) 253 Final; European Commission: Brussels, Belgium, 2013.
6. *Directive 2012/27/EU of the European Parliament and of the Council of 25 October 2012 on Energy Efficiency*; L315/1-56; Official Journal of the European Union: Brussels, Belgium, 2012.
7. *Directive 2009/28/EC of the European Parliament and of the Council of 23 April 2009 on the Promotion of the Use of Energy from Renewable Sources*; L140/16-62; Official Journal of the European Union: Brussels, Belgium, 2009.
8. *Directive 2002/91/EU of the European Parliament and of the Council of 16 December 2002 on the Energy Performance of Buildings*; L1/65; Official Journal of the European Union: Brussels, Belgium, 2003.
9. *Directive (EU) 2018/844 of the European Parliament and of the Council of 30 May 2018 Amending Directive 2010/31/EU on the Energy Performance of Buildings Directive 2012/27/EU on Energy Efficiency*; L156/75; Official Journal of the European Union: Brussels, Belgium, 2018.
10. Verbeke, S.; Ma, Y.; Van Tichelen, P.; Bogaert, S.; Gómez Oñate, V.; Waide, P.; Bettgenhäuser, K.; Ashok, J.; Hermelink, A.; Offermann, M.; et al. *Support. for Setting Up a Smart Readiness Indicator for Buildings and Related Impact Assessment, Final Report*; Study Accomplished under the Authority of the European Commission DG Energy 2017/SEB/R/1610684; Vito NV: Mol, Belgium, 2018.
11. Verbeke, S.; Aerts, D.; Rynders, G.; Ma, Y.; Waide, P. *3rd Interim Report of the 2nd Technical Support. Study on the Smart Readiness Indicator for Buildings*; Study Accomplished under the Authority of the European Commission DG Energy ENER/C3/2018-447/06; Vito NV: Mol, Belgium, 2020.
12. Janhunen, E.; Pulkka, L.; Säynäjoki, A.; Junnila, S. Applicability of the Smart Readiness Indicator for Cold Climate Countries. *Buildings* **2019**, *9*, 102. [CrossRef]
13. Vigna, I.; Pernetti, R.; Pernigotto, G.; Gasparella, A. Analysis of the Building Smart Readiness Indicator Calculation: A Comparative Case-Study with Two Panels of Experts. *Energies* **2020**, *13*, 2796. [CrossRef]
14. Søren Østergaard, J.; Madsen, H.; Lopes, R.; Junker, R.G.; Aelenei, D.; Li, R.; Metzger, S.; Lindberg, K.B.; Marszal, A.J.; Kummert, M.; et al. *Annex 67: Energy Flexible Buildings, Energy Flexibility as a Key Asset in a Smart Building Future Contribution of Annex 67 to the European Smart Building Initiative*; Position Paper of the IEA Energy in Buildings and Communities Program (EBC) Annex 67 "Energy Flexible Buildings"; IEA: Paris, France, 2017.
15. Vigna, I.; Pernetti, R.; Pasut, W.; Lollini, R. New domain for promoting energy efficiency: Energy Flexible Building Cluster. *Sustain. Cities Soc.* **2018**, *38*, 526–533. [CrossRef]
16. Märzinger, T.; Österreicher, D. Supporting the Smart Readiness Indicator—A Methodology to Integrate A Quantitative Assessment of the Load Shifting Potential of Smart Buildings. *Energies* **2019**, *12*, 22. [CrossRef]
17. Bott, H.; Grassl, G.C.; Anders, S. *Nachhaltige Stadtplanung, Konzepte für nachhaltige Quartiere*; Detail: München, Germany, 2013; ISBN 978-3955531935.
18. Concerted Action Energy Performance of Buildings Directive. Available online: https://epbd-ca.eu (accessed on 14 April 2020).
19. Fokaides, P.A.; Polycarpou, K.; Kalogirou, S. The impact of the implementation of the European Energy Performance of Buildings Directive on the European building stock: The case of the Cyprus Land Development Corporation. *Energy Policy* **2017**, *111*, 1–8. [CrossRef]

20. Lopez-Ochoa, L.M.; Las-Heras-Casas, J.; Lopez-Gonzalez, L.M.; Olasolo-Alonso, P. Towards nearly zero-energy buildings in Mediterranean countries: Energy Performance of Buildings Directive evolution and the energy rehabilitation challenge in the Spanish residential sector. *Energy* **2019**, *176*, 335–352. [CrossRef]
21. Brunsgaard, C.; Dvorakova, P.; Wyckmans, A.; Stutterecker, W.; Laskari, M.; Almeida, M.; Kabele, K.; Magyar, Z.; Bartkiewicz, P.; Op't Veld, P. Integrated energy design—Education and training in cross-disciplinary teams implementing energy performance of buildings directive (EPBD). *Build. Environ.* **2014**, *72*, 1–14. [CrossRef]
22. Li, Y.; Kubicki, S.; Guerriero, A.; Rezgui, Y. Review of building energy performance certification schemes towards future improvement. *Renew. Sustain. Energy Rev.* **2019**, *113*, 13. [CrossRef]
23. Wigginton, M.; Harris, J. *Intelligent Skins*; Butterworth-Heinemann: Oxford, UK, 2002.
24. Omar, O. Intelligent building, definitions, factors and evaluation criteria of selection. *Alex. Eng. J.* **2018**, *57*, 2903–2910. [CrossRef]
25. Österreicher, D. Methodology towards integrated refurbishment actions in school buildings; Special Issue: Addressing Sustainable Building Refurbishment: A Journey through Energy Optimization and Structural Retrofit. *Buildings* **2018**, *8*, 42. [CrossRef]
26. Conci, M.; Schneider, J. A District Approach to Building Renovation for the Integral Energy Redevelopment of Existing Residential Areas. *Sustainability* **2017**, *9*, 747. [CrossRef]
27. Apanaviciene, R.; Vanagas, A.; Fokaides, P.A. Smart Building Integration into a Smart City (SBISC): Development of a New Evaluation Framework. *Energies* **2020**, *13*, 2190. [CrossRef]
28. Building Intelligence Quotient. Available online: http://www.building-iq.com/biq/index.html (accessed on 30 May 2020).
29. Honeywell and Ernst & Young LLP. *Smart Buildings Make Smart Cities; Honeywell Smart Building Score; Green. Safe. Productive*; Honeywell International Inc.: Gurgaon, India; Available online: http://smartbuildings.honeywell.com/hsbs_home (accessed on 30 May 2020).
30. Mattoni, B.; Nardecchia, F.; Bisegna, F. Towards the development of a smart district: The application of an holistic planning approach. *Sustain. Cities Soc.* **2019**, *48*, 17. [CrossRef]
31. CityCalc. Available online: http://citycalc.com (accessed on 19 April 2020).
32. Smutny, R.; Österreicher, D.; Sattler, S.; Treberspurg, M.; Battisti, K.; Gratzl, M.; Rainer, E.; Staller, H. Low-tech solution for Smart Cities—Optimization tool CityCalc for solar urban design. In *Proceedings of the International Conference on Urban Planning and Regional Development in the Information Society GeoMultimedia, Hamburg, Germany, 22–24 June 2016*; Schrenk, M., Popovich, V., Zeile, P., Elisei, P., Beyer, C., Eds.; Real Corp 2016 Proceedings: Hamburg, Germany, 2016; ISBN 978-3-9504173-0-2.
33. City Energy Analyst. Available online: https://cityenergyanalyst.com (accessed on 19 April 2020).
34. Fonseca, J.A.; Schlueter, A. Integrated model for characterization of spatiotemporal building energy consumption patterns in neighborhoods and city districts. *Appl. Energy* **2015**, *142*, 247–265. [CrossRef]
35. Fonseca, J.A.; Nguyen, T.A.; Schlueter, A.; Marechal, F. City Energy Analyst (CEA): Integrated framework for analysis and optimization of building energy systems in neighborhoods and city districts. *Energy Build.* **2016**, *113*, 202–226. [CrossRef]
36. Eicker, U.; Weiler, V.; Schumacher, J.; Braun, R. On the design of an urban data and modeling platform and its application to urban district analyses. *Energy Build.* **2020**, *217*, 19. [CrossRef]
37. Ferrari, S.; Zagarella, F.; Caputo, P.; Bonomolo, M. Assessment of tools for urban energy planning. *Energy* **2019**, *176*, 544–551. [CrossRef]
38. IEA EBC Annex 75—Cost-Effective Building Renovation at District Level Combining Energy Efficiency & Renewables. Available online: http://annex75.iea-ebc.org (accessed on 19 April 2020).
39. Rifkin, J. *The Green New Deal: Why the Fossil Fuel Civilization Will Collapse by 2028, and the Bold Economic Plan. to Save Life on Earth*; St. Martins Press: New York, NY, USA, 2019; ISBN 978-1250253200.
40. Aichele, C.; Doleski, O.D. *Smart Market—Vom Smart Grid zum intelligenten Energiemarkt*; Springer Vieweg: Wiesbaden, Germany, 2014. [CrossRef]
41. Darivianakis, G.; Georghiou, A.; Smith, R.S.; Lygeros, J. The Power of Diversity: Data-Driven Robust Predictive Control for Energy-Efficient Buildings and Districts. *IEEE Trans. Control Syst. Technol.* **2019**, *27*, 132–145. [CrossRef]
42. Hakimi, S.M.; Hasankhani, A.; Shafie-khah, M.; Catalão, J.P.S. Demand response method for smart microgrids considering high renewable energies penetration. *Sustain. Energy Grids Netw.* **2020**, *21*, 100325. [CrossRef]

43. De Paola, A.; Angeli, D.; Strbac, G. Distributed Control of Micro-Storage Devices with Mean Field Games. *IEEE Trans. Smart Grid* **2015**. [CrossRef]
44. Zhang, C.; Wu, J.; Zhou, Y.; Cheng, M.; Long, C. Peer-to-Peer energy trading in a Microgrid. *Appl. Energy* **2018**, *220*, 1–12. [CrossRef]
45. Fleiter, T.; Steinbach, J.; Ragwitz, M.; Arens, M.; Aydemir, A.; Elsland, R.; Fleiter, T.; Frassine, C.; Herbst, A.; Hirzel, S.; et al. *Mapping and Analyses of the Current and Future (2020–2030) Heating/Cooling Fuel Deployment (Fossil/Renewables)*; European Commission Directorate-General for Energy: Brussels, Belgium; Luxembourg, 2016; Available online: https://ec.europa.eu/energy/sites/ener/files/documents/mapping-hc-excecutivesummary.pdf (accessed on 27 April 2020).
46. Loibl, W.; Stollnberger, R.; Österreicher, D. Residential Heat Supply by Waste-Heat Re-Use: Sources, Supply Potential and Demand Coverage-A Case Study. *Sustainability* **2017**, *9*, 250. [CrossRef]
47. Lichtenwoehrer, P.; Erker, S.; Zach, F.; Stoeglehner, G. Future compatibility of district heating in urban areas—A case study analysis in the context of integrated spatial and energy planning. *Energy Sustain. Soc.* **2019**, *9*, 12. [CrossRef]
48. Huang, P.; Copertaro, B.; Zhang, X.X.; Shen, J.C.; Lofgren, I.; Ronnelid, M.; Fahlen, J.; Andersson, D.; Svanfeldt, M. A review of data centers as prosumers in district energy systems: Renewable energy integration and waste heat reuse for district heating. *Appl. Energy* **2020**, *258*, 20. [CrossRef]
49. Brand, L.; Calven, A.; Englund, J.; Landersjo, H.; Lauenburg, P. Smart district heating networks—A simulation study of prosumers' impact on technical parameters in distribution networks. *Appl. Energy* **2014**, *129*, 39–48. [CrossRef]
50. Lichtenegger, K.; Wöss Halmdienst, C.; Höftberger, E.; Schmidl, C.; Pröll, T. Intelligent heat networks: First results of an energy information-cost-model. *SEGAN* **2017**, *11*, 1–12. [CrossRef]
51. Lichtenegger, K.; Leitner, A.; Märzinger, T.; Mair, C.; Moser, A.; Wöss, D.; Schmidl, C.; Pröll, T. Decentralized heating grid operation: A comparison of centralized and agent-based optimization. *Sustain. Energy Grids Netw.* **2020**, *21*, 100300. [CrossRef]
52. Kilkis, S. Energy system analysis of a pilot net-zero exergy district. *Energy Convers. Manag.* **2014**, *87*, 1077–1092. [CrossRef]
53. Bischi, A.; Taccari, L.; Martelli, E.; Amaldi, E.; Manzolini, G.; Silva, P.; Campanari, S.; Macchi, E. A detailed MILP optimization model for combined cooling, heat and power system operation planning. *Energy* **2014**, *74*, 12–26. [CrossRef]
54. Bracco, S.; Delfino, F.; Ferro, G.; Pagnini, L.; Robba, M.; Rossi, M. Energy planning of sustainable districts: Towards the exploitation of small size intermittent renewables in urban areas. *Appl. Energy* **2018**, *228*, 2288–2297. [CrossRef]
55. Österreicher, D.; Märzinger, T. Assessing the Load Shifting Potential in Buildings—Application of a Methodology for the Smart Readiness Indicator on a Theoretical Use Case in the City of Vienna. In Proceedings of the SBE 19 Malta Sustainable Built Environment; International Conference on Sustainability and Resilience, Valetta, Malta, 21–22 November; ISBN 978-99957-1-613-4 (ebook).
56. Available online: https://www.wien.gv.at/flaechenwidmung/public/start.aspx (accessed on 13 April 2020).
57. Google Maps. Available online: https://www.google.com/maps/place/Tivoligasse (accessed on 13 April 2020).
58. Available online: http://episcope.eu/building-typology/tabula-webtool/ (accessed on 13 April 2020).
59. OIB Austrian Institute of Construction Engineering. *OIB Guideline 6, Energy Saving and Heat Insulation*; OIB-330.6-026/19; Austrian Institute of Construction Engineering: Vienna, Austria, 2019.
60. Garcia-Ayllon, S.; Miralles, J.L. New Strategies to Improve Governance in Territorial Management: Evolving from "Smart Cities" to "Smart Territories". *Procedia Eng.* **2015**, *118*, 3–11. [CrossRef]
61. Sharifi, A. A typology of smart city assessment tools and indicator sets. *Sustain. Cities Soc.* **2020**, *53*, 15. [CrossRef]

MDPI
St. Alban-Anlage 66
4052 Basel
Switzerland
Tel. +41 61 683 77 34
Fax +41 61 302 89 18
www.mdpi.com

Energies Editorial Office
E-mail: energies@mdpi.com
www.mdpi.com/journal/energies

www.ingramcontent.com/pod-product-compliance
Lightning Source LLC
LaVergne TN
LVHW071619170726
843515LV00009B/2429

* 9 7 8 3 0 3 9 4 3 8 4 9 5 *

(<1976), SH2 (1976–2002), SH3 (2003–2009), and SH4 (>2010). Each age category has tighter thermal insulation requirements than the previous. The ventilation system for SH1 was natural ventilation and for SH2 it was mechanical exhaust ventilation without heat recovery. SH3 utilized mechanical supply and exhaust ventilation with heat recovery as did SH4, but with a higher heat recovery efficiency. The details of the reference building properties are shown in Table 1.

Table 1. Properties of the reference single-family houses [34].

	Building Age Class	SH1	SH2	SH3	SH4
	Construction years	–1975	1976–2002	2003–2009	2010–
U-values of envelope (W/(m^2·K))	External wall	0.58	0.28	0.25	0.17
	Floor	0.48	0.36	0.25	0.16
	Ceiling	0.34	0.22	0.16	0.09
	Doors	1.4	1.4	1.4	1.0
	Windows	1.8	1.6	1.4	1.0
Total solar heat transmittance (g)		0.71	0.59	0.46	0.46
Direct solar transmittance (ST)		0.64	0.52	0.39	0.39
Air tightness	n_{50}, (1/h)	6	4	3.5	2
	q_{50}m^3/(h m^2)	15.6	10.4	9.1	5.2
Ventilation	Type	Natural ventilation	Mech. E. vent.	Mech. S.&E. vent.	Mech. S.&E vent.
	Heat recovery temp. eff.	0	0	0.55	0.65
	Ventilation rate (L/s/m^2)	0.30	0.33	0.36	0.36
	Total air exchange rate (1/h)	0.41	0.46	0.5	0.5
	SFP (kW/m^3/s)	0	1.5	2.5	2
	Heating setpoint (°C)	22	22	21.5	21

In addition, the buildings were divided according to the main heating system in use in the building: district heating (DH), wood/oil boiler, direct electric heating, or ground-source heat pump (GSHP). The share of different heating systems in the building stock within each building age category is shown in Figure 2.

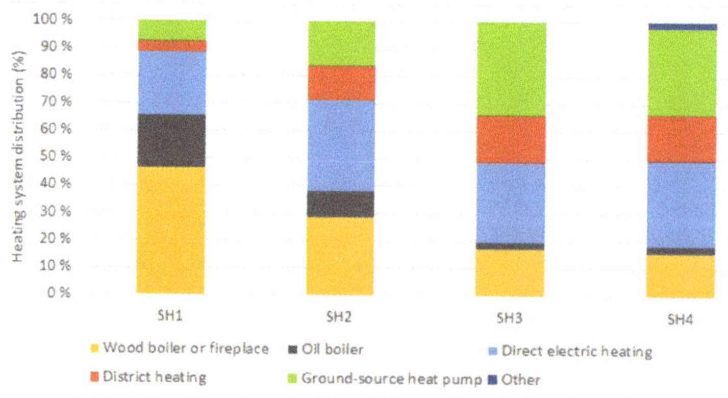

Figure 2. Distribution of main heating systems in the building stock within each building age category.